Advances in
Affective and
Pleasurable Design

Advances in Human Factors and Ergonomics Series

Series Editors

Gavriel Salvendy

Professor Emeritus
School of Industrial Engineering
Purdue University

Chair Professor & Head
Dept. of Industrial Engineering
Tsinghua Univ., P.R. China

Waldemar Karwowski

Professor & Chair
Industrial Engineering and
Management Systems
University of Central Florida
Orlando, Florida, U.S.A.

3rd International Conference on Applied Human Factors and Ergonomics (AHFE) 2010

Advances in Applied Digital Human Modeling
Vincent G. Duffy

Advances in Cognitive Ergonomics
David Kaber and Guy Boy

Advances in Cross-Cultural Decision Making
Dylan D. Schmorrow and Denise M. Nicholson

Advances in Ergonomics Modeling and Usability Evaluation
Halimahtun Khalid, Alan Hedge, and Tareq Z. Ahram

Advances in Human Factors and Ergonomics in Healthcare
Vincent G. Duffy

Advances in Human Factors, Ergonomics, and Safety in Manufacturing and Service Industries
Waldemar Karwowski and Gavriel Salvendy

Advances in Occupational, Social, and Organizational Ergonomics
Peter Vink and Jussi Kantola

Advances in Understanding Human Performance: Neuroergonomics, Human Factors Design, and Special Populations
Tadeusz Marek, Waldemar Karwowski, and Valerie Rice

4th International Conference on Applied Human Factors and Ergonomics (AHFE) 2012

Advances in Affective and Pleasurable Design
Yong Gu Ji

Advances in Applied Human Modeling and Simulation
Vincent G. Duffy

Advances in Cognitive Engineering and Neuroergonomics
Kay M. Stanney and Kelly S. Hale

Advances in Design for Cross-Cultural Activities Part I
Dylan D. Schmorrow and Denise M. Nicholson

Advances in Design for Cross-Cultural Activities Part II
Denise M. Nicholson and Dylan D. Schmorrow

Advances in Ergonomics in Manufacturing
Stefan Trzcielinski and Waldemar Karwowski

Advances in Human Aspects of Aviation
Steven J. Landry

Advances in Human Aspects of Healthcare
Vincent G. Duffy

Advances in Human Aspects of Road and Rail Transportation
Neville A. Stanton

Advances in Human Factors and Ergonomics, 2012-14 Volume Set:
Proceedings of the 4th AHFE Conference 21-25 July 2012
Gavriel Salvendy and Waldemar Karwowski

Advances in the Human Side of Service Engineering
James C. Spohrer and Louis E. Freund

Advances in Physical Ergonomics and Safety
Tareq Z. Ahram and Waldemar Karwowski

Advances in Social and Organizational Factors
Peter Vink

Advances in Usability Evaluation Part I
Marcelo M. Soares and Francisco Rebelo

Advances in Usability Evaluation Part II
Francisco Rebelo and Marcelo M. Soares

Advances in
Affective and
Pleasurable Design

Edited by
Yong Gu Ji

CRC Press
Taylor & Francis Group
Boca Raton London New York

CRC Press is an imprint of the
Taylor & Francis Group, an **informa** business

CRC Press
Taylor & Francis Group
6000 Broken Sound Parkway NW, Suite 300
Boca Raton, FL 33487-2742

© 2013 by Taylor & Francis Group, LLC
CRC Press is an imprint of Taylor & Francis Group, an Informa business

No claim to original U.S. Government works

International Standard Book Number: 978-1-4398-7118-8 (Hardback)

Visit the Taylor & Francis Web site at
http://www.taylorandfrancis.com

and the CRC Press Web site at
http://www.crcpress.com

Table of Contents

Section III. Ergonomics and Human Factors

Section V. Human Interface in Product Design

Section VIII. Diverse Approaches: Biosignals, Textiles, and Clothing

Section IX. Novel Devices, Information Visualization, and Augmented Reality

Preface

This book focuses on a positive emotional approach in product, service, and system design and emphasizes aesthetics and enjoyment in user experience. This book provides dissemination and exchange of scientific information on the theoretical and practical areas of affective and pleasurable design for research experts and industry practitioners from multidisciplinary backgrounds, including industrial designers, emotion designer, ethnographers, human-computer interaction researchers, human factors engineers, interaction designers, mobile product designers, and vehicle system designers.

This book is organized in nine sections which focus on the following subjects:

I: Designing for Diversity
II: Cultural and Traditional Aspects
III: Ergonomics and Human Factors
IV: Product, Service, and System Design
V: Human Interface in Product Design
VI: Emotion and UX Design
VII: Design and Development Methodology
VIII: Diverse Approaches: Biosignals, Textiles, and Clothing
IX: Novel Devices, Information Visualization, and Augmented Reality

Sections I through III of this book cover special approaches in affective and pleasurable design with emphasis on diversity, cultural and traditional contexts, and ergonomics and human factors. Sections IV through VII focus on design issues in product, service, and system development, human interface, emotional aspect in UX, and methodological issues in design and development. Sections VIII and IX handle emotional design approaches in diverse areas, i.e. biosignals, textiles, and clothing, and emerging technologies for human interaction in smart computing era. Overall structure of this book is organized to move from special interests in design, design and development issues, to novel approaches for emotional design.

All papers in this book were either reviewed or contributed by the members of Editorial Board and Interaction Design Lab at Yonsei University. For this, I would like to appreciate the Board members listed below:

A. Aoussat, France
A. Chan, Hong Kong
S. B. Chang, Korea
L. L. Chen, Taiwan
Y. C. Chiuan, Singapore
G. S. Cho, Korea
S. J. Chung, Korea
D. A. Coelho,Portugal
O. Demirbilek, Australia
Q. Gao, China
R. Goonetilleke, Hong Kong
B. Henson, UK

W. Hwang, Korea
C. Jun, China
M. C. Jung, Korea
S. R. Kang, USA
H. Khalid, Malaysia
H. Kim, Korea
J. Kim, Germany
Y. K. Kong, Korea
K. Kotani, Japan
O. Kwon, Korea
G. Kyung, Korea
H. Lee, Korea

I. Lee, Korea
Y. K. Lim, Korea
K. Morimoto, Japan
M. Ohkura, Japan
Y. W. Pan, Korea

P-L. P. Rau, China
S. Schutte, Sweden
H. Umemuro, Japan
A. Warell, Sweden
M. H. Yun, Korea

This book is the first approach in covering diverse approaches of special areas and including design and development methodological researches and practices in affective and pleasurable design. I hope this book is informative and helpful for the researchers and practitioners in developing more emotional products, services, and systems.

April 2010

Yong Gu Ji
Yonsei University
Seoul, Korea

Editor

Section I

Designing for Diversity

CHAPTER 1

Emoticons: Cultural Analysis

Myung Hae Park

California State University
Sacramento, USA
myung@csus.edu

ABSTRACT

As computer-mediated communication (CMC) continues to replace face-to-face (F2F) interaction, the nature of communication has changed. Emoticons (emotional icons) are facial expressions pictorially represented by text and punctuation marks. Emoticons have become substitutes for the visual cues of F2F communication in CMC, and have been used for many years to diversify communication in informal text messages. This study compares and analyzes syntactic typographic structures and variables between two different cultural emoticons, American and Korean, and examines whether emoticons can be culturally neutral.

Keywords: computer-mediated communication, emotion and typographic emoticon, typographic syntactic variables

1   INTRODUCTION

All over the world people connect through CMC. According to The Radicati Group, in the Email Statistics Report, 2011–2015, "the number of worldwide email accounts is expected to increase from an installed base of 3.1 billion in 2011 to nearly 4.1 billion by year-end 2015". Instant Messaging (IM) is also continuing to growth in popularity, especially the younger generation. "Worldwide IM accounts are expected to grow from over 2.5 billion in 2011 to more than 3.3 billion by 2015" (Radicati Group Inc, 2011). Short messaging service (SMS) has become an important mode of communication throughout the world and its use is increasing rapidly (Global Mobile Statistics, 2011).

As CMC replaces some forms of F2F interaction, the nature of communication has also changed. As one is unable to view the other person in CMC, there is a lack of nonverbal cues, such as facial expressions and body gestures. This lack of nonverbal information means certain information cannot be fully transferred (McKenna & Bargh, 2000). As a result, utilizing other ways of expressing intended emotions in CMC becomes important. Using icons to express emotions (emoticons) has become a substitute for nonverbal cues used in F2F interactions. Walther and D'Addario (2001) defined emoticons as graphic representations of facial expressions that are embedded in electronic messages. These often include alphabetic characters and punctuation marks to create emotional expressions. Frequently used typographic (e.g., text-based) emoticons include facial expressions representing happy, sad, angry, etc, as shown in Table 1.1. Recently, graphic emoticons have been introduced in IM, resulting in an improved visual language for expressing human emotion.

Many researchers have noted the importance of emoticons to convey meaning in CMC. Emoticons help accentuate meaning during development and interpretation (Crystal, 2001). Lo's study (2008) showed that most internet users cannot perceive the correct emotion, attitude, and attention intent from pure text without emoticons. Adding emoticons significantly improves the receiver's perception of a message. They not only carry the warmth of F2F communication, but also add breadth to the message (Blake, 1999).

Basic human facial expressions are not learned, but are universal across cultures (Ekman in Matsumoto, 1992). Based on this reasoning, this author hypothesizes that the basic emotions in emoticons are also culturally neutral. This paper reviews two types of studies that: (1) analyze syntactic structure and variables in two different cultural emoticons, and (2) examine whether emoticons can be culturally neutral. This study focuses only on typographic emoticons.

Typographic Emoticons	:-)	:-(	:P	:<	:O
Graphical Emoticons					

Table 1.1 Typographic Emoticons and Graphic Emoticons

2   METHOD

The study consisted of 60 American undergraduate student participants majoring in graphic design. Female participants comprised about 60% and male participants 40% of the sample. About 90% of participants were aged in their 20s, and approximately 40% had been exposed to Eastern (e.g., Korean, Japanese) style emoticons (Table 2.1). This group is called the exposed group and the other group the non-exposed group in the analysis of the results. The study used a combination of fixed-response (i.e., structured) and closed-ended (i.e., non-structured) questions. The fixed-response questions were: (1) demographic information, (2) frequency of emoticon usage, (3) media usage of emoticons, (4) frequency of typographic and graphic emoticon usage, (5) attitude toward emoticons, (6) difficulty in understanding emoticons, and (7) experience in Eastern emoticons. The closed-ended questions were: (1) commonly used emoticons, and (2) perceived emotions on both types of cultural emoticons.

	Participants	10s	20s	30s	40s	Have been exposed to Eastern style emoticons?
Male	25	–	22	2	1	9
Female	35	1	33	–	1	17
Total	60	1	55	2	2	26

Table 2.1 Participant Demographics

3   RESULTS

3.1   Typographic Elements and Structure

The typographic emoticons in Korea are made up of Korean 'Hangul' characters with punctuation marks (e.g., asterisk, tilde, grave accent) in a similar way to American emoticons, which use alphabetic characters with punctuation marks (e.g., colon, round bracket, slash). Countless emoticons can be formed using different combinations of characters in both types of cultural emoticon. The most popular American emoticons include punctuation marks such as the colon - : - representing the eyes, and brackets - ( ) - representing the mouth. The most popular Korean emoticons include characters ㅅ or ㅜ for the eyes, and – for the mouth.

Orientation of the marks is seen as a significant difference in the formation of the typographic emoticons for both cultural emoticons. American emoticons have horizontal orientation; in other words the eye is on the left and the mouth is on the right. This is the traditional way to write in English; from left to right, the way one reads and writes, and the side-by-side way letter characters are formed. However, Korean emoticons have vertical orientation; the eye is topmost and the mouth is bottommost. This is the traditional way to write in Korean and also the way characters are formed; top to bottom. Examples are shown in Table 3.1.1.

Emotions	American Emoticons		Korean Emoticons	
happy/smile	:)	:D	^ ^	^o^
sad/cry	:(	:-(	ㅜ_ㅜ	ㅠ_ㅠ
flirtatious	;)	;-)	^.~	^_~
angry	:\	>:\	`_´	.V.

Table 3.1.1 Structural Difference Between American and Korean Emoticons

3.2   Syntactic Variables

The face as a whole indicates human emotion. Specific emotional modes, such as happiness or sadness, are expressed through a combination of five different facial features: eyebrows, eyes, nose, cheeks, and mouth. The enormous complexities of physiognomy have been reduced to the bare essentials through emoticons. The human face has been simplified; two eyes become dots and a mouth becomes a line. Emoticons are made up of typographic facial motifs to represent emotions. For example, the closing round bracket -) - represents a smiling mouth to indicate happiness, whereas (represents a downturned mouth to indicate unhappiness, and / indicates confusion. Therefore, facial expressions in emoticons rely on typographic syntactic variables (i.e., formal mode of visual signs) such as shape, size, proportion, direction, and orientation of the five facial features. The unambiguous typographic syntactic variables correspond with certain facial expressions, used to effectively convey intended emotions.

Table 3.2.1 lists facial features from the most popular American emoticons known to participants. Interestingly, a wide variety of syntactic variables are found for the mouth, followed by those for the eyes. Interestingly, for Korean emoticons, the author found more syntactic variables for the eyes, followed by those for the mouth (Table 3.2.2).

Features	Facial Syntactic Variables in American Emoticons						
eyebrow	>						
eye	:	— —	> <	=	**		
	;	,	i i	X			
nose	-						
cheek	'	*					
mouth	)	D	P	|	\	[	>
	(	O	T	_	/	]	<
	))	X	u	—	.	3	*
	((	{	#	@			

Table 3.2.1 American Emoticons: Facial Syntactic Variables

Features	Facial Syntactic Variables in Korean Emoticons					
eyebrow						
eye	^ ^	ㅜㅜ	^ ~	` `	**	= =
	> <	ㅠㅠ	^ _	` '	@ @	z z
	V	- -	oO	;;		
nose						
cheek	**					
mouth	~	o	–	.		

Table 3.2.2 Korean Emoticons: Facial Syntactic Variables

Figure 3.2.1 indicates culture is a determining factor in the formation of emoticons when representing emotions. In American emoticons, a wide variety of distinguishing variables are found for the mouth. For example, the closing round bracket in the :-) representing happiness can be replaced with the opening round bracket to form :-(to represent sadness. Likewise, the emotion statement can be increased by replacing) with D to form :-D, representing great happiness. It may be that the change in the mouth of an emoticon can be generally understood as a difference in facial expression in American emoticons. Therefore, American emoticons are largely reliant on the syntactic variables of the mouth.

Conversely, a greater variety of distinguishing variables in Korean emoticons are found for the eye (Figure 3.2.1). For example, ^ ^ are used for smiling eyes, representing happiness, and can be replaced with ㅜㅜ for crying eyes, representing sadness, or ` ' for vicious eyes which represent anger. For this reason, Korean emoticons are more reliant on syntactic variables of the eye. This suggests American emoticons use visual stimuli from the mouth to express emotions, whereas Korean emoticons use the eyes.

Other facial features, such as eyebrows, nose, and cheeks are not significant in the formation of emotions in either culture's emoticons. Some features are excluded from an emoticon such that the remaining items gather importance. Eyebrows and noses rarely contribute to different emotional expressions in Korean emoticons, compared to American emoticons. For instance, a smiling face ^ ^ excludes a nose and mouth. The only expressive features, the eyes, become more prominent. This indicates that American emoticons rely on a variety of facial features to convey emotional expression, whereas Korean emoticons express emotions using a minimal number of facial features.

8

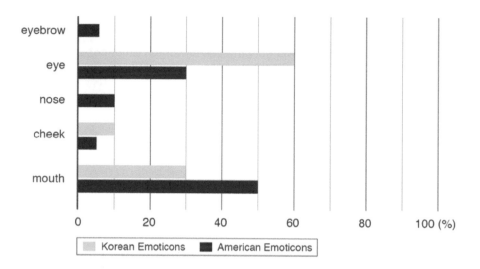

Figure 3.2.1 Frequency of Facial Syntactic Variables

3.3   Culture and Emoticons

Facial expressions are a form of nonverbal communication which convey emotional states during F2F communication. Most anthropologists believe facial expressions are learned, and therefore vary from culture to culture (Jack et al., 2009). On the other hand, Ekman showed facial expressions of emotion are universal across cultures (Ekman in Matsumoto, 1992). If this is the case, regardless of culture, the same emotion from different cultural emoticons should be recognized. Every emoticon bears visual resemblance to facial expressions. The author hypothesizes that the probability of correct interpretation of other culture's emoticons is high as people can derive meaning from the iconic illustration, regardless of culture.

When shown several popular Korean emoticons representing the basic emotions of happiness, sadness, anger, and flirtatiousness, participants were asked to give their perceived emotions on each emoticon (See Table 3.3.1).

Table 3.3.1 Korean Typographic Emoticons

Some participants accurately interpreted a number of emoticons, regardless of their knowledge of Eastern style emoticons. Others, however, found interpretation difficult. 62% of the non-exposed group accurately identified emoticons of smiling eyes (^ ^ and ^o^) and 75% of the exposed group correctly identified them. With the 2x2 Chi-Square calculation, the Chi-Squared value χ^2 =0.8181. It leads to the two-tailed p-value of 0.3657 (p>0.05). This difference between the exposed and non-exposed group is not statistically significant. Therefore, the smiling emoticons (happiness) are culturally neutral. For the emoticon of winking eyes (^_~), the same percentages are observed from the both groups as for the smiling emoticons. Therefore the emoticon of winking eyes (flirtatiousness) is also culturally neutral. For the emoticons of crying eyes (ㅜ_ㅜ and ㅠ_ㅠ), 18% of the non-exposed group accurately identified them and 71% of the exposed group accurately identified them. With the 2x2 Chi-Square calculation, the Chi-Squared value χ^2 = 16.62. It amounts to the p-value of 0.001 (p<0.05). By conventional criteria, this difference between the two groups is considered to be extremely statistically significant. Therefore, the emoticons of crying eyes (sadness) are culturally dependent. The reason for this cultural dependency is clear when they are made up of Korean 'Hangul' characters. For the emoticon of angry eyes ('_'), 53% of the non-exposed group and 58% of the exposed group accurately identified it. With the 2x2 Chi-Square calculation, the Chi-Squared value χ^2 = 0.165. It leads to the two-tailed p-value of 0.6846 (p>0.05). This difference between the exposed and non-exposed group is not statistically significant. Therefore, the emoticon of angry eyes is culturally neutral.

The study concluded that certain emoticons correspond with facial expressions, irrespective of cultural background, and are therefore culturally neutral unless the emoticons are made up of the culturally oriented characters.

4   CONCLUSIONS

Typographic emoticons are made up of typographic syntactic variables. American emoticons focus on variables of the mouth, whereas Korean emoticons emphasize variables of the eyes. The orientation of constructing emoticons is another factor in distinguishing between two cultural emoticons: American uses horizontal format, while Korean uses vertical. Even if emoticons between two cultural backgrounds are in the different orientation focusing on different facial features, there is a great probability that two different cultural groups can recognize emoticons that bear visual resemblance to facial expressions. The study concluded that certain emoticons are in general culturally neutral unless they are made up of the culturally oriented characters.

This study compared and analyzed syntactic typographic structures and variables between two different cultural emoticons, and examined whether emoticons could be culturally neutral. Further studies into cross-cultural comparisons of a more extensive number of emoticons are required to validate the level of cultural neutrality, and to investigate whether the gender variable plays a meaningful role in the interpretation of emoticons.

REFERENCES

Blake, Gary. "E-mail with feeling." *Research Technology Management* 42 (6) Nov/Dec 1999: 12–13.

Crystal, David. *Language and the internet*. Cambridge, UK: Cambridge University Press. 2001.

"Email Statistics Report, 2011–2015." *Radicati Group, Inc.* Accessed January 5, 2012 <http://www.radicati.com/wp/wp-content/uploads/2011/05/Email-Statistics-Report-2011-2015-Executive-Summary.pdf>.

"Global Mobile Statistics 2011." *MobiThinking*. Accessed January 2012 <http://mobithinking.com/mobile-marketing-tools/latest-mobile-stats>.

"Instant Messaging Market 2011–2005." The *Radicati Group, Inc.* Accessed January 2012 <http://www.radicati.com/wp/wp-content/uploads/2011/11/Instant-Messaging-Market-2011-2015-Executive-Summary.pdf>.

Jack, Rachael E, Caroline Blais, Christoph Scheepers, Philippe G. Schyns & Roberto Caldara. "Cultural Confusions Show that Facial Expressions Are Not Universal." *Current Biology*, 19 (18) 13 August 2009: 1543-1548.

Lo, Shao-Kang. "The Nonverbal Communication Functions of Emoticons in Computer-Mediated Communication." *CyberPsychology & Behavior* 11 2008: 595–597.

Matsumoto, David. "More Evidence for the Universality of a Contempt Expression." *Motivation and Emotion*, 16 (4) December 1992: 363–368.

McKenna, K. Y. A., & Bargh, J. A. "Plan 9 from cyberspace: The implications of the Internet for personality and social psychology." *Personality and Social Psychology Review* 4 2000: 57–75.

Walther, J. B., & D'Addario, K. P. "The impacts of emoticons on message interpretation in computer – mediated communication." *Social Science Computer Review* 19 2001: 324–347.

CHAPTER 2

Designing Spaces for Aging Eyes

Kimberly Melhus Mitchell

Northern Arizona University
Flagstaff, Arizona USA
Kimberly.Melhus@nau.edu

ABSTRACT

If the least conservative estimates are used, by the year 2040 the average life expectancy of older people could increase by 20 years. Some projections are that by the middle of the 21st century, there will be 16 million Americans over 85 years of age. The sensory, cognitive and motor abilities decline as we age. With a rapidly aging population, design for the elderly is going to have to be given greater consideration than it has in the past. Often times it is the architects that design the interior spaces of the assisted living environments, and quite often these four key legibility variables have not been factored in to the design. The purpose of this research is to produce some preliminary guidelines for the wayfinding, organization, and experience design of assisted living environments. The study will establish critical legibility factors related to aging vision and designing supportive environments that enhance comfort, safety and independent functioning. This study involves an in-depth look at an assisted living facility in Ames, Iowa. The methodologies will consist of overall observation and one-on-one interviews with the staff. The case study will reveal how the overall experience in assisted living environments can be improved.

Keywords: universal design, wayfinding, experience design, vision, aging

1   INTRODUCTION

The sensory, cognitive and motor abilities decline as we age. With a rapidly aging population, design for the elderly is going to have to be taken in greater consideration than it has in the past. If the least conservative estimates are used, by the year 2040 the average life expectancy of older people could increase by 20 years. Some projections are that by the middle of the 21st century, there will be 16 million Americans over 85 years of age. Prognosticators also say that the average 65-year-old will spend 7.5 years of his or her remaining 17 years living with some functional disability (Spirduso, Francis & MacRae, 2005). When looking at many designed spaces today it does not seem that the needs of the aging population are even being considered.

The need for sensitivity to usability issues will only become more pressing in the coming decades as user populations become more diverse. One significant trend is the increasing longevity of the human race, worldwide. Another factor is improved medical technologies that allow more critically injured and seriously ill people to survive (Story, Mueller and Mace, 1998).

The ability to perform the activities of daily living (ADLs) becomes more challenging as one ages. As these ADLs become more difficult, many older adults have turned to assisted living environments. The Assisted Living Federation of America, which was created in 1990, defines assisted living as "a long-term care option that combines housing, support services and health care, as needed" (alfa.org). Assisted living is designed for individuals who "require assistance with everyday activities such as meals, medication management or assistance, bathing, dressing and transportation." In addition, "some residents may have memory disorders including Alzheimer's, or they may need help with mobility, incontinence or other challenges" (IBID).

The following content reveals a case study involving an in-depth look of an assisted living facility in Ames, Iowa. The methodologies used consisted of overall observation and one-on-one interview sessions with the staff. The two aspects of design researched in the study were: wayfinding and experience design. The goals of the study were to find how the overall experience of assisted living environments, particularly the assisted living facility in Ames, Iowa, and the overall universal design codes could be improved.

2   THE BACKGROUND OF THE ROSE OF AMES

The Amenities

The Rose of Ames is a fifty-six unit, one and two-bedroom private senior-living apartment complex that offers assisted living services as well as 24-hour monitoring. The facility offers social programming, coin-free laundry facilities, a nurse's office, and dining room meal service with a private chef. According to the Rose of Ames website, each apartment unit features a "fully equipped kitchen, mini-blinds on all windows, a shower with a built-in seat, and a large in-unit storage closet." The property features a beauty salon, a mini-theater, a small computer room, activities room, a library and fireplace, a whirlpool, front and back porches with patio furniture and a guest suit.

Pricing

Like most rental units, there are several application fees. Rent varies from $546 to $716 for 1-bedroom apartments and 2-bedroom apartment units are $849/month with gas and electric included in the unit rent cost. There is an income guideline that the seniors cannot exceed during move-in. A household size of one person cannot make more than $21,680/year, while a household of two people cannot make more than $37,200/year. The pricing is very competitive when compared to the national average. According to the American Association of Homes and Services for the

Aging, "the average daily cost for a private room in a nursing home is $214.00 ($6,390 per month and $77,745 annually), while the average monthly cost of assisted living facilities is $2,969.00 or $35,628 annually" (Caregiverlist.com).

The Location

Located in Ames, Iowa, approximately 30 minutes from Iowa's state capital, Des Moines. Ames is also less than a day's drive from Minneapolis, Kansas City, Omaha, Chicago, St. Louis and Milwaukee. According to the City of Ames website, in 2002, Ames was ranked one of the "Best 20 places in America to Live and Work" by BestJobsUSA.com. Also in 2002, it was ranked 20th on the "Best Place to Live in America" by *Men's Journal* magazine. Ames, Iowa is also home to Iowa State University.

The Rose of Ames is conveniently located, as it is less than a block to Iowa State University's CyRide bus stop. This transportation system runs every single day of the week during many hours of the day and all throughout the city of Ames.

The Development and Purpose of the Rose of Ames

According to an article published in the *Nursing Homes* magazine, "In 2001, a statewide study by the Iowa Finance Authority indicated that more than 50% of the elderly aged 75 years and older could not afford what was currently available on the assisted living market" (http://www.ltlmagazine.com/). At that time, the average monthly costs apartments within assisted-living facilities ranged between $1,272 and $2,517. Yet, according to the article, "one in four seniors statewide aged 75 and older had a monthly income at or below $884 (from the U.S. Census Bureau, 2000), and 50% of the annual median area income ranged from $17,300 to $23,550 (from the U.S. Department of Housing and Urban Development)" (IBID).

This is where the Rose of Ames comes into play: they wanted to close the gap between what was *needed* and *what people could afford.* Their goal was to respond to this critical need by providing an "assisted living community that would not only offer an affordable option for moderate-to-low-income seniors," but, "also would maintain the same quality and scope of housing and services as those available in market-rate assisted living facilities" (IBID).

In order to create an environment at such low-costs, their plan has to separate the housing from the services by making the services optional and purchased separately. Seniors may rent apartments in the facility and either obtain the optional services from the building owner or its affiliates, or obtain them from any provider they choose. The same is true with their meals: residents may purchase flexible meal plan or cook on their own (IBID).

2.1   LITERATURE REVIEW

Wayfinding

According to author, licensed architect and Professor of Architecture, Arvid E. Osterberg, "signs that provide directions to rooms or spaces and to accessible means of egress need to be accessible." Signage requirements, provided by the *Americans with Disabilities Act Accessibility Guidelines* (ADAAG), are listed in his book, *Access for Everyone*. They include:

> • Signs that are not required to be accessible are the building directories, menus, occupant names, building addresses, and company names and/or logos (Access for Everyone). Though they are not required to meet the accessibility requirements, Osterberg and Kain, recommend making the information readily available to all people whenever it is possible.

> • Signs provide important information about locations and services, including information about accessible locations and services. All people should have access to all types of information provided by signs. To assist the greatest number of people, signs should be placed at appropriate locations and heights, contain characters and backgrounds that meet specific requirements for readability, and use symbols that have been adopted internationally to indicate accessible locations and features.

> • The design and placement of signs should be uniform in and around the buildings and sites. People will be able to find and use signs more easily and quickly if the placement and heights of the signs are consistent.

> • Accessibly signs may include tactile characters (such as raised characters, Braille, and/or pictograms), visual characters, or both visual and tactile features. Where signs are required to be both tactile and visual, you may install one sign that includes both types of information or you may install two signs, one visual and the other tactile (Osterberg and Kain, 2005)

Experience Design

The second portion of the study focused on experience design, or the overall comfort level of the Rose of Ames. Experience design is the practice of designing products, processes, services, events and environments based on the consideration of an individual or group's needs, desires, believes, knowledge, skills, experiences and perceptions. It involves emotions and memories as well as overall feelings of satisfaction or disgust (Diller, Shedroff, Rhea, 2005).

Experience design is not driven from a single design discipline; rather, it's a cross-discipline perspective. It considers many aspects of the brand/business/environment/experience- from product, packaging and retail environment to clothing and attitude of employees (Diller, Shedroff, Rhea, 2005). It involves all human senses.

2.2   ASSESSMENTS

The outside of Rose of Ames is quite inviting because of the large, white covered porch (Figure 1). The benches and seats on the porch seem to get a lot of attention from the residents. Trees outline the vicinity of the porch, giving the resident a sense of being on a patio of their own home. Bird feeders are seen hanging from the nearby trees, which aides in a form of entertainment for the residents as well. Inside felt much more home-like than institution-like with the warm, earth tone colors all throughout the facility. The tones were rich and deep, giving a sense of being in a home, not an apartment complex or an assisted living facility. The carpets were deep green and red, and natural colored wood was everywhere. Facilities such as a hair salon, laundry on each floor, a computer room with two computers, candy and soda machines, a small movie theater and a whirlpool room facilitated in entertainment and made it a place where residents wanted to be. The reading room felt very peaceful with the nice wooden framed paintings overhead the fireplace. The plush couch and chairs and the nice selection of books gave it a sense of warmth. The décor throughout the facility was rich and really quite lavish. Framed pictures outline all of the walls, and plants were seen in areas such as the reading room and dining area.

Finding *The Rose of Ames* facility was quite easy because of the large, detailed sign directly in front of the parking lot (Figure 2). The facility is located amongst several other apartment complexes in a residential community. The sign reads, "The Rose of Ames Senior Residencies". From the outside view, there is no confusion that it is a living facility is for seniors.

Before entering *The Rose of Ames,* one is greeted by a white covered porch with wooden benches to sit upon (Figure 3). The main entrance is not clearly marked, but with the help of the automatic doors, it is fairly obvious that is the main entrance. No doors have any markings, except for one, which is the handicap accessible main entrance (Figure 4). The handicap accessible button to push the door open was quite far (about 3 feet) from the actual door to get in. If the button is pushed, the door will open slowly and allow the person to walk in. If the person is able to pull the door, he or she will find it to be quite heavy. The door was surprisingly narrow for a main entrance to an assisted living facility, but it is feasible with a wheelchair. There was minimal to no threshold under the door, which would make passing over with a wheeled chair very easy.

Once inside the main building, it was very difficult to tell where to go. Straight ahead was a beautiful wooden staircase with green, carpeted stairs (Figure 5), and to the immediate left was a hallway that led into another much larger room (Figure 6). There was no signage directing one to what was upstairs, how to get upstairs if one was not capable of climbing, or what was around the corner. The staircase would be an impossible feat for someone in a wheelchair or who fell short of breath easily.

After walking around the corner, a map became visible (Figure 7). The map was clearly printed by one of the staff members, as it was not professional and was enclosed within an 8.5" x 11" glossy sheet protector. The glare on the paper was troublesome at times and the text was very small and hard to read for someone with 20/20 vision. This was a very bad hazard. Around the corner and into the hallway into the actual facility was an easily assessable staff office. Nearby was a fireplace with chairs (Figure 8) and to the left of the office was a large dining room offering plenty of natural light (Figure 9).

The exit signs were clearly marked, lit, and easy to find. In addition, all of the rooms had signs posted flush against the wall right outside the door to let residents know what was inside each room. Each sign also had Braille (Figures 10 & 11). Unfortunately the signage was a very similar color to that of the painted walls. For someone with 20/20 vision this was not a concern, however, it would be very hard to distinguish for someone with limited vision.

Each floor had color-coded entryways to the residents' rooms, which was extremely helpful to differentiate the floors from one another. In addition, every resident's room was also clearly marked with his or her name outside the door. Each resident was able to decorate the area right outside their door with pictures, stuffed animals, shelving units– whatever they so desired as long as it did not come into the hallway (Figures 12 & 13). This helped to aid in wayfinding, in addition to allowing the residents' to customize their own entryway. Some residents had doorbells installed outside their doors as well.

The hallways were long, and although there were wooden handrails on each side, there was not an area for someone to sit and rest during his or her walk (Figure 14). The location of the elevators was not marked as clearly as it could have been. There was a seating area right outside the elevator so someone could sit to wait (Figure 15). Once inside the elevator, the buttons were nicely marked and lower to the ground, which is helpful to people in wheelchairs.

2.3   RECOMMENDATIONS

"The Rose of Ames" could be screen printed onto the main entrance door so that it was very clear which door was the main entrance. The doorway itself could actually be extended so that it was two doorways wide. This would also help people realize that this was the main entrance because of the emphasis on the large doorway. Also, the handicap press for the door could be closer to the door as well.

A sign is needed to let people know what is upstairs and what is around the corner right when entering. Also, the map needs to seem more important and be larger. The map could also be closer to the main entrance, so that way people know immediately where they are and where they need to go.

All of the rooms were marked just fine, but since the hallways were dark, the signs seemed to blend in with the walls. Some signs do not need to stick out, such as the maintenance closet, but some, like the laundry facility, or the elevator, could stick out up above the doorway. This would help people who have a difficult time walking far distances to see how far they actually have to go. This would also help emphasize the importance of that room. Some people might walk past the room because they were not looking at each sign as they walked by. Color contrast within each sign might also help show importance. Signage around corners, such as arrows pointing to which room numbers were down that particular hallway would be helpful, too

The patterned carpet was very nice, however, it may be too dark for the residents. A non-patterned carpet would have been a better solution. The staircases were especially dark. In fact, because there was a window at the end of each hallway (Figure 16), and the staircases were also at the end of each hall, there was quite a bit of a difference between the lighting of the end of the hallway and then once one enters inside the staircase area (Figure 17). It takes the elder's eyes a bit more to adjust to light changes, and so this was a definite hazard.

3   CONCLUSIONS

Overall the experience and wayfinding signage at the Rose of Ames was very nice. The facility was very inviting, intimate and private. The designer's paid special attention to small details that made the facility feel home-like to the residents' that lived there. From the outside covered porch to the small movie theater, residents' had a sense of community. The warm earth-tone colors of the green, patterned carpet and wooden doors made the facility feel very inviting. Little touches such as allowing the residents' to customize their entryways really made the facility have a friendly feeling.

There were potential hazards, however. The dim lighting in the hallways and the bright natural lighting coming from the windows could be a potential hazard. The signage needed some distinguishing features to differentiate the different rooms, as well.

As visual communicators, it is our responsibility to consider first the needs, and then the wants of society. We also need to understand just how influential our designs become; beyond printed matter and digital interfaces, designers can actually assist people in remaining active, independent individuals in society.

FIGURES

Figure 1 & 2. Rose of Ames facility and signage

Figure 3. Main entrance porch   **Figure 4.** Main entrance

Figure 5 & 6. Immediately inside   **Figure 7**. Interior Map

Figure 8. Fireplace   **Figure 9**. Dining Hall

Figures 10 & 11. Signage

Figure 12 & 13. Residents' doors

Figure 14. Hallway   Figure 15. Bench outside elevator

Figure 16. Hallway   Figure 17. Stairway

REFERENCES

"A Breakthrough in Creating Affordable Assisted Living." *Nursing Homes*. 2005.
Retrieved February 25, 2012. http://www.ltlmagazine.com/article/breakthrough-creating-affordable-assisted-living

"Ames Iowa." *Community Information*. Retrieved June 11, 2007. http://www.ames.ia.us/

Baumeister, R.F., & Leary, M.R. The need to belong: Desire for Interpersonal Attachments as a Fundamental Human Motivation. *Psychological Bulletin 117*, 497-529. 1995.

Berger, Craig M. *Wayfinding: Designing and Implementing Graphic Navigational Systems.* Switzerland: RotoVision SA. 2005.

Diller, Stephen, Nathan Shedroff, Darrel Rhea. *Making Meaning: How Successful Businesses Deliver Meaningful Customer Experiences*. New Riders Press. 2005.

"Heartland Senior Services." Retrieved June 11, 2007.
http://www.heartlandseniorservices.com

"Home Healthcare." *Lutheran Services in Iowa*. Retrieved June 11, 2007. http://www.lsiowa.org/home_healthcare.asp

"How to Choose a Senior Assisted Living Community." Retrieved January 20, 2012. http://www.caregiverlist.com/AssistedLiving.aspx

Lidwell, William, Kritina Holden, Jill Butler. *Universal Principles of Design: 100 Ways to Enhance Usability, Influence Perception, Increase Appeal, Make Better Design Decisions and Teach Through Design*. Massachusetts: Rockport Publishers. 2003.

"The Rose of Ames." *Evergreen Real Estate Development Corporation*. Retrieved June 11, 2007. http://www.evergreenredc.com/roseofames.php

Noble and Bestley, *Visual Research: An Introduction to Research Methodologies in Graphic Design*. London: AVA Publishing. 2005.

Osterberg, Arvid and Donna Kain. *Access for Everyone*. Ames, Iowa: Iowa State University. 2005.

"Senior Living Options." Retrieved January 20, 2012.
http://www.alfa.org/alfa/Assisted_Living_Information.asp

Story, M. F., Mueller, J. L., & Mace, R. L. (1998). A Brief History of Universal Design. In *The universal design file: Designing for people of all ages and abilities*. North Carolina: The Center for Universal Design. Retrieved June 7, 2007.

Spirduso, W., Francis, K., & MacRae, P. (2005). Physical Dimensions of Aging (Second Edition). Champaign, IL: Human Kinetics.

CHAPTER 3

New Concept for Newspaper Kiosk through Understanding Users' Behavior

Yassaman Khodadadeh, Asma Toobaie

University of Tehran, Iran
Email: khodadade@ut.ac.ir
Asma_toobaie@yahoo.com

ABSTRACT

This study is concerned with obtaining design specifications for generating a new concept of newspaper kiosk, located in the streets of Tehran. Despite of the importance of newspaper kiosk in Persian culture, they have many shortcomings. In order to obtain a better understanding regarding newspaper kiosks' problems, there was a need for conducting a study. For this aim a precise observation was carried out and thirty users were interviewed regarding their feeling and interactions with kiosk. Samples were chosen randomly among the people who were interacting with the newspaper kiosk. They were from different social groups and were various in gender and age. As a result of this study, four key words were obtained. The customers were asked to mention their opinions regarding each of these key words. The data were analyzed statistically with Excel software and key words were classified due to the customers' priorities. The Quality Function Deployment (QFD) method was used to translate the voice of customer to the design parameters. The outputs of QFD can help to generate new concept for newspaper kiosk.

Keywords: QFD, User behaviour, Kiosk

1   INTRODUCTION

Modern urban life involves a variety of elements for different activities. Sometime urban life is so complicated that people hardly can find their ways and also has no understanding about different spaces. The spaces in cities are both physical and non physical. The non physical spaces are the sole of the city, which are affected by design of different elements. In order to provide better understanding of urban life for people, there is a need for clearly defining urban spaces and structures (Ho, Wang and Lee, 2007). Also for improving the quality of urban life, adequate urban elements and equipments are required. Street furniture is the common term for these equipment and facilities. A classic publication by Design Council (1979: 5) stated: "Public and open spaces are essential parts of people's living space. Poorly designed environment and facilities, including street furniture and open space facilities, can be a nightmare for residents and visitors". Moreover in recent years a lot of designers and researchers such as Rapuano (1994), Orr (2002) and Siu (2008) recognized that public spaces not only provide the functional needs of city users, but also fulfill their social, cultural, psychological and ideological needs.

Devereux (2007) noticed that "user and their views and needs are very important for companies and designers. They are seeking more innovative ways to understand their users' needs and desires". In case of street furniture, the main goal is to harmonize the form, scale, materials and placement to obtain beauty, accessibility and safety through understanding users' needs.

Sometimes users have difficulty for describing design problems and, in many cases, are not even consciously aware of them (Yanagisawa et al., 2009). Therefore, observing people while using a product, allows designers to understand how they interact with a product and the environment. This can provide an understanding for unexpected design problems. Sohan and Nam (2009) stated that the behavior patterns are helpful for analyzing a design problem. Also, the unconscious human behaviors could help to find design solutions.

The newspaper kiosk is one of the street furniture that is very popular in Tehran streets. It has been recognized as the main center of selling the newspapers and magazines in Persian culture. Therefore they are frequently visited by people of different age and backgrounds. Due to the close relationship of people with newspaper kiosks, the quality of these products plays an important role in improvement of urban life quality. According to general observation, it seems that the current kiosks have some problems during the using process and they can not sufficiently provide users needs and desires. These problems are not limited to the end users, and include all people that live and work in the city such as pedestrians and vendors. Therefore, deeper investigation into these problems and their reasons seems necessary. In order to find users needs and desires regarding newspaper kiosks in urban space, there is a need for performing a study. Therefore, a study was carried out and will be explained as fallow.

2   METHODS

The study was carried out by observation and interview to collect data. Quality Function Deployment (QFD) method was used to translate the voice of customer to design specifications.

Eight different kiosks were chosen in Tehran for this study. Four of them are located in very crowded streets in downtown and the other four are placed in less crowded streets. The newspaper kiosks, which were studied, have similar appearance however they are different in dimensions. The main structure of the studied kiosks, are made of iron, which is covered by fiberglass composite panels. The color of the panels is white. The studied kiosks have cubic form and their dimensions are about 400*200*300 or 300*200*250 centimeters. Some of them have a little colorful plastic canopy for protection against sunlight, although, it is not big enough to protect people and newspapers from sun.

2.1   Observational study

Observation is included gathering high quality data through looking and listening very carefully. In this method the particular information about users' behavior is discovered. It can help to study people in their natural ways without their behavior being influenced by the presence of a researcher. For this study reaction and behavior of different people including end users, vendors and pedestrians who pass the kiosk were observed and recorded through videotaping and photography. The users were various in age, gender and social classes. The data were collected, considering five key points of Who, What, Where, When and How.

2.2   Interviews

Interview can assist investigating users' psychology and opinions. It helps to find out the problems that users are experienced during the process of use. For this study thirty people between ages seventeen to seventy years old were selected randomly among users. As the users of the newspaper kiosk are mostly male, in this study seventy percent of samples (21 persons) were male and the rest thirty percent (9 persons) were female. The questions were about what they think about kiosks' appearance and performance. In addition they were asked to explain their feeling during the process of use and describing their opinions about their interactions with kiosk. Through these methods the user's needs and expectations were found out and by investigating them it was tried to identify the shortcomings.

2.3   Results

As a result of the study, the problems were identified. The results were classified in four key words to make them comparable statistically. As table 1 show the key words are form, color, material and newspapers placement, which the most important one is the newspapers placement. Twenty seven people out of thirty interviewees believed that the main problem is putting the large number of newspapers on the floor. In order to choose or take a newspaper, the users need to bend in an unsafe posture or even sit on their knees (figure1).

Twenty six of thirty people complained about the form of these kiosks. They believe that the cubic form of kiosk is boring and unattractive. Also they stated that the newspaper kiosk is unrecognizable and has no identification. Twenty five of thirty people said that they had problem to access the newspapers due to the incorrect placements of them (figure2). They also believe that the white color of kiosk is not recognizable in the crowded public space and is unattractive. The result of photography proves many problems regarding newspapers placements, users and pedestrians. As it can be seen in figure 3 pedestrian pathways is properly blocked by the users of kiosk while they read or buy newspapers. Excessive accumulation in front of kiosk is also a big problem, especially when people intend to talk to vendor or pay for a newspaper (figure 4).

Table 1 data collection via study on users' behavior

keyword	Voice of customer	persons
newspapers placement	Excessive accumulation in front of kiosk	22
	Putting the large number of newspapers on the floor	27
	Lack of access to the newspaper	25
	Very messy appearance of kiosk	14
	Block up pavement	24
form	Unattractive and un recognizable form	26
	lack of safety against wind and rain	15
	Shortage of sun protection shade	17
color	Unattractive and un recognizable color	25
material	Dirty and stainable body	10

Seventeen people mentioned that the canopy of kiosk is not appropriate to protect the newspapers and the customers against the sun (figure 5).

Also fifteen people were complained about lack of facilities for protection against wind and rain, which results to wet and damaged newspapers. At this moment vendors put a piece of stone or metal on the newspapers to protect them against wind (figure 6). It causes difficulty in the process of choosing and taking newspapers. Fourteen people believed that the appearance of kiosk is very messy. Ten people out of thirty had a bad feeling about the kiosks, due to dirtiness of appearance.

Figure 1 Unsafe posture

Figure 2 Difficult access to newspapers

Figure 3 Blocked pedestrian pathway

Figure 4 Excessive accumulation

Figure 5 Lack of protection by canopy

Figure 6 Stone for saving newspapers

The results of data collection were analyzed by excel software, which are presented in figure 7.

After analyzing the problem and determining the keywords, it is necessary to translate these problems to design parameters. To perform this translation, the quality function deployment was employed.

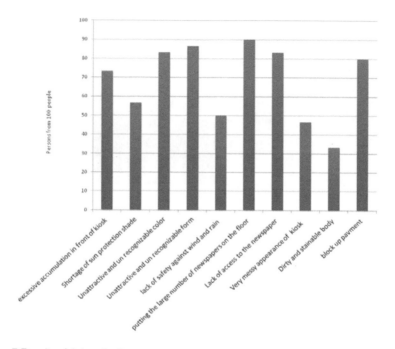

Figure 7 Results of data collection

2.4 Quality Function Deployment (QFD)

Quality Function Deployment is an exhaustive method for translating customer requirements to engineering characteristics or design parameters. This process is applied to extract design parameters by focusing on quality of product (ReVelle, 1998). In fact QFD is a strong tool, which transforms voice of customer into engineering characteristics and design specifications. QFD method works with a matrix for mapping the voice of customer to the voice of engineer or designer. This matrix is called House of Quality, which is shown in figure 8.

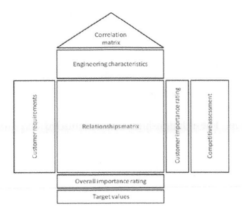

Figure 8 Typical House of Quality Matrix (QFD)

QFD has typically seven steps (Cross, 2000), which were applied for this study and are explained as fallow.

1. **Identifying customer' requirements in terms of product attribute**
   Customer requirements contain customer wants, needs and expectation for the product. They are establishing a clear understanding of all customers' needs that are called "voice of customer". In this study voice of customer was determined by translating the results of data collection to words and clear design instructions.
2. **Determining the relative importance of the attributes**
   In this step voice of customer is rated to determine relative importance of requirements. This rating was performed based on the data which were collected and presented in figure 7.
3. **Evaluating the attributes of competing products**
   Customers often make judgments about product attributes. Therefore, determining the competitors' position in the market and evaluate their attribute is a good way of finding the weaknesses and shortcomings of a product that needs to be developed. For this aim a market research was performed to find the best kiosks in Iran's market. The result of conducting market research showed that the newspaper kiosk has no variety in Iran. Therefore, the available samples of other countries were chosen for the benchmark. Technical specifications and design parameters of newspaper kiosks were obtained through design books and manuals. These specifications were classified to several items such as; color, form, material, texture, canopy color, canopy form, canopy material, canopy dimensions, canopy texture, stand color, stand form, stand dimensions, and overall dimensions. The summery of benchmark table is presented in table 2.
4. **Drawing a matrix of product attributes against design characteristics**
   In this step a matrix was prepared by placing the voice of customer in the column and design specification which were obtained through benchmark in the raw.

5. Identifying the relationship between design characteristics and product attributes

The relationships between customer attributes and design characteristics have been assessed as strong, medium or weak in the matrix. The strength of the relationships was indicated by symbols. The orange bold point with a weight of five was used for strong relationship. The orange circle with a weight of three was used for medium relationship and a green triangle with a weight of one for weak relationship. The relationships were identified and sum of each column was calculated and placed in overall importance rating that was shown as EC importance in the bottom of house of quality matrix. Importance ratings can help the trade off decision making process through identifying which design characteristics has more influence on customers' overall perception of the products.

Table 2 the Benchmark Table

design Characteristic	Kensington kiosk	Pausanias kiosk
material	Brass plywood	Stainless steel Fiberglass panel
form	curved	rectangular
color	Yellow and ocher	silver
texture	opaque	opaque
dimension	3*3*4m	3*3.50*2m
Roof material	Stainless steel	Stainless steel
Form of roof	curved	rectangular
Color of roof	silver	silver
Texture of roof	opaque	opaque
Window material	plexiglass	-
Form of window	rectangular	-
Canopy material	-	Stainless steel
Form of canopy	-	rectangular
Color of canopy	-	silver
Texture of canopy	-	opaque
Dimension of canopy	-	3*1m

Kensington kiosk

pausanias kiosk

6. Identifying any relevant interactions between design characteristics specifications

It's also called correlation matrix. Correlation matrix is used to recognized how design specifications support or conflict with one another. The roof matrix of house of quality provides this check. These rating are usually quantified between 2 and -2. In the roof of the matrix, interaction between design characteristics has been identified by numbers. The strong support was shown by 2 and strong conflict by -2. Also 1 indicates weak support and -1 shows weak conflict.

7. **Setting target figures to be achieved for the design characteristics**

Another part of QFD is the targets, which can be set for the measurable parameters of the design characteristics in order to satisfy customer requirements. For setting the targets, reference is required. Reference is made through the comparisons with two competitors, given on the right-hand side of the matrix. The analysis of two competitors helped to find the strength and the weakness of the competitors' products in the market. Two kiosks, Kensington and Pausanias, were compared with a five point scale rating. As it can be seen in figure 9, Kensington kiosk has got better rate in all items (voice of customer) against Pausanias kiosk except the canopy.

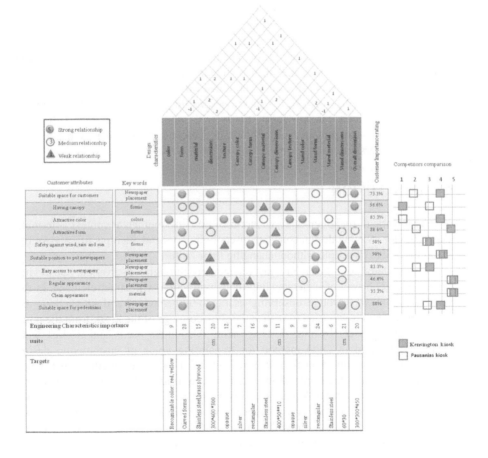

Figure 9 performing QFD Matrix

3   DISCUSSION

As the result of study shows, the most important item for the customer is putting the large number of newspapers on the floor. 90% of people who were interviewed complained about the placement of newspapers. Also the result of observation showed that people have difficulty in choosing and taking the newspapers. As the main function of the kiosk is presenting the newspapers, there is a need for an appropriate way of presenting them. Therefore, designing a suitable place such as good stand is vital. Considering user behaviors, Iranian users' have the habit of reading the titles of newspapers before taking them. In fact, most of the users choose their newspapers by reading the titles and even some of them just read the titles without buying any newspaper. As a result of this behavior, many people gather in a very small place in front of the kiosk, which cause difficulty in having access to the newspapers and the vendor. According to QFD matrix, suitable position of newspapers has a strong relationship with form of stands. The second important problem is the main form of newspaper kiosk. Eighty six percent of people believed that the form of kiosk is not attractive. As QFD matrix explains the attractive form has a strong relationship with the form of kiosk, canopy and stands. In the other hand, EC importance in QFD matrix shows, that the form of kiosk with score of 28 is the most important factor. The next score belongs to the form of stand with the score of 24. Thus, the form of the kiosk is the first important item in design, which should be considered. Even it can influence the existence and form of stand for newspapers, which earlier mentioned as the most important factor for the main function of the kiosk. Obviously the second important design item is the form of stand, which can provide and appropriate placement for the newspapers and organize the users.

4   CONCLUSION

The historical, social, cultural, economical and environmental context determine user behavior. A good design considers users behaviors and attempts to improve the social behaviors of people without changing their culture.

In this study user behavior was recognized through observation. Observing people is a good way of understanding their interaction with street elements in the particular context. Through observing users, their behaviors and interactions are determined. Then it was tried to realize the shortcomings of the kiosk from the point of view of the users by interviewing them. Also their needs and desires were obtained and used in QFD method for translating voice of customer to design specifications. Quality Function deployment is a powerful method that helps designer to make decisions about product's attributes by considering users' requirements. Through QFD matrix it was found that firstly, suitable position for putting newspapers was the most important factor from customers' point of view, which has a strong relationship with form of stand. Secondly, the forms of kiosk and stand are the most important design characteristics that should be considered in new concept of newspaper kiosk. Thirdly, through competitors' comparison it was found what the competitors achieved regarding design characteristics of their products. The outputs of House of Quality matrix can be used for concept generation and evaluation.

REFERENCES

Cross, N. 2000. *Engineering Design Methods: Strategies for Product Design.* 3rd Ed. England: John Wiley & Son Ltd.

Design Council and the Royal Town Planning Institute. 1979. *Street ahead.* London: Design Council.

Devereux, C. 2007. *People power: Designing the perfect costumer,* report for CNN.

Ho, Y., Wang, H. and Lee, R. 2007. The Sustainable Value of Urban Design. *Proceeding of second IASDR (International Association of Societies of Design Research).* Hong Kong, 12-15 November.

Orr, D. W. 2002. *The nature of design: Ecology, culture, and human intention.* Oxford: Oxford University Press.

Rapuano, M., Pirone, P. P., and Wigginton, B. E. 1994. *Open space in urban design: A report* (Revised ed.). Cleveland, OH: Cleveland Development Foundation.

ReVelle, J. B., Moran J. W., and Cox, C. A. 1998. *The QFD handbook.* New York: John Wiley and Sons Inc.

Siu, K. W. M. 2008. Better design quality of public toilets for visually impaired persons: An all-round concept in design for the promotion of health. *The Journal of the Royal Society for the Promotion of Health,* 128(6), 313-319.

Sohn, M. and Nam, T. 2009. Design Method for Sustainable Interaction-Understanding and Applying Unconscious Human Behaviors in Design. *Proceeding of third IASDR (International Association of Societies of Design Research).* Seoul, Korea, 18-22 October.

Yanagisawa, H., Kozuka, Y., Matsunaga, M. and Murakami,T. 2009. Observation support system for recording, reviewing and sharing observed design problems. *Proceeding of third IASDR (International Association of Societies of Design Research).* Seoul, Korea, 18-22 October.

CHAPTER 4

Connectivity Model: Design Methods for Diverse Users

Sunghyun Kang, Debra Satterfield

Iowa State University
Ames, Iowa, USA
Shrkang@iastate.edu, Debra815@iastate.edu

ABSTRACT

When we design a product or an artifact, the design solution should consider the user's social, cultural, and emotional needs as well as their physical constrains. This paper introduces a framework of the connectivity model and demonstrates how this model can be applied to the design process of products and services for diverse users. The connectivity model is a design and evaluation method based on the mental trilogy, *kansei* engineering, and activity theory. This new model contextualizes usability in terms of its social and emotional appropriateness as well as its physical, cognitive, and cultural contexts.

The framework of the connectivity model has been used in product/artifact design development and evaluation especially targeted to diverse user groups including older adults and those with physical and cognitive disabilities. To demonstrate how the connectivity model can be applied to the design process, a design development process model is also introduced in this paper.

Keywords: design methods, connectivity model, diverse users

1 INTRODUCTION

In the field of design, experience design treats the designed artifact (i.e., the IT product or the *environment* being designed) as part of a holistic system; that is, it considers the artifact in the context of its use. From this perspective, the social and emotional needs of the user are considered along with the physical, organizational, and social constraints of the project. In our previous work (Kang and Satterfield, 2009), we proposed the connectivity model as a framework for design and evaluation (figure 1), and presented a case study on how the Connectivity Model can be adopted as a design and evaluation process. The Connectivity Model demonstrates the fact that design solutions should only be considered as optimal solutions when they encompass not only the local but also these global constraints and requirements. Of course, most projects that consider these elements – the social, emotional, and physical constraints – will have a plethora of optimal solutions; nevertheless, the framework offers a useful model for designing and evaluating those factors that will be instrumental in constructing design solutions that address both local and global solution requirements.

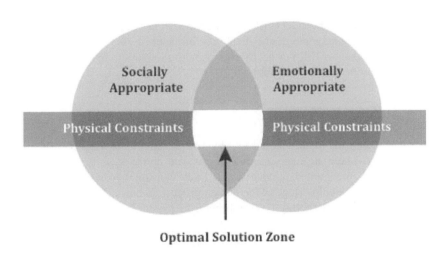

Figure 1 Optimal Solution Zone based on the intersection of Physical Constraints and spheres of Socially and Emotionally Appropriate Behaviors

The Physical constraints are related to the design elements, size, shape, weight, color, material, and so on. Also they relate to the target audience's physical ability including vision, mobility, cognitive ability, and sensory recognition. As we stated in our previous paper, the framework of *Kansei* engineering methods are well developed to understand people's emotions and experiences as they relate to physical constraints (Kang and Satterfield, 2009). We have adopted *Kansei* engineering methods in the design evaluation process to understand people's emotion and experiences towards a product/artifact.

Social appropriateness is considered in the context of gender, age, organization, and society. Activity theory (Vygotsky, 1978) sought to understand people's activities based on the social and cultural context.

Emotional appropriateness is considered how the artifact empowers the individual to meet needs, goals or concerns. A cane, for example, is an assistive tool for working but some people may be embarrassed by using a cane. The testimonial, "I realize that being ashamed of using a cane has kept me from doing things, and using the cane helps me do more, safely (aota.org)," shows that emotion encompasses physiological, affective, behavioral, and cognitive components (Brave and Nass, 2003). In the cane case however, the motivation to work safely encourages the user to overcome the emotional feeling of embarrassment. LeDoux (2002) defined motivation as "neural activity that guided us toward goals, outcomes that we desire and for which we will exert effort, or ones that we dread and will exert to prevent, escape from or avoid." Finding motivating factors is very important in the design development process. Fogg (2009) suggests that there are three motivator factors with three pairs; pleasure/pain, hope/fear, and social acceptance/rejection. According to Fogg, hope is most powerful motivation factor in his Behavior Model. Motivation, ability and triggers are the three factors that Fogg includes in his Behavior Model (FBM). Someone who has high motivation and high ability will most likely increasing their performance to the targeted behavior. However, the behavior will not occur even though motivation and ability are high without an appropriate trigger (2009).

2   DESIGN INQUIRIES

Understanding the target audience's physical constraints is very critical in the design process. There are many cases that physical ability is fully functional but cognitive ability are limited or vice versa. Each constraint in context should be addressed in the beginning stage of the design process and also evaluated at the end of the design process. Usability is an important factor. It makes the artifact easy to use while desirability is an important factor that enhances the motivation of the target audience to use the artifact.

Also understanding how the society accepts the concept or perceives behavior.

We have addressed following questions in the design each development and evaluation process.

2-1. Identifing Target Audience's Needs

Fogg states that "increasing motivation is not always the solution" and he suggests "increasing ability (making the behavior simpler) is the path for increasing behavior performance" (2009). As illustrated in figure 1, understanding target audience's physical constraints are backbone of the Connectivity model.

The following three key areas to understanding the target audience's needs should be answered in the design process:

Physical needs: Do the physical aspects of this design meet the physical needs of the target audience in terms of size, weight, material choices, interaction qualities, sensory qualities, and other visceral considerations?

Social needs: Is this design appropriate to the social needs of the target audience? Is it age appropriate? Is it appropriate to the gender and culture of the target audiences?

Emotional needs: Does this design have a quality of empowerment or desirability that appropriately meets the needs of the target audience?

2-2. Identifying Target Audience's Motivating Factors

Fogg et al. (2008) state that "creating successful human-computer interaction requires skills in motivating and persuading people." But how do we trigger the target audience's motivation? Satterfield (2009) has used an approach through ethnographic observation to identify the audience's primary motivational factors especially for children with cognitive and development disabilities. Satterfield (2009) suggested that following questions should be answered to identify motivation factors:

> **Identify primary motivating factors in human behavior:** Have the primary motivating factors of the various target audiences been identified and incorporated into a meaningful design solution?
> **Identify the role of emotion in human interaction design:** Have the primary motivating factors of the various target audiences been identified and incorporated into a meaningful design solution?
> Does the solution allow the user to negotiate for reinforcement by allowing the user to tailor the experience?
> **Identify the role of human interaction in communication**: Can this design be a catalyst or facilitation tool for human interaction between children with autism, epilepsy, and cerebral palsy and their neurologically typical peers?

2-3. Identifying Communication Methods

It is important to identify what is the best way to communicate the functionality of the interface and interaction with the targeted artifact. Satterfield (2009) suggested that following questions should be answered to enhance communication (2009):

> **Identify multi channel sensory systems that can enhance communication**: Have the sensory systems of the body been researched and analyzed? Was information from fields such as perceptual psychology, occupational therapy, and neurology used in order to inform the design, interaction, educational, and communication decisions?
> **Identify micro and macro sensory experiences:** Are the Communication experiences designed to make use of both fine and gross motor involvement for the user? Has research in body movement, spatial orientation and tactile responses been used as a basis for developing effective educational experiences?
> **Effectively utilize multiple learning styles**: Does this solution recognize and effectively incorporate strategies that accommodate multiple learning styles of target audience?

2-4. Identifying Design Elements

This is step the final step to identify design elements that can motivate the target audience group. The look and feel of the interface can create an immediate connection with the social, emotional, and cultural identity of the various target audiences. Visual interface design should answer following areas:

Color: Does the dominant color in this design set the appropriate emotional tone for the target audience? Is there an identifiable color palette? Does the color palette support the emotional tone expected by the target audience? Does it have appropriate color relationships and an effective visual hierarchy that allows the design to effectively communicate its intentions and functions?

Typography: Does the choice of typography effectively create the correct visual ethos for this design? Does it allow for effective typographic hierarchy and information design?

Style and material selection: Does the use of style and material selection create the correct visual ethos for this design? Do the material choices support the style and function of the design in appropriate ways? Do the material combinations create an effective visual hierarchy that enhances the style and usability of the design?

3   RESEARCH METHODS

Connectivity Model uses various research methods including *Kansei* engineering to measure audience's emotional feeling and their perceptions, combined with ethnographical research methods. A focus group study, interviews, surveys are also adopted in the design development and evaluation process. Through interviews and a focus group study qualitative data sets are collected. And via a survey and *Kansei* engineering methods quantitative data sets are collected. Being able to understand the role and scope of qualitative and quantitative data sets and how to work between these two sets is critical to understanding the problem to be solved and seeing the possible solutions. Qualitative data has a sensory richness and has many possible solutions. Quantitative data, on the other hand, has a more exact or limited set of solutions. In essence, for the designer, quantitative data indicates what the problem is and qualitative data indicates how the problem could be solved. It is crucial that the designer have access to both sets of data and knows how to work with them. Quantitative data identifies who has a problem and what the problem is like. Qualitative data helps designers to understand the social and emotional aspects of their target audiences and reinforces the idea that the target audience is real group of people with real needs. By identifying with the people and the needs, a greater sense of urgency is created, thus making problem solving timelier and more relevant. It also gives the designer an evaluation tool by providing a situation that can serve to determine if the proposed design is in fact a viable and creative solution to their specific problem.

Usability studies allow us to identify how well the interface design communicates its function through the implementation of appropriate physical and visual affordances. Through the usability study we also collect qualitative data as well quantitative data sets.

3 APPLICATION OF CONNECTIVITY MODEL TO DEVELOP A DECISION AID TOOL FOR REPRODUCTIVE HEATH CARE

The Connectivity Model has been applied for various projects. Figure 2 presents how the Connectivity Model was adopted in the design and evaluation process to develop a decision aid tool for reproductive health care targeted at college student ages from 18 to 25 years old.

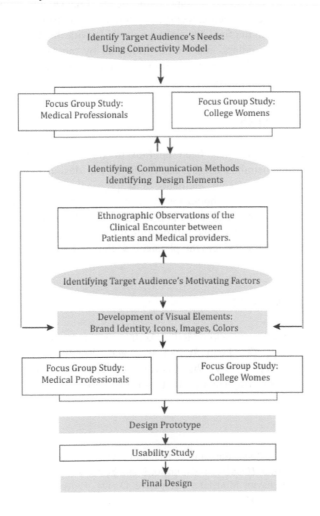

Figure 2 Design and Evaluation Process of Designing a Decision Aid Tool for Reproductive Health Care for College Women with Connectivity Model

The goal of the decision aids tool is to provide information that young women can use to communicate their needs with medical professionals. As the first step, we conducted a study of existing material to examine the needs for this target audience. At the same time we conducted a focus group study with college women and another focus group with medical professionals to identify and understand both target audience's perceptions on existing materials and needs for future designs. Ethnographic observations were used to observe the interaction between patients and medical provides during the clinical encounter to identify the motivating factors of both groups. The clinical encounter observation provides information on how to prepare the communication strategies for the final design (Satterfield et al., 2011).

With the analysis data from two focus group data sets and observation of the clinical encounter, we developed a brand name, icons, images, and color palettes. With six design variables in each category, we conducted focus group studies with young women. We are in the final design stage working with the design preferences in brand identity, icons, colors, and images indicated by our target audience. The final design will be a web-based intervention communication tool that works for both smart phones and computers.

4   CONCLUSIONS

The Connectivity model has been applied to various projects. This model considers physical constraints as the backbone in the design development and evaluation process. The design elements and physical constraints should be acceptable to society and they should be emotional appropriate to target audience group. This model is a very flexible method to adopt into the design and evaluation process for any target audience. The ultimate goal of the Connectivity Model is to provide a design tool that considers the physical, social, and emotional needs for all people.

ACKNOWLEDGMENTS

The authors would like to acknowledge Andrea Quam, Nora Ladjahasan, Cyndi Wiley, Brandon Alvarado, Leah Willadesen, and Whitney Farrell who have participated in the reproductive health care projects that have adopted the Connectivity Model as a design and evaluation framework.

REFERENCES

Brave, S., and Nass, C., 2008. Emotion in Human-Computer Interaction. In *The Human Computer Interaction Handbook*, eds. Jacko, J. and Sears, A. CRC Press.

Fogg, B.J., 2009. A Behavior Model for Persuasive Design. *Proceedings of Persuasive '09*, Claremont, California, USA.

Fogg, B.J, Cuellar,G., and Danielson, D., 2008. Motivating, Influencing, and Persuading Users: An introduction to Captology, In *The Human Computer Interaction Handbook*, eds. Jacko, J. and Sears, A. CRC Press.

Kang, S, R., and Satterfield, D., 2009. Connectivity Model: evaluating and designing social and emotional Experiences. *Proceedings of IASDR*, Seoul, Korea.

Satterfield, D., 2009. Designing Social and Emotional Experiences for Children with Cognitive and Developmental Disabilities. *Proceedings of the Interactive Creative Playwith Disabled Children workshop*, The 8th International Conference on Interaction Design and Children, Como, Italy.

Satterfield, D., Kang, S. R., Bruski, P., Malven, F., Quam, A., and Ladjahasan,. 2011. Developing a Reproductive Health Care Decision Aid for Women Ages 18-25 and Their Medical Providers. *Proceedings of IASDR*, Delft, Netherland.

The American Occupational Therapy Association, Occupational Therapy Helps Prevent Decline in Seniors, Accessed Feb. 20, 2012.
http://www.aota.org/News/Consumer/Well-Elderly.aspx,

Vygotsky, L. S., 1978. Mind in society: *The development of higher psychological Process*, Cole, M., John-Steiner, V., Scribner, S., & Souberman, E. (Eds.), Cambridge, Mass. and London, England: Harvard University Press.

CHAPTER 5

Educational Play Experiences for Children with Cognitive and Physical Disabilities

Debra Satterfield, Sunghyun Kang

Iowa State University
Ames, Iowa, USA
debra815@iastate.edu, shrkang@iastate.edu,

ABSTRACT

Many children are currently diagnosed and living with cognitive and developmental disabilities such as autism, epilepsy, and cerebral palsy. The Center for Disease Control and Prevention (CDC) defines developmental disabilities as a diverse group of severe chronic conditions that are due to mental and/or physical impairments. Individuals with developmental disabilities have problems with major life activities such as language, mobility, learning, self-help, and independent living. Therefore, there is a critical need to develop effective and motivating educational experiences that will mediate social interactions between typical children and children with cognitive or developmental disabilities such as autism, epilepsy and cerebral palsy. The purpose of this research to determine how to design educational play experiences that effectively engage and facilitate children with cognitive and physical disabilities and their neurologically typical peers into effective collaborative learning situations. As a research method, focus group studies with parents, educators, and caregivers of persons with cognitive and physical impairments such as autism, epilepsy, and CP were recruited to identify how designers can best create effective learning experiences that meet the social, emotional, behavioral, and motivational needs of all of these constituent groups. The research data and its implications for evidence-based design will be discussed.

Keywords: educational play, children, cognitive and physical disabilities

1   INTRODUCTION

For children with autism there are many forms of therapy, but fewer activities and outlets for socialization and play. In addition, families with children with autism spectrum disorders (ASD) often struggle with many aspects of socialization and communication as they pertain to peer relationships and making friends for kids with ASD. Therefore identifying those aspects of autism that form the greatest barriers to socialization and communication with typical peers is of critical importance in designing effective educational play experiences that can facilitate better peer relationships.

In order to identify those behaviors and issues that cause the greatest problems for ASD kids, a survey was conducted with parents and caregivers. The survey examined the physical, social, emotional, motivational and behavioral aspects of autism as they impact daily living, communication and socialization.

The ultimate goal of this research to use the data from this survey to inform the design of educational play experiences that effectively facilitate communication and socialization between children with cognitive and physical disabilities and their neurologically typical peers. In this preliminary study, we collected and analyzed information from parents, educators, caregivers, and medical providers through these focus group studies.

2 RESEARCH METHOD

The Connectivity Model (Kang and Satterfield, 2009) is a design and evaluation methodology based on; the mental trilogy, activity theory, and *Kansei* Engineering. It also incorporates ethnographic research strategies from information architecture and educational strategies from Dewey and Vygotsky that are specifically designed to understand and work with the idiosyncratic nature of cognitive and developmental disabilities (Satterfield 2010). These research strategies are used to inform data collection through focus group studies and ethnographic observations. This information is then combined with information design analysis theories to create a comprehensive design and evaluation framework for creating information tools to support decision-making (Satterfield et al, 2011).

In our preliminary study we are using the Connectivity Model to examine the impact of autism on communication and socialization based on the criteria of physical, cognitive, behavioral, social, emotional, and motivational aspects of peer interactions. Through focus group studies, this preliminary research is examining the following questions:

- How long and in what capacity have you known or interacted with children with physical and cognitive disabilities?
- What are the main social skills that negatively impact a child with physical and cognitive disabilities in typical peer social interactions?

- What are the main communication skill issues that impact a child with physical and cognitive disabilities in typical peer social interactions?
- What are the main behavioral issues that impact a child with physical and cognitive disabilities in typical peer social interactions?
- What are other factors that negatively impact the social relationships of children with physical and cognitive disabilities from their peers?
- What assistive or augmentative technologies do you use for social or communication purposes with children with physical and cognitive disabilities?
- How effective are these technologies and what are their main strengths and weaknesses?
- What types of situations or conditions best facilitate social interactions between children with physical and cognitive disabilities and their neurologically typical peers?
- How important are social relationships with their peers to children with physical and cognitive disabilities?
- What other information would you like to add with regard to the needs of children with physical and cognitive disabilities?

3   FINDINGS AND DISCUSSION

This preliminary study is to examine the role of autism and autistic features as they impact peer interactions, communication, socialization, and quality of life for children with autism. The outcomes of this research will be used to determine how to best target communication strategies to enhance educational play experiences for children with ASD and their neurologically typical peers for the goal of improved socialization. Our focus group studies seek to identify the most problematic issues of autism in the areas of socialization, communication, and behavior as determined by parents, caregivers, educators, and medical providers. The outcomes of this study will give insight with regard to how to best identify and manage autism and autistic features with regard to their impact on the social and emotional lives of children with autism and their families. This preliminary research also brings together mixed groups of stakeholders in focus group settings and encourages them to explore these questions from their own viewpoints. In the focus groups, stakeholders are allowed to interact with members of other stakeholder groups. These combined groups encourage better understand of the priorities of members of other stakeholder groups. From this preliminary research we hope to gain valuable knowledge about how to effectively combine stakeholder groups in ways that allow them to interact and identify their differing priorities.

In this ongoing study, this paper will discuss three focus group sessions with a total of 5 parents and one educator. Some participants have more than one child with varying degrees and types of autistic features. These 5 parents have a total of 8 children, seven boys and one girl. One educator who participated in the focus group study has over 20 years of experience working with autism and ADHD (Attention deficit hyperactivity disorder). The parents have children ages 5 to 12 years with ASD and ADHD. Inclusion in the focus groups was based on self-reported diagnosis of autism or a related cognitive impairment with autistic features. Some questions from the focus groups are combined into one response in the analysis due to the similarity and relatedness of the questions and answers.

3-1 Social skills that negatively impact a child with physical and cognitive disabilities in typical peer social interactions

Most parents expressed that their children do not know what are personal boundaries and how to express themselves verbally. For example, the children lack a firm understanding of issues such as when to say things and what to say. For example they lack a certain level of sensitivity to the feelings of others and may make factual, yet inappropriate statements such as, "you are fat." In most cases, participants noted that the children exhibited language and behaviors that are not age appropriate. Also they are not doing well in transitioning from one play to the other. For example, going from recess back into school might trigger a tantrum or serious resistance to leaving one activity for another. Also, turn taking among peers posed a problem for many kids.

3-2 Main communication skill issues that impact a child with physical and cognitive disabilities in typical peer social interactions

Similar to their issues with personal boundaries, many children with autism don't understand innuendos, sarcasm, tone of voice as they change the meaning of spoken words. One parent noted that nothing is intuitive to these children. For one non-verbal boy, his parent expressed that "his number one issue is being able to successfully communicate...it's an absolute nightmare and a struggle." Some parents have used Picture Exchange Communication System (PECS) as a method of communication. However, picture exchange systems may be limited in what they can communicate and how they are organized for access and retrieval. Several parents noted that children with autism experience difficulty in discussing their feelings and may either avoid the subject, internalize their feelings, or act out as a way of expressing themselves.

3-3 Main behavioral issues that impact a child with physical and cognitive disabilities in typical peer social interactions

The educator said that "any behavior issues are going to have social impact. So the behavior issue then becomes a social issue because they've just had a meltdown now in front of their classmates." Most participants expressed that calming an autistic child in public places is very hard and frustrating. Many children have inappropriate behaviors such as making inappropriate noises, inconsolable tantrums, and the need for quiet or private spaces to regain composure. It was noted that in some cases, rather than removing the autistic child from a class or group during these episodes, all of the other children have been coached in how to leave the area where the autistic child is acting out.

Other parents noted that lack of boundaries with regard to ownership is a problem and understanding when they can and cannot touch objects. In an extreme case, another parent described sensory related touching behaviors involving toileting and smearing of fecal matter.

3-4 Assistive or augmentative technologies for social or communication purposes with children with physical and cognitive disabilities

The educator mentioned b-Calm, Kush balls, and PECs communication as an assistive tool to communicate with children. Most parents mentioned that they have used PECs. Some other tools have been used among participants are Dynavox, PODS, and social stories. They expressed that some tools such as PODS and Dynavox are not easy to use.

Some participants suggested i-pads are useful for social interaction because children can read a story together and do music or art activities together. Most participants mentioned that they've tried many different gadgets but most worked only for a very short time. The teacher noted a preference for paper based solutions like note cards or games.

3-5 Types of situations or conditions can best facilitate social interactions between children with physical and cognitive disabilities and their neurologically typical peers

The educator suggested that building social relationships between children with physical and cognitive disabilities and typical kids is a good way to facilitate socialization. Some ideas are being adopted in education as a way to form social bonds such as recess buddies, lunch buddies, and school bus buddies. Keeping children with ASD engaged in a variety of activities would be a good way to facilitate social interaction. One parent noted that her daughter prefers to be active and entertained by her peers. Also, it was noted that preventing bullying will also help children with autism be better accepted. From this discussion the question arose how do you raise a kid to be more accepting of people with disabilities? One parent noted that she had approached the school to discuss kids with autism and bullying as a way to prevent it. It was also no two kids with autism are alike and what works with one child may or may not work with another. They also noted that what works today may not work tomorrow even with the same child.

4   CONCLUSION AND FUTURE STUDY

An observation from these focus groups was the change in situations that occurs for children as they grow older with ASD. This is especially true for the higher functioning children in the Asperger's Syndrome diagnosis. One parent said, "now it's becoming a big deal. When he was in third, fourth, fifth grade... You know it hurt me more than it hurt him, when he had lack of friends or if I saw kids roll their eyes. I mean the mother bear in me would come out and I would want to smack them upside the head. But for him now at this age, at twelve, sixth, seventh grade, it's really starting to be a big deal to him because everything is intensified. I mean you're already at that age. You're already self-conscious. You think you're a dork." The teacher added, "as a teachers, that's one of our main jobs with these kids is to help build those peer relationships at a very young age, knowing that they may change by the time they get out there to middle school because suddenly this person doesn't want to hang out with them any more." For the lower functioning children the socialization problems seem to be less pronounced. They seem to be more insulated from the social expectations and accepted for their own abilities and strengths. This seems to stem from the fact that the higher functioning children seem to be just "normal enough" that the social expectations are still high for them among their classmates.

There was strong agreement that socialization among peer groups is important for all children with ASD. There was also a strong consensus that odd or disruptive behaviors can negatively impact the quality of social relationships for children with autism. Things such as inappropriate touching, loud noises, lack of age appropriate language, the inability to understand boundaries, and the lack of an ability to interpret innuendo, sarcasm, or other affective aspects of language greatly impedes social relationships for children with autism. There was also a strong consensus that these social and behavioral issues are extremely frustrating and embarrassing for families and caregivers, thus greatly increasing the stress levels experienced by these people.

The preliminary findings of this research suggest a strong need to connect design to the social, behavioral and sensory needs of children with ASD. By using this data to design socialization activities and learning experiences, it is hoped that designers can create design solutions for children with autism and their families that have a greater connection to the real world issues of autism.

ACKNOWLEDGMENTS

The authors would like to acknowledge Hannah Hunt and Anson Call for their contribution to this research and the In•Site Research Group at Iowa State University.

REFERENCES

Kang, S, R., and Satterfield, D., 2009. Connectivity Model: evaluating and designing social and emotional Experiences. *Proceedings of IASDR,* Seoul, Korea

Maenner, M and Durkin, M., 2010. Trends in the Prevalence of Autism on the Basis of Special Education Data. *Pediatrics 2010*; 126; e1018; originally published online October 25, 2010; DOI: 10.1542/peds.2010-1023.

Satterfield, D., 2010. Play•IT: A Methodology for Designing and Evaluating Educational Play Experiences for Children with Cognitive Disabilities, *Proceedings of The 7th Design and Emotion Conference*, Como, Italy.

Satterfield, D., Kang, S. R., Bruski, P., Malven, F., Quam, A., and Ladjahasan,. 2011. Developing a Reproductive Health Care Decision Aid for Women Ages 18-25 and Their Medical Providers. *Proceedings of IASDR,* Delft, Netherland.

CHAPTER 6

Universal Product Family Design for Human Variability and Aesthetics

Jesun Hwang[1], Seung Ki Moon[2], Yun Ho[2], and Changsoo Noh[1]

[1]Samsung Electronics
Suwon, South Korea
Jesun.hwang, chsnoh @samsung.com

[2]Nanyang Technological University
Singapore
skmoon, @ntu.edu.sg, HOYU0011@e.ntu.edu.sg

ABSTRACT

The present research is motivated by the need to create specific methods for universal product family design based on human variability. Product family design is a way to achieve cost-effective mass customization by allowing highly differentiated products to be developed from a common platform while targeting products to distinct market segments. In this paper, we extend concepts from mass customization and product family design to provide a basis of universal design methods in product design. The objective of this research is to propose a method for identifying a platform for the families of universal products in economical feasible design concepts and integrating human variability into the design process to improve usability and performance as well as aesthetics. We generate platform design strategies for a universal product family based on performance utilities to reflect human variability and aesthetics. A coalitional game is applied to model a strategic design decision-making problem and evaluate the marginal profit contribution of each strategy for determining a platform design strategy in the families of universal products. To demonstrate implementation of the proposed method, we use a case study involving a family of consumer products.

Keywords: universal design, product family and platform design, human variability, aesthetics

1   INTRODUCTION

Universal design is a recently suggested term for designing for persons with a disability (Mace, 1985). Universal design specifically suggests the concepts of equity and social justice. Also, in the context of separate is not equal, universal design suggests the design of solutions that simultaneously and equally serve both the fully able and not fully able. Design of new products for everyone requires numerous functions for many individuals and groups often separated by capabilities and limitations due to age and disabilities (Preiser and Ostroff, 2001). Innovative companies that generate a variety of products and services for satisfying customers' specific needs are invoking and increasing research on mass-customized products, but the majority of their efforts are still focused on general consumers who are without disabilities (Beecher and Paquet, 2005). In a highly competitive market, universal design can be considered as appropriate marketing strategies by providing the broadest market segment.

For mass customization, companies are increasing their efforts to reduce cost and lead-time for developing new products and services while satisfying individual customer needs. Mass customization depends on a company's ability to provide customized products or services based on economical and flexible development and production systems (Silveria et al., 2001). By sharing and reusing assets such as components, processes, information, and knowledge across a family of products and services, companies can efficiently develop a set of differentiated economic offerings by improving flexibility and responsiveness of product and service development (Simpson, 2004). Product family design is a way to achieve cost-effective mass customization by allowing highly differentiated products to be developed from a common platform while targeting products to distinct market segments. A product family is a group of related products based on a product platform, facilitating mass customization by providing a variety of products for different market segments cost-effectively. A successful product family depends on how well the trade-offs between the economic benefits and performance losses incurred from having a platform are managed (Simpson et al., 2005; Moon et al., 2008).

The present research is motivated by the need to create specific methods for universal product family design based on human variability. In this paper, we extend concepts from mass customization and product family design to provide a basis of universal design methods in product design. The objective of this research is to propose a method for identifying a platform for the families of universal products in economical feasible design concepts and integrating human variability into the design process to improve usability and performance as well as aesthetics. We generate platform design strategies for a universal product family based on performance utilities to reflect human variability and aesthetics. One approach to universal design can be to focus on diversity in creating products, services, and environments, which the design is facilitated to a range of customers' needs. To determine a platform strategy that consists of common modules, we will investigate which functional modules will be more contributions in a universal product family. A coalitional game is applied to model a strategic design decision-making problem and evaluate the marginal profit contribution of each strategy for determining a platform design strategy in the families of universal products. Game theoretic approaches provide a rigorous framework for managing and evaluating strategies to achieve players' goals using their information and knowledge (Osborne and Rubinstein, 2002).

The remainder of this paper is organized as follows. Section 2 describes the proposed design method for developing a universal product family using a coalitional game. Section 3 gives a case study using a family of TV remote controls. Closing remarks and future work are presented in Section 4.

2. UNIVERSAL PRODUCT FAMILY DESIGN

Figure 1 shows the proposed process for developing a universal product family based on the top-down and module-based approaches in product family design. The proposed method consists of three phases: (1) generate design strategies, (2) identify design preference, and (3) determine a design strategy. Customer needs can be collected through surveys and market studies. The market study begins by establishing target markets and customers. In the initial phase, customer needs based on human variability and aesthetics are analyzed to develop market segments for a universal product family. The customer needs are also used to identify required product functionality for individual products and across a range of products. In universal product design, customers' preference is determined by information related to customers' accessibilities or functional limitations. Product reference information can help develop market segmentation for universal product family design by identifying an initial platform based on functional requirements. During conceptual design, products can be designed based on functional requirements, and their functional modules can also be determined. In particular, a family of universal products can be first configured by defining a product platform. A product platform consists of several common modules that can be shared across a family of product. Then, platform design strategies are generated by module-based design concepts. After evaluating different platform design strategies using universal design principles and a game theoretic approach, a platform design strategy is determined to generate universal product family concepts according market segmentations and design constraints.

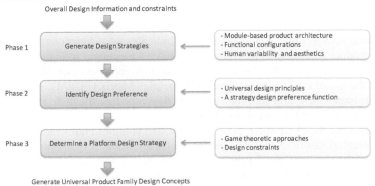

Figure 1: The Proposed Process of Developing a Universal Product Family

2.1 Phase 1: Generate Design Strategies

The universal product platform framework is built on representing the product space in terms of five different modules: common modules, variant modules, universal modules, accessible modules, and typical modules (Moon and McAdams, 2010). The notion of common and variant modules is generally a well understood concept in product family design. Common modules are those shared across the product family regardless of the module's characterization with respect to typical and accessible products. In general, these common modules are suitable candidates for establishing the product platform. Variant modules refer to the differing elements used to introduce variety into a range of products in the family. The common elements plus the variant elements combined create a product family. The framework used to design a product family here is modular, but the notions of common and variant need not be limited to a modular framework.

A module based product family strategy allows for the design and production of economically viable universal product families. Specifically, modules for universal design can be categorized into: 1) universal; 2) accessible; and 3) typical modules. Universal modules are those that are the same in function and form for both typical and disabled users. Accessible modules provide specific functionality or form solutions for persons with limitations due to age and disabilities. Typical modules contain functional and form solutions, or both, that are not suitable for user with a disability. For generating product family configuration concepts, accessible and typical modules can be used to build the product platform with respect to the economy of scale in product development. For example, anthropometric data can provide designers with design alternatives when determining the dimension of an accessible or typical module. Dimensions of modules based on body measurements lead to the modules that are better suited for the intended users' anthropometry. The next section introduces a product preference model for evaluating preference and performance in a universal product.

2.2 Phase 2: Identify Design Preference

To evaluate and measure preference of a product, we use a strategy quality function that is positively related to functional accessibility level (FL) and usability level (UL) as follows (Moon and McAdams, 2010):

$$Q=f(FL, UL) \tag{1}$$

The functional accessibility level represents the interaction of product functionality and product accessibility: it is a measure that indicates what functions are needed to make a product accessible to individuals who have a functional limitation as defined in the ICF (WHO, 2001). To determine the functional accessibility level, we propose the use of the Function-Universal Principles Matrix (FUPM). This matrix is based on impairment and usability measure developed in the ICF and the seven principles of universal design (Connell, 1997).

Table 1 shows a FUPM template. The first two columns enumerate and then list all the potential functions that may be needed by all products in the product family. Across the header row, the 7 principles of universal design are recorded. The last two columns contain the functional accessibility level and the usability level.

Table 1: The proposed Functions-Universal Principles Matrix

No	Function	a	Universal principle 1	Universal principle 2	...	Universal principle 6	Universal principle 7	Functional accessibility level value	Usability level
1	F1	a_1	$a_1\lambda_1up_1$	$a_1\lambda_2up_2$	...	$a_1\lambda_6up_6$	$a_1\lambda_7up_7$	fl_1	ul_1
2	F2	a_2	$a_2\lambda_1up_1$	$a_2\lambda_2up_2$	...	$a_2\lambda_6up_6$	$a_2\lambda_7up_7$	fl_2	ul_2
...	...	...	...	...	...	...	...	...	...
i	F_i	a_i	$a_i\lambda_1up_1$	$a_i\lambda_2up_2$	...	$a_i\lambda_6up_6$	$a_i\lambda_7up_7$	fl_i	ul_i
...	...	...	...	...	...	...	...	...	...
n	F_n	a_n	$a_n\lambda_1up_1$	$a_n\lambda_2up_2$	...	$a_n\lambda_6up_6$	$a_n\lambda_7up_7$	fl_n	ul_n
	λ		λ_1	λ_2	...	λ_6	λ_7	-	-

In the FUPM, the values, fl_i ($i = 1,2,...,n$), of the functional accessibility level for each function can be calculated as follows:

$$fl_i = \sum_i^n a_i \lambda_p up_{i,p}$$ (2)

where a_i is the degree of importance for ith function in terms of accessibilities and the degree is determined based on the accessibility of impairment as follows:

$$a_i = \begin{cases} 1 & \text{for No impairment} \\ 2 & \text{for Mild impairment} \\ 3 & \text{for Moderate impairment} \\ 4 & \text{for Severe impairment} \\ 5 & \text{for Complete impairment} \end{cases}$$ (3)

And, λ_p is the importance weight of pth universal principle in terms of functions ($p = 1,2,...,7$) and can be determined by the Analytical Hierarchy Process (AHP) or group decision-making methods based on product's characteristics and utilization. up_p is a binary variable (0, 1) for indicating the dependence between functions and the pth universal principle.

For the usability level, ul_i, we categorize the usability of a function into five levels based on the difficulty of using the function with respect to impairment and capacity limitation (WHO, 2001): (1) No, (2) Mild, (3) Moderate, (4) Severe, and (5) Complete difficulties. The value of the usability level can be determined as follows:

$$ul_i = \begin{cases} 5 & \text{for No difficulty} \\ 4 & \text{for Mild difficulty} \\ 3 & \text{for Moderate difficulty} \\ 2 & \text{for Severe difficulty} \\ 1 & \text{for Complete difficulty} \end{cases}$$ (4)

The expected strategy quality, Q_i, for function i can be estimated by an expected quality function: $f^i : FL \times UL \mapsto \Re$. Hence, the real number of $f^i(FL, UL)$ represents the quality of strategy i having accessibility level FL for usability level *UL*. For example, the expected quality for strategy *i* can be determined as:

$$f^i(FL,UL) = fl_i \times ul_i \qquad (5)$$

The proposed strategy quality function will be applied to measure accessibility for determining product qualities in terms of platform design strategies. The next section discusses a coalitional game model for determining a platform design strategy.

2.3. Phase 3: Determine a Platform Design Strategy

A coalitional game is designed to model situations wherein some of players have cooperation for seeking a goal in a game (Osborne and Rubinstein, 2002). A coalitional model focuses on the potential benefits of the groups of players rather than individual players. In the coalitional model, the sets of payoff vectors are used to represent value or worth that each group of individuals can achieve through cooperation. In this paper, we employ a coalitional game to model module sharing situations regarding human variability and solve the functional module selection problem in given universal product family design. To determine modules for a platform, we decide which functional modules provide more benefit when in the platform based on the marginal contribution of each module.

We assume that each module in a product can be modeled as a player. Then, consider the following module selection problem for platform design. Each group of players (coalition) can be represented as a platform design strategy for a universal product family and be independent on the remaining players. To determine modules for platform design, we consider the set of all possible coalitions and evaluate the benefits of coalitions based on individuals' preferences.

In order to formulate the proposed scenario as a coalitional game, we must first identify the set of all players, N, and a function, v, that associate with every nonempty subset S of N (a coalition) (Osborne and Rubinstein, 2002). A real number $v(S)$ represents the worth of S and the total payoff that is available for division among the members of S. And, v satisfies the following two conditions: (1) $v(\varnothing) = 0$, and (2) (superadditivity) If $S, T \subset N$ and $S \cap T = \varnothing$, then $v(S \cup T) \geq v(S) + v(T)$. Based on the definition of the coalitional game, the proposed game can be defined as:

- N: players who represent (variant) modules
- $v(S)$: the benefit of a coalition, $S \subset N$

where a coalition, S, represents a platform design strategy that consists of several modules. In this research, we use the Shapley value to analyze the benefits of family design and determine modules for platform design (Shapley, 1971). The Shapley value is a solution concept for coalitional games and is interpreted as the expected marginal contribution of each player in the set of coalitions.

Based on the results of marginal contributions for variant modules, we can determine a platform strategy according to market segmentations and design constraints. The selected platform strategy provides a guideline for generating universal product family design concepts. A successful universal product family depends balancing the tradeoffs between economic benefits and accessibilities incurred from having a platform. In the next section, the proposed method is applied to determine a platform design strategy using a case study involving a family of TV remote controls.

3. CASE STUDY

To demonstrate implementation of the proposed method, we investigate a family of TV remote controls that are operated by a touch-screen. These products offer the opportunity to create a product family that is able to accommodate to the wide spectrum of user needs. The remote control design provides a good example of common and different sizes for remote controls related to human variability. We define the seven principles of design when reinventing a remote control as shown in Table 2.

The objective in this case study is to determine a platform design strategy represented by usability for a touch-screen subject to hand sizes. This case study focuses on how to determine the marginal contributions of designs related to the dimension of the touch screen for the new platform design of a remote control family using the proposed game at the conceptual design stage of development.

Table 2: Universal design principles for a TV remote control

Universal Design Principles	Description
Equitable Use	A remote control design buttons should be within easy reach of most users in terms of comfort and ease. The number of menu icons and arrangement of the touch-screen phones are usually determined on a basis of button size and space of buttons.
Flexibility in use	The design of a remote control should also fit nicely into the palm of the users and can be manipulated by both the left and right hand users.
Simple and Institutive Use	The design of a remote control should be user friendly and easy to use even for a novice, regardless of user's experience, knowledge and language skills.
Perceptible Information	Remote control design communicates essential information effectively to the users, regardless of ambient conditions or the user's sensory abilities.
Tolerance for Error	The design of a remote control should be able to minimize hazards and the adverse consequences of accidental and unintended actions. Alternatively, the remote control could hibernate automatically to save power if it is not in use after some lapse of time.
Low Physical Effort	The design of a remote control should facilitate the use such that it is efficient and comfortable and with minimum effort. It is necessary to understand the negative impact of the different controls when using the thumb for touch screen devices.
Size and Space for Approach and Use	Appropriate size and space is to provide for approach, reach, manipulations and use regardless of user's body size, posture or mobility.

3.1 Phase 1: Generate Design Strategies

According to different hand widths and lengths, we can generate the dimension of remote controls that have various dimensions. In this paper, we use anthropometric data to determine the dimensions of the remote controls for generate design strategies. Based on the anthropometric data (Tilley and Dreyfuss, 2001), the hand width and length of the 95 % man are 73 mm and 161 mm, respectively. While the hand width and length of the 95 % woman are 70 mm and 155 mm, respectively. And, the hand width and length of the 95 % children under 12 are 59 mm and 126mm, respectively. The standard dimension of the remote control selected is about 115 x 58.6 x 9.3mm. The dimension is similar to the size of the iPhone. The main reason why we selected the dimension is because the size of the iPhone had been well received by the public. The end user is able to control the screen with one hand and both the left and right handers had no problem manipulating the phone at all. We define the seven design strategies of the dimensions of a touch-screen for a TV remote control based on the standard dimension and the anthropometric data of hand dimensions as shown in Table 3.

Table 3: Seven design strategies of a touch-screen

Touch-screen	Design #1	Design #2	Design #3	Design #4	Design #5	Design #6	Design #7
Dimension (mm)	125 x 60 x 5.2 mm	95 x 50 x 5.2 mm	75 x 35 x 5.2 mm	115 x 54 x 5.2 mm	85 x 42.5 x 5.2 mm	95 x 46.8 x 5.2 mm	95 x 48 x 5.2 mm
Coalition	Man	Woman	Child	M+W	M+C	W+C	M+W+C

3.2 Phase 2: Identify Design Preference

To identify design preferences for the touch-screen of a remote control, we performed a survey that was participated by 5 respondents from each different group of people, Man, Woman and Child (aged 12 and below). We also considered two type colors (cool and warm) for aesthetic of the products. Every respondent is asked to rank from 1 to 6 what are the most important principles they look for when using the remote control. Based on the seven design strategies, the expected strategy qualities for the products can be calculated by the functional accessibility level and the usability level as mentioned in Section 2.2. The Functions-Universal Principles Matrix was used to determine the functional accessibility level for the dimensions of the products as shown in Table 4. The degree of important (a) for touch-screen accessibility and the weight of universal principles (λ) for the touch-screen were determined by characteristics related to the products as shown in Table 4. We assume that the values of the usability levels for the products including the platform strategies are 5.

3.3 Phase 3: Determine a Platform Design Strategy

To determine a platform design strategy for a remote control, we used the performance of a touch-screen, which a user can easily access to the area of the touch-screen by his/her right or left thumb with free grip posture. Based on the proposed seven design strategies, twelve participants took part in the experiment, aged between 22 and 46 to determine the performance. The results from the experiment, the performances of the proposed designs are shown in Table 5. The proposed coalitional game was applied to obtain the marginal contributions of human variability based on the performances and preferences of the seven design strategies.

Table 4: Functions-Universal Principles Matrix for a touch-screen

Touch-screen	a	Equitable use	Flexibility in use	Simple and intuitive use	Perceptible information	Tolerance for error	Low physical effort	Size and space for use	Accessibility level	Usability level	Preferred color
Man	3	4	6	1	3	0	2	5	249	5	Cool
Woman	3	3	2	4	5	0	1	6	237	5	Warm
Child	3	2	4	3	6	0	1	5	237	5	Warm
☐	-	3	4	3	3	1	5	5	-	-	

Table 5: Performances of the proposed touch-screen designs (%)

Touch-screen	Design #1	Design #2	Design #3	Design #4	Design #5	Design #6	Design #7
Performance (average)	77	95.1	100	90.1	99.6	96.9	97.9

The game between three players (man, woman, and child) for platform design of this product family is defined as the proposed coalitional game that is described in Section 2.3. Seven coalitions for the design strategies were defined as Table 3. To determine marginal contributions for each player, the coalitional benefits of the design strategies were calculated by multiplying the performance and the preference. Therefore, the payoff vector of the game is v(0, 958.65, 1129.305, 1185, 1085.705, 1200.18, 1147.08, 1179.695). To determine the marginal contribution of each player, we used the Shapley value as mentioned in Section 2.3. The Shapley values of the players are (325.685, 384.463, 469.548).

Based on the marginal contributions of players, we can decide a platform design strategy for a TV remote control family according to the dimension of Child's hands with the preferred color.

4. CLOSING REMARKS AND FUTURE WORK

In this research, we have introduced a method for developing a universal product family based on human variability and aesthetics through a game theoretic approach Module-based design was applied to allow a range of trade-off in determining the specific function configuration for a platform at a conceptual design phase. We proposed a quality function to evaluate and measure accessibility of a product using the Functions-Universal Principles Matrix. Through the case study, we demonstrated that the proposed coalitional game could be used to determine a platform strategy by considering human variability that provide more benefits with respect to hand sizes in remote control design. Future research efforts will be focused on improving the efficiency of the proposed method, developing product cost models and design strategies for various universal product family environments, and comparing to the proposed game with other decision-making methods for determining a design strategy in a universal product family.

REFERENCES

Beecher, V. and Paquet, V., 2005, "Survey instrument for the universal design of consumer products," *Applied Ergonomics*, vol. 36, no. 3, pp. 363-372.

Connell, B.R., M. Jones, R. Mace, J. Mueller, A. Mullick, E. Ostroff, J. Sanford, E. Steinfeld, M. Story, and G. Anderheiden, 1997, *Center for Universal Design*, North Carolina State University, Raleigh, North Carolina.

Demirbilek, O. and Demirkan, H., 2004, "Universal product design involving elderly users: a participatory design model," *Applied Ergonomics*, vol. 35, no. 4, pp. 361-370.

Mace, R., 1985, *Universal Design: Barrier Free Environments for* Everyone. Los Angeles, CA: Designers West.

Moon, S. K., Park, J., Simpson, T. W., and Kumara, S. R. T., 2008, "A Dynamic Multi-Agent System Based on a Negotiation Mechanism for Product Family Design," *IEEE Transactions on Automation Science and Engineering*, vol. 5, no. 2, pp. 234-244.

Moon, S.K. and D.A. McAdams, 2010, "A Platform-based Strategic Design Approach for Universal Products," *International Journal of Mass Customization*, Vol. 3, No. 3, p. 227-246.

Osborne, M. J. and Rubinstein, A., 2002, *A Course in Game Theory*, MIT, Massachusetts, MA.

Preiser, W. F. E. and Ostroff, E., 2001, *Universal Design Handbook*, McGraw-Hill Inc., United States.

Shapley, L. S., 1971, "Cores of Convex Games," *International Journal of Game Theory*, vol. 1, no. 1, pp. 111-129.

Silveria, G. D., Borenstein, D., and Fogliatto, F. S., 2001, "Mass Customization: Literature review and research directions," *International Journal of Production Economics*, vol. 72, no. 1, pp. 1-13.

Simpson, T. W., 2004, "Product Platform Design and Customization: Status and Promise," *Artificial Intelligence for Engineering Design, Analysis, and Manufacturing*, vol. 18, no. 1, pp. 3-20.

Simpson, T. W., Siddique, Z., and Jiao, J., 2005, *Product Platform and Product Family Design: Methods and Applications*. Springer, New York, YN.

Tilley, A. R. and Dreyfuss, H., 2001, *The Measure of Man and Woman: Human Factors in Design*, Wiley, New Jersey.

WHO, 2001, International Classification of Functioning, Disability and Health: ICF Short version, World Health Organization, Geneva.

Section II

Cultural and Traditional Aspects

CHAPTER 7

Comparison of Evaluation of Kawaii Ribbons between Gender and Generation

Michiko Ohkura, Tsuyoshi Tomatsu, Somchanok Tivatansakul,

Shibaura Institute of Technology
Tokyo, Japan
ohkura@sic.shibaura-it.ac.jp

Siitapong Settapat

Thomson Reuters
Bangkok, Thailand

Saromporn Charoenpit

Thai-Nichi Institute of Technology
Bangkok, Thailand

ABSTRACT

"Kawaii" is a Japanese word that represents an emotional value; it has positive meanings, such as cute, lovable, and small. In the 21st century, the emotional values of industrial products become very important. However, since not many studies have focused on the kawaii attributes, we focus on a systematic analysis of kawaii interfaces themselves, that is kawaii feelings caused by the attributes such as shapes, colors, and materials. We have already performed some experiments for abstract objects in virtual environment and obtained some interesting tendencies on kawaii attributes such as kawaii shapes and kawaii colors. This article introduces our trial that dealt with combinations of attributes, including colors and patterns, and applied them to an actual product, a ribbon, using a web questionnaire system. From analysis of the questionnaire results, we compared the selection tendencies of kawaii ribbons by genders and generation.

Keywords: Kansei value, kawaii, cute, ribbon, gender, generation, color, pattern

1   INTRODUCTION

Recently, the kansei value has become crucial in the field of manufacturing in Japan. The Japanese Ministry of Economy, Trade and Industry (METI) has determined that it is the fourth most important characteristic of industrial products after function, reliability, and cost. METI believes that it is important not only to offer new functions and competitive prices but to also create a new value to strengthen Japan's industrial competitiveness.

Several years ago, we began new research to apply our previous research results to the systematic creation of the kansei values of artificial products, especially the large export surpluses of such digital contents as Japanese games, cartoons, and animations (JEITIA, 2002). One of the main reasons for the success of the digital content is the existence of "kawaii" characters and their highly sensitive techniques (Belson and Bremner, 2004). Therefore, we selected kawaii as a crucial kansei value of artificial products.

Kawaii is an adjective in the Japanese language. Recent works (Belson and Bremner, 2004) (Yomota, 2006) (Koga, 2009) (Makabe, 2009) (Sakurai, 2009) have recognized the following common attributes of kawaii:

- An emotional value of Japanese origin.
- Such positive meanings as cute, lovable, and small.

Because the Japanese word kawaii is not exactly the same as "cute" or "lovable," and its use has become international (Sakurai, 2009), we use it directly, both as an adjective and a noun.

Even though such Japanese kawaii characters as Hello Kitty and Pokemon have become popular worldwide, few studies have focused on kawaii attributes; therefore, we systematically began to analyze kawaii interfaces themselves: kawaii feelings evoked by shapes, colors, and materials. Our objective is to describe a method for constructing a kawaii interface from research results.

We previously performed experiments with abstract objects in a virtual environment and obtained interesting tendencies on such kawaii attributes as shapes and colors (Ohkura and Aoto, 2007) (Ohkura et al., 2009) (Ohkura and Aoto, 2010).

This article introduces our new trial, in which we dealt with combinations of attributes, colors and patterns and applied them to an actual product: a ribbon. In addition, we employed a web questionnaire system.

2   METHOD

2.1   Outline Web Questionnaire System

One of the authors developed a web questionnaire system that is accessible through such web browsers as Internet Explorer and Google Chrome using Apache, PHP, and My SQL under a Windows environment (Charoenpit and Ohkura, 2012).

The system was modified and used for this research.

Although the participants of this questionnaire research were only Japanese, we planed to provide it for other countries in the future. Thus, the first several web pages were described in English. The screen shot of the top page is shown in Figure 1. After selecting Japanese as the participant's nationality, the web pages contained both in English and Japanese.

Figure 1 Top page the web questionnaire system

2.2   Candidates of Kawaii Ribbons

The shape and patterns of the ribbon candidates were selected from a reference (MdN Editorial Office, 2010). We chose one shape and three patterns. Six colors were selected from the results of our research, and an achromatic color was added. Thus, the total number of target ribbons became 21 (Figure 2).

2.3   Structure of Questionnaire System

The following is the structure of the questionnaire system:

(1) Top page: Explanation of questionnaire
(2) Selection of participant's attributes: Selection of gender, age group, and nationality
(3) Explanation of selection of ribbons: election of more kawaii ribbons from the displayed two ribbon images in every five seconds. The selection results should be answered using the keyboard's arrow keys.
(4) Selection of ribbons. The number of the compared pairs of 21 ribbons is 210. Thus, comparisons should only be made between pairs of different patterns of

62

the same color and different colors of the same pattern, which reduces the number of compared pairs from 210 to 84. Each ribbon appears to compare 8 times. An example of the comparison display is shown in Fig. 4.

(5) Questionnaire for selection reasons: after the comparison, a 5-scale questionnaire was performed for the selection reason: (5: strongly agree, 4: agree, 3: neutral, 2: disagree, 1: strongly disagree)

Q1: Patterns were kawaii.
Q2: Colors were kawaii.
Q3: Whitish ribbons were kawaii.
Q4: Total impression
Q5: First impression

(6) Questionnaire results: The selection and reason results were saved as a database. Necessary results were extracted by My SQL

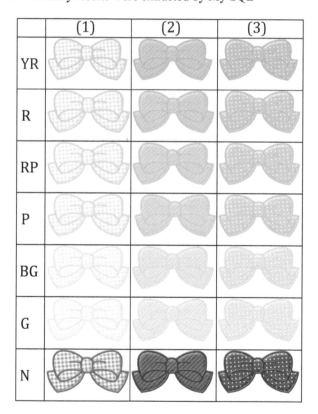

Figure 2 Candidates of kawaii ribbons

Figure 3 Selection of gender

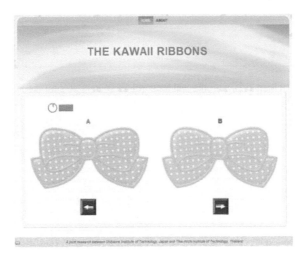

Figure 4 Example of comparisonr

3   RESULTS

3.1   Questionnaire and Participants

We collected data two times. 45 participants, 35 male and 10 female students in their 20's, volunteered for the first time collection. 20 participants, 10 women in their 20's and 10 women in their 40's or 50's, volunteered for the second time.

3.2　Cumulative data for each participants group

Each participant was randomly shown 84 pairs of ribbons (Figure 4) and asked to select which pair is more kawaii within five seconds. The cumulative data for the first time and the second time are shown in Figure 5.

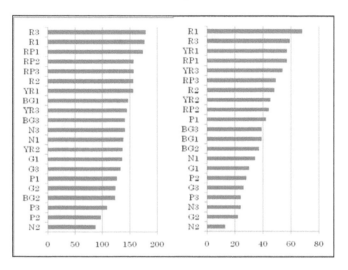

(a) Men in 20's for the first time　(b) Women in 20's for the first time

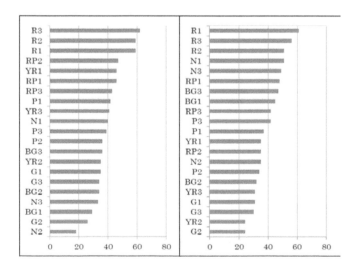

(c) Wome in 20's for the second time　(d) Women in 40's and 50's in the second time

Figure 5　Cumulative data of the numbers being selected for each ribbon. The total numbers of appearance for comparison of each ribbon are 280 in (a), and 80 in (b), (c), and (d).

In each figure of Figure 5, the horizontal axis is the numbers of selected, and the vertical axis is the kinds of ribbons shown in Figure 2.

3.3   Results of 2-factor analysis of variance

All cumulative data shown in Figure 5 were analyzed separately. Pattern and color of ribbons were served as the factors for the 2-factor analysis.

From the analysis of the results for the first time, the patterns and colors had main effects at 1% significance for both men and women. In addition, the results of the lower level tests showed as follows:

- Significant differences between Patterns (1) and (3) and Patterns (2) and (3) for men.
- Significant differences in all combinations of the three patterns for women.

The order of the kawaii degree was basically as follows:

Pattern (1) > Pattern (3) > Pattern (2)

As for colors, not all the combinations had significant differences in the lower level tests. The tendencies of the averaged scores were basically as follows:

R >= RP >= YR >= BG >= G >= N >= P (Men)
R >= YR >= RP > BG > P >= G >= N (Women)

These results show that the tendencies of the men and women were almost the same. However, the orders of YR (Orange) and P (Purple) for women were higher than those for men.

From the analysis of the results for the second time, the patterns and colors had main effects at 1% significance only for women in their 40's and 50's. Only colors had main effect at 1% significance for women in their 20's. The results of the lower level tests showed similar tendencies for patterns to the first time. As for colors, not all the combinations had significant differences in the lower level tests. The tendencies of the averaged scores were basically as follows:

R > RP >= YR >= P > N >= BG >= G (Women in 20's)
R > N >= RP >= BG > P >= YR >= G (Women in 40's and 50's)

In total for the results of the first time and the second time, the followings were obtained:

- R (Red) and RP (Red purple) were selected often regardless of gender and generation.
- P (Purple) was selected more often by women than men.
- Women in their 40's and 50's selected N (achromatic color) more often and selected YR (Orange) less often than men and women in their 20's.

3.4   Analysis of selection reasons

Figure 6 shows the histograms of Q1 to Q5 by gender and generation. From the comparison between (a) and (b), the scores of the female participants were generally higher. In addition, the ratio was relatively high for the male participants who put higher importance on their first impressions and lower importance on their total impression. On the other hand, the ratio was relatively high for female participants who put higher importance on their total impression. This was the biggest difference between genders. Figures (c) and (d) support the above tendency.

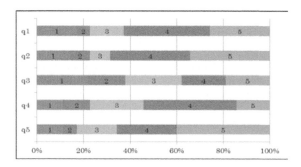

(a) The histograms of Q1 to Q5 for men in their 20's of the first time

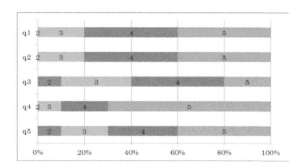

(b) The histograms of Q1 to Q5 for women in their 20's of the first time

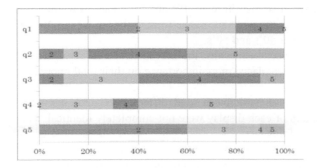

(c) The histograms of Q1 to Q5 for women in their 20's of the second time

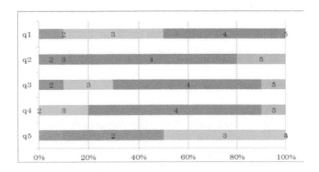

(c) The histograms of Q1 to Q5 for women in their 40's and 50's of the second time

Figure 6 Results of reasons of selection

4   DISCUSSION AND CONCLUSIONS

This article introduces our new trial in which we dealt with combinations of attributes, colors, and patterns and applied them to an actual product, a ribbon, employing a web questionnaire system. The most important feature of this study is to evaluate the difference between ribbons which is considered kawaii in general and moreover set their colors evaluated kawaii in our previous study (Komatsu and Ohkura, 2011). It is because we wanted to clarify the differences between genders and generations.

From the analysis of the accumulated data, we obtained the following results:

- For kawaii ribbons, the effects of patterns and colors were significant.
- The selection tendencies were almost the same between men and women in their 20's. However, the ranks of orange and purple were higher in women than in men.
- The selection tendencies were almost the same between women in their 40's and 50's and women in their 20's. However, the ranks of achromatic color

were higher in women in their 40's and 50's than in women in their 20's. On the other hand, the ranks of orange were higher in women in their 20's than in women in their 40's and 50's.

● As for the selection reasons, the averaged scores of women were relatively high and women placed more importance on their total impression.

Note that because participants used different PCs for our web questionnaire system, the colors of each display were not completely identical. The above results should consider of this fact.

In addition, the participants of this research were limited to Japanese people. Research with more participants including different nationalities is future work.

ACKNOWLEDGMENTS

The authors would like to acknowledge all the participants to serve as volunteers and Ms. Somachanok Tivatansakul for helping data collection.

REFERENCES

Belson, K. and B. Bremner 2004. *Hello Kitty: The Remarkable Story of Sanrio and the Billion Dollar Feline Phenomenon*, New Jersey: John Wiley & Sons.

Charoenpit, S., and M. Ohkura. "The kansei research on the price labels of shoes." Paper presented at AHFE International Conference, jointly with 12th International Conference on Human Aspects of Advanced Manufacturing (HAAMAHA), San Francisco, CA, 2012.

Koga, R. 2009. *"Kawaii" no Teikoku.* Tokyo: Seidosha. (in Japanese)

Japan Electronics and Information Technology Industries Association. "Statistics of Exports/Imports of Software in 2000." Accessed Dec. 10, 2012, http://it.jeita.or.jp/statistics/software/2000/index.html. (in Japanese).

Makabe, T. 2009. *Kawaii Paradigm Design Kenkyu.* Tokyo: Heibonsha. (in Japanese)

MdN Editorial Office 2010. *Kawaii Materials: Hearts and Ribbons.* Tokyo: MdN Corporation. (in Japanese)

Ohkura, M.,and T. Aoto. 2007. Systematic Study for "Kawaii" Products, *Proceedings of the 1st International Conference on Kansei Engineering and Emotion Research 2007 (KEER2007)*, Sapporo, Japan.

Ohkura, M., S. Goto, and T. Aoto 2009. Systematic Study for 'Kawaii' Products: Study on Kawaii Colors Using Virtual Objects, *Proceedings of the 13th International Conference on Human-Computer Interaction*, 633-637, San Diego, CA.

Ohkura, M.,and T. Aoto 2010. Systematic Study of Kawaii Products: Relation between Kawaii Feelings and Attributes of Industrial Products, *Proceedings of the ASME 2010 International Design Engineering Technical Conference & Computers and Information in Engineering Conference (IDETC/CIE 2010)*,DETC2010-28182, Montreal, Canada.

Komatsu T. and M. Ohkura 2011. Study on Evaluation of Kawaii Colors Using Visual Analog Scale, *Human-Computer Interaction, Part I*, 103-108, Orlando, FL.

Sakurai, T. 2009. *Sekai Kawaii Kakumei*, Tokyo: Ascii Media Works. (in Japanese)

Yomota, I. 2006. *Kawaii Ron*, Tokyo: Chikuma Shobo. (in Japanese)

CHAPTER 8

Assessment of Material Perception of Black Lacquer

Tsuyoshi Komatsu, Michiko Ohkura

Shibaura Institute of Technology,
3-7-5, Toyosu, Koto City, Tokyo, Japan
ma11068@shibaura-it.ac.jp

Tomoharu Ishikawa, Miyoshi Ayama

Utsunomiya University
7-1-2, Yoto, Utsunomiya, Japan

ABSTRACT

Material perception is a crucial factor in product design. We can feel material perception not only by vision but also by several other senses including our tactile sense. Therefore, material perception has been assessed under vision-only, tactile-only or visuo-tactile conditions. Based on its assessment, many studies exist on such physical characteristics as surface roughness or hardness, but few studies have addressed such emotional characteristics as nostalgia or pleasure.

Therefore, we focused on the emotional responses of participants to the characteristics of black lacquers with different degrees of gloss to clarify their effects in material perception.

First, we experimentally obtained indexes to assess the material perception of black lacquer boards and clarified how material perceptions of them differed based on sensory modalities.

Keywords: material perception, kansei value, emotion, lacquer

1   INTRODUCTION

Material perception is a crucial factor in product design. For example, the beauty of black lacquer products made reflects not only their shape and color but also such material perceptions as gloss, glaze, and depth. We can experience material perception by vision and sensation modality including touch. Material perception has been assessed under vision-only, tactile-only or visuo-tactile conditions. However, in assessing material perception, many studies exist on physical characteristics surface roughness or hardness, but studies on such characteristics as nostalgia or pleasure are scarce. We focused on the emotional responses of participants to the characteristics of black lacquer boards with different degrees of gloss to clarify their effect in material perception.

2   PRELIMINARY EXPERIMENT

2.1   Experiental Method

We experimentally obtained indexes to assess the material perceptions of black lacquer boards and to clarify whether they are affected by types of light sources. The participants were a female and nine males in their 20s.

We employed three black lacquer boards with different gloss ratios, and a black lacquer board that was polished for glazing. Table 1 shows the four black lacquer boards. We employed two experimental environments with different light sources. Table 2 shows the illuminance measurement results.

Table 1 Four black lacquer boards

Kinds	Descriptions
B#0	low-gloss ratio
B#5	middle-gloss ratio
B#10	high-gloss ratio
B#11	polished for glazing

Table 2 illuminance measurement results

Experimental environments	illuminance	x	y
Meeting room	898 lx	0.429	0.411
Japanese-style room	342 lx	0.470	0.425

In this experiment, participants looked at, touched, picked up, and assessed each board's material perceptions in the following experimental procedure:
  i.   Participants subjectively ranked the four boards.
  ii. They explained their arrangement criteria of the four boards.

iii.   They assessed each board for each criterion and preference from 0 to 100.

The participants repeated the above procedures for different experimental environments.

2.3   Experiental results

Tables 3 and 4 show the subjective order of the four boards. Table 3 shows the result in the meeting room. Table 4 shows the result in the Japanese-style room.

The following were obtained from tables 3 and 4.

- The most preferred black lacquer boards differed by individual. The black lacquer board with a high-gloss ratio and one polished for glazing were more preferred.
- The results in the two experimental environments were about the same.

From the results of these experiments, we also obtained the following major criteria for the assessment of the material perception of black lacquer boards:

- Gloss: (pikapika, glazed, glassy, glossy, burnish etc.)
- Slick: (friction, grainy, slick, velvet, rough etc.)
- Sensuous comfortable tactile: (comfort, pleasing to the touch etc.)
- Luxurious: (luxury, like expensive, fancy etc.)

Table 3 Order in meeting room

	Arrangement of four boards by preferences			
Participant	No.1	No.2	No.3	No.4
1	B#0	B#11	B#10	B#5
2	B#5	B#10	B#11	B#0
3	B#0	B#5	B#10	B#11
4	B#11	B#0	B#10	B#5
5	B#0	B#5	B#11	B#10
6	B#10	B#5	B#0	B#11
7	B#11	B#10	B#5	B#0
8	B#11	B#10	B#5	B#0
9	B#10	B#11	B#5	B#0
10	B#10	B#5	B#11	B#0

Table 4 Order in Japanese-style room

	Arrangement of four boards by preferences			
Participant	No.1	No.2	No.3	No.4
1	B#0	B#11	B#10	B#5
2	B#11	B#10	B#5	B#0
3	B#0	B#5	B#11	B#10
4	B#11	B#10	B#0	B#5
5	B#5	B#0	B#11	B#10

6	B#10	B#11	B#5	B#0
7	B#10	B#11	B#5	B#0
8	B#10	B#11	B#5	B#0
9	B#11	B#10	B#5	B#0
10	B#10	B#5	B#11	B#0

2.4   Discussion

In this experiment, we obtained indexes to assess the material perception of black lacquer boards. However, the age groups and gender were limited because the participants included just one female and nine males in their 20s. Also, the black lacquer boards were heavy because the objects covered with black lacquer weren't wood or wood powder, but acrylic. Such situations might have affected the assessment of their material perceptions.

The results in the two experimental environments tended to be the same. However, we received such comments as the following because the illuminance of the light source in the meeting room was too bright.

·   The lights were dazzling.
·   I don't like the black lacquer boards on which I can see a reflection of my own face.

Based on such comments, we conclude that meeting rooms aren't appropriate experimental environments for assessing black lacquer boards.

3   EXPERIMENT

3.1   Experiental set-up

We performed experiments to obtain indexes to assess the material perception of black lacquer boards to clarify that assessments of the material perception of black lacquer boards differ by sensory modalities.

We employed the same four black lacquer boards as in our preliminary experiment. Participants are the staffs and the students of Shibaura Institute of Technology included 10 females and 7 males from 20 to 50.

First, participants answered questionnaires before the assessment experiments. Next they assessed the material perceptions of the four black lacquer boards under vision-only, tactile-only and visuo-tactile conditions. Participants didn't pick up these black lacquer boards, and used a Japanese-style room as an experimental environment.

3.2   Questionnaire

The questionnaire was comprised of the following items:
Q1. Did you see, touch or use products (bowls, multi-tiered boxes etc.) made of lacquer last New Year holidays? Are you familiar with such products?

Q2. Are you interested in art or crafts (paintings, sculptures, handcrafts, calligraphy etc.)? Have you been exposed to art and crafts? If you answered "yes" to Q2, respond to Q3.

Q3. What art or crafts are you interested in?

3.3   Assessment procedure

Participants assessed the material perceptions of the four black lacquer boards under vision-only, tactile-only and visuo-tactile conditions. They touched the black lacquer boards with the index finger of their dominant hand under the tactile-only and visuo-tactile conditions. Figure 1 shows the experimental environment for the vision-only and visuo-tactile conditions. Figure 2 shows it for the tactile-only condition in which the black lacquer boards were covered.

Each experimental procedure was the same under the vision-only, tactile-only and visuo-tactile conditions. The following is the experimental procedure:

i.   Participants instructed the experimenter to change the order of the four boards based on their preferences.

ii.   Participants explained their subjective arrangement criteria of the four boards as in the preliminary experiment.

iii. Participants assessed each board for each criterion of judging, gloss and preference with a visual analog scale (VAS).

Participants repeated the above procedures for different conditions.

Figure 1 Experimental environment under vision-only and visuo-tactile conditions.

Figure 2 Experimental environment under tactile-only condition.

3.4   Experimental result

3.4.1   Assessment of gloss

Figure 3 shows an example of the results under the vision-only, tactile-only and visuo-tactile conditions. The vertical axis is the average of the scores of the gloss degrees with the kinds of black lacquer boards shown in the horizontal axis. The error bars indicate the standard deviations. We obtained the following:

- The scores of B#10 and B#11 tended to be the same under the vision-only condition.
- The difference between the scores of B#10 and B#11 under the tactile-only and visuo-tactile conditions tended to be higher than under the vision-only condition.

    We obtained the following results from an analysis of variance (ANOVA) with six factors: gender, age group, Q1, Q2, kind of black lacquer board, and sensation modality.

- The main effect of the kinds of black lacquer boards is significant at a 0.1% level.
- The interaction effect between the kinds of black lacquer boards and the sensation modality is significant at a 5% level.

We also obtained the following from the results of multiple comparisons.

- The assessed degrees of gloss aren't significantly different between B#10 and B#11, but they are among the other black lacquer boards under the vision-only and visuo-tactile conditions.
- The assessed degrees of gloss aren't significantly different among each black lacquer board under the tactile-only condition.

Figure 4 shows the number of participants who assessed B#11 as glossier than B#10. Those who assessed B#11 as glossier than B#10 tended to be less under the vision-only condition but more under the tactile-only and visuo-tactile conditions.

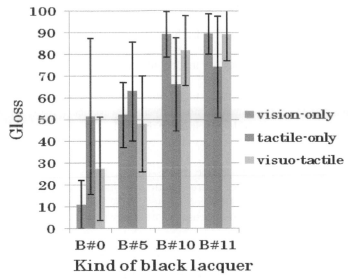

Figure 3 Average scores of gloss degrees by sensation modality

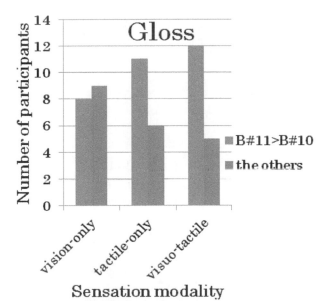

Figure 4 Participants who assessed B#11 as glossier than B#10

3.4.2 Assessment of preferences

Figure 5 shows the results by sensation modality. The vertical axis shows the average scores of the preference degrees by the kinds of black lacquer boards shown in the horizontal axis. The error bars indicate the standard deviations. We obtained the following from Figure 5:

- The B#11 scores tended to be higher under the vision-only and visuo-tactile conditions.
- The B#0 and B#11 scores tended to be higher under the tactile-only condition.
- The standard deviations were high under each sensation modality condition.

We obtained the following from an analysis of variance (ANOVA) with six factors: gender, age group, Q1, Q2, kind of black lacquer board, and sensation modality. The main effect of the kind of black lacquer boards and Q1 was significant at 5% level.

Figure 6 shows an example of the results of Q1. The vertical axis shows the average scores of the preference degrees with the kinds of black lacquer boards shown in the horizontal axis. The scores assessed by participants who answered "yes" on Q1 were higher under each sensation modality condition.

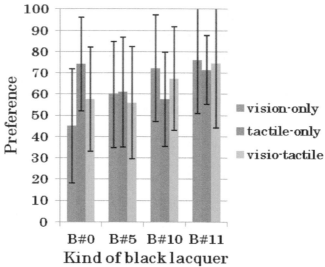

Figure 5 Average scores of preference degrees by sensation modality

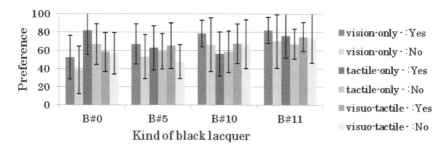

Figure 6 Average scores of preference degrees of Q1

3.5 Discussion

3.5.1 Assessment of gloss

The results obtained from Figure 3 and the ANOVA indicate that distingishing B#10 from B#11 is difficult under the vision-only condition. The number of participant who assessed B#11 as glossier than B#10 tended to increase under the visuo-tactile condition. This indicates that both seeing and touching are important for perceiving an object's gloss. However, the difference isn't significant among every black lacquer board under the tactile-only condition in the results of multiple comparisons. We found that many participants distinguished the material perception of the black lacquer boards under the tactile-only condition, but had difficulty judging the gloss of the black lacquer boards.

3.5.2 Assessment of preference

The scores assessed by participants who answered "yes" on Q1 were higher under each sensation modality, suggesting that the assessment of preference for black lacquer boards reflects familiarity with lacquer products.

The results obtained from Figure 5 also show the following regardless of answers to Q1.

- The B#11 score tended to be higher under the vision-only and visuo-tactile conditions.
- The B#0 and B#11 scores tended to be higher under the tactile-only condition.

CONCLUSIONS

We focused on the emotional responses of participants to the characteristics of black lacquer boards with different degrees of gloss and aim to clarify the effect of the differences of material perception.

First, we experimentally obtained indexes to assess the material perception of black lacquer boards and clarified that the assessment of the material perception of black lacquer boards differed by sensory modalities. We obtained the following:

- Both seeing and touching are important for perceiving an object's gloss.
- Many participants distinguished among the four black lacquer boards, but had difficulty judging the gloss of the black lacquer boards.
- Subjective assessments for the black lacquer boards reflect familiarity with lacquer products.
- The B#11 (a black lacquer board that was polished for glazing) score tended to be higher under vision-only and visuo-tactile conditions.
- The B#0 (a black lacquer board with low-gloss ratio) and B#11 (a black lacquer board that was polished for glazing) scores tended to be higher under tactile-only condition.

ACKNOWLEDGMENTS

We express gratitude to the staffs and the students of Shibaura Institute of Technology for their cooperation.

REFERENCES

T. Komatsu and M. Ohkura. 2011. Study on Evaluation of Kawaii Colors Using Visual Analog Scale. Human-Computer Interaction, Part I, HCII2011, LNCS 6771: 103-108.

SB. Bird and EW. Dickson. 2001. Clinically significant changes in pain along the Visual Analog Scale. ANNALS OF EMERGENCY MEDICINE, Vol. 38, No. 6: 639-643.

X. Chen 1, C.J. Barnes, T.H.C. Childs, B. Henson, F. Shao. 2009. Materials' tactile testing and characterisation for consumer products'affective packaging design. Materials and Design 30: 4299-4310.

K. Drewing, A. Ramisch, and F. Bayer. 2009. Haptic, visual and visuo-haptic softness judgments for objects with deformable surfaces. Third Joint Eurohaptics Conference and Symposium on Haptic Interfaces for Virtual Environment and Teleoperator Systems: 640-645.

M. Ayama, T. Eda and T. Ishikawa. 2010. Studies on blackness perception, The Institute of Electronics, Information and Communication Engineers, Vol.93, No.4: 316-321

CHAPTER 9

Analysis of Search Results of Kawaii Search

Kyoko Hashiguchi, Katsuhiko Ogawa

Keio University
Kawagawa, Japan
Kyon@sfc.keio.ac.jp

ABSTRACT

Kawaii Search is a blog search engine used to search for sundry items that are popular among Japanese women. Kawaii Search categorizes blog articles into five categories according to their visual appearances "cute" (cute), "yurukawa" (mellow), "kirei" (beautiful), "omoshiro" (amusing), and "majime" (conservative). We analyzed the search results to evaluate the search behaviors of users. The results revealed the characteristics of the five kawaii categories.

Keywords: Impression, Blog search engine, Text formatting, Japanese blogosphere, Information retrieval

1   INTRODUCTION

The word "kawaii" in Japanese indicates the degree of cuteness, especially in the context of Japanese culture. The term kawaii is utilized widely as a mainstream concept in fashion, cosmetic, and sundry industries in Japan. As a result, many Japanese women write blogs on these topics. However, searching for specific blog articles is difficult. To this end, on June 29, 2011, we developed and released a blog search engine called "Kawaii Search" on goo Lab, a website developed by NTT laboratory group for showcasing advanced technologies. The purpose of Kawaii Search is to search for blog articles on sundry items that are popular among Japanese women.

Conventionally, blog articles are classified using different approaches, for example, according to the age of the writer, the organization that the writer may belong to, or the magazine the writer may be employed with. However, our search engine classifies blog articles into five types of kawaii categories by analyzing their contents.

In this paper, we analyze the search results of our search engine to evaluate the search behaviors of users while using Kawaii Search.

2   INTERFACE *OF KAWAII SEARCH*

Figure 1 shows the Kawaii Search interface, and Figure 2 shows search results for the keyword "knit."

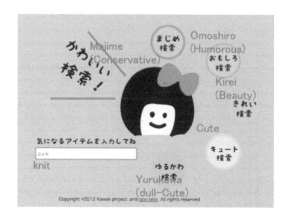

Figure 1 Kawaii Search interface

Figure 2 Search results of Kawaii Search for "knit"

2.1 USAGE SCENARIO

In this section, we describe the usage scenario. The search is carried out as follows:

1. Enter the keyword.
For example, if you want to search for the keyword "knit," enter it in the text box as shown in Figure 1.

2. Select the kawaii category.
As shown in Figure 1, click on one of the five buttons representing the following kawaii categories: mellow (yurukawa), cute (cute), beautiful (kirei), amusing (omoshiro), and conservative (majime).

3. Results are displayed on the interface.
Search results, along with the corresponding titles and thumbnails, are displayed at the bottom of the Kawaii Search interface. If you find a blog article interesting, click on the corresponding link (Figure 3).

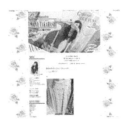

Figure 3 Example of blog article returned by Kawaii Search for "knit"

2.2 Five Kawaii patterns

The placement of text in blogs written by the Japanese can reveal their personality. For example, some writers leave a large space between lines or use hieroglyphics and slang or abbreviations such as "gal." Further, Japanese words can consist of four types of characters: kanji, hiragana, katakana, and kana. Different combinations of these features indicate different personalities. Thus, blog readers can not only read the blog but also interpret the writer's personality. On the basis of these features, we propose a new search algorithm specifically for blogs written by Japanese writers.

As described above, the blog articles are categorized into five distinct categories according to their contents: mellow (yurukawa), cute (cute), beautiful (kirei), amusing (omoshiro), and conservative (majime). A "yurukawa" type of blog generally contains many Yuru-Smileys. (Yuru-Smiley is an animated GIF.It is similar to Smiley, however Yuru-Smileys are all originally created by the users.)It also contains few more than an average number of hiragana characters, and the spaces between lines are fairly large. A "kawaii" type of blog contains more smileys than the other types of blogs. Further, it contains a large number of hiragana characters, and the spaces between lines are very wide. A "kirei" type of blog contains many pictographs and a few smileys. A "majime" type of blog contains many letters, with several strokes in each letter, and words. It contains few hiragana characters, and the spaces between lines are narrow. An "omoshiro" type of blog is characterized by features such as many symbols that are used to draw an image.

Figure 4 Five Kawaii patterns

3 ANALYSIS OF USERS' SEARCH BEHAVIOR

In this section, we discuss the analysis results for the search behaviors of users.

3.1 Analysis data

Data was obtained from 8000 users using Kawaii Search from July to November 2011.

3.2 Search Behaviors

From the above data, we analyzed the search behaviors of users as follows (Figure 5):

1. Most users carried out their search in the order yurukawa, cute, kirei, omoshiro, and majime because of the order of appearance of the corresponding buttons on the search interface (Figure2).
2. Second most Some users carried out their search in the order view   "majime", "omoshiro" , and "majime", or "majime", "omoshiro" , and   "kirei" in that order.
3. Some users carried out their search in the order yurukawa, cute, and yurukawa or yurukawa, cute, and kirei.

Cases 2 and 3 indicate that maximum number of users searched for combinations of majime and omoshiro, and yurukawa and kawaii types of blogs.

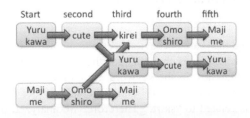

Figure 5 Search behavior of users in Kawaii Search

3.3 Users' Purpose for using Kawaii Search

Next, we attempted to understand the users' purpose of using Kawaii Search. We extracted the first 500 search keywords from the number of page views for each kawaii category (2500 search keywords in total). If the duration of visit of a user to a blog article is long, it is considered that the user reads the article including the search keyword in detail, indicating that the search result is satisfactory. Therefore, in this study, we analyzed the search results for which the mean sojourn time was more than 46 s. Table 1 shows the 35 keywords searched under all the kawaii categories. It can be seen that words such as clothes and autumn wear accounted for more than 70% of the keywords. In addition, users mainly searched for generic terms such as cosmetics, skirts, and bags, and not specific products (e.g., dotted miniskirts). These results suggest that many users searched for kawaii items and a wide range of fashion items such as clothes and accessories rather than specific items. The non-fashion items included terms such as Kyoto, Korea, animals, characters, and miscellaneous goods; however, these terms appear frequently in women's magazines. Thus, it was found that users use Kawaii Search for the purpose of searching for kawaii type of blog articles.

Table 1 Keywords used In Kawaii Search

Fashion items	Non-fashion items
洋服(Clothes) 秋服(autumn wear)	イ・ビョンホン(Lee Byung Hun)
カラコン(colored contact lens)	雑誌(magazine)　京都(Kyoto)
スカート(skirt)　マキシ(maxiskirt)	韓国(Korea)　スマフォ(smartphone)
ニット(knit) ワンピース(dress)	リラックマ(Rilakkuma)　恋愛(love)
トップス(tops) 指輪(ring)	犬(dog)　スタバ(Starbucks)
ヘアスタイル(hairstyle)	プレゼント(gift)
bag　かごバック(basket bag)	
髪(hair) つけまつげ(false eyelashes)	
化粧品(cosmetics) ブーツ(boots)	

3.4   Search Behavior based on Differences in Five Kawaii Categories

Next, we analyzed the search behaviors of users on the basis of the differences in the five Kawaii categories. It is thought that the keyword and the kawaii pattern under which a user searches for the keyword are closely related to each other. For example, when a user looks for "skirt" under the category of kawaii (cute), he/she thinks that "cute" is related to "skirt." Therefore, we analyzed different keywords for each kawaii category (Table 2). The proportion of fashion or sundry items searched in each kawaii category was as follows: yurukawa, 48%; kawaii, 50%; kirei, 35%; omoshiro, 8%; and majime, 23%.

In addition to fashion terms such as "cosplay," "camisole," and "knitting," many users searched for "2ch ", "game," "Power Stone," "Oracle," "Neutrino," "shogi," and "Doraemon" under the category of omoshiro. Thus, we can infer that the user wants to look for interesting contents.

Under majime, 23% of the searched keywords were fashion terms. Most users looked for luxury brands. In addition, users searched for electronic items using keywords such as "android," "LED," "3DS, and "cameras" and names of universities such as "Keio," "Kanagawa," and "Meiji." Because the frequency of kanji characters and the number of sentences in majime type of blog articles are high, it is thought that an informative and difficult keyword would be required for an effective search.

Table 2. Keywords used in five kawaii categories

Yurukawa	Kawaii
ネックレス(necklace) ポーチ(porch) ダイエット(diet) アート(art) 手作り (handicraft) パンプス(pumps) goo	リボン(ribbon) 鞄(milly bag)長靴(boots) ファンデーション(foundation) 彼氏(boyfriend) マスカラ(mascara)

Kirei	Omoshiro
ドレス(dress) パンプス(pumps) リング(hand ring)キャミ (camisole) チョコレート(chocolate) カラー(color)	コスプレ(cosplay) 将棋(shogi, 2ch) キャミ (camisole) どらえもん(Doraemon) 編み物(knitting) 楽しい(happy) ゲーム(game)

Majime	
ネックレス(necklace) むくみ(swelling) ヘアアレンジ(hair arrangement) 明治大学(Meiji University)	ニュース(news, android)足(foot spa) 慶應(Keio University)

3.5 Characteristics of blog articles searched under five Kawaii categories

Next, we analyzed the blog articles searched by the users. First, we analyzed the background color and then the type of blog article.

1. Background colors of blog articles
The first ten high-ranked blog articles read by the users in each category (total 50) were analyzed. Figure 6 shows the classification of the blog articles in each category, according to their background colors. It can be seen that blog articles with pink and white background colors were often chosen under yurukawa. On the other hand, the category of kawaii mostly included blog articles with pink or multicolored background. Blog articles with a white background were often chosen under kirei. Under omoshiro, blog articles of various background colors, including blue and yellow, were chosen. Under majime, the background colors were mostly beige and white.

A blog writer who uses emoji, large spaces between lines, and many hiragana letters tends to choose background colors such as pink and white, and these articles belong to the category of yurukawa.

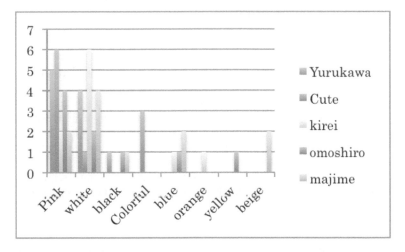

Figure 6. Classification of blog articles according to their background colors

2. Types of blog articles
Next, we investigated whether users selected blog articles according to their title or the related image (Figure 3).

Figure 7 shows the classification of blog articles in each category, selected according to the title or image. It can be seen that many users selected articles in yurukawa, kawaii, kirei, and omoshiro because of the images displayed in the search results. On the other hand, articles under majime were often selected for their title. This was because, in this category, the users were more interested in the content than the images.

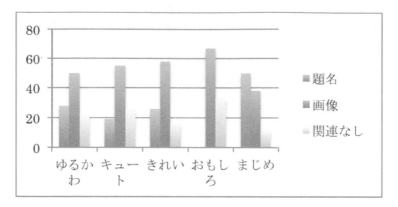

Figure 7. Classification of blog articles according to title and image

4   CONCLUSIONS

We analyzed the search results of our blog search engine, Kawaii Search, and found the following four characteristics.

1. Many users searched for **kawaii** items such as fashion items generally found in women's magazines.
2. All the users showed similar search behaviors for all the five **kawaii** categories.
3. The five Kawaii categories showed several unique features. Most users searched for fashion items under yurukawa and kawaii. Majime was used when users wanted to look for difficult words. Omoshiro was used when the users wanted to search for blogs for recreational purposes, such as those written on comics, movies, etc.
4. When users wanted to know the contents of blog articles, they selected the majime category. On the other hand, when users wanted to read articles according to the images related to the keyword, they selected the other categories.

On the basis of these results, we hypothesize that the contents and background colors of blog articles influence the type of kawaii pattern selected by the user. We will try to prove this hypothesis in our future work. In particular, it is necessary to develop an algorithm by considering the characteristics of each kawaii category.

REFERENCES

Kyoko Hashiguchi, Katsuhiko Ogawa, 2011. Proposal of the Kawaii Search System Based on the First Sight of Impression, HCI International.
goo Lab," http://labs.goo.ne.jp/.
Kawaii Search, http://kawaii-search.jp/.

CHAPTER 10

3D Character Creation System Based on Sensibility Rule Extraction

Takuya Ogura, Masafumi Hagiwara

Department of Information and Computer Science, Keio University
3-14-1 Hiyoshi, Kohoku-ku, Yokohama, 223-8522, Japan
ogura@soft.ics.keio.ac.jp, hagiwara@soft.ics.keio.ac.jp

ABSTRACT

3D characters have been widely used in various fields such as advertizing and entertainment. This paper proposes an automatic 3D character creation system based on sensibility rule extraction. Since the proposed system employs an interactive evolutionary computation (IEC) mechanism, it can automatically learn the users' preference. What the users have to do is only evaluation of displayed 3D characters created by the system. The proposed system can create 3D characters with high diversity. A lot of evaluation experiments were carried out and the superiority is demonstrated.

Keywords: 3D Character, Rough Sets Theory, Genetic Algorithm, *Kansei* Engineering

1 INTRODUCTION

Nowadays, various kinds of digital contents have been developed such as social network services (SNSs), games and so on(Digital Content Association of Japan, DNP Media Create Co., Ltd., Ministry of Economy, and Trade and Industry, Commerce and Information Policy Bureau, 2011). There are many studies on these digital contents. (Nishikawa, Mashita, and Ogawa, et al, 2011)(Toma, Kagami, and Hashimot, 2011)(Chen, 2009).

3D characters have been widely used as an avatar in those contents of various fields such as advertizing and entertainment. Because an avatar has a personal role expressing, opportunities to create 3D character in individuals increase. However, It is difficult to create 3D character in individuals(Mizuno, Kashiwazaki, and Takai, et al, 2008)(Igarashi and Hughes, 2002)(Ando and Hagiwara, 2009).

On the other hand, sensibility called *Kawaii* originates in Japan attracts attention worldwide. There are many studies on *Kawaii* (Ohkura, 2011)(Mitake, Aoki, and Hasegawa, et al, 2011)(Sugahara, 2011). A Special issue on *Kawaii* was published by Japan Society of Kansei Engineering (JSKE).

There are a method using Genetic Algorithm (GA) (Sakawa and Tanaka, 1995.) and a method using Rough Sets Theory (RST) (Inoue, Harada, and Shiizuka, et al, 2009) as 3D character creation to reflect a sensibility of a user such as *Kawaii* (Ito and Hagiwara, 2005)(Ando and Hagiwara, 2009)(Gu, Tang, and Frazer, 2006). Ito et al studied the combinatorial search of parts in consideration of a sensibility of a user using a frame of GA (Ito and Hagiwara, 2005). Ando et al studied a reflection of a sensibility of a user by extracting a sensibility rule using RST(Ando and Hagiwara, 2009). What the users have to do is only evaluation of the displayed 3D characters created by the proposed system. However, there are problems such as quality of the 3D characters created by the system becomes estranged largely with that created by the commercially available game.

This paper proposes an automatic 3D character creation system based on sensibility rule extraction. Since the proposed system employs an interactive evolutionary computation (IEC) mechanism, it can automatically learn the users' preference. What the users have to do is only evaluation of displayed 3D characters created by the system.

This paper is organized as follows. In section 2, we detail our proposed 3D character creation system. Section 3 describes the experiment and examples of 3D Character created by the proposed system are shown in section 4. Finally, we conclude the paper in section 5.

2   3D CHARACTER CREATION SYSTEM

2.1   System Outline

Since the proposed system employs an interactive evolutionary computation (IEC) mechanism, it can automatically learn the users' preference. What the users have to do is only evaluation of displayed 3D characters created by the system.

Figure 1 shows the flow of the whole the proposed system.

Processes used in the proposed system are as follows.

1) The system creates 3D characters.

2) The 3D characters are shown to the user.

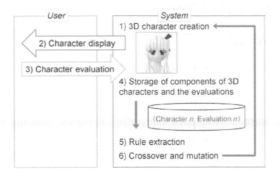

Figure 1 Flow of the whole the proposed system

3) The user evaluates the characters.

4) The component of the 3D characters and the evaluations are accumulated.

5) Rules are extracted from the accumulated data.

6) The component of the 3D characters is regarded as a gene, and genetic operations, crossover and mutation are carried out.

7) 1)-6) are repeated until a satisfied character is provided.

In 1), based on the rules extracted by a preliminary experiment, N initial individuals are created. In 5), Rough Sets Theory (RST) is employed to extract sensibility rules. In 6), crossover, mutation of Genetic Algorithm (GA) and revision by rules are carried out. Then, processing returns to 1) again.

2.2 COMPONENT OF THE 3D CHARACTER IN THE PROPOSED SYSTEM

First, components of the 3D character in the proposed system are explained. The components are divided into 5 attributes and 11 parts. Table 1 shows attributes and the values. Table 2 shows parts and the values. The attributes are abstract, and the parts are concrete. Linguistic expression is employed for the attributes so that the proposed system can be easily applied to the other systems. The values of the attributes and the parts are set in reference to research(Kawatani, Kashiwazaki, and Takai, et al, 2010.) and *Moe* attribute category of Wikiopedia.

2.3 SENSIBILITY RULE EXTRACTED BY THE PROPRSED SYSTEM

The rule extracted by the proposed system is expressed as follows.

Antecedent part If A_i is a_i and A_j is A_j and $\cdots\cdots$

Consequent part Then sensibility word(***Kawaii***, *Kawaikunai*)

Table 1 Attributes and the values

Attributes	Values
Kind of eyes	[droopy eyes, up-angled eyes, narrow eyes, cat like eys] 4 kinds in total
Hairstyle	[hairstyle such as the intake, straight bangs, hime cut, butch haircut, sweptback hair, afro, mohawk, sausage curls, bunches, ponytail, braid, side topknot] 12 kinds in total
The length of the hair	[medium, semi long, long, very short] 4 kinds in total
Color of the hair	[green, pink, silver, red, purple, mazarine, orange, brown, black, yellow]10 kinds in total
Outline of the face	[normal, round, sharp] 3 kinds in total

Table 2 Parts and the values

Parts	Values
Vertical ratio of face	5 kinds from small to big
Height of the nose	5 kinds from flat nose to long nose
Texture of eyes	16 kinds of texture
Color of eyes	9kinds of texture
Size of eyes	5 kinds from small to big
Color of cheeks	2 inds of texture
Forelock	38 kinds of 3D model
Back hair	34 kinds of 3D model
Sideburns	11 kinds of 3D model
Interval between eyes	6 kinds from far to near
Position of mouth	6 kinds from high to low

A_i is the attribute and a_i is the value of the attribute. The attribute corresponds to that of 3D character's component. Sensibility word expresses the impression that a user has towards 3D character. Kawaii and Kawaikunai (not Kawaii) are used as the sensibility word in the proposed system.

2.4   INITIAL 3D CHARACTER CREATION

Attributes and parts are set in the initial 3D character creation. Attributes are set at random, and corrected by the initial sensibility rules. Figure 2 shows an example of correction by rules. The initial sensibility rules are sensibility rules using RST in a preliminary experiment. This combination of GA and RST refers to researches(Shijie Dai, He Huang, Fang Wu, Shumei Xiao Ting Zhang, 2009)(Zhang Liangzhi, He Minai, Zhang Mengmeng, 2010). Afterwards, parts are set as far as it is not against the attributes. For example, if attribute *Hairstyle* is *twin tail*, parts *Back hair* is selected from 3D models to meet *twin tail*.

2.5   CHARACTER DISPLAY

Figure 3 shows an example of character display in the proposed system. 3D characters are created according to the components and displayed to the user.

2.6   CHARACTER EVALUATION

The examination about the evaluation method is important in a study of the Kansei engineering. A preliminary experiment was carried out, and a pair comparison method is selected because it was easy to evaluate 3D characters most. In the method, individuals are sorted according to the evaluation of the user, and a high score is given in descending order.

2.7   STORAGE OF COMPONENTS OF 3D CHARACTERS AND THE EVALUATIONS

Components of 3D character are stored with the evaluation of the user. The Components of 3D character are attributes and parts. The evaluation value is a score given by the user.

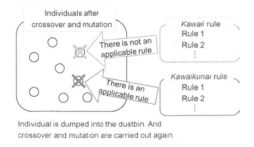

Figure 2 An example of correction by rules.

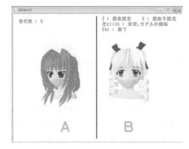

Figure 3 An example of character display.

2.8   RULE EXTRACTION

Kawaii rule and *Kawaikunai* rule are extracted using RST.

2.9   CROSSOVER AND MUTATION

At first, 2 individuals are selected from the M individuals with high score. Next, attributes are determined by Uniform Crossover. Then, parts are determined by Uniform Crossover as far as it is not against the attributes. Next, mutation is carried out. Attributes and parts change with a predetermined probability in every generation. Parts change as far as they are not against the attributes. Afterwards, attributes are corrected by sensibility rules like Figure 2 in the same way as the initial 3D character creation. This combination of GA and RST refers to researches (Shijie Dai, He Huang, Fang Wu, Shumei Xiao Ting Zhang, 2009)(Zhang Liangzhi, He Minai, Zhang Mengmeng, 2010). When an individual is dumped into the dustbin, crossover and mutation are carried out again. It is repeated until N individuals are created.

3   EXPERIMENTS

Experimental objective
Experiments were carried out to compare the proposed system with a method to choose the parts of the proposed system, an existing system(Ando and Hagiwara, 2009) and a commercially available game.

Experimental method
The procedures of the experiment are as follows. Examinee used the four systems. Afterwards, questionnaires were carried out for each system.

Table 3 shows items of the questionnaire. All the evaluations are five-level. The examinees are 20 men and women in their twenties. About (3), only the proposed system and the existing system were carried out.

Table 4 shows the values of various parameters used by this experiment. Initial sensibility rules were extracted from 1200 characters evaluated by the examinees of 12 men and women in their twenties.

Systems used for the comparison
The existing system is the system of our former research(Ando and Hagiwara, 2009). The commercially available game is Phantasy star portable 2 infinity("Official Website of Phantasy Star Portable 2 Infinity"). In this game, firstly a user creates a character. Because a variety of parts exist in this character creation, and it was thought that the quality is high, it was used as a comparison for reference.

Result and consideration
Table 5 summarizes the result. (1)-(8) in table 5 correspond (1)-(8) in table 3.

Table 6 shows result of Wilcoxon signed-rank test(Ueda, 2009.). *Significant difference* shows there is significant difference. *None* shows there is no significant difference.

There are significant differences in (5) and (6) between the proposed system and a method to choose the parts of the proposed system. The result shows 3D character can be easily created by using the proposed system than a method to choose the parts of the proposed system.

There is significant difference in all items of the questionnaire between the proposed system and the existing system. The result shows the proposed system is better than the existing system.

Accoding to Tables 5 and 6, it has shown that quality of the 3D characters created by the proposed system is comparable to thoes created by the commercially available game.

4   EXAMPLES OF 3D CHARACTER CREATED BY THE PROPOSED SYSTEM

Figure 4 shows examples of 3D Character created by the proposed system.

Table 3 items of the questionnaire

(1) Were you satisfied with a system?	[1: aggrievement—5: satisfaction]
(2) Were you able to make favorite character?	[1: aggrievement—5: satisfaction]
(3) Did you feel that preference was reflected?	[1: not feel—5: feel]
(4) Did you enjoy?	[1: not enjoy—5: enjoy]
(5) Were you tired?	[1: fatigable—5: not fatigable]
(6) Was it easy to operate it?	[1: not easy—5: easy]
(7) Hau about the variety?	[1: small variety—5: large variety]
(8) Was the quality high?	[1: low—5: high]

Table 4 Parameters

Parameters	Values
Population of one generation N	5
Number of higher individuals M	3
Mutation rate	60∼10%(decreased by every generation)

Table 5 Result

	The proposed system	A method to choose the parts of the proposed system	The existing system	A commercially available game(reference)
(1)	3.65	3.60	2.10	3.95
(2)	3.85	4.15	2.20	4.25
(3)	4.20	---	2.50	---
(4)	4.10	3.70	2.85	3.75
(5)	3.95	3.30	2.65	2.75
(6)	4.65	3.15	3.35	3.05
(7)	3.60	3.75	2.75	4.20
(8)	3.55	3.50	1.95	4.75

Table 6 Result of Wilcoxon signed-rank test (5%)

	A method to choose the parts of the proposed system	The existing system	A commercially available game(reference)
(1)	×	○	×
(2)	×	○	×
(3)	---	○	---
(4)	×	○	○
(5)	○	○	○
(6)	○	○	○
(7)	×	○	○
(8)	×	○	○

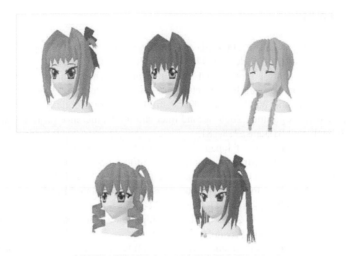

Figure 4 Examples of 3D Character created by the proposed system.

5   CONCLUSIONS

3D characters have been widely used in various fields such as advertizing and entertainment. This paper proposes an automatic 3D character creation system based on sensibility rule extraction. Since the proposed system employs an interactive evolutionary computation (IEC) mechanism, it can automatically learn the users' preference. What the users have to do is only evaluation of displayed 3D characters created by the system. The proposed system can create 3D characters with high diversity.

A lot of evaluation experiments were carried out and the superiority is demonstrated. In addition, it has shown that quality of the 3D characters created by the proposed system is comparable to that created by the commercially available game.

For the future work, there is creation the whole body as well as a face and there is improvement of quority or variety by entrusting an expert with the making of parts. In addition, there is engineered analysis about *Kawaii*.

ACKNOWLEDGMENTS

The authors are grateful to examinee of the evaluation experiment.

REFERENCES

Ando, M. and Hagiwara, M. 2009. 3D Character Creation System Using Kansei Rule with the Fitness Extraction Method. FUZZ-IEEE 2009: 1507-1512.

Chen, X. 2009. Analysis for digital content industry value chain. IEEE International Conference on Network Infrastructure and Digital Content, 2009: 349-352.

Dai, S., Huang, H., and Wu, F., et al. 2009. Path Planning for Mobile Robot Based on Rough Set Genetic Algorithm. 2009 Second International Conference on Intelligent Networks and Intelligent Systems: 278-281.

Digital Content Association of Japan, DNP Media Create Co., Ltd., Ministry of Economy, and Trade and Industry, Commerce and Information Policy Bureau. 2011. Digital Content Hakusyo 2011. Tokyo, Japan: Digital Content Association of Japan.

Gu, Z., Tang, M. X., and Frazer, J.H. 2006. Capturing Aesthetic Intention during Interactive Evolution. Computer- Aided Design 38: 224-237.

Igarashi, T. and Hughes, J. F. 2002. Clothing Manipulation. 15th Annual Symposium on User Interface Software and Technology: 91-100.

Inoue, K., Harada, T., and Shiizuka, H., et al. 2009. Rough Sets are applied to Kansei Engineering. Tokyo, Japan: Kaibundo.

Ito, H. and Hagiwara, M. 2005. Character-agent automatic creating system reflecting user's kansei. Journal of Japan Society of Kansei Engineering, Vol.5, No.3: 11-16.

Kawatani, H., Kashiwazaki, H., and Takai, Y., et al. 2010. Feature Evaluation by Moe-Factor of ANIME Characters Images and its Application. IEICE Technical Report, ITS, Vol.109, No.414: 113-118.

Liangzhi, Z., Minai, H., and Mengmeng, Z. 2010. Study on Road Network Bi-level Programming under the Traffic Flow Guidance. 2010 International Conference on Measuring Technology and Mechatronics Automation: 631-634.

Mitake, H., Aoki, T., and Hasegawa, S., et al. 2011. Research on motion generation of reactive virtual creatures. Journal of Japan Society of kansei Engineering, Vol.10, No.2: 79-82.

Mizuno, K., Kashiwazaki, H., and Takai, Y., et al. 2008. Human Motion Estimation System for 3D Character Animation. IPSJ SIG Notes GCAD, Vol. 2008, No.80: 45-48.

Nishikawa, T., Mashita, T. and Ogawa, T., et al. 2011 : A Context-Sensitive Prediction Method for Ordering Multimedia Content in a Mobile Environment. IEICE Trans. D, Vol.94, No.1: 147-158.

"Official Website of Phantasy Star Portable 2 Infinity." Accessed February 21, 2012, http://phantasystar.sega.jp/psp2i/.

Ohkura, M. 2011. Systematic Study on Kawaii Products. Journal of Japan Society of kansei Engineering, Vol.10, No.2: 73-78.

Sakawa, M. and Tanaka, M. 1995. Genetic Algorithm. Tokyo, Japan: Asakura Publishing Co., Ltd.

Sugahara, T. 2011. Considerations of the geometrical features of the smiling face creating cuteness. Journal of Japan Society of kansei Engineering, Vol.10, No.2: 96-98.

Toma, K., Kagami, S., and Hashimot, K. 2011. 3D Measurement Using a High-Speed Projector-Camera System with Application to Physical Interaction Games. Transactions of the Virtual Reality Society of Japan, Vol. 16, No.2: 251-260.

Ueda, T. 2009. How to Solve Statistical Test and Estimates to Learn from 44 Exercises. Tokyo, Japan: Ohmsha.

CHAPTER 11

Shaboned Display: An Interactive Substantial Display Using Expansion and Explosion of Soap Bubbles

Shiho Hirayama, Yasuaki Kakehi

Keio University
Kanagawa, JAPAN
{hirayama, ykakehi}@sfc.keio.ac.jp

ABSTRACT

We propose an interactive substantial display named "Shaboned Display" consisting of bubble film array. In this system, each bubble does not float away and it functions as a pixel by expanding and contracting at the same position. As a result, this system can show various kinds of image or motion by controlling the action of each bubble. In addition, we succeeded in detecting an explosion of the bubble film using an electrical approach and utilizing this is as an input. Using such an ephemeral material as an interface, this touch interface stimulates humans to touch and break them. Moreover, we can create unpredictable changes in an artistic representation. We herein report the design, implementation, and evaluation of a display system and touch interaction using a bubble film array along with its application examples.

Keywords: Bubbles, Substantial display, Touch Interaction, Ephemeral material

1. INTRODUCTION

So far, many researchers and artists, mainly from the field of media art, have made several attempts to create substantial displays so that users can directly touch the materials forming pixels. As one of characteristics of the substantial display, these pixels can sometimes be affected by environmental factors and conditions such as lighting, temperature, humidity, time and actions of users/audiences. Normally, such kinds of contingent factors are not desirable for showing information precisely. However, some artists and researchers have willingly involved them in the output of their displays mainly as art expressions. In this research, we also adopt this approach and propose a novel substantial display, which can show information affected by ambient factors. Furthermore, in this system, we propose a method for utilizing the contingencies not only for output but also input, and interactions between the system and the users or ambient factors.

As a material for forming pixels, we focused on soap bubbles. Bubble seems beautiful and many people have pleasant memories to play it. Soap bubbles have several characteristics, for example, it can be blown into precise sphere. However bubbles are very fragile and can lose their shape easily without a piece. Certainly, the surface of a bubble seems transparent at first because it is as thin as visible light wavelength. However, gravity causes a difference in the surface thickness resulting in an extremely thin liquid film, and the bubble appears extremely iridescent. This surface color changes incessantly under the influence of the environment. The beauty and variety in the appearance of a bubble is as important as its fragility.

Our research proposes the bubble interface using those characteristics. First, we present an interactive substantial display named "Shaboned Display" (see Figure 1). The bubble film functions as a pixel by expanding and contracting at the same position. Second, we present a sound interactive art named "Shaboned Chime" with haptic input obtained by detecting bubbles' explosion. This system can generate various sounds according to the explosions.

Figure 1 Shaboned Display

2. RELATED STUDIES

So fur, substantial displays have been studied in many projects; of these include Wooden mirror [1], WATERLOGO [2], and Shade Pixel [3]. Thus, really various materials, liquid or solid, natural or artificial, have been examined, and unique representations have been realized. On the other hand, Sandscape [4] and Khronos Projector [5] realized input functions making use of material characteristics of sand or clothes. Although these systems, which can be touched and controlled directly, provide intuitive interaction, they cannot function without optical devices such as a projector.

In contrast, our research aims to realize touchable input and substantial display using material characteristics simultaneously. As an example adopting the similar approach, Super Cilia Skin [6] is a substantial display, which consists of an elastic membrane and an array of cilia actuators. When the surface of this membrane is stroked, this system detects the position of each actuator as input and this information is used to modify figure in three dimensions as output. While this system adopted a kinetic approach, in our research, we explore a novel display technology with haptic and ambient interactions using soap bubbles and air controls.

As for the soap bubbles, they are popular for children's game. Moreover, this unique and beautiful material attracts scientists [7][8] and artists. As typical previous display systems utilizing bubbles as image pixels, Kashiwagi and his colleagues realized a 3D substantial display [9]. In this system, static electricity attracts bubbles and coordinates to establish their position. Therefore a 3D-image can be formed but that system does not have functions for interaction. On the other hand, many project have paid attention to the explosion of soap bubbles for input events while the position of the bubbles are not controllable. In Bubble cosmos [10] and Sylvester's research [11], a camera detects the position and size of a bubble with smoke and applies the results to interaction. Ephemeral melody [12] employs electricity instead of camera. When scattered bubbles hits electric copper pipes, power is turned on and sound is generated.

In comparison to these previous researches, we explore to realize both of the two functions simultaneously: a substantial display and interaction with bubble explosions. Our system can show concrete representations by controlling expansion and contraction of soap bubbles and can provide interactions by detecting touching and bursting of bubbles.

3. SUBSTANTIAL DISPLAY WITH TOUCH INPUT USING SOAP BUBBLES

3.1. Overview

As stated above, we proposed an interactive substantial display consisting of a bubble film continually expands and contracts. In this project, we realize that the following two aims. First, by controlling scale, position and timing of each bubble, this system can output visual images envisaged by an artist. Unlike in a conventional automatic bubble machine, the bubbles in our system function as a pixel. Second, a touch interface is implemented by detecting bubble explosion. The fragility of bubbles motivates us to touch or burst them, and it encourages audiences to interact more effectively.

We will explain the design of a system that realizes the following three functions.

- Continual expansion and contraction of the bubbles film.
- Generation of the bubble film
- Detection of bubble explosion.

3.2. System design to control bubble movement

Figure 2 shows the overview of the system design. We attached bubble vents on a bottom surface of a black acrylic box filled with soap liquid. Then, we put a tube made of soft vinyl and sponge that always absorb enough soap liquid at the tip of each vent (see Figure 3). In addition, to create soap bubbles automatically, we attached a valve mechanism with a solenoid for pinching the tip of the vent (see Figure 4). After the valve pinches the vent, it naturally opens due to the elasticity of vinyl so that a soap film is generated. Each vent is connected with a small air pump under the acrylic box. And there is a hole of 2 mm between the vent and the pump to allow escape of air. Thus, by controlling the on/off timing of the pump, the system can expand and contract the bubble film.

In the current implementation, we use P54A02R pumps （Max pressure 90 kPa ， Quantity of flow 1500 cc/min） manufactured by of Okenseiko Co. Ltd. To turn pixel ON, the pump blows on the bubble film for 1 second to form bubbles of 4 cm across. To contract the film, the pump is switched off for a minimum 2 second. We designed the size of the bubble so that it is large enough to display a representation clearly and suitable to maintain expansion and contraction. In addition, we set 5 cm for the distance between each vent to prevent the bubbles from touching each other.

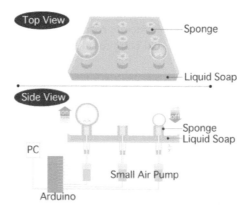

Figure 2 System for expansion and contraction bubbles.

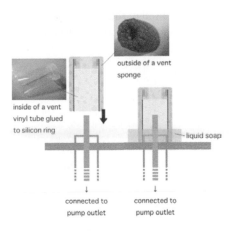

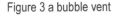

Figure 3 a bubble vent

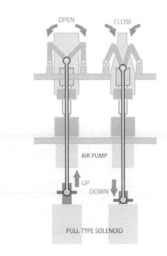

Figure 4 a valve mechanism with a solenoid
to reproduce bubbles film automatically

3.3. System design for detection of bubbles' explosion

Here, we described the system for detection of bubble explosion. First of all, we need note that a bubble is an electrically-conductive material. Thus, in this system, we adopted this material characteristic for detecting explosion. Concretely, we attached two electrodes near the vent (see Figure 5). While one is hung on the vent, the other one is attached 2cm above the vent so that a bubble reaches it when expanding fully. These electrodes are part of the same electric circuit with a pull-down switch and are able to sense a change in voltage (Figure 6). In this system, a voltage value is sent to a computer via a micro-controller called Arduino. Figure 7 shows an abrupt drop in voltage between the electrodes when a bubble explosion occurs. The vertical axis represents voltage and the lateral axis represents time. Comparing the current input value with the value in the previous frame, if the difference exceeds a threshold we set in advance, the system regard a bubble explosion occurs.

We perform the following experiment to confirm the accuracy of the sensing system. The procedure of the experiment is as follows:
1). Inflate a bubble film on a vent using a pump for a second.
2). The bubble becomes the maximum size, 4cm across.
3). Break this bubble when it expanded at the maximum.

We repeated this set 100 times in a room of following condition: temperature 21 ℃, the humidity 25%. Figure 8 shows a graph representing a histogram of potential differences between the electrodes at the moment of explosion. In this experiment, threshold was set up 0.15 V, which is an empirically derived value. Through this experiment, the success rate of the system to detect was 96%. In most of the failure cases, we observed that the bubble could not reach the electrode fully since it expands diagonally affected by the vent conditions. In the future, we plan to improve the success rate more by setting different threshold to each vent according to the condition.

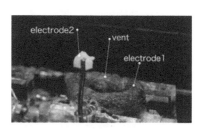

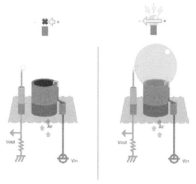

Figure 5 a vent with electrodes.

Figure 6 System to detect bubble explosion using electricity.

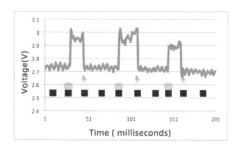

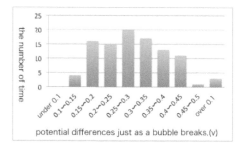

Figure 7 Variation in potential depending on presence of bubble film.

Figure 8 the histogram of potential difference when system detects an explosion.

4. REPRESENTATIONS USING SOAP BUBBLES

In this section, we describe a specific representation realized by Shaboned Display and Shaboned Chime.

4.1. Infomational representation on Shaboned Display

The Shaboned Display can output specific images that consist of pixels using bubbles. Figure 8 shows alphabet images using 5 x 5 bubbles. Moreover, this system can output the various representations for example a movement from center like ripples spread across the water, a switching movement like a hound's-tooth check pattern. Figure 9 shows a movement like a wave; ten bubbles in a row expand for a second from left to right. At this time, the Shaboned Display requires some time to change images because bubble behaves slowly.

Figure 9
Representation examples 01:
Drawing alphabets "H""O" and "N"

Figure 10
Representation example02:
motion like a wave
from left to right

4.2. Shaboned Chime: interaction with detection of explosion

Shaboned Chime has 8 vents in arrow, and it generates an individual sound when a bubble is exploded. The electrodes arranged on each vent measure the voltage numerically and send them as signals for Processing. The system plays a different sound depending on which bubble is broken because each vent is associated with an individual sound effect.

In this system, the audiences can play it intentionally like playing a piano. In addition, they can also hear ambient melody created by unintentional bubble explosion due to ambient conditions. These sounds assigned to each vent are high-pitched are electronic to match the image of bubble explosion. While we cannot hear the actual sound of the bubble explosion directly, we can enjoy the explosions exaggerated by these synthesized sounds.

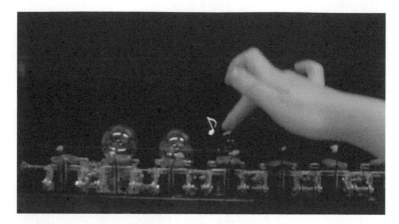

Figure 11 Shaboned Chime

4.3. Discussion

We have demonstrated Shaboned Display and Shaboned Chime in several exhibitions including SIGGRAPH 2010 Emerging Technologies and Keio University Open Research Forum 2010.

Through the exhibitions, various people irrespective of age or gender were interested in these systems and the experiment. As for the Shaboned Display, many audiences gave us their impressions as typified by following comments: ``beautiful'', ``Fun to watch. I want to watch it for a long time'' and ``It's looks like a creature breathing.'' In our system, the bubble pixels itself change their look incessantly, and we never observe identical situations. Many factors such as ambient light, wind, audience behaviors, surrounding objects can influence bubbles. Furthermore, the bubbles keep being generated and bursting. We assume that such kind of characteristics affected the feedbacks of the audiences.

Further, some people walk around the display and tried to watch it from several points of view. The bubble pixels are substantial and three-dimensional spheres. So the appearance of the display can be changed depending on the viewing positions. In addition, their iridescent surface has a structural color. Therefore, the color of each pixel can be changed depends on the distance and angle of view. In addition, other typical comment of audiences was that they felt pleased when some bubbles could keep existing without bursting. We assume this aspect can also gather attentions of audiences to the displayed images by deriving feelings of KAWAII.

In case of the Shaboned Chime, many audiences touch and broke bubbles at their own initiative without special explanations about the usage. We believe that the material of the bubble film provides a certain affordance or an appeal itself to the audiences so that they tend to feel and break. When a sound was generated at the time of the explosion, some audiences burst into laughter, and some tried to touch them more to listen and confirm the sound assigned to other vents. As for notable behaviors, some audiences tried to input by not only touching the bubbles by hand, but also blowing them or shaking them with wet fingers. We think that such actions of audiences can be important design factors for our system. In the near future, we plan to improve the hardware to detect various actions of audiences as input and realize much more various interactions.

5. CONCLUSION

The soap bubble is very beautiful and made of daily used material. In addition, this delicate sphere burst easily without a piece, because of various influences from environmental factors in this real world such as a wind. This fragility motivates us want to touch and break it positively. By using these unique characteristics, our research developed two systems. One is a substantial display named Shaboned Display. This system can make ephemeral images consisting of bubble pixels. The other is an audio interactive art named Shaboned Chime. This system can detect the bubble explosion and therefore, we utilize effectively lack of pixels and make an unintentional art. Through some exhibitions, these systems interested the audiences with beauty and fluid representation. Further, many audiences tried breaking bubbles and enjoyed it without explanation, it can be said that our systems are intuitive and entertainment.

In the future, we should deepen representation utilizing bubble characteristics more effectively and depend that as platform for art representation, which has both controlled description and unintentional changes

REFERENCES

[1]. Rozin D. 2000. : "Wooden mirror". *ACM SIGGRAPH 2000 Art Gallery.*

[2]. Atlier OMOYA : WATERLOGO: Accessed 25 February, 2012 http://atelieromoya.jp/wl09.html

[3]. Hyunjung K., Woohun L.2008.: Shade Pixel: Interactive Skin for Ambient Information Displays". *SIGGRAPH ASIA 2008, Posters, 2008.*

[4]. H Ishii, C Ratti, B Piper, Y Wang, A Biderman and E Ben-Joseph.2004. : "Bringing clay and sand into digital design—continuous tangible user interfaces," *BT Technology Journal*, Vol.22 No.4,

[5]. Cassinelli A. and Ishikawa, M.2005: "Khronos projector". *ACM SIGGRAPH 2005 Emerging Technologies.*

[6]. Hayes R., Mitchell W. J. James T.2003.: Super Cilia Skin: An Interactive Membrane. CHI2003, *Conference on Human Factors in Computing Systems.*

[7]. Charles V. B.; Translated by Hiroshi Noguchi1987. Soap Bubbles: Their Colours and the Forces Which Mould Them", TokyoTosho Co.

[8]. Hachiro T. 1951-1958. : On the Electrification of Soap Bubble By Expansion Part1-5. *the faculty of liberal arts of Iwate University, The annual reportVol.3 Part2- Part Vol.13 Part2*

[9]. Go K. Tomohiro T. Michitaka H. 2006: Material-based 3D Display by Static Electricity, Entertainment Computing Collected Papers, pp.23-30, 2006.

[10]. M. Nakamura, Go Inaba Jun T., Kazuto S., Junichi H.2006: Mounting and Application of Soap Bubble Display Method, *Transactions of the Virtual Reality Society of Japan, Vol.11, No.2, pp.339-349.*

[11]. Axel S. Tanja D. Albrecht S. 2010. : Liquids, Smoke, and Soap Bubbles Reflections on Materials for Ephemeral User Interfaces, *TEI '10 Proceedings of the fourth international conference on Tangible, embedded, and embodied interaction.*

[12]. Risa S. Taro S. Makoto I. Chuuichi A. 2008. : Slow Interactive Media Art Using Soap Bubbles, *1A5-3, 13th roceedings of the Virtual Reality Society of Japan annual conference.*

CHAPTER 12

Study on Kawaii in Motion -Classifying Kawaii Motion Using Roomba-

Shohei SUGANO [†], Haruna MORITA [††], Ken TOMIYAMA [†††]

The Department of Advanced Robotics, Chiba Institute of Technology
Narashino, Chiba, Japan 275-0016
[†] mad.hatter.teaparty3173@gmail.com
[††] yumemiru_yuuki@yahoo.co.jp
[†††] tomiyama.ken@it-chiba.ac.jp

ABSTRACT

In this study, we attempt to categorize Kawaii-ness in motion using Roomba as a testing bed. Kawaii, a Japanese word expressing favorable characteristics such as pretty, adorable and fairly like, is a part of Japanese original concept Kansei (sensibility, sensitivity, feeling, emotion, etc.) and becomes known to many people in relation to Japanese anime. Researchers have started studying Kawaii-ness and results on Kawaii-ness in shape and color have been published but not in motion. We, therefore, focus our study on Kawaii-ness in motion. We chose Roomba for our study because Roomba is not particularly Kawaii in shape or color and that helps us evaluating Kawaii-ness in motion. Furthermore, Roomba shows variety of motions using combinations of three basic motions consisting of straight motion, turn, and spin. This makes it easy to segment and categorize its motion according to Kawaii-ness contents. First, natural motion of Roomba was shown to several subjects and asked them to identify parts of motion that they found Kawaii. We asked to identify Kawaii motions using arrows of various shapes, such as a circular arrow for the circling motion. We then segmented and categorized the motion of Roomba according to the identified Kawaii blocks. Randomly chosen 10 prepared video segments were shown to another set of subjects to ask Kawaii-ness content of those segmented motions. For every Roomba motion a subject found Kawaii, we asked the subject to answer the questionnaire consisting of pairs of adjectives prepared according to the SD method. The representative physical features of categorized Roomba motions are correlated with the Kawaii-ness based on the result of the questionnaire. We report our findings in detail in this article.

Keywords: *Kawaii*, Roomba, Motion Classification, Kansei values.

1   INTRODUCTION

In this study, we investigate Kawaii-ness in motion. Kawaii, a Japanese original concept, is spreading around the world and became known to people in many countries.   Kawaii is one of the Kansei categories that describes favorable characters such as pretty, adorable, fairy-like, and cute.   It is among the Kansei values that have become important in manufacturing in Japan. The Japanese Ministry of Economy, Trade and Industry (METI) has been conducting a project called the "Kansei Value Creation Initiative" from 2007 (METI, 2008 and Araki, 2007). The main objective of the project is to introduce a new value axis in addition to the conventional value axes of reliability, performance, price and so on), which are recognized in the manufacturing sector. METI held a Kansei value creation fair, called the "Kansei-Japan Design Exhibition," at Les Arts Decoratifs (museum of decorative arts) at Musée du Louvre, Paris in December 2008.   Launched as an event of the "Kansei Value Creation Years," the exhibition had more than 10,000 visitors during the ten-day period and was received quite favorably (METI, 2009). We have started studying Virtual Kansei for robots where robots can understand human emotions, can generate own (virtual) emotion and can express own emotion while performing given tasks (Miyaji, 2007 and Kogami, 2009).

Recently, researchers have started studying Kawaii-ness features from various directions. For example, colors and shapes have been studied by Ohkura and Murai (Ohkura *et al.*, 2008 and Murai *et al.*, 2008). Those results show that curved shapes such as a torus and a sphere are generally evaluated as more Kawaii than straight-lined shapes. Ohkura also studied the relationship between Kawaii-ness in color and physiological signals (Ohkura *et al.*, 2009). Other researchers such as Nittono adopted a behavioral science approach to study Kawaii artifacts (Nittono, 2011). Mitake and others reported a motion generation technique for reactive virtual creatures from Kansei point of view (Mitake *et al.*, 2011). Mori and others reported a series of studies on motion generation of interactive robot taking human Kansei into consideration. They studied the relationship between boredom and impression of robot motion and proposed an algorithm for robot motion generation that does not make human bored (Mori *et al.*, 2004 and Mori *et al.*, 2004a).

However, there is little work on Kawaii-ness in motion itself. Here, we focus on Kawaii-ness in ordinary-life motion observed in daily life. This is because we are not interested in intentional motions such as over acting theatrical plays. As a first step of our study, we adopted Roomba as a devise to show motions. This is because, unlike dolls and animals, Roomba does not have particularly Kawaii shape and color and, therefore, evaluators can concentrate on motion itself. This paper describes our findings on classifying Kawaii motion using Roomba.

It is noted that we have been concentrating our Virtual Kansei research to emotion so far. Here we intend to expand our study to Kansei proper, and choose Kawaii-ness as the first target.

2   CLASSIFYING KAWAII MOTION

As we stated above, we chose Roomba, model 537, in Fig. 1 to categorize Kawaii-ness in motion. Roomba is a vacuum cleaning robot created and marketed by iRobot company (iRobot, 2012). The followings are the reasons why we chose Roomba.

- Shape and color do not particularly induce Kawaii-ness.
- Motions are combinations of three simple motion elements.
- Motion is in 2D.
- Roomba is the most familiar autonomous mobile robot used in home electrical appliances category.

Basic motions of Roomba consist of straight motion, left and right turns, and left and right spins. Roomba shows variety of motions using combinations of those motions. For example, the spiral motion is a combination of spin and forward motions. Thus Roomba is ideal to categorize basic Kawaii motions. We first conducted two preliminary experiments.

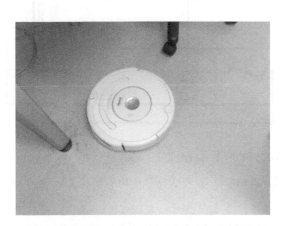

Figure 1   The picture of Roomba used in the questionnaire.

2.1   FIRST PRELIMINARY EXPERIMENT

The objective of the first preliminary experiment was to find how to segment the Roomba motion. We showed the Roomba in auto mode cleaning a designated area in one of the college classrooms to subjects over ten minutes. Figure 2 shows the floor plan of the classroom and the designated area used in experiment. We asked subjects to describe any part of motions that they felt Kawaii. We asked to identify those Kawaii motions using arrows of various shapes, such as a circular arrow for circling motion. We also video recorded the motion of Roomba for later processing.

2.2   SECOND PRELIMINARY EXPERIMENT

The objective of the second preliminary experiment was to obtain the human's natural view angle. This was necessary to determine camera angle to take video footages for the main experiment. First, we asked the subjects to stand at a distance natural for them to watch Roomba motion. Then, we measured the distance from Roomba to the subjects. Those distances and the heights of the subjects were used to calculate the view angles at which the subjects watched the Roomba (Figure 3). The heights of the eyes of the subjects were calculated by subtracting 10 (cm) from the heights of them.

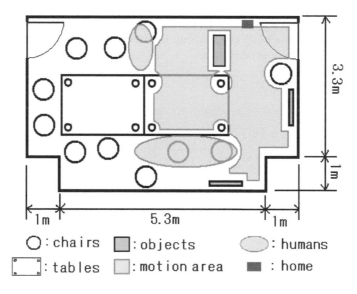

Figure 2   The floor plan and the designated area of the classroom used in the first preliminary experiment.

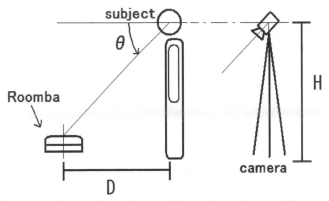

D: Distance from Roomba to the observing subject
H: Distance from the ground to subject's eyes
θ: The view angle at which the subject watched the Roomba

Figure 3 The image of second preliminary experiment to determine the camera angle.

2.3   RESULTS OF PRELIMINARY EXPERIMENTS

The subjects of the first experiment were three female and six male student volunteers all in twenties. Figure 4 shows examples of arrows indicating Kawaii motions that the subjects drawn. Based on those arrows of motions, we segmented the recorded video of Roomba's motion into small individual footages. The length of the footages of the segmented motions became 2 sec to 30 sec long.

Second experiment was performed by three female and seven male student volunteers in twenties. The measured average distance was about 155 centimeter. The average height of the subjects was also about 155 centimeter. As a result, the obtained average view angle was approximately 45 degrees (Table 1).

Based on the results of two preliminary experiments, we are ready to prepare video footages to show to subjects in the main experiment.

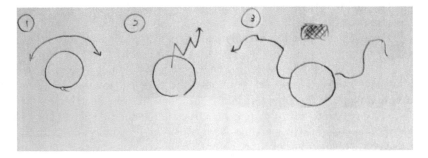

Figure 4   An example of arrows indicating Kawaii motions that the subjects drawn.

2.4 MAIN EXPERIMENT

We produced ten types of motion video footages based on the findings in preliminary experiments. Table 1 shows the names and the lengths of the motion footages. The ten types of motions are as follows:

- Spin: Roomba spins around like a top spinning at a spot.
- Back: Roomba moves backward as it leaves the home.
- Spiral: Roomba moves along an outward spiral.
- Zigzag: Roomba moves forward in sharp angles.
- Bounce: Roomba changes the direction of travel after colliding with an object.
- Home: Roomba slowly moves forward when entering home.
- Turn: Roomba changes the direction of travel using rotating motion.
- Dizzy: Roomba wonders at nearly constant speed.
- Pattern: Roomba repeats similar motion.
- Fuzzy: Trial and error type motion of Roomba including stops between.

We conducted a survey to evaluate Kawaii-ness using those Roomba motions. The experiment was carried out in two or three people at a time, and was carried out by projecting the image of a PC using a big screen so the subjects can see clearly the motion of the video. The experimental procedures are as follows:

1. The subjects sat approximately at 5 (m) from the screen.
2. The experimenter described the experiment.
3. The subjects answered three questions relevant to Roomba.
4. The experimenter showed a photo of Roomba.
5. The subjects answered the question "Do you think Roomba is Kawaii?"
6. The experimenter showed the video of the selected motion to the subjects.
7. Each subject answered a questionnaire on the motion only if they felt it Kawaii. Otherwise, they skipped the questionnaire.

Three questions in Step 3 are as follows:

1. Do you know Roomba?
2. Have you ever seen Roomba?
3. Have you ever seen Roomba in motion?

Figure 1 in Chapter 2 is the photo image presented to the subjects at step 4. Steps 6 and 7 were repeated ten times for ten different motions, which were displayed in random order. Answers in the questionnaire were analyzed by the SD method. Table 2 shows the 20 pairing adjectives in the questionnaire. They were listed with 5 point scale. The markings were converted to numbers of -2 to 2 from the left adjective to the right one.

113

Table 1   The name and the time of the motion.

Type	Time [sec]	Type	Time [sec]
Spin	7	Home	11
Back	6	Turn	11
Spiral	30	Dizzy	14
Zigzag	8	Pattern	11
Bounce	9	Fuzzy	10

Table 2   The pairing adjectives used in the questionnaire.

No	Adjective pairs	No	Adjective pairs
1	Obedient – Rebellious	11	Toddling – Brisk
2	Complicated – Simple	12	Precocious – Childish
3	Familiar – Unknown	13	Smart – Stupid
4	Artificial – Natural	14	Relaxed – Nervous
5	Yielding – Stubborn	15	Serious – Funny
6	Long – Short	16	Smooth – Ragged
7	Quick – Slow	17	Hasty – Gradual
8	Elder – Young	18	Busy – Idle
9	Strong – Weak	19	Regular – Irregular
10	Straight – Curved	20	Clear – Ambiguous

2.5   RESULT OF MAIN EXPERIMENT

There were 9 female and 9 male participants for the evaluation experiment. The age demography of them are 1 teens, 6 twenties, 1 forties, and 1 seventies females and 8 twenties and 1 sixties male. 90% of all the participants answered that they know Roomba and had seen it moving. The answers to the Kawaii-ness of the shape of Roomba are summarized in Figure 5. It is noted that 45% of female answered no and that is much higher than 22% of male. Table 3 lists top three motions that are classified Kawaii. All male subjects chose Dizzy as Kawaii motion whereas female subjects chose Spiral and Bounce as tied top. An interesting observation here is that although about half of female subjects answered the shape of Roomba not Kawaii but 2/3 of them found Spiral and Bounce motions Kawaii.

We also checked consistencies in the choice of adjectives by analyzing standard deviations (SD) in markings of pairs of adjectives. By looking at those motions with more than 9 subjects chose Kawaii, we found three adjective pairs with less than 0.8 SD. They are

- <u>Simple</u> / Complicated (SD: 0.53, Average: -1.5)
- <u>Smooth</u> / Ragged (SD: 0.52, Average: -1.5)
- <u>Regular</u> / Irregular (SD: 0.70, Average: -1.4)

Note that negative averages indicate that the subjects chose the right adjectives (underlined).

The fact that there are those adjective pairs with low SD indicates that there are some underlining common features that contribute to the Kawaii-ness in motion. For example, Kawaii-ness in primitive motions of babies may be tied to the choice of Simple (Average: -1.5). Smoothness (Average: -1.5) suggests physical parameters such as acceleration may be associated with Kawaii-ness. Regular (Average: -1.4) means repetition of simple motion and that may contribute to Kawaii-ness because it may remind unintentional repeated trials of an unskilled.

(a) All   (b) Male   (c) Female

Figure 5   Response to the question "Do you think Roomba is Kawaii?"

Table 3   Rankings of the responses.

(a) Female

Rank	Score (%)	Motion type
1st	6 (67%)	Spiral
	6 (67%)	Bounce
3rd	5 (56%)	Fuzzy

(b) Male

Rank	Score (%)	Motion type
1st	9 (100%)	Dizzy
2nd	7 (78%)	Bounce
3rd	6 (67%)	Pattern

3   CONCLUSIONS

In this study, we investigate Kawaii-ness in motion using segmented motions of Roomba. We first conducted two preliminary experiments for finding how to divide Roomba's continuous motion into identifiable segments. Based on the findings of two preliminary experiments, we were able to categorize ten separate motions. We showed those motions to subjects and asked them to evaluate those motions they found Kawaii by marking between pairs of adjectives.

It was observed that, although 45% of female subjects answered the shape of Roomba not Kawaii In main experiment, 2/3 of them found motions named Spiral and Bounce Kawaii. We consider this indicating that Roomba was very appropriate equipment in studying Kawaii-ness in motion, not in shape or in color. We further analyzed the consistency in answers for questionnaire by looking at smallness of SDs of numerated markings. It was found that subjects tended to mark simple, smooth, and regular as qualities associated with Kawaii-ness. This indicates the possibility of finding association between Kawaii-ness and physical quantities.

It is our immediate task to increase the number of subjects and investigate if the findings here are generalizable. A more important and interesting task is to extract physical characteristics from the categorized motions of Roomba and investigate relationship between them and Kawaii-ness in motion.

4   ACKNOWLEDGEMENTS

We thank to the students and their families of Chiba Institute of Technology who contributed to this research and served as volunteers.

REFERENCES

Araki, J. 2007. "Kansei and Value Creation Initiative," Journal of Japan Society of Kansei Engineering, Vol. 7(3), pp.417-419, 2007. (in Japanese).

"iRobot corporation". 2011. Roomba official site. Accessed January 25, 2012, http://store.irobot.com/category/index.jsp?categoryId=3334619&cp=2804605.

Kogami, J., Y. Miyaji, and K. Tomiyama. 2009. "Kansei Generator using HMM for Virtual KANSEI in Caretaker Support Robot," KANSEI Engineering International, pp.83-90, 2009.

METI. 2008. "Kansei Value Creation Initiative" http://www.meti.go.jp/english/information/downloadfiles/PressRelease/080620KANSEI.pdf.

METI. 2009. Announcement of the "Kansei Value Creation Museum" http://www.meti.go.jp/english/press/data/20090119_02.html.

Mitake, H., T. Aoki, and S. Hasegawa. 2011. "Research on motion generation of reactive virtual creatures," Journal of Japan Society of Kansei Engineering, Vol. 10(2), pp.79-82, 2011. (in Japanese).

Miyaji, Y., and K. Tomiyama. 2007. "Virtual KANSEI for Robots in Welfare." International Conference on Complex Medical Engineering Proceedings (IEEE/ICME 2007), 2007.

Mori, Y., K. Ota, and T. Nakamura. 2004. "Motion Generation of Interactive Robot Considering KANSEI of Human (The First Report) -Relationship between Complexity and Boredom, and Impression of Robot Motion-," Journal of Japan Society of Kansei Engineering, Vol. 4(1), pp.17-20, 2004. (in Japanese).

Mori, Y., K. Ota, and T. Nakamura. 2004a. "Motion Generation of Interactive Robot Considering KANSEI of Human (The Second Report) –Motion Algorithm for Robot that does not tire Human-," Journal of Japan Society of Kansei Engineering, Vol. 4(1), pp.21-26, 2004. (in Japanese).

Nittono, H. 2011. "A behavioral science approach to research and development of Kawaii artifacts," Journal of Japan Society of Kansei Engineering, Vol. 10(2), pp.91-95, 2011. (in Japanese).

Ohkura, M., A. Konuma, S. Murai, and T. Aoto. 2008. "Systematic Study for "Kawaii" Products (The Second Report) -Comparison of "Kawaii" colors and shapes-," Proceedings of SICE2008, Chofu, Japan.

Murai, S., S. Goto, T. Aoto, and M. Ohkura. 2008. "Systematic Study for "Kawaii" Products (The Third Report) -Comparison of "Kawaii" between 2D and 3D-," Proceedings of VRSJ2008.(in Japanese).

Ohkura, M., S. Goto, T. Aoto. 2009. "Systematic Study for "Kawaii" Products (Fifth Report) -Relation between Kawaii Feelings and Biological Signals-," ICCAS-SICE2009, Fukuoka, Japan.

CHAPTER 13

Holistic Analysis on Affective Source of Japanese Traditional Skills

K. Morimoto and N. Kuwahara
Kyoto Institute of Technology
Matsugasaki, Sakyo-ku, Kyoto, JAPAN
morix@kit.ac.jp, nkuwahar@kit.ac.jp

ABSTRACT

Traditional industries supported by skillful craftsmanship have being narrowed fields of activity in Japan, because their advanced and affective skills are not transformed to the next generation. Skills of traditional craftsmanship are unique in Japan, and the viewpoint will be changed to promote the industry relevant to their skills. The characteristics of the traditional craftsmanship have to be solved and utilized it in a frontier industry. It is required to know what is the tips and intuition of them, and to harness it in proper fields. If a machine can be made to learn their tips and intuition, we will be able to apply it to various manufacturing processes. We propose and discuss new approach of skill research based on overall viewpoint to relief and express the elements that compose tacit knowledge of traditional craftsmanship.

Keywords: craftsmanship, affective skill, proficiency

1   INTRODUCTION

The traditional craftsmanship authorization system in Japan started in 1974 targeting the expanding demand of traditional handicraft industry that has hung low by the shortage of successors, etc. However, the narrowing trend of the traditional industries of Japan backed by skillful traditional craftsmanship does not stop, and the number of engineers called "master" and "craftsman" are also decreasing in proportion to it. For the reason, the technology that currently extinguished was considered to be a cultural heritage and many trials have been performed to save affective skills in a certain form the cultural and academic viewpoint. It is important to understand the master's technique in various feelings in many cases, because they use characteristic words on teaching traditional skills. He points out to steal his technique to catch his know-how. The technique that can transmit efficiently the knowledge such as the vital point and tips, which cannot be expressed by nonverbal methods, is required for preservation and succession of his great skills. And it is required to know and to harness tips and intuition that is called inherent tacit knowledge of traditional craftsmanship. To accomplish the purpose the medium for expressing the knowledge (tacit knowledge) that is fountainhead of tips or intuition has to be identified. In this paper we propose how to analyze synthetically the proficiency for traditional arts that constitutes tacit knowledge.

2   PROFICIENCY FOR TRADITIONAL ARTS

2.1   Tacit Knowledge and Proficiency of Experts

Skill is the capability acquired through culture and training based on talent. In traditional performing arts in Japan, traditional handicrafts, and traditional craftsman, as mentioned above, the skills have been handed down by the method of seeing and stealing master's work. The pupils have repeated the exercise to catch up with his know-how, because it is difficult to transfer tacit knowledge of masters to their pupils by means of writing it down or verbal manner. In this paper, the tacit knowledge caught up by working experience will be called the proficiency for traditional arts in order to express and transform by analyzing the level of skills.

2.2   Ellements of Proficiency for Traditional Arts

Shirato et al. (2007) showed the difference between expert and novice operator in kitchen knife grinding used to make traditional kyoto-cooking in Japan, by three-dimensional motion analysis and electromyogram analysis. Moreover, Ohnishi et al. (2009) searched on feature of kyoto-confections and clarified the great differences the aesthetic form of the Japanese traditional sweets depending on grasping power of craftsman. Cheng et al. (2012) reported on nurses' skills through observing her usual manner and task analysis. These approaches only do not tend to carry out video record of the operation, and to explain skill by measured data of operation or physiological measured value. In order to express elements of proficiency for traditional arts, it is required not only physiological (brain waves, electromyogram, blood flow etc.) and physical measurement (three-dimensional body movement, hand power etc.) but the text data obtained from sound data, voice data, interview data, etc. From those data analysis from various sides elements of the proficiency for traditional arts is clarified, and it will be available to transmit traditional techniques and to extract the essence for education of traditional skills. Then mutual data integration is performed to find deep manner of experts getting high level of traditional skills.

3   HOLISTIC MODEL ON ANALYSIS OF PROFICIENCY

As pointed out above, it is clear that measurement of skill elements that compose on base of skills is required. So far, various methods have been carried out in measuring movement of body and hands, EMG, ECG, EOG, brain wave, changes of physiological data, thinking process, etc. However there was one important blind spot. It means that relation between degrees of master's operation for modeling a form or stats and changes of the object is insufficient. Moreover, although changes of the form in work process are reproducible, but it is difficult to reproduce and transmit the changes of master's thinking process. Therefore, it is also required to solve what is the intuition and tips that are essence of traditional craftsmanship through measuring and analyzing the characteristics of material and product including   measurement of skillful technique. In other words, it is just insufficient to measure master's stats and form of product. Therefor we need data on changes of physical data of contact portion with target product and the changes of feeling or thinking during make up the product, because the masters are always making products by perceiving the target states with high precision. Moreover, when they use a tool, measurement data in interfaces between a tool and work article is also required.

We propose the holistic model on analysis of proficiency for traditional arts in figure 1. In master's side, there are measurement data about interface portion between his manner and states of his tool, such as pressure and tactile sense, in addition to measurement of making up and thinking processes, and changes of physiological states. In work products side, the data of environmental states, such as temperature and humidity, is also required in addition to form of product. Moreover, there is also the data of interface portion between tool and product. We consider that it is possible to express the proficiency for traditional arts by unifying these data and specifying the correlation among these data.

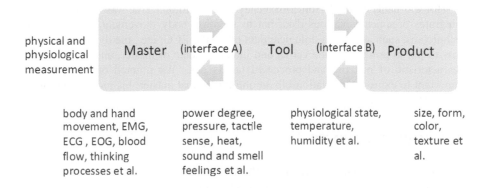

| physical and physiological measurement | Master | (interface A) | Tool | (interface B) | Product |

| body and hand movement, EMG, ECG , EOG, blood flow, thinking processes et al. | power degree, pressure, tactile sense, heat, sound and smell feelings et al. | physiological state, temperature, humidity et al. | size, form, color, texture et al. |

Figure 1 The holistic model on analysis of proficiency for traditional arts.

4   DEPLOYMENT OF HOLISTIC ANALYSIS ON PROFICIENCY

Holistic analysis on proficiency for traditional arts make it possible to archive the master's skills that focused on various aspects of work process instead of technological articles, such as target traditional-handicrafts and art work for the conventional digital archives.   Moreover, the various data can also be indicated simultaneously and mutual comparison of physiological data and physical data can be done easily.   It also becomes to catch the technology that can also grasp temporal correlation, and to characterize the component technique by static data.

We can imagine the possibility to change study method of traditional skills.   For instance, it is possible to display the virtual teacher as avatar who expresses an ideal teacher using the measured three-dimensional data on master's movement.   By piling it on video images novice learner can study the differences between him and the master visually.   About learning on traditional skills Araki et al.(2010) discussed on the framework that creates an operations acquisition supporting system.   He introduced the multi-modal information presentation system using the framework. In order to teach skills proper word and instruction are demanded between not only person to person, but also person to robot for future interactive system on learning traditional skills.   Sakamoto et al. (2010) challenged and proposed the technique of teaching a robot some operations using abstract language, and reported the result of the evaluation on teaching effects. And Ohira et al. (2011) proposed the schema graph for introducing specialization and generalization to a graph-based data model in order to systematize and reuse knowledge effectively.   The concept of holistic analysis on proficiency will contribute to develop not only for preservation of skill but also for education to take over, or research on human-human interaction and human-machine interaction.

5   CONCLUSIONS

Holistic analysis model on proficiency for traditional arts was proposed that integrates organically various data, not to mention body movement of traditional technicians, such as the physiological data of EMG, ECG, etc., the dynamic data of how to put in power etc., the verbal and interview date of master, and the physical characteristic of material and processed goods etc.   We pointed out that it was important to express the relationship among traditional master, his tool and product by measurement data.

ACKNOWLEDGMENTS

The authors would like to thank JSPS Grant-in-Aid for Scientific Research (B)(23300037) that made it possible to develop this study.

REFERENCES

Araki, M. and Hattori, T. 2010, Proposal of a Practical Spoken Dialogue System Development Method: Data-management Centered Approach, In Minker, W. et al. (eds) Spoken Dialogue Systems Technology and Design, Springer, 187-211.

Cheng, M., Kanai, P. M., Kuwahara, N., Itoh, O. H., Kogure, K., Ota, J. 2012, Dynamic Scheduling-based Inpatient Nursing Support: Applicability Evaluation by Laboratory Experiments, Int. J. Autonomous and Adaptive Communications Systems, 5, 1.

Hochin, T. 2011, Decomposition of Graphs Representing the Contents of Multimedia Data, Journal of Communication and Computer, 7, 4, 43 -49.

Ohira, Y., Hochin, T., and Nomiya, H. 2011, Introducing Specialization and Generalization to a Graph-Based Data Model, Lecture Notes in Computer Science, Springer, 6884, 1-13.

Ohnishi, A. et al. 2009, Evaluation of the Technique and Skills of Making "Kyo-gashi" Sweets by Analyzing Finger Motions and Weight and Forms of Sweet. J. Science of Labour, 85, 3, 108-119.

Sakamoto, Y., Araki, O., Uemura, T., Saratani, K., Honda, T., Ozeki, M. Oka, N., 2010, Abstract Instructions to a Robot: Experimental Study on Teachability, Annual Meeting of Information Processing Society of Japan Kansai Branch, C-05 (in Japanese).

Shirato, M. et al. 2007, Comparison of Body Movement and Muscle Activity Patterns during Sharpening a Kitchen Knife between Skilled and Unskilled Subjects. J. Science of Labour, 84, 4, 139-150.

CHAPTER 14

Representation and Management of Physical Movements of Technicians in Graph-Based Data Model

Teruhisa Hochin, Yuki Ohira, Hiroki Nomiya

Kyoto Institute of Technology
Goshokaidocho, Matsugasaki, Sakyoku, Kyoto, Japan
{hochin, nomiya}@kit.ac.jp

ABSTRACT

The physical movements of skilled people are tried to be represented in a graph-based data model. This representation is in the form of a kind of directed graph. It is proposed that the graph representing the pause, which is considered to be inevitable in representing the movements of the skilled people, should be explicitly included in the representation of the movements. The representation of a video stream in the form of a kind of graph and that of its contents are also discussed.

Keywords: Graph-based data model, Content representation, Movement, Pause

1   INTRODUCTION

In recent years, content retrieval of multimedia data has extensively been investigated. One approach of addressing to this issue uses feature values of multimedia data. For example, when a picture is given as a desired one, similarity between the picture and the one in a database is calculated by using feature values. Pictures having high scores of similarity are presented to the user as the query result. Another approach uses graphs representing the contents of multimedia data. Petrakis *et al.* have proposed the representation of the contents of medical images by using directed labeled graphs (Petrakis, 1997). Uehara *et al.* have used the semantic network in order to represent the contents of a scene of a video clip (Uehara, 1996). Jaimes has proposed a data model representing the contents of multimedia by using four components and the relationships between them (Jaimes, 2005). Contents of video data are represented with a kind of tree structure in XML (Manjunath, Salembier, Sikora, 2002).

We have proposed a graph-based data model, the *Directed Recursive Hypergraph data Model* (DRHM), for representing the contents of multimedia data (Hochin, 2006, Hochin and Nomiya, 2009, Hochin, 2010, Ohira, Hochin and Nomiya, 2011, Hochin, Ohira and Nomiya, 2012). It incorporates the concepts of directed graphs, recursive graphs, and hypergraphs. An *instance graph* is the fundamental unit in representing an instance. A *collection graph* is a graph having instance graphs as its components. A *shape graph* represents the structure of the collection graph. Shape graphs may change when instance graphs are inserted, or modified. As the existence of instance graphs affects shape graphs, DRHM is said to be an *instance-based* data model. Although DRHM has been proposed for representing the contents of multimedia data, image stills are the targets of the contents representation in DRHM. Video data are not treated yet. The movements of skilled people are required to be represented for deriving their skills and teaching them to novices.

This paper tries to represent the contents of video data including the movements of skilled people in DRHM. In this paper, the tea ceremony is used as an example of the movements of the skilled people. First, the representation of videos is discussed. Several methods of representing videos are shown. The representation of a representative picture is included in the discussion. Next, the representation of the contents of a video stream is examined. The discussion includes positional relationships, actions, and the abstraction. Lastly, the representation of the movement of a skilled person is discussed. Introducing the instance graph representing the pause is proposed in order to precisely represent the movement.

This paper is organized as follows: Section 2 briefly explains the structure in DRHM by using an example. Section 3 describes the tea ceremony used as an example of the movements of a skilled person. Section 4 describes the representation of a video stream. The representation of the contents of a video stream is examined in Section 5. The representation of the movements of the skilled people is discussed in Section 6. Lastly, Section 7 concludes this paper.

2 DIRECTED RECURSIVE HYPERGRAPH DATA MODEL

The structure of DRHM is described through an example. The formal definition is included in our previous work (Hochin, 2010). In DRHM, the fundamental unit in representing data or knowledge is an *instance graph*. It is a directed recursive hypergraph. It has a label composed of its identifier, its name, and its data value. It corresponds to a tuple in the relational model.

Consider the representation of the picture shown in Fig. 1(a). An ornament is on a floor. The ornament consists of three bags of rice and a tassel. Fig. 1(b) represents

124

the contents of this picture in DRHM. An instance graph is represented with a round rectangle. For example, *g1* is an instance graph. An edge is represented with an arrow. A dotted round rectangle surrounds a set of initial or terminal elements of an edge. For example, *g5* and *g6*, which are surrounded by a dotted round rectangle, are the initial elements of the edge *e2*. When an edge has only one element as an initial or terminal element, the dotted round rectangle could be omitted for simplicity. The instance graph *g4*, which is the terminal element of the edge *e2*, is an example of this representation. An instance graph may contain instance graphs and edges. For example, *g1* contains *g2*, *g3*, *e1*, and *e4*.

A set of the instance graphs having similar structure is captured as a *collection graph*. A collection graph is a graph whose components are instance graphs. The components are called *representative instance graphs*. A collection graph corresponds to a relation in the relational model.

The structure of a collection graph is represented with the graph called a *shape graph*. It corresponds to a relation schema in the relational model. The collection graph, whose structure the shape graph represents, is called its *corresponding collection graph*.

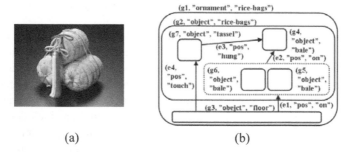

(a)   (b)

Figure 1 A picture of an ornament (a) and an instance graph representing its contents (b).

3   TEA CEREMONY AS MOVEMENTS OF SKILLED PEOPLE

In the tea ceremony, a host serves cups of tea for guests. There are strict rules on how to prepare, make, and serve tea. The tea ceremony consists of many steps including many actions. The actions may depend on the tools used in the tea ceremony. A fireplace in a floor is used in winter, while the one on a floor is used in summer. The former (the latter, respectively) is called "Ro" ("Furo") in Japanese. The fireplace "Furo" is used in the scenes shown in Fig. 2. Other tools depend on the kind of fireplace.

A scene of the tea ceremony is shown in Fig. 2. This scene is of Fukusa Sabaki, where a sheet of cloth called *the fukusa* is elegantly folded. The fukusa is used for cleaning the instruments used in the tea ceremony.

The host appearing in Fig. 2 does not put Japanese clothes on because of capturing his motion precisely.

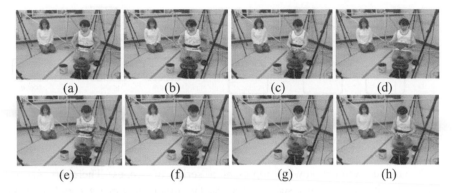

Figure 2 Snapshots of a scene of a video stream of tea ceremony. The scenes are of Fukusa Sabaki, where the fukusa, which is a sheet of cloth, is elegantly folded.

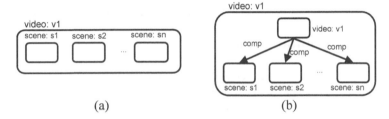

Figure 3 A video stream is represented with a series (a), or a tree (b).

4 REPRESENTATION OF VIDEOS

4.1 Representation

Here, how to represent a video stream is examined. In DRHM, it is natural that a video stream is represented with a representative instance graph rather than a set of ones because a representative instance graph is a unit of representing an object. A video stream, which consists of many scenes, could be represented with a series of scenes, or a tree, whose nodes represent scenes. In the former approach, scenes are represented with instance graphs, and are included in a representative instance graph (Fig. 3(a)). From now on, identifiers of the labels of elements are omitted for brevity as shown in Fig. 3. The name of a label is followed by the character ":" and the value. For example, the string "video: v1" shown in Fig. 3(a) means that the name of the label is "video," and its value is "v1." In the latter approach, the instance graph of representing a whole video stream is introduced. This instance graph is a root node of a tree. Scenes are represented with instance graphs, and are of the child nodes of the root node (Fig. 3(b)). The edges of the tree represent the composition relationships.

126

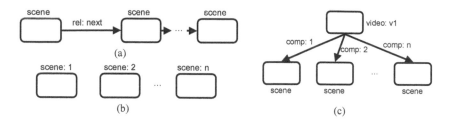

Figure 4 Order of scenes is represented with links (a), serial numbers (b), or edges (c).

Next, we examine how to represent a series of scenes. There are several approaches in representing a series of scenes. The first is the *link approach*. A scene is connected to the next scene (Fig. 4(a)). Scenes are usually linked in a line. The next is the *serial number approach*. A serial number is given to a scene. The order of scenes is represented with serial numbers (Fig. 4(b)). The third is the *serial edge approach*. Scenes could be represented as the child nodes of the tree representing a video stream. In this case, scenes could be serialized by ordering edges (Fig. 4(c)).

4.2   Representative Pictures

Next, we examine the representation of a representative picture. A representative picture is the representative of a video stream. This picture should clearly represent the contents of a video stream. That is, the contents of a video stream should be captured through the representative picture.

A video stream has the hierarchical structure. That is, a video stream includes one or more scenes. A scene consists of several shots. A representative picture of a video stream is usually a shot in the scene. If the content representation of the shot of a representative picture is changed, that of the video stream as well as that of the scene must be changed. The representation of a representative picture should be strong for such modification.

There are several approaches in representing a representative picture. The first is the *pointer approach*. The pointer contains the address of the instance graph of the shot, which is of the representative picture. The next is the *edge approach*. The instance graph representing a representative picture is connected to an instance graph representing a video stream through an edge. The relationship between a video stream and a representative picture is represented as a part of a graph representing the structure of the video stream.

5   REPRESENTATION OF CONTENTS

5.1   Positional Relationships

The representation of relationships among objects appearing in a video stream is discussed. In DRHM, relationships are represented with edges. Various relationships exist in a video stream, and may change in every second. We focus on the positional relationships, which are considered to be important in representing the movements.

Two types of positional relationships between objects should be considered. The one is the positional relationship between the centers of objects. The relationship between C_A and C_B, which are the centers of the objects A and B shown in Fig. 5, is an example of this type of relationship. The other is the positional relationship between the nearest points of objects. In Fig. 5, the corner P_A (P_{B1}, respectively) is the nearest point of the object A (B) to the object B (A). This relationship could represent whether objects are touched, or are close to each other.

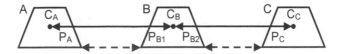

Figure 5　Positional relationships.

Here, the representation of the snapshot shown in Fig. 2(d) is examined. Especially, both hands, the body, and the fukusa are focused on. In Fig. 2(d), the fukusa is held with both hands in front of the body.

It is convenient that the positional relationship is represented in the polar coordinates, where the relative position is represented with the distance and two angles from an original point to a target point. Table 1 shows the relative positions between the centers of two objects of a pair selected from both hands, the body, and the fukusa. A triplet in a field represents the distance and two angles. The first angle is in a horizontal plane, and the second one is in a perpendicular plane. In Table 1, the angles are represented with zero degree when the distance is zero. Table 2 shows the relative positions between the nearest points of two objects. As the fukusa is held with both hands, the distance between the fukusa and the right hand, and that between the fukusa and the left hand are equal to zero.

The values of both types of relative positions themselves could be used as the values of edges. An example of this usage is shown in Fig. 6(a). Here, we use one direction out of two directions opposite to each other. For example, the direction from the left hand to the body is the opposite one of that from the body to the left hand. The latter direction is adopted in Fig. 6(a). In Fig. 6(a), the edges whose distances are larger than 450 are omitted. The positional relationships between the nearest points whose distances are equal to zero are only adopted.

Positional literals could be derived by using the relative positions. An example of the positional literal is "left" representing that the fukusa is left of the right hand. It is also used for the case that the left hand is left of the fukusa. An example of the usage of positional literals is shown in Fig. 6(b). Here, the literal "right_ant" ("left_ant", respectively) means *right (left) anterior*. The positional literals shown in Fig. 6(b) are derived from the relative positions shown in Fig. 6(a). Please note that *right (left,* respectively) *anterior* is the opposite direction of *left (right) posterior*.

128

Table 1 Relative positions between the centers of two objects.

Target	Left Hand	Right Hand	Body	Fukusa
Left Hand	(0, 0d, 0d)	(457, 1d, -1d)	(446, -36d, 17d)	(237, 4d, -10d)
Right Hand	(457, -179d, 1d)	(0, 0d, 0d)	(317, -114d, 27d)	(226, 177d, -8d)
Body	(446, 144d, -17d)	(317, 66d, -27d)	(0, 0d, 0d)	(340, 112d, -31d)
Fukusa	(237, -176d, 10d)	(226, -3d, 8d)	(340, -68d, 31d)	(0, 0d, 0d)

Table 2 Relative positions between the nearest points of two objects.

Target	Left Hand	Right Hand	Body	Fukusa
Left Hand	(0, 0d, 0d)	(457, 1d, -1d)	(367, -79d, 10d)	(0, 0d, 0d)
Right Hand	(457, -179d, 1d)	(0, 0d, 0d)	(369, -88d, 11d)	(0, 0d, 0d)
Body	(367, 101d, -10d)	(369, 92d, -11d)	(0, 0d, 0d)	(420, 66d, -14d)
Fukusa	(0, 0d, 0d)	(0, 0d, 0d)	(420, -114d, 14d)	(0, 0d, 0d)

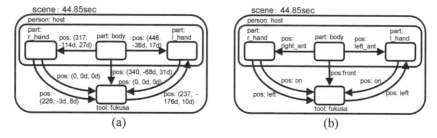

Figure 6 Positional relationships represented with relative positions (a), and positional literals (b).

5.2   Action and Abstraction

A scene, which is a sequence of shots, may represent an action. The scene of opening the fukusa is represented with instance graphs as shown in Fig. 7. The upper instance graph represents the scene, while the bottom ones represent the shots in the scene. This scene starts at holding the fukusa with both hands in front of the body. The left bottom instance graph represents this shot. Then, the distance between both hands gradually increases. Lastly, the fukusa is completely spread out. The right bottom instance graph represents this shot. The action could be captured as a kind of abstraction. Several shots are put together to an action scene.

Several actions are also put together to a more general action. An example of this abstraction is shown in Fig. 8. The action "fukusa_sabaki" is composed of a series of actions: take the fukusa, open it, put it between fingers, hold it with the right hand, and fold it. In the same way, serving a cup of tea is also represented with a series of actions. These relationships constitute a hierarchy. The root of the hierarchy represents the whole of a video stream. The leaves of the hierarchy are the shots of the video stream.

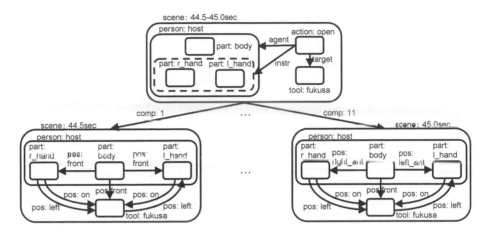

Figure 7   Several shots are summed up to the action "open."

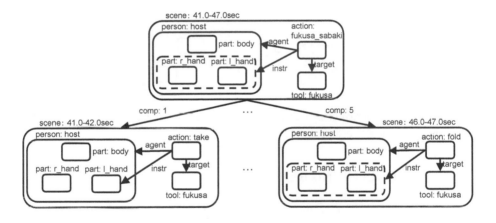

Figure 8   Primitive actions, e.g., *take* and *fold*, are summed up to an action *fukusa_sabaki*.

6   REPRESENTATION OF SKILLED PEOPLE'S MOVEMENTS

The representation method described until here is general in describing the movements. It is not considered to be sufficient to describe the movements of skilled people. Here, the method of describing the movements of the skilled people is discussed.

We could feel fluency and smoothness from the movements of the skilled people. Fluency and smoothness is the moving appearance without hesitation. This means constant speed, or constant acceleration. This may mean that speeds and/or accelerations are not different one another. From this observation, the average speed, the average acceleration, the standard deviation of speeds, and that of accelerations had better be derived in order to examine fluency and smoothness of the movement.

Smoothness is, however, different from sluggishness. The movement of the skilled person is not in a sluggish manner. On the contrary, we feel sharpness. What makes the movement sharp?

One is that the strong part could clearly be distinguished from the weak part. The clear difference between the strong and the weak parts results in the sharpness of the movement. This could be measured by the magnitude of the movement.

Another may be caused by the pause. The pause is called "Ma" in Japanese. Its importance is revealed and is emphasized (Nakamura, 2009). The pause means no movement. It is the temporary stop. It has the shots having no movement. Examples of the similar kind of action are *keep, hold, stay,* and *stop*. Please note that this kind of action is not explicitly represented usually. It is included in the previous or next action. It is considered to be important for the action without any movement to be explicitly represented. An instance graph representing this kind of action should be introduced in the representation of the movement. In Fig. 9, the action *keep* follows the action *open*. The action *keep* represents the pause. The duration of the pause may be important information of the movements of the skilled people.

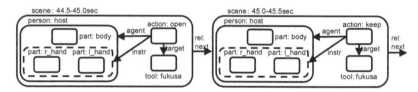

Figure 9   The instance graph including the action "*keep*" is introduced for the purpose of the explicit representation of the pause.

7   CONCLUDING REMARKS

This paper tried to represent the movements of skilled people in DRHM. The representation of a video stream in the form of a kind of graph is examined. After several methods of representing videos were shown, positional relationships, actions, and the abstraction in representing the contents of a video stream were examined. It is shown that the graph representing the pause should be introduced because the pause is important in representing the movements of the skilled people.

Confirming the efficiency of the introduction of the pause to the movement representation is in future work. This paper uses the tea ceremony as an example of the movements of skilled people. Representation of other kinds of movements is also in future work. Precise information used in representing positional relationships may be preferred for the precise retrieval. It, however, needs large amount of storage, and wastes time in processing queries. Concise representation may be better for the storage and the query processing, while it could not provide the precise retrieval capability. The decision of the kind of information used for the positional representation is also in future work.

ACKNOWLEDGEMENTS

This research is partially supported by the Ministry of Education, Science, Sports and Culture, Grant-in-Aid for Scientific Research (B), 23300037, 2011-2014.

REFERENCES

Hochin, T. 2006, Graph-Based Data Model for the Content Representation of Multimedia Data, *Proc. of 10th Int'l Conf. on Knowledge-Based Intelligent Information and Eng. Systems (KES2006)*, 1182–1190.

Hochin, T., and Nomiya, H. 2009, A Logical and Graphical Operation of a Graph-based Data Model, *Proc. of 8th IEEE/ACIS Int'l Conference on Computer and Information Science (ICIS2009)*, 1079–1084.

Hochin, T. 2010, Decomposition of Graphs Representing the Contents of Multimedia Data, *Journal of Communication and Computer*, 7(4): 43–49.

Hochin, T., Ohira, Y., and Nomiya, H. 2012, Incremental Representation and Management of Recursive Types in Graph-based Data Model for Content Representation of Multimedia Data, Accepted in *5th Int'l Conference on Intelligent Interactive Multimedia Systems and Services (KES-IIMSS2012)*).

Jaimes, A. 2005, A Component-Based Multimedia A Data Model, *Proc. of ACM Workshop on Multimedia for Human Communication: from Capture to Convey (MHC'05)*, 7–10.

Manjunath, B. S., Salembier, P., and Sikora, T. (eds.) 2002, *Introduction to MPEG-7*, John Wiley & Sons, Ltd.

Nakamura, T. 2009, Psychological Study of 'Ma' (a synonym of 'pause') in Communication, Journal of the Phonetic Society of Japan, 13: 40–52 (in Japanese).

Ohira, Y., Hochin, T., and Nomiya, H. 2011, Introducing Specialization and Generalization to a Graph-Based Data Model, *Lecture Notes in Computer Science*, Springer, 6884(*Proc. of 15th Int'l Conf. on Knowledge-Based Intelligent Information and Eng. Systems (KES2011)*): 1–13.

Petrakis, E. G. M., and Faloutsos, C. 1997, Similarity Searching in Medical Image Databases, *IEEE Trans. on Know. and Data Eng.*, 9: 435–447.

Silberschatz, A., Korth, H., and Sudarshan, S. 2002, *Database System Concepts (4th ed.)*, McGraw Hill.

Sowa, J. F. 1984, *Conceptual Structures - Information Processing in Mind and Machine*, Addison-Wesley.

Tanaka, K., Nishio, S., Yoshikawa, M., Shimojo, S., Morishita, J., and Jozen, T. 1993, Obase Object Database Model: Towards a More Flexible Object-Oriented Database System, *Proc. of Int'l. Symp. on Next Generation Database Systems and Their Applications (NDA'93)*, 159–166.

Uehara, K., Oe, M., and Maehara, K. 1996, Knowledge Representation, Concept Acquisition and Retrieval of Video Data, *Proc. of Int'l Symposium on Cooperative Database Systems for Advanced Applications*, 218–225.

CHAPTER 15

Multimodal Motion Learning System for Traditional Arts

Masahiro Araki

Kyoto Institute of Technology
Kyoto, Japan
araki@kit.ac.jp

ABSTRACT

The present paper describes an interactive multimodal motion learning system for traditional arts. This system uses several modalities, such as synthesized speech, video, and agent actions, to instruct a user. Input modality uses speech modality to control the system considering a hands-busy situation in motion learning. This system is developed using the proposed multimodal interaction system framework, which realizes easy construction of learning contents from high-level data modeling. As an example to illustrate the use of this framework, we apply the multimodal learning system to the Japanese tea ceremony.

Keywords: multimodal dialogue system, motion learning

1   INTRODUCTION

In Japan, various traditional arts are in danger of being lost because of a shortage of interest in the younger generation. In order to ensure that these arts are not lost, it is important to realize a digital library of traditional arts that records the making process of traditional products or the procedures of traditional arts. However, recording such information is insufficient. In order to spread knowledge on traditional arts, this digital library data should be used as educational contents. For this purpose, we designed a framework for a motion learning system using the contents of a digital library of Japanese traditional arts.

The proposed framework enables easy prototyping of a multimodal motion learning system that uses several modalities, such as synthesized speech, video, and agent action, as output and speech modality as input, considering the hands-busy situation in motion learning. The main part of the development process involves the creation of semantically annotated contents. Therefore, troublesome multimodal control description and/or complicated state transition definition can be avoided in implementing the dialogue system. As a result of this content-based development method, we can apply this system to several types of motion learning problems easily.

The remainder of the present paper is organized as follows. Section 2 surveys previous research on multimodal interaction systems and toolkits for constructing multimodal interaction systems. Section 3 explains the proposed data-driven framework for the multimodal interaction system development. Section 4 describes an example of the implementation of the multimodal motion learning system for the Japanese tea ceremony. Conclusions are presented in Section 5 along with a discussion of future research.

2   RELATED RESEARCH

A number of multimodal interactive systems (e.g., Johnston and Bangalore, 2004; Wahlster, 2006) have been proposed that realize rich human-computer interaction through mechanisms such as voice, touch, and gaze. A large part of the research has concentrated on the problems of modality fusion for input and modality fission for output. Since the target tasks of previous systems were primarily information-seeking dialogue, which is common to existing Web applications, software engineering problems, such as ease of prototyping, reusability, and extendibility, tend to be out of the scope of the research. In order to implement several types of educational multimodal interactive systems from the contents of a digital library, an established framework for multimodal interaction is necessary.

The CSLU toolkit (McTear, 2004) and Galatea toolkit (Kawamoto et al., 2004) are flexible frameworks for multimodal interaction, especially with virtual agents. A virtual agent component is essential for the present purpose because we must deal with human motion as one of the learning contents. However, both of these toolkits provide only a face agent. Therefore, it is difficult to apply these toolkits to the proposed system.

MMDAgent[1] is the most up-to-date toolkit for implementing a multimodal dialogue system using a full-body agent. MMDAgent realizes rich human-computer interaction using speech recognition, speech synthesis, and an animation agent. The disadvantage of this toolkit is the difficulty in writing a dialogue scenario because this toolkit is based on state transition description. Therefore, the goal is to design a framework that will help to implement a multimodal interaction system using MMDAgent that is specific to the purpose of motion learning.

---

[1] http://www.mmdagent.jp/

3   FRAMEWORK OF MULTIMODAL INTERACTION SYSTEMS

The key concept of the proposed development framework is automatic generation of dialogue controller code and data management code from a data model definition. We previously proposed Mrails (Araki and Mizukami, 2011; Araki and Hattori, 2010), which follows the concept of the Rails framework for generating speech and GUI Web applications. The slot-filling type task and the database search type task are covered by Mrails. However, the explanation type task, which is a target task of the present paper, is not suitable for automatic generation of dialogue controller code from the data definition because, unlike the other two types, which depend on the task type, the dialogue controller code of the explanation type task depends largely on the data model definition.

Therefore, we construct a new framework for automatic code generation of the explanation type task. We fix the data type for learning contents and implement a generator of the state-based dialogue scenario for MMDAgent. MMDAgent is in charge of speech recognition of user's input, speech synthesis for learning process management and explanation of the contents, and motion display. Other explanation materials, such as explanation by character and video display, are written in HTML5 code and being called from MMDAgent. An overview of our framework is shown in Figure 1.

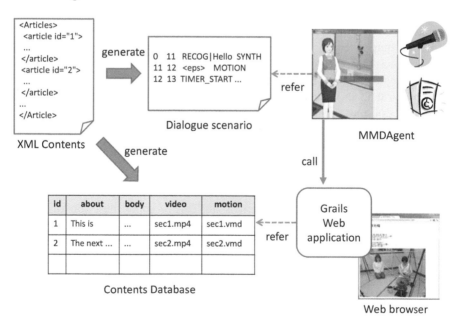

Figure 1 Overview of the proposed framework

As data type for learning contents, we use the Article class of Schema.org[2]. Schema.org is a collection of schemas, i.e., HTML tags, that webmasters can use to markup their pages in ways recognized by major search providers. Since major search engine providers agree on this schema, the use of Schema.org over the Web is expected to become widespread. Article class[3] is a Schema.org class and has several properties that are suitable for representing learning contents.

The development process begins by making HTML pages that have semantic annotations following Schema.org elements or by making an XML file representing the equivalent contents of the HTML pages. An example of an XML file is shown in Figure 2. Table 1 shows a description of each element.

```
<Articles>
  <article id="1">
    <about>
       This section explains ...
    </about>
    <articleBody>
       At first, let's look at ...
    </articleBody>
    <video>
       http://example.org/tea/sec1.mp4
    </video>
    <associatedMedia>
       http://example.org/tea/sec1.vmd
    </associatedMedia>
  </article>

  <article id="2">
  ...
  </article>

...
</Articles>
```

Figure 2 Example of an XML file for learning contents

---

[2] http://schema.org/
[3] http://schema.org/Article

136

Table 1 Description of elements in the Article class

Element	Description
article	One unit of interaction. Typically, one long explanation is divided into some articles. The id attribute is required.
about	The content of this element expresses the outline of this article. It is used for opening message of the interaction generated from this article.
articleBody	The content of this element expresses the detailed explanation of this article. It is used for explanation parallels with video play.
video	The content of this element indicates URL of the video data of this article.
associatedMedia	The content of this element indicates URL of the motion data of this article which is played by the virtual agent.

From the markup contents shown in Figure 2, the proposed framework generates (1) a finite state transition description file of MMDAgent, which handles speech and motion interaction, and (2) a Grails application, which manages the video handling.

In generating a finite state transition description, each article element corresponds to one large state, which has smaller states that correspond to an opening message, a detailed video explanation, awaiting user input, and response to specific user input. The state transition model is shown in Figure 3.

Each large state includes a speech control script that includes commands such as replay the video, show the motion, change the camera angle, and go to the next article. When a motion demonstration is required by the user, the motion is performed by the virtual agent.

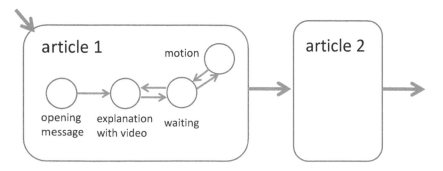

Figure 3 State transition model of the motion learning system

Video data is managed in a Grails application with an additional explanation in HTML5 format, which is called from MMDAgent script. The server mechanism is realized by a Grails application. Grails components, such as model part which handles Article class, controller part which realizes dialogue flow of explanation task, and view codes which displays learning contents, are implemented and adjusted to the motion learning task.

4   MOTION LEARNING SYSTEM FOR JAPANESE TEA CEREMONY

Using the framework explained in Section 3, we implemented a motion learning system for the Japanese tea ceremony. From a digital library data of tea ceremony, which was recorded for the purpose of detailed motion capture, we construct an XML file following the definition shown in Table 1. A screenshot of the interaction is shown in Figure 4.

Figure 4 Screenshot of the motion learning system for the Japanese tea ceremony

In this article, the virtual agent first voices an opening message. A detailed video explanation is then presented. When the user commands the agent to perform the motion, the agent performs the motion following the motion definition file. The user can control the camera angle by speech commands, as shown in Figure 5.

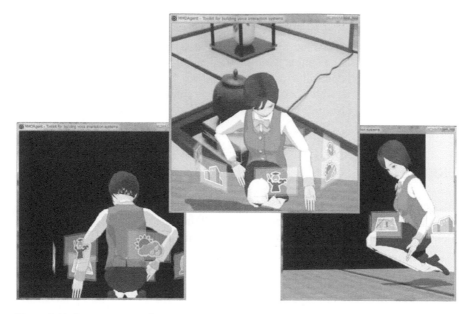

Figure 5 Various camera angles controller by speech command

5   CONCLUSIONS

The present paper described an interactive multimodal motion learning system for the Japanese tea ceremony that uses several modalities, such as synthesized speech, video, and agent action. This system is developed using the proposed multimodal interaction system framework that realizes easy construction of the learning contents from high-level data modeling.

In the future, we intend to develop a motion capture component for input and for the realization of an adaptive control of the learning process following the score of motion imitation.

ACKNOWLEDGMENTS

The present research was supported in part by the Ministry of Education, Science, Sports, and Culture through a Grant-in-Aid for Scientific Research (B), 23300037, 2011.

REFERENCES

Araki, M and T. Hattori, 2010. Proposal of a Practical Spoken Dialogue System Development Method: Data-management Centered Approach, In W. Minker et al. (eds.) Spoken Dialogue Systems Technology and Design, Springer-Verlag, 187-211.

Araki, M and Y. Mizukami, 2011. Development of a Data-driven Framework for Multimodal Interactive Systems, In Proceedings of the Paralinguistic Information and its Integration in Spoken Dialogue Systems Workshop, Springer-Verlag, 91-101.

Johnston, M. and B. Srinivas, 2004. MATCHkiosk: A Multimodal Interactive City Guide. *The Companion Volume to the Proceedings of 42st Annual Meeting of the Association for Computational Linguistics*, 222-225.

Kawamoto, S., H. Shimodaira, T. Nitta, T. Nishimoto, S. Nakamura, K. Itou, S. Morishima, T. Yotsukura, A. Kai, A. Lee, Y. Yamashita, T. Kobayashi, K. Tokuda, K. Hirose, N. Minematsu, A. Yamada, Y. Den, T. Utsuro, S. Sagayama. 2004, Galatea: Open source software for developing anthropomorphic spoken dialog agents, in Life-Like Characters, H. Prendinger and M. Ishizuka (eds.), Springer-Verlag, 187-212.

McTear, M. 2004, Spoken Dialogue Technology, Springer-Verlag.

Wahlster, W. (ed), 2006. SmartKom: Foundations of Multimodal Dialogue Systems, Springer-Verlag.

CHAPTER 16

Characteristics of Technique or Skill in Traditional Craft Workers in Japan

Masashi Kume, Tetsuya Yoshida

Kyoto Bunkyo Junior College,
Kyoto, JAPAN
Email address: kume@po.kbu.ac.jp
Kyoto Institute of Technology
Kyoto, JAPAN
Email address: yoshida@kit.jp

ABSTRACT

The present research aims to clarify the work processes and biomechanical characteristics related to the techniques and skills of Japanese traditional craft workers, and through electromyograms and motion analysis, considers the differences in motion and muscles used at work by top-notch and beginning crafts workers involved in "Kyoto confectioneries", "Knife sharpening" and "Mud wall plastering". The results were that the work of top-notch craft workers in "Kyoto confectioneries" and "Knife sharpening" was highly reproducible and rhythmical. Moreover, the top-notch craft workers in "Mud wall plastering" worked with better posture and less physical burden than beginners. The characteristics of these techniques and skills may relate to the raising of quality in Japanese traditional crafts.

Keywords: three-dimensional motion analysis, manufacturing process, traditional craft

1   INTRODUCTION

There are many traditional crafts and traditional industries in Japan. Most of these involve making things by hand, without the use of machinery. However, at present, few young people work in traditional crafts, so it will probably be difficult for the next generation of young people to inherit the techniques and skills of traditional crafts (Chen, 2002; Yamaguchi, 2002). In order to become skillful at traditional techniques and skills, a considerable amount of time and experience is necessary. Hence, analyzing the techniques and skills of top-notch craft workers may contribute to young people having an interest in traditional crafts and industries, and to the early learning of those skills and techniques (Yoshida,2008).

Three-dimensional motion analysis and electromyogram analysis are used to evaluate the skills and techniques of traditional craft workers at work (Shirato, 2007; Shibata, 2009; Ota, 2009). The present research aims to clarify the work processes and biomechanical characteristics related to the techniques and skills of Japanese traditional craft workers, and through electromyograms (EMG) and motion analysis, considers the differences in motion and muscles used at work by top-notch and beginning crafts workers involved in "Kyoto confectioneries", "Knife sharpening" and "Mud wall plastering".

2   METHOD

2.1   Kyoto Confectioneries

One male expert with fourteen years carriers performed the making of sweets under two conditions; 1) regular making method (condition A), 2) making sweets with slightly more pressure with both hands to give the disturbance (condition B). Three processes of making sweets was divided; wrapping from the bottom to the middle position of sweets (Phase 1), wrapping from the middle to the top position of sweets (Phase 2), and the finalization of details (Phase 3). The subject carried out 49 trials in two conditions, however, data sampled six trials randomly from each condition. The weight, height and diameters of the sweets were measured, and two diameters were defined as peak values of vertical (diameter A) and lateral (diameter B) directions on a transverse plane (Figure 1). The data were calculated mean value and standard deviation.

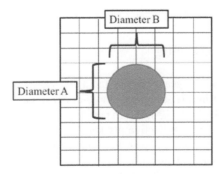

Figure 1 Measurement positions of two diameters in sweets.

2.2   Knife Sharpening

Two skilled (cooking technical college teacher with 30 and 19 years carriers, E1 and E2) and one unskilled subject (cooking technical college student, N) were participated in this study. Body motion was measured using three-dimensional motion analysis with four cameras, which allows optical real-time motion capture. The 19 reflective markers attached to the skin of the subjects with double-sided tape. Each marker placed on the head, right and left shoulder, elbows, wrists, hand, and hip (Figure 2). EMG sensors attached on the left and right deltoid muscle, triceps muscle, biceps muscle, and flexor carpi ulnaris (Figure 2). Three-dimensional motion analysis (MAC 3D SYSTEM) were performed four times; first day in both skilled and unskilled subject, and two days, one week, and three weeks later in unskilled subject who was performed practice of sharpening kitchen knife for 5 days per week.

Figure 2 Attached position of reflective markers and electromyogram sensors.

2.3   Mud Wall Plastering

An expert plasterer with 26 years carriers and a non-expert plasterer with 4 years carriers were participated in this study. The three-dimensional motion analysis was performed with a MAC 3D SYSTEM. Seven cameras were set up to capture motion of the body (Figure 3). The 22 reflective markers were attached to the skin of the subjects with double-sided tape. Markers were placed on the head, shoulder, right and left elbows, wrists, hand, greater trochanter, knees, ankles, and toes. A trial is defined as a vertical motion of the right hand from bottom to top while plastering a wall. Each subject was asked to perform three trials of plastering a clay wall (Figure 4). After compensating for defective data, frames were obtained showing the coordinates for each marker.

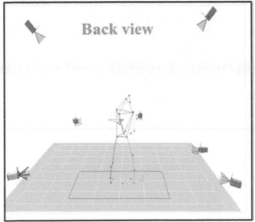

Figure 3 Seven cameras setting to capture body motion.

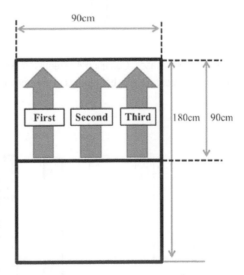

Figure 4 Measurement of motion during the plastering a clay wall.

3   RESULTS AND DISCUSSION

3.1   Kyoto Confectioneries

Figure 5 shows the making time of sweets in three phases. The mean time of six trials for phase 3 was about 1.5 seconds longer under condition B (5.0±0.8sec) than under condition A (3.5±1.1sec), however, the time for phase 1 (condition A, 4.8±0.6 sec; condition B, 4.7±0.5sec) and phase 2 (condition A, 3.0±0.2sec; condition B, 3.4±0.4sec) was similar between conditions. Thus, phase 3 may be important to make the sweet's form, and the time for phase 3 increased when an expert made sweets with slightly more pressure with both hands. The height of sweets was about 2mm higher under condition B (36.02±1.10mm) than under condition A (34.25±0.86mm), however, diameter A (condition A, 51.25±1.03mm; condition B, 50.83±0.87mm) and diameter B (condition A, 50.22±0.87mm; condition B, 50.94±1.15mm) was similar between conditions (Figure 6). These results demonstrate that the sweets less than 1mm diameters of the error have made by expert in spite of the disturbance during handwork. Thus, the high reproducibility to make sweats independent of the disturbance was found in expert.

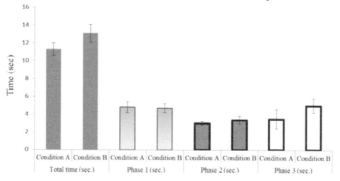

Figure 5 Making time of sweets in three phases. Condition A is regular making method. Condition B is making sweets with slightly more pressure with both hands to give the disturbance. Value are mean ± S.D.

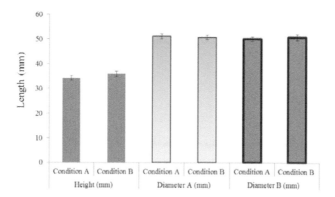

Figure 6 Comparison of sweets height and two position diameters on two conditions. The two diameters were defined as peak values of vertical (diameter A) and lateral (diameter B) directions on a transverse plane. Values are mean ± S.D.

3.2   Knife Sharpening

Figure 7 shows a typical motion of anteroposterior position of knife during sharpening kitchen knife by one subject. The kitchen knife of anteroposterior motion in the skilled subjects had rhythmical and stability. Figure 8 shows the relative anteroposterior position of the knife of each stroke by all subjects. The start position of each stroke was arranged to origin. During the sharpening a kitchen knife, motion duration was shorter, motion pathway was longer, and reproducibility in each stroke was higher in the skilled subjects than in the unskilled subject. Figure 9 shows the muscle activity during sharpening a kitchen knife in each subject. In each stroke, greater EMG activities in flexsor carpi ulnaris and posterior deltoid muscle were observed in the skilled subjects (E1 and E2), but not in the unskilled subject after two days of practice (N_2).

Thus, body motion or muscle activities during the sharpening a kitchen knife differed between the skilled and an unskilled subject, suggesting that rhythmical and stability motion observed in skilled subjects may be related with flexsor carpi ulnaris and posterior deltoid muscle activities.

Figure 7 Typical motion of anteroposterior position of knife during sharpening by E1.

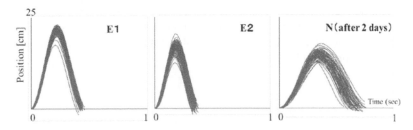

Figure 8 Relative anteroposterior position of the knife of each stroke by all subjects. The start position of each stroke was arranged to origin. E1and E2, skilled subjects; N, unskilled subject after two days of practice.

146

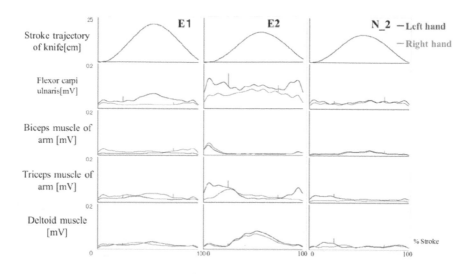

Figure 9 Muscle activity during knife sharpening in each subject. Eight electromyogram (EMG) sensor were attached to the skin of the subjects with double-sided tape. Sensors were placed on the Left and right deltoid muscle, triceps muscle, biceps muscle, and flexor carpi ulnaris. E1and E2, skilled subjects; N_2, unskilled subject after two days of practice.

3.3   Mud Wall Plastering

Body movements of plastering a clay wall in both the expert and non-expert plasterer are shown in Figure 10. Movement of the shoulder, greater trochanter, and knee was greater in the non-expert plasterer than that in the expert plasterer. Therefore, shoulder inclination were analyzed by the drawing straight line from the left to right marker fixed at the shoulder. Since the position of the right shoulder marker was higher than that of a left shoulder marker during all trials, the slope (angles) of the shoulder inclination was calculated as the difference between the shoulder and a horizontal line.

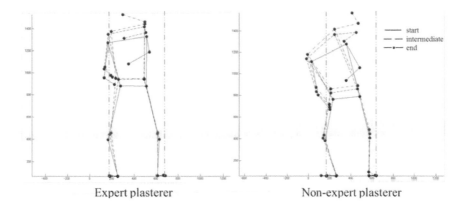

Expert plasterer   Non-expert plasterer

Figure 10 Typical motion of mud wall plastering in expert and non-expert (back view). Stick picture in both subjects were obtained coordinate data from 14markers including the shoulder, greater trochanter, knee, ankle and toe on the right and left sides during plastering a wall. Three trials of measurement were performed, and typical trial indicate start, intermediate and end of plastering a wall with clay lengthwise.

The shoulder inclination was steep in the non-expert than the expert during all trails (Figure 11). These results suggest that movements of the shoulder and knee may be important factors in increasing stability and technique while plastering a wall with clay.

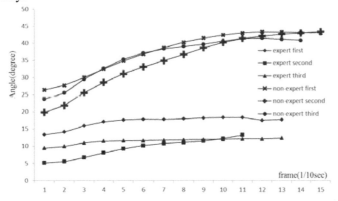

Figure 11 Changes in the shoulder inclination (indicated as angles) during the all trials of mud wall plastering by expert and non-expert plasterers.

4   CONCLUSIONS

The results were that the work of top-notch craft workers in "Kyoto confectioneries" and "Knife sharpening" was highly reproducible and rhythmical. Moreover, the top-notch craft workers in "Mud wall plastering" worked with better posture and less physical burden than beginners. The characteristics of these techniques and skills may relate to the raising of quality in Japanese traditional crafts.

REFERENCES

Chen G., Yamashita A., Shibata K., and Tanaka C. 2002. Three-dimensional analysis of the work motion using woodworking tools 1 Journal of wood science, 48(2): 80-88.

Shibata K., Nasu S., Kume M., Nakai A., and Hamada H. 2009. Effects of natural adhesive "Nibe" for drawing kyoto bow: ASME 2009 International mechanical engineering congress and exposition (IMECE2009): 111-115.

Shirato M., Ohnishi A., Kume M., Maki M., Nakai A., Yamashita Y., Yoshida T. 2007. Biomechanical analysis of the technique for producing "shirabeo" string on expert: 10th Japan International SAMPE Symposium & Exhibition (JISSE-10): CD-ROM TC-1-1.

Yamaguchi T., Kitamura K., Uenishi T., Azuma H., Takahasi A., Akamatsu M. 2002. Comparative analysis of skilled and unskilled behaviors in sewing-machine operation Biomechanism 16: 207-220.

Ota T., Kume M., Iue M., Hamasaki K., Yoshida T. 2009. Motion Analysis of the "Temae" in the Way of Tea: 11th Japan International SAMPE Symposium & Exhibition (JISSE-10): TC2-2.

Yoshida T., Ohnishi A., Shirato M., Kume M., Nakai A. 2008. Characteristics of "TAKUMINO-WAZA" in Japanese traditional craft Materials integration, 21(8): 20-25.

CHAPTER 17

A Study on the Traditional Charm of Natural Dyes

- Focusing on pre-printing mordant-

Kyu-Beom KIM [1], Min-Ju KIM [2], Gun-Ha Choi [3]

[1]Kyungnam National University of Science and Technology
Jinju, Korea
[2]Kyoto Institute of Technology, Japan
[3]Ryulsan Art, Korea
http://cms.gntech.ac.kr/user/gntech/

ABSTRACT

Weakness of preprocessing the mordant before multi-color dye in natural dye can be solved by following the visible benefits deduction.

Applied mordant with air contact caused change on standing effect after printing but between 6 hours to 1 week it became stabilized. Used jointly mordant's greatest characteristic is there are no changes in color tones when copper is mixed with mordant. Result of lowest change in color tone was from Pagoda tree and iron mordanting showed maximum absorption wavelength shifted to bathochromic shift when multi color pigment was processed with preprocessing mordant. From sharpness of outline effect, preprocessing with tin showed low result where iron with used jointly caused unlevel dyeing occurrence. Preprocessing mordant for individual mordant was replaced by color fastness where mordant used jointly with copper showed lower color fastness result comparing to individual mordant.

Keywords: natural dye, mordant, printing

1   INTRODUCTION

Comparing natural dyes with artificial dyes, natural dyes provide more attractive and delicate natural color tones. Environment preservation is important nowadays and natural dyes are attracting attention as it does not harm the environment.

Natural dye provides less harm to the human body as it's formulated with natural ingredients, however there are few negative facts about the natural dyes as well. Firstly, in general natural dye is lower in fastness rating in than artificial dyes. Secondly, to gain the desired color tone, dying process require repetition comparing to artificial dyes, and the color range given from natural dyes are limited compared to artificial dyes. In addition, textile printing from natural dyes provided limited printing patterns as the concentration of color liquid from natural dyes seem low and homeopathically clear. Thus, natural dyes are less practical in use and not useful for design than synthetic dyes (B. I. Jun and J. H. Hwang, 2003).

Pre-printing techniques with natural dyes are usually used to use traditional way of leader cloth which can only get a simple pattern fabric. Moreover, as the concentration of solution of dye extracted from natural dyestuff is low it is difficult to have hyperchromic effect when mixing with thickening agent. Also, due to concentration of thickening agent become watery sharpness of outline, to express clear line get worse. In addition to textile printing technique for mordant of natural dye is that the mordant and thickening agent mixed into woven fabrics form a predetermined pattern and then, various multi-color and single-color functional dyestuff is dyed along with common way of solid dyeing so that a range of patterns and designs are added.

Textile printing techniques with natural dyes are usually used to use traditional way of leader cloth which can only get a simple pattern fabric (B. I. Jun and J. H. Hwang, 2003), (Y. H. Jang and J. B. Lee, 1999). Moreover, as the concentration of solution of dye extracted from natural dyestuff is low it is difficult to have hyper-chromic effect when mixing with thickening agent (K. S. Kim, D. W. Jeon, J. J. Kim, 2005), (E. K. Kim and J. H. Chang, 2003), (M. Y. Park, H. J. Kim, and M. C. Lee, 2003). Also, due to concentration of thickening agent become watery sharpness of outline, to express clear line get worse. In addition to textile printing technique for mordant of natural dye is that the mordant and thickening agent mixed into woven fabrics form a predetermined pattern and then, various multi-color and single-color functional dyestuff is dyed along with common way of solid dyeing so that a range of patterns and designs are added (N. H. Shin, S. Y. Kim, and K. R. Cho, 2005).

In general, previous researched textile printing of natural dyes was use of concurrent mordancy technique in order to show color. As result, chemical change is occurred among thickening agent, dyestuff and mordant for elapsed time. Also, in the case of textile printing is carried out with concentrated dye or powdered dye causes lowing of workability, loss of degree of exhaustion because of lignin elution or inner maintenance happened during the solution of dye process. It was also difficult to decide appropriate moment for printing to maintain sharp line or sharp dividing line for design and there are several change factor raised in the part of mordant and miscibility on the boundary surface.

This research is designed for study textile printing of mordant used in natural dyes to improve the usage of low and homeopathically clear concentration of color and sharpness. Preprocessing the pre-printing mordant mixed with multi colors will provide more color tone options, shapes and designs. In particular, trend of colors and designs of chromatin water can be widely changed to adopt into commercialization as well as differentiate the product with artificial with artificial with functionality, aesthetic, appreciation and quality for desired customers' need.

After printing by number of pattern along with pre-printing mordant, dye process using multi color dyestuff is used to examine stable adaptability of printing, to select appropriate room for natural dye, discharge printing agent and binder, and to examine the condition of performance evaluation, degree of exhaustion and color fastness improvement. As result, in order to avoid not only defect of natural dyes with mordant textile printing which is simplicity and but also to study dyeing process to improve a variety of colors and sharpness textile printing with mordant is used first multi-color natural dye stuff is dyed later that result in to gain sharpness and to show textile printing effect.

2   Samples and experiment process
2.1 Samples and reagent

(1) Silk fabrics

Textile Printing fabric was Crepe de chine(chintz: 21 d/3fly, twistless120d/fly, weft: 21 d/3 fly, 2700 S/Z, 96ply/inch). Sample's characteristic is provided in Table 1.

Table 1 Characteristics of fabrics

Fabric	weave	denier		density (threads/inch)		surface color		
		warp	weft	warp	weft	H	V	C
Silk	plain	21d /3fly	21d /3fly	120	96	5.1Y	9.4	0.1

(2) Ingredient

Pagoda tree and Sappan wood was used for vegetable dye. Animal dyes are Gallapple, Lac and Cochineal purchased from drug medication.

(3) Reagent

Table 2 will demonstrate 4 main mordant. Aluminium potassium sulfate, copper(II) sulfate, Ferrous sulfate, tin(II) chloride are first class reagent (Duksan pure Chemical Co, Ltd).

Natural reagent was mixture of sodium alginate($C_6H_7NaO_6$) and indalca pa-30.

Table 2 Chemical structure and name of mordanting agent

Mordanting agent	Chemical name	Chemical structure
Al	Aluminum potassium sulfate	$AlK(SO_4)_2.2H_2O$
Cu	Copper(II) sulfate	$CuSO_4.5H_2O$
Fe	Ferrous sulfate	$FeSO_4.7H_2O$
Sn	Tin(II) chloride	$SnCl_2.2H_2O$

2.2 Textile printing
(1) Textile printing unit production

This was designed to measure the ratio of colors produced from pre-printing unit and backing production.

Textile printing R1(Rylusan) was designed to gather color realization by using Acetic acid and ordinary salt and 2% of vegetable oil. R2was designed to measure the fixation of mordancy and inflammation prevention by using alginic soda indalca pa-30 ratio. The results are shown in Table 3.

Table 3 Pre-treatment condition by each spec

specimen	mordant ratio(%)	R-1(%)	R2(%)	pH
1	Cu 4.0	0.5	-	4
2	Cu 2.0, Sn 2.0	-	2.5	3.5
3	Fe 2.0	0.7	-	4
4	Al 4.0	-	-	3.5
5	Sn 3.0	-	2.5	3.5
6	Fe 1.0, Al 1.0	0.5	-	4
7	Fe 1.0, Cu 1.0	1.0	-	3.5
8	Al 2.0, Cu 2.0	1.0	-	3.5
9	Fe 1.0, Al 1.0, Cu 1.0	0.3	0.06	3.5
10	Fe 1.0, Sn 1.0	0.25	3.0	3.5

(2) Printing

Printing was adopted screen-printing by hand to measure the effect of the mordancy of mordant and sharpness. Especially study of tinting on print effects was carried out by designing the patterns to spread widely.

(3) Steaming

Defensive was measured by acid steaming at 90℃ for an hour then rinsing in water.

(4) Dye extraction

20g/ℓ Gallapple concentration was washed in clean water to remove bugs then boiled in water for 30 minutes.

Pagoda tree and Sappan wood of 30g/ℓ was also boiled in water for 30 minutes.

Animal nature of Lac and Cochineal, especially Lac was dampened for over 2 hours to gain color stabilization and Cochineal was boiled for 30 minutes mixed with 1g/ℓ acetic to gain purple characteristic before extraction.

(5) Dye

Textile printing mordancy samples were each dyed at 1:80 ratio for an hour before rinse and dry. The dye process is the following.

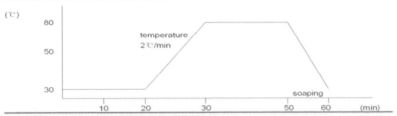

Figure 1 Dyeing process of natural dyestuff.

2.3 Color tone assessment

(1) Exhaustion concentration

Degree of exhaustion was measured by Computer color matching system (Gretag Macbath 7000-A USA) to measure reflexivity for each wave and to compute the surface reflexivity's maximum absorption impact using Kebelaka-Munk formula.

$$KS = \frac{(1 - R)^2}{2R}$$

In the formula, R is the surface reflexivity's maximum absorption impact ($0 < R < 1$). K is extinction coefficient, S is scattering coefficient.

(2) Mordant and color tone per mordant textile-printing

Computer color matching system was used to measure CIE Lab color measurement committee's $L^* a^* b^*$ values. To gather color tone data, each measurement was done per each tone.

2.4 Performance evaluation

(1) Fade test

To measure the color fade tone under sun light, KS K 0700's Fade-O-Meter (Shimadzu, XF-15FN, Japan) was used.

(2) Color fastness to water test

To measure the dyed color's resistibility in water, KS K 0645's Perspirometer Method was used for best results.

(3) Color fastness to perspiration test

Dyed color's resistibility with perspiration from human body was measured by KS K 0715's Perspirometer Method.

3   Result and consideration

3.1 Multi-color dye data

Purpose of adopting printing on mordant textile-printing was to gather required color tones with multi color dyes. In addition, polygenetic dye data is required to gather information for each color tones' dye concentration to show printing effect. Therefore, as previous dying process of mordant pre-printing only gave limited shapes, designs and colors synthetic dye was used in three primary color to show a range of color after printing by number of pattern along with pre-printing mordant dye yellow, green, purple and other color tones is chosen from multi color dyes and the colors was analyzed.

154

Using above dye data, the best dyestuff was searched and appropriate room for natural dyes, discharged printing agent, binder and other factors for selection and performance test was carried out. Also, tests for dying affinity and color concentration were examined. Following the research result, comparing to existing natural dyes printing mordant gave better and clear boundary line in color tones. Therefore, this research has diversified the usage of natural dyes and improved the dip dying which lead to clear in colors as well as creating clearer boundary lines and patterns. When use of concurrent mordancy technique the color tone does not fade over time. Thus reproducibility of color tone was improved. At the same time, testing color concentration in different situations and finding out the color tone does not fade in sun light, water and perspiration circumstances, reproducibility can be advanced.

Fig. 2 and Fig. 3 demonstrate that a variety of color with clear line can be presented by color chip made from multi-color dyes from yellow (Pagoda tree), green (Sappan wood, Lac), purple (Cochineal) and other (Gallapple).

Figure 2   Color chip of Multi-colored natural dyes preprocessed from pre-printing mordant (Pagoda tree, Sappan wood, Lac, Cochineal and Gallapple).

As you can see in the photos, colors shown from pre-printing mordant was similar to ordinary dip dyes.

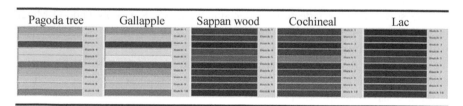

Figure 3   Color chip of Multi-colored natural dyes preprocessed from pre-printing mordant (Pagoda tree, Sappan wood, Lac, Cochineal and Gallapple).

Fig. 4's color tones were studied by Computer Color Matching (CCM) to measure the fadedness of different test circumstances with value of K/S (Kebelaka-Munk formula).

Vegetable dye shows higher color tone by mixing iron and annotation, animal nature dye showed high color tone with aluminium mordancy. Color tones shifted

| Pagoda tree | Gallapple | Sappan wood | Cochineal | Lac |

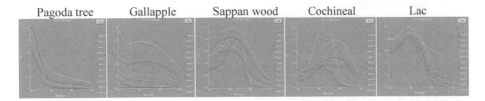

Figure 4 K/S value by textile pre-treatment condition of dye.

greatly to bath-chromic from maximum absorption impact caused from Gallapple used of iron mordanting. Pagoda tree of yellow color tone did not show much change in maximum absorption impact but in Sappan wood annotation mordancy circumstances, great shift was shown.

When Cochineal mixed with steel mordanting, Lac with steel only mordanting showed great difference in absorption impact. Mixed mordancy's greatest characteristic was shown when mordant mixed with copper met with other color tones; color tones did not change greatly, therefore influence from copper mixing with mordant is greatly important.

3.2 Difference between thickening agent and mordant and its compatibility analysis

(1) Color tone change due to printing condition

Thickening agent and metallic compound of mordant used in printing was tested for 1 week after its mixed together to check the compatibility. The change on standing was researched.

Mixed thickening agent made little change in the beginning but it was steady after 6 hours till 1 week period, it seems air room temperature was stabilized.

Fig. 5 was created using Table 3's components; samples were dyed after printing mordant previously. Same with natural dye's dip dye, the most similar color tone given by the test was from sample 4's (Pic in grey) aluminum mordanting's condition. Highest color density was from sample 3 (Pic in blue).

And least affected color tone from mixed mordant was dyestuff and nutgall, most affected color tone was from Pagoda tree. Other dyestuff created similar color tone depending on the mordant, but Pagoda tree was affected the most from mordant.

| Pagoda tree | Gallapple | Sappan wood | Cochineal | Lac |

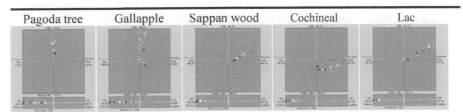

Figure 5 Color difference by processing condition.

156

(2) Sharpness of outline

Fig. 6 was designed to evaluate sharpness of outline. Line designs were printed then dyed with dyestuff.

Photos seems difficult to tell the difference, but pre-mordanting spec dye showed the result that design's sharpness of outline was great but spec 2 and spec 5's borderline was very muddy. It seems to be the association and disposition from mordant by scapolite mordanting process.

Iron and scapolite's mixed mordanting created imbalanced dye.

| Pagoda tree | Gallapple | Sappan wood | Cochineal | Lac |

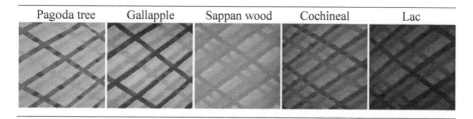

Figure 6   Preprocess per spec sharpness of outline evaluation.

(3) The effect of color fastness improvement

Natural dye's mordant preprocessing is shown in Table 4. Color fastness was replaced by dip dye method, vegetable dyes
replaced by color fastness to washing, and animal nature dye replaced by color fastness to light.

Acid dye from synthetic dyestuff was replaced by color fastness when used on silk fabric was shown.

For individual mordant cases, Sappan wood 's Brazilian used 2 types of -OH mordant and complex on dihydropyran to get the effect of molecular weight and mordant printing preprocess by steaming and color fixing.

Especially color fastness to light effect from Cochineal mixed with iron and copper's individual mordant showed great results but from tin and aluminum's individual mordant mixed with copper showed low results.

Other dyes showed similar results, therefore when copper is mixed with other mordant and used joint will cause color fastness to light effect to reduce its tone. Color fastness to water and color fastness to perspiration also showed great results but copper mixed with other mordant and used joint showed the same result. Therefore, copper is useful when only mixing with individual mordant.

Natural dye products attraction is aesthetic appreciation and psychological stability, it can be used to follow new trends to meet customers' needs and wants. However, stability of color tone and handling treatment is difficult for mass production.

Recently, to solve these issues, natural dyes in powder form has been created but comparing to actual dye, concentration is lower. This research result shows if mordant preprocessing is used on printing effect will ensure control and change of standing in color tones, also depending on purpose and design, pre-operation is possible for printing effect for not only single color but to change the point of view on natural dye.

4   CONCLUSIONS

Weakness of preprocessing the mordant before multi-color dye in natural dye can be solved by following the visible benefits deduction.

1. Applied mordant with air contact caused change on standing effect after printing but between 6 hours to 1 week it became stabilized.
2. Used jointly mordant's greatest characteristic is there are no changes in color tones when copper is mixed with mordant.
3. Result of lowest change in color tone was from Pagoda tree and iron mordanting showed maximum absorption wavelength shifted to bathochromic shift when multi color pigment was processed with preprocessing mordant.
4. From sharpness of outline effect, preprocessing with tin showed low result where iron with used jointly caused unlevel dyeing occurrence.
5. Preprocessing mordant for individual mordant were replaced by color fastness where mordant used jointly with copper showed lower color fastness result comparing to individual mordant.

Table 4 Color fastness by various condition

Fastness Condition		Light	Water			Perspiration					
kind of dye	spec.						Acid			Alkaline	
			Fade	stain		fade	stain		fade	stain	
				silk	cotton		silk	cotton		silk	cotton
Pagoda tree	1	4	2	3	1	1	3	2	2	2	1
	2	3	1	4	4	2	4	4	2	3	2
	3	3	2	4	4	2	4	4	3	4	3
	4	1	2	4	3	2	4	4	3	4	2
	5	1	3	4	3	4	4	3	3	2	1
	6	2	2	4	4	3	4	4	2	3	2
	7	1	1	4	4	2	4	4	3	2	1
	8	1	1	4	4	3	3	3	3	1	1
	9	1	1	4	4	3	4	3	3	1	1
	10	2	2	4	4	3	4	4	2	3	2
Gallapple	1	3	3	4	3-4	3	3	1	1	4	2
	2	1	4	4	4	4	3	2	3	4	3
	3	3	4	4-5	4-5	3	4	2	1	3	2
	4	1	4	4-5	4	4	4	3	3	3	2
	5	1	4	4-5	4	4	4	3	4	4	2
	6	2	3	4-5	4-5	3	4	2	1	4	2
	7	2	4	4	4	1	4	3	1	4	3

kind of dye	spec.	Light	Fade	Water (stain)		fade	Acid (stain)		fade	Alkaline (stain)	
				silk	cotton		silk	cotton		silk	cotton
	8	2	3	4-5	4	3	4	3	2	3	2
	9	2	4	4	4	2	4	2	2	4	2
	10	2	4	4-5	4-5	2	4	2	1	3	2
	1	3	3	3	1	1	1	1	2	2	1
	2	3	3	4	1	3	2	1	2	1	1
	3	3	3	4	3	3	2	1	3	3	1
	4	2	3	4	2	2	2	1	1	2	1
Sappan wood	5	2	3	4	2	2	4	3	2	3	1
	6	2	2	3	2	2	2	1	1	2	1
	7	2	3	3	3	1	1	1	1	1	1
	8	2	2	3	2	1	1	1	2	1	1
	9	2	2	2	2	1	1	1	1	1	1
	10	3	1	3	2	2	3	1	1	3	2
	1	4	4	1	1	1	1	1	1	1	1
	2	4	4	1	1	1	1	1	1	2	1
	3	4	4	1	2	2	2	2	2	3	2
	4	2	3	1	1	2	1	1	2	2	1
Cochineal	5	2	3	1	1	1	1	1	1	1	1
	6	4	3	1	1	2	2	1	2	1	1
	7	2	4	1	1	1	1	1	1	2	1
	8	2	3	1	1	2	1	1	1	1	1
	9	3	2	1	1	1	1	1	1	1	1
	10	4	3	1	1	1	2	1	1	1	1
Fastness Condition		Light		Water			Perspiration (Acid)			(Alkaline)	
kind of dye	spec.		Fade	stain		fade	stain		fade	stain	
				silk	cotton		silk	cotton		silk	cotton
	1	5	4	1	1	2	1	1	1	1	1
	2	5	4	3	1	2	1	1	1	1	1
	3	5	4	2	1	4	1	1	2	1	1
	4	5	4	2	1	4	1	1	1	1	1
Lac	5	5	4	2	1	3	1	1	1	1	1
	6	5	4	3	1	1	1	1	1	1	1
	7	5	4	1	1	1	1	1	1	1	1
	8	5	4	1	1	2	1	1	1	1	1
	9	5	4	1	1	1	1	1	1	1	1
	10	5	4	3	1	1	1	1	1	1	1

REFERENCES

B. I. Jun and J. H. Hwang. 2003. "Studies on the Printing with Natural Dyes on Sappan Wood", *J. Korean Scociety of Industrial Application*, Vol. 6, No. 3, 239-245.

B. I. Jun and J. H. Hwang. 2003. "Studies on the Printing with Natural Dyes Two phase printing method", *J. Korean Scociety of Industrial Application*, Vol. 6, No. 3, 247-252..

Y. H. Jang and J. B. Lee. 1999. "Printing Mehtod Natural Dyes", J. *Korean Scociety of Craft*, Vol. 2, No. 2, 161-174.

K. S. Kim, D. W. Jeon, J. J. Kim. 2005. "Effect of the Dye Bath and Mordants on the Dyeing of Silk Fabric using Cochineal", *J. Korean Home Economic Association*, Vol. 43, No. 7, 109-116..

E. K. Kim and J. H. Chang. 2003. "Dyeability of Cotton and Silk Fabrics Printed with Cochineal", *Human life Science*, Vol. 6, No.-, 233-242.

M. Y. Park, H. J. Kim, and M. C. Lee. 2003. "Dyeabilites of Lac extract onto the silk and wool fabrics(Ⅱ)-Effect of mordanting methods and various mordants-", *J. Korean Scociety of Cloting and Textiles*, Vol. 27, No. 9/10, 1134-1143.

D. W. Jeon, J. J. Kim, and H. S. Shin. 2003. "The Effcet of Chitosan Treatment of Fabrics on the Natural Dyeing using Japanese Pagoda Tree(I)", *The Research Journal of the Costume Culture*, Vol. 11, No. 3, 423-430.

N. H. Shin, S. Y. Kim, and K. R. Cho. 2005. "A Study on Using Gray Color Dyeing from Gallapple", *J. Kor. Soc. Cloth. Ind.*, Vol. 7, No. 5, 547-552..

CHAPTER 18

Effect of Culture Interdependency on Interpersonal Trust

– A New Challenge for Information System and Service Design

Jun Liu, Pei-Luen Patrick Rau

Institute of Human Factors and Ergonomics,
Tsinghua University
Beijing, China
rpl@mail.tsinghua.edu.cn

ABSTRACT

Interpersonal trust is a crucial issue in collaborative information technology and service design. The study aims at exploring the effect of cultural interdependency on interpersonal trust in computer-mediated team collaborations. A social dilemma experiment with 20 Chinese students and 20 German students were conducted to compare interdependent with independent cultures. Results found that when facing team vs. personal benefit contradictions, interdependent people (i.e. Chinese) have more trust with their teammate than independent people (i.e. Germans). Also the reduction of trust in non-face-to-face virtual communication is less suffered by interdependent people than by independent people. The results support that interdependent culture build trust based on intimate relationships, while independent culture establish trust on the basis of previously agreed rules and regulations. The findings provide implications for cross-cultural management, and information system and service design.

Keywords: cultural interdependency, interpersonal trust, self-construal, computer-mediated communication, virtual team

1   INTRODUCTION

How do you trust Amazon online store sellers? How to find a trustable teammate on Warcraft? Will you allow a Facebook friend request from a stranger? All these questions come to a crucial issue in the information technology world – interpersonal trust. While enjoying the convenience of collaboration with information systems and services, people have to face new challenges of interpersonal trust. They have to learn whom to trust, how to build trust, and how to appear trustable in the new forms of social interaction.

People's process of learning and building trust is fundamentally influenced by their cultural background. One of the most influential cultural variables is interdependency, which shapes individuals' beliefs and perceptions of their relationship with others. In interdependent cultures, such as China and Japan, people view themselves as interconnected with related others, and build trust on the basis of shared social identity and reciprocal behavior. On the contrary, in independent cultures, such as West Europe and the United States, people view themselves as distinct individuals, and build trust on the basis of regulation and credit systems. These cultural differences make it important to explore the cultural effects on interpersonal trust in collaboration with information system and service.

This chapter introduces a study exploring the effect of cultural interdependency on interpersonal trust in computer-mediated team collaboration. The study compares interdependent Chinese culture and independent German culture through an experiment. The chapter also provides implications for cross-cultural management and information system and service design. The rest of the chapter includes four parts: first, a general review of related literatures on cultural interdependency and interpersonal trust; second, the hypothesis and design of the experiment; third, an analysis of the experimental results; and forth, discussions and implications of the study.

2   LITERATURE REVIEW

2.1   Interpersonal Trust

Most definitions describe interpersonal trust as the expectation of certain behaviors of the interacting partner. The definition, probably being the most important for empirical research on trust dates back to Rotter (1967) and refers to exactly these expectations about the other person. Rotter says that trust is based on the expectation of a person or a group to rely on another given promise, orally or written, of another person or group, no matter if it is positive or negative (Rotter 1967). After that, Cook & Wall's (1980) defined trust on the expectation of good intentions when interacting with a partner. They stated that trust refers primarily to the degree to which someone is willing to ascribe good intentions to other persons and thus to trust their words and actions (Cook & Wall 1980).

Depended on the level of social trust existing in a society, Fukuyama (1995) distinguishes between "low-trust societies" and "high-trust societies". In the first one, highly personalized social relations and intimate social networks are existing while in the second one, formal networks and functional relationships, so called social capital, is dominating.

According to Fukuyama (1995), China is a low-trust society. This results from the ethical philosophy of Confucianism. This philosophy puts the ties of a family above everything else in Chinese society. For that reason, the trust into not family members or, in case of a group, out-group members is very low or even not existing. A typical low-trust experience you can make in China is during shopping. The people are extremely critical until you paid the money because they are all the time afraid that you might try to cheat them. Exemplary, Chinese people check the 100 Yuan bill very carefully and always ask you for smaller money if possible. In contrast to this, German people don't have to worry that much about being scammed. Life is very regulated by law and the punishment if trying to rip someone of is quite hard. Consequently, Germany is a high-trust society. Family or in-group relationships are not a necessary requirement to build up trust (Fukuyama 1995).

2.2 Cultural Interdependency

The concept of cultural interdependency evolved from a comparison of Western and Eastern conceptualizations of the self – self-construal (Markus & Kitayama 1991). Self-construal is an image of oneself, including independent and interdependent types. It has been found to be a better predictor of communication in interpersonal and small group settings than cultural individualism-collectivism (Oetzel & Bolton-Oetzel, 1997).

In independent cultures, people construct themselves as unique individuals. Their basic characteristics are (a) internal abilities, feelings, and thoughts; (b) being unique and expressing the self; and (c) realizing internal attributes and promoting one's goals (Markus & Kitayama 1991).

The interdependent self-construal people believe that they are connected to other group members and have the motivation to fit in with others, act appropriately, promote others' goals, and value conformity and cooperation (Markus & Kitayama 1991).

Cultural interdependency affects the need to maintain cohesion. An interdependent self-construal individual tends to avoid engaging in conflicts (Zhang et al. 2006). A possible indication from this fact is that interdependence might have a positive impact on relationship outcome. The fact that interdependent self-construal people are more concerned about relationship outcomes (Oetzel & Bolton-Oetzel, 1997) also indicates this opinion above.

3. HYPOTHESIS AND METHODOLOGY

3.1 Hypothesis

This study focuses on the effect of cultural interdependency on interpersonal trust in computer-mediated team collaboration. In most collaboration, there are contradictions between team benefit and personal benefit. The current study adopted a social dilemma to simulate the team vs. self benefit contradiction in team collaborations.

From previous literatures, German culture is a highly independent culture, and Chinese culture is highly interdependent. Chinese people value team cohesion more importantly than Germans. Consequently, they may build the common trust to others' willingness to collaborate rather than to compete. So we hypothesize that:

H1: When facing team vs. self benefit contradiction, Chinese participants have more trust with their teammate than German participants.

In general, virtual communication causes lower interpersonal trust than face-to-face communication (Gibson & Cohen, 2003). However, when we take culture into consideration, as interdependent culture build trust on relationship, the different communication media may have reduced influence on interpersonal trust, especially when teammates have already built an interdependent relationship. But for independent people, the relationship issue is less essential in their trust consideration, so they may be more influenced by different communication media. Therefore, we hypothesize that:

H2: Chinese participants' trusts with their teammates are less influenced by virtual vs. face-to-face communication media than German participants.

3.2 Experiment design and participants

A 2*2 between-subject experiment was designed to test the hypothesis. The two independent variables were: communication media (i.e. face-to-face or online chatting) and culture (i.e. Chinese or German). Forty participants took part in the experiment in pairs: ten pairs of Chinese and ten pairs of Germans. All of them were graduate students in Tsinghua University. Within each culture, five pairs participated in the face-to-face condition and the other five pairs in online chatting. Each pair of participants had already known each other as friends before they came to the experiment.

3.3 Task and Measure

The team collaboration task had three steps. First, each pair of participants was given two different scientific articles separately. They were asked to read it and remember the facts listed on it. Second, after the learning period, they were allowed to share their own knowledge with the teammate in a limited time. Finally, the reading materials were taken back. Participants took a quiz of the facts.

The social dilemma of team vs. personal benefit contradiction was designed with the final quiz. The quiz involved facts on both articles with half-to-half ratio. Both participants must get at least 50% correct answers to pass it and to earn a reward. Besides, the participant with higher score in the team can get a higher reward. Participants learnt this rule before the task. Therefore, participants can choose to share with the partner more facts to fulfill the team benefit, or to share less (or even wrong facts) to fulfill the personal benefit.

In face-to-face conditions, both participants sit in the same room side by side. In virtual communication conditions, they were separated into two rooms and finish the second phase of sharing through Skype chatting on computers. The time of each period was identical between the two conditions.

After the task, a questionnaire of interpersonal trust was handled to the participants. The questionnaire involves 6 items in a 7-point Likert scale from previous literatures (McKnight etc., 2002; Cook & Wall, 1980) asking about trust to the teammate's benevolence as well as confidence of the teammate's action. Finally, the participants were interviewed separately on their trust with their teammate.

4. RESULTS

The two hypotheses were tested using T-test algorithm. Hypothesis 1 predicts that when facing team vs. self benefit contradiction, Chinese participants have more trust with their teammate than German participants. The T-test results show that culture has significant influence on interpersonal trust ($F=4.84$, $p<0.05$). Chinese (Mean=6.04, SD=0.77) scored higher interpersonal trust than Germans (Mean=5.53, SD=0.76). Therefore the first hypothesis was supported.

Hypothesis 2 states that Chinese' trusts with their teammates are less influenced by virtual vs. face-to-face communication media than Germans. To test it, the media effects on the two samples were tested separately. With the Chinese sample, T-test shows non-significant media differences of interpersonal trust ($F=3.24$, $p=0.08$). But with the German sample, T-test shows significant differences: German participants trust their teammates more ($F=6.54$, $p=0.02$) in face-to-face than in virtual communication. Hypothesis 2 is also supported.

The interview about participants' trust with their teammates revealed some reasons of the different effects. Fourteen out of twenty Chinese participants mentioned that they trust the teammates' benevolence because they are friends. One of them said "I believe she tried her best to share the facts with me, because she is always responsive when I need her." To compare, eight out of the ten German participants in virtual teams reported uncertainty on their teammates' collaboration. To quote one of them: "I think some of the facts he shared with me may not be accurate enough."

Also there was an interesting phenomenon observed by the experimenters: at the beginning of the sharing phase, all the German pairs started with setting a rule of sharing in turn, like you share one, I share one. But only two pairs of Chinese participants did so. All the other Chinese started sharing directly. Chinese face-to-face pairs also shared in turn, but they didn't need the prior discussion. Chinese virtual pairs didn't share in turn; they typed all the time and tried to typing in as many facts as possible.

5. DISCUSSION

Information technology brings new opportunities of collaboration, but also introduces new challenges of interpersonal trust in a collaborative team. The challenges are different across cultures. In interdependent cultures (e.g. China), trust are built on interpersonal relationships. People share the same identity or within intimate relations trust each other no matter what is the media of communication. In independent cultures (e.g. German), trust is more influenced by communication media and prior setting of rules and regulations. According to Gibson & Cohen (2003), virtual teams have more misunderstanding problems than face-to-face teams. But this is a less serious problem in interdependent cultures, where interpersonal relationship can act as lubricant for trust.

There are several implications for cross-cultural team management. First, to enhance social trust in interdependent culture, the dominating issue is to build interdependent in-group relations and shared identity within a team. Second, in independent culture, the first thing to do is to establish rules and to regulate responsibilities among team members. Third, to manage a virtual team of independent culture, it is essential to make sure the communication of rules and regulations are successful, even though it may take extra time.

For collaborative information technology and service design, the study suggests to support relationship building and shared identity in interdependent cultures by emphasizing social features, such as social media or personal chat rooms. But in independent cultures, information systems or services should always support a rule setting process before work.

6. CONCLUSION

Interpersonal trust is a new challenge for information technology and service design. The study explored the effect of cultural interdependency on interpersonal trust in computer-mediated team collaborations. It found that when facing team vs. personal benefit contradictions, interdependent people (i.e. Chinese) have more trust with their teammate than independent people (i.e. Germans). Also the negative effect of virtual communication on trust happens only in independent culture, but not in interdependent culture. The results support that interdependent culture build trust based on intimate relationships, while independent culture establish trust on the basis of previously agreed rules and regulations. The findings provide implications for cross-cultural management, and information system and service design.

166

ACKNOWLEDGMENTS

The authors would like to acknowledge Nico Wendler and Na Chen from Tsinghua University for their help with the data collection. This study was funded by a National Science Foundation China grant 70971074 and a National Science Foundation China grant 71031005.

REFERENCES

Cook, J. and Wall, T. 1980. New work attitude measures of trust, organizational commitment, and personal need fulfillment. *Journal of Occupational Psychology,* 53: 39–52.

Fukuyama, F. 1995. *Trust. The social virtues and the creation of prosperity.* New York: Free Press.

Gibson, C. B. and Cohen, S. G. 2003. *Virtual teams at work.* San Francisco: John Wiley & Sons.

Markus, H. R., and Kitayama, S. 1991. Culture and the self: Implications for cognition, emotion, and motivation. *Psychological Review*, 98: 224-253.

McKnight, D. H., Choudhury, V., and Kacmar, C. 2002. Developing and Validating Trust Measures for e-Commerce: An Integrative Typology. *Information Systems Research,* 13 (3): 334–359.

Oetzel, J. G., and Bolton-Oetzel, K. D. 1997. Exploring the relationship between selfconstrual and dimensions of group effectiveness. *Management Communication Quarterly,* 10: 289-315.

Rotter, J. B. 1967. A new scale for the measurement of interpersonal trust. *Journal of Personality*, 35 (4): 651–665.

Zhang, Y., Feick, L., and Price, L. J. 2006. The impact of self construal on aesthetic preference for angular versus rounded shapes. *Personality and Social Psychology Bulletin*, 32: 794-805.

CHAPTER 19

Exploration on the Relationship between Chinese Characters and Ergonomic Affordances

wei-han,chen 1 yu-ju,lin 2

National Taiwan University of Arts 1
Taipei College of Maritime Technology2
Taiwan
aska199@yahoo.com.tw
naralin@mail.tcmt.edu.tw

ABSTRACT

The process through which people understand the world tends to begin with their surrounding objects and activities, and often involves becoming aware of the function and usage of their hands. The idea of "hand" constantly enters a person's thoughts and consciousness, as it can be seen as a necessity of life, a culture connected to thoughts, and a symbol of representation. In Chinese writing systems including Chinese characters, the oracle bone script and Chinese bronze inscriptions, many characters are constructed partially with the symbol of手 (sou), which expresses thoughts and concepts richly associated with the hand. In the developmental history of Chinese characters, the hand has been continuously made more abstract and transformed into various symbols. Nevertheless, characters such as手 (shou), 爪 (zhua), 又 (you), 勺 (shao) and characters containing these parts share many characteristic features.

This paper seeks to discuss the meanings behind the common understanding of "images or situation models contained in Chinese characters," and how these meanings are connected to the development of creativity in arts and design. The research process utilizes philology materials, assisted by semiotic studies and the affordance theory in ergonomics, to further describe and examine the topic. The research attempts to trace the various sources of ideas relating to "hand" in Chinese characters, and discuss these in terms of ergonomic affordances with the aim to explore the corresponding relationships within a situation model. The research finds that the construction of Chinese characters often contains the conceptual application of "representation and non-representation" as well as "similarity" in semiotic characteristics. The study explores possibilities of applying these findings to design and creativity, and provides further discussions and explanations on the works of the artists and designers and the experiment of the writer.

The goal of this study not only explains the cognitive awareness people have of how pictograms and ideograms function in Chinese characters, but also describes how this understanding can be transformed and applied concretely to the development of creativity in arts and design. It also verifies that the situation under which a Chinese character is constructed can be used creatively by designers, in ways such as extracting a visual element from a traditional, cultural image, or interpreting a text through cultural knowledge and thereby transforming it into a source or method for producing a situation model. In addition, this study presents a diachronic and comparative analysis, and further examines the process through which the "hand" radical in Chinese characters was developed, replaced, and mixed with other characters.

Keywords: Chinese characters, Motion Economy consists , hand radical

1.CHINESE CHARACTERS

The goal of this study not only explains the cognitive awareness people have of how pictograms and ideograms function in Chinese characters, but also describes how this understanding can be transformed and applied concretely to the development of creativity in arts and design. It also verifies that the situation under which a Chinese character is constructed can be used creatively by designers, in ways such as extracting a visual element from a traditional, cultural image, or interpreting a text through cultural knowledge and thereby transforming it into a source or method for producing a situation model. In addition, this study presents a diachronic and comparative analysis, and further examines the process through which the "hand" radical in Chinese characters was developed, replaced, and mixed with other characters.

〈 Chinese characters are applied to the planar design example 〉

2. Characters containing the hand (手) radical (手部形)

The long history of Chinese civilization has resulted in the development of nearly 50,000 characters. Each radical within a character provides a significant symbol associated with its meaning.

The process through which people understand the world tends to begin with their surrounding objects and activities, and often involves becoming aware of the function and usage of their hands. The idea of "hand" constantly enters a person's thoughts and consciousness, as it can be seen as a necessity of life, a culture connected to thoughts, and a symbol of representation. In Chinese writing systems including Chinese characters, the oracle bone script and Chinese bronze inscriptions, many characters are constructed partially with the symbol of手 (sou), which expresses thoughts and concepts richly associated with the hand. In the developmental history of Chinese characters, the hand has been continuously made more abstract and transformed into various symbols. Nevertheless, characters such as手 (shou), 爪 (zhua), 又 (you), 勺 (shao) and characters containing these parts share many characteristic features.

	hand (手) radical (手部形)	Motion Economy consists
手 (sou)		1.Level One: Finger motions　　　Explanation: the lowest level; the fastest speed; precise motions
手 (shou)		2.Level Two: Finger motions + Wrist motions　　　Explanation: The upper arm and forearm remain unmoved; motions are limited to fingers and wrist
爪 (zhua)		3.Level Three: Finger motions +Wrist motions + Forearm motions (elbow motions)　　　Explanation: Motions are limited to below the elbow; the upper arm remains unmoved

又 (you)	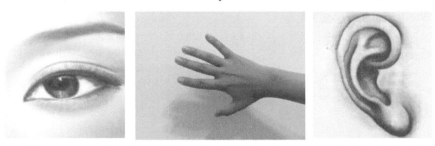	4.Level Four: Finger motions +Wrist motions + Forearm motions + Upper arm motions (shoulder motions) Explanation: The object or tool is farther from the body and therefore cannot be obtained by Level Three motions; requires the motion of "extending the arm"
勺 (shao)		5.Level Five: Finger motions +Wrist motions + Forearm motions + Upper arm motions + Body motions Explanation: The highest level; requires the most energy; the slowest speed; motions involve the entire body

In my study of Chinese characters, I discovered that the development of characters is very closely related to many types of human activities. There are many characters and pictograms related to our five senses and various body parts, and these characters share a close relationship to our behavior and movements.

(There are many characters and pictograms related to our five senses and various body parts)

3.Motion Economy consists

By studying these characters I found that they are closely related to varying degrees of precision in movements. By looking at how these characters are composed, we see that they are intricately linked to principles of motion economy in ergonomics.

Motion Economy consists of:
1.Level One: Finger motions
    Explanation: the lowest level; the fastest speed; precise motions
2.Level Two: Finger motions + Wrist motions
    Explanation: The upper arm and forearm remain unmoved; motions are limited to fingers and wrist

3.Level Three: Finger motions +Wrist motions + Forearm motions (elbow motions)

Explanation: Motions are limited to below the elbow; the upper arm remains unmoved

4.Level Four: Finger motions +Wrist motions + Forearm motions + Upper arm motions (shoulder motions)

Explanation: The object or tool is farther from the body and therefore cannot be obtained by Level Three motions; requires the motion of "extending the arm"

5.Level Five: Finger motions +Wrist motions + Forearm motions + Upper arm motions + Body motions

Explanation: The highest level; requires the most energy; the slowest speed; motions involve the entire body

4. Relation between Chinese character and Motion Economy consists

In the process of researching this topic, and with the help of my advisor I have gathered a great deal of information and reference materials. I have reorganized the acquired data and made a table categorizing these five levels of motions. For each level I provide an example of a Chinese character to discuss the relationship between the character and the motions involved.

1.   Level One: Finger motions
Explanation: the lowest level; the fastest speed; precise motions

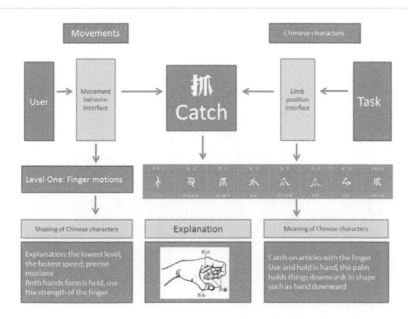

2.Level Two: Finger motions + Wrist motions

Explanation: The upper arm and forearm remain unmoved; motions are limited to fingers and wrist

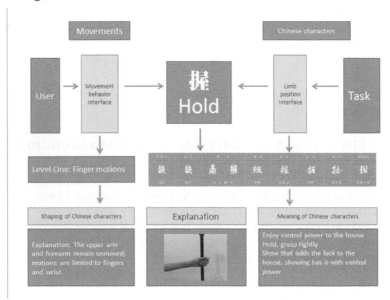

3.Level Three: Finger motions +Wrist motions + Forearm motions (elbow motions)

Explanation: Motions are limited to below the elbow; the upper arm remains unmoved

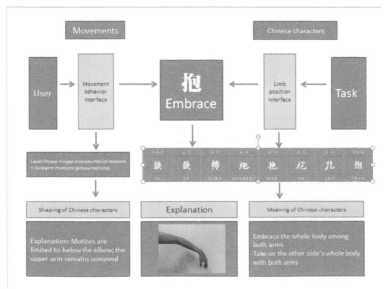

4.Level Four: Finger motions +Wrist motions + Forearm motions + Upper arm motions (shoulder motions)

Explanation: The object or tool is farther from the body and therefore cannot be obtained by Level Three motions; requires the motion of "extending the arm"

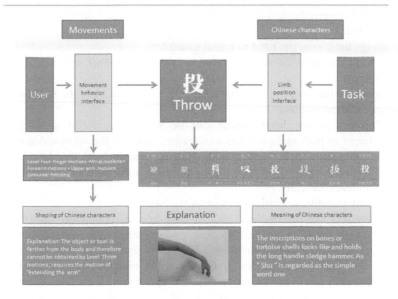

5.Level Five: Finger motions +Wrist motions + Forearm motions + Upper arm motions + Body motions

Explanation: The highest level; requires the most energy; the slowest speed; motions involve the entire body

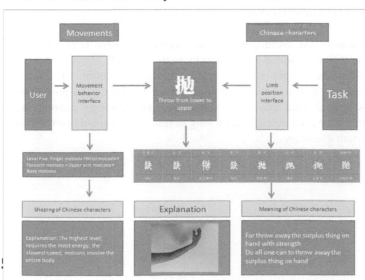

5. CONCLUSIONS

1.The use of the hand (手) radical in Chinese characters

As a body part, the hand is one of the body parts most understood by mankind early on in history.

2.Correlations between the development of human behavior and the motions and directions of the hand

Human beings have used the hand as a tool to explore surrounding objects both close and far from the body, as well as meanings that are abstract and concrete, simple and complex.

3.The relationship between the meanings of Chinese characters and the behavior mechanisms of different motions

The development of the hand radical can be seen as a useful cognitive mechanism in terms of both metaphor and metonymy. Through the relationship of the hand with concrete, abstract, and spatial domains, people have developed a series of words and expressions linked to the hand, hence expanding and enriching the system of Chinese characters.

4.The perceptive (tactile, sensual) extension of the hand

Chinese characters that use the hand radical have become commonly used by Chinese speakers in their daily life, but few people notice the metaphorical meanings associated with these characters. This does not mean the role of metaphors has diminished in our everyday language, but rather, it could point to new developments and potentials for metaphorical usage. Expressions that contain metaphors which cannot be easily recognized without careful analysis actually serves as ample proof that metaphors exist because they serve as a cognitive mechanism in our daily life.

5.Hand (radical) of Chinese characters Left and right sides type attitude with movement principle

Hands are as to index of Chinese characters, it is one of the radicals. The hand belongs to and draws radicals four times.

This study discusses the correlation between certain Chinese characters and hand motions. It is still in the initial stage of preliminary research. I hope that in the future I can offer more insight and exploration on the evolution of Chinese characters and behavior and movements in ergonomics. Exploring the process by which metaphors have entered into our everyday language through customs and habits which people have accumulated over a long period of time should prove to be a novel and important discovery.

REFERENCES

Art Directors Club. (n.b.) Archive: 1993 Hall of Fame: Yusaku Kamekura. Retrieved November
20, 2009 from http://www.adcglobal.org/archive/hof/1993/?id=214

Bolton, K. (2000). Part I: Language in Context. The Socialinguistics of Hong Kong and the space
for Hong Kong English. World Englishes, Vol. 19(3), p. 265-285.

British Council. (2006). International Young Design Entrepreneur of the Year award 2006.
Retrieved November 10, 2009 from http://www.britishcouncil.org/arts-aad-design-crossdisciplinary-

Bustler. (2009). New Design Educational Program in Shanghai and Rotterdam. Retrieved
November 22, 2009 from

Fung, A. S., & Lo, A. C.Y. (2001). Design Education in China: New proposals to address
endemic problems. Journal of Arts and Design Education, Vol. 2(2),

Kim, C., Yang, Z. & Lee, H. (2009). Cultural Differences in Consumer Socialization: A
Comparison of Chinese-Canadian and Caucasian-Canadian Children. Journal of Business
Research, 62: 955-962.

Knight, N. (2006). Reflecting on the Paradox of Globalisation: China's Search for Cultural
Identity and Coherence. China: An International Journal, Vol. 4(1)

Alvarez-Filip, L., N. K. Dulvy, and J. A. Gill, et al. 2009. Flattening of Caribbean coral reefs:
Region-wide declines in architectural complexity. *Proceedings of the Royal Society, B*
276: 3019–3025.

Boyd, J. and S. Banzhaf. 2007. What are ecosystem services? *Ecological Economics* 63:
616–626.

Section III

Ergonomics and Human Factors

Evaluation of Customers' Subjective Effort and Satisfaction in Opening and Closing Tail Gates of Sport-Utility Vehicles

Taebeum Ryu, Byungki Jin, Myung Hwan Yun, Wonjun Kim

Hanbat National University
Daejeon, South Korea
tbryu75@me.com

ABSTRACT

The user's effort of opening and closing tail gates of sport-utility vehicles (SUV) is an important factor in the affective quality of them. The purpose of this study is to analyze customers' effort and satisfaction in closing and opening SUV tail gates with related mechanical characteristics of them. About 100 participants evaluated their subjective effort and satisfaction in closing and opening the tail gates of 42 SUVs. ANOVA of mixed-factors design was conducted to find significant mechanical properties of tail gates affecting customers' effort and satisfaction. Three properties of the force-angle graph of tail gates such as middle force more than initial force and range of angle of steady force affected customers' perceived effort and satisfaction in closing SUV tail gates. Two properties such as initial force and entire angle range of force-angle graph did them in opening. Based on these significant mechanical properties, customers' subjective effort and satisfaction on closing and opening SUV tail gates were modeled in a statistical manner, and the explanation power of the models were fairly high.

Keywords: customers' effort and satisfaction, opening and closing tail gates, sports-utility vehicles, affective design, affective quality control

1 INTRODUCTION

The affective quality of passenger vehicles such as the look and feel is becoming important factor in customers' purchase decision. When making purchase decisions, more customers are placing high importance on driving comfort, availability of convenience features (add-on features for convenience, such as automatic headlight on/off, anti-lockout device, and underseat storage), luxuriousness of materials, and quality of finish rather than engine power and fuel consumption rate (Jindo and Hirasago, 1997; White, 2001; Welch, 2002).

Affection is defined as a customer's psychological response to the perceptual design details of the product (Demirbileck and Sener, 2003). Product designs that do not consider customers' affection may essentially be weakened (Helander and Tham, 2003). Kansei engineering has been well recognized as a technique of translating consumers' psychological feelings about a product into perceptual design attributes (Nagamachi, 1996). Six technical styles of Kansei engineering methods were proposed with applications such as the automobile industry, cosmetics, house design, and sketch diagnosis (Nagamachi, 2002). Other product design practitioners in a variety of areas have utilized these approaches to translate a consumer's perceptions into design elements (Jindo and Hirasago, 1997; Nakada, 1997; Yun et al., 2001; Ou et al., 2004; Yun et al., 2003; Schutte, 2005; You et al., 2006).

However, few studies have been conducted on the satisfaction of opening and closing the tail gates of SUV. Automotive manufacturers put a lot of attention on the customers' effort of opening and closing the tail gates and are trying to get better Initial Quality Score (IQS) related to it. While many Kansei engineering studies have focused on the visual design and touch feel, few empirical studies exist focusing on the customers' affect on their behavioral performance.

This study attempted to analyze customers' satisfaction and perceived effort of closing and opening tail gates of SUV. To do this, this study conducted following tasks: 1) an evaluation form was developed to evaluate customers' affect on closing and opening the tail gates, 2) an experiment was performed to evaluate customers' satisfaction and perceived effort, 3) statistical analysis was conducted to find significant mechanical properties in force curves of tail gates and 4) the models of customers' satisfaction and perceived effort were developed.

2 EVALUATION OF EFFORT AND SATISFACTION OF TAIL GATE OPENING AND CLOSING

2.1 Affective Variables

Satisfaction and perceived effort were selected to evaluate the affect in closing and opening the SUV tail gaits. Because this study focus on the ease or hardness of closing and opening the tail gates, customers perceived effort was selected as a behavioral affect, their satisfaction as a reflective affect according to Norman (2005). We confirmed if the variables can be used with a pilot-test in which five students participated and the evaluation experience was reflected on preparing evaluation form.

2.2 Evaluation form

The developed questionnaire to evaluate customers' satisfaction and effort on closing and opening the tail gate consisted of 1) the questions for basic information of customers, 2) explanation script of the evaluation method, and 3) the question to rate customer's affect. The questions for basic information of customers were included to obtain the demographic and anthropometric data, driving experience and the experience of usage about tail gates. In the explanation script of evaluation method, a detailed task for evaluator to perform, and the explanation of the affective variables were included. In the form, customers' satisfaction is rated by using seven-points semantic differential scale and perceived effort is rated by Borg's CR-10.

2.3 Participants and vehicles

A total of 100 males participated in the evaluation experiment of SUV tail gates. Of the participants 64 were members of SUV clubs and 36 were general customers. Among them, 16, 45, 28, and 11 were in their 20s, 30s, 40s and 50s respectively. Most of them had more than the driving experience of six years. In respective of height and weight, they were normally distributed.

The present study used 42 SUVs from varying car markers in Europe, Korea, Japan, the United States. They included all of compact, mid-sized, large-sized and luxury SUVs. The vehicles were placed at a yard of an auto manufacturing company. It was difficult and unreasonable for a participant to evaluate all of them; we grouped the vehicles into three groups (14 vehicle in one group) and made one participant to evaluate one group. The SUVs were distributed evenly as possible in car makers and size (see Table 1).

2.4 Experiment procedure

The evaluation experiment in the study consisted of two sessions: introduction, satisfaction evaluation. At the introduction session, the purpose and method of evaluation were explained to the participants, and the basic questions in the questionnaire were answered. Then, in the evaluation session, participants evaluated their affect on closing and opening each SUV tail gate by following a predetermined order (the evaluation orders of the vehicles were randomized by the balanced Latin-square design to counterbalance the carryover effects). To do this, the evaluation order of SUVs was given to participants and a code was placed on the rear parts of SUVs (see Figure 1)

Table 1 Grouping of 42 SUVs

Size	Group1			Group2			Group3		
	Maker	Name	Code	Maker	Name	Code	Maker	Name	Code
Luxury	FORD	EXPLORER	☐	TOYOTA	SIENNA	1	TOYOTA	4-RUNNER	A
	V.W	TOUAREG	☐	HYUNDAI	VERACRUZ	2	KIA	GRAND CARNIVAL	B
	KIA	MOHAVE	☐						
Large-sized	BMW	X-5	☐	NISSAN	ROGUE	3	FORD	FREESTYLE	C
	VOLVO	XC-90	☐	TOYOTA	HIGHLANDER	4	NISSAN	MURANO	D
	AUDI	Q5	☐	HONDA	CR-V	5	HONDA	PILOT	E
	HYUNDAI	SANTAFE	☐	PEUGEOT	4007	6	CHEVROLET	EQUINOX	F
				KIA	SORENTO R	7			
Mid-sized	CHRYSLER	COMPASS	☐	V.W	TIGUAN	8	BUICK	ORLANDO	G
	TOYOTA	COROLLA VERSO	☐	FORD	KUGA	9	RENAULT	GRAND SCENIC	H
	OPEL	ZAFIRA	☐	CITROEN	PICASSO C-4	10	MAZDA	MAZDA 5	I
	NISSAN	QASHQAI	☐	RENAULT SAMSUNG	QM5	11	V.W	TOURAN	J
	KIA	SPORTAGE R	☐	HYUNDAI	TUCSON ix	12	KIA	SPORTAGE	K
							HYUNDAI	TUCSON	L
Compact	DAIHATSU	TANTO	☐	SUZUKI	SX-4	13	TOYOTA	URBAN CRUISER	M
	SUZUKI	PALETTE	☐	CITROEN	C3	14	KIA	SOUL	N

Figure 1 A picture of tail gate evaluation experiment

3 DEVELOPMENT OF SATISFACTION AND EFFORT MODEL OF SUV TAIL GATE

3.1 Effect of customer variables on satisfaction and effort of closing/opening tail gate

Analyzing the effect of each customer variable such as age, height, driving experience on the two affective variables with one-way ANOVA, all of the variables had no significant effect on them (at 95% confidence level). For example, the satisfaction and perceived effort at closing scored similarly between club members and general customers (Figure 2).

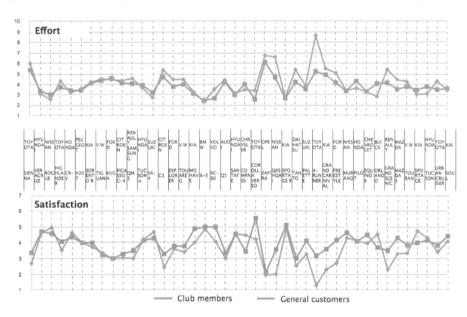

Figure 2   Satisfaction and effort scores between club members and general customers

3.2 Extracting design variables from tail gate force curve

Conducing the mixed-factor ANOVA with SUV group, tail gate (nested in the group), customer type, subject (nested in customer type), customer's satisfaction and effort were significantly different only by tail gate itself (95% of confidence level). Thus the force to be required to open and close tail gates was measured by angles with a force gauge, and a force graph was obtained for the 42 SUVs.

We compared and grouped the force graphs based on the satisfaction and effort scores using a post-hoc test (Student-Newman Test), and found a relationship between the score and some properties in the force graph. In the force graphs of closing tail

184

gates, the smaller the initial closing force, the better the customers' affect, and the initial force should be smaller as the initial close angle is larger. In addition, if there is a force in the middle of closing more than the initial force, customers' affect was worse. Last, the longer the steady state of closing force, the worse the customers' affect.

Likewise, in the force graphs of opening tail gates, the smaller the initial opening force, the better the customers' affect. In addition, the longer the range of angle of the force graph, the worse the customers' affect

From this analysis, we selected some mechanical properties affecting customers' satisfaction and effort. At closing, three variables were selected: initial force and its angle, maximum force more than the initial force and the angle range of steady state (Figure 4 [a]). At opening, two were selected: initial force and the angle range of the force graph.

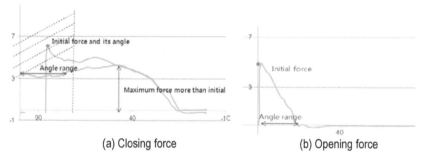

(a) Closing force (b) Opening force

Figure 3   Design variables affecting customers' satisfaction and effort

3.3 Modeling customer's satisfaction and effort

A customers' satisfaction and perceived effort model were developed with the selected properties at closing and opening tail gates. The equation of the models was developed with multiple regression modeling. All of the four models (customers's satisfaction and perceived effort at closing and opening) had the coefficient of determination (R^2) more than 0.5. The details of the equations were omitted with confidential reasons.

4 DISCUSSION AND CONCLUSIONS

As expected, customer's satisfaction and effort of closing and opening tail gates were significantly affected by the force required to operate them, while the effect the customer variables was insignificant. The mechanical designer s of tail gates have recognized the some properties of the force graph as important factors affecting IQS related to closing and opening tail gates. This study proved them and found some new variables in an empirical manner. Most of the designers also have thought that customer's satisfaction and effort of closing and opening tail gates were affected by their height and weight. This study rejected the misunderstanding empirically and passed the difficult issue of customizing tail gates cause by it.

The developed models explained customer's satisfaction and perceived effort with fairly high fitness. The total variation of customer's satisfaction and perceived effort by tail gates was much reduced by the selected properties from the force graph. This may be because the selected affective variables were closed related to the performance of closing and opening tail gates and were relatively easy to measure participants' own affective state. Contrast to simple affect, model of complicated affect showed relative low coefficient of determination like You et al. (2006).

This study modeled customers' satisfaction and perceived effort in closing and opening SUV tail gates with the properties of their force-angle curves. In the future, we will model IQS related to opening and closing tail gates with the affective variables. Although this model can not explain the complex mechanism of ISQ scoring, the part of the IQS which is determined by the effort in closing and opening will be covered with this model.

ACKNOWLEDGEMENT

This research was supported by Basic Science Research Program through the National Research Foundation of Korea(NRF) funded by the Ministry of Education, Science and Technology(2012-0003457).

REFERENCES

Demirbilek, O., and Sener, B. 2003. Product Design, Semantics and Emotional Response. *Ergonomics* 46(13/14): 1346–1360.
Helander, M. G., and Tham, M. P. 2003. Hedonomics-Affective Human Factors Design. *Ergonomics* 46(13/14): 1269–1272.
Jindo, T., and Hirasago, K., 1997, Application Studies to Car Interior of Kansei Engineering. *International Journal of Industrial Ergonomics* 19(2): 105-114.
Nagamachi, M. 1996. Introduction of Kansei Engineering. Tokyo: Japan Standard Association.
Nagamachi, M. 2002. Kansei Engineering in Consumer Product Design. *Ergonomics in Design*10(2): 5-9.
Nakada, K. 1997. Kansei Engineering Research on the Design of Construction Machinery. *International Journal of Industrial Ergonomics* 19: 129-146.

Norman, D. A. 2005. *Emotional design: why we love (or hate) everyday things*. New York: Basic Books.

Schütte, S., & Eklund, J. 2005. Design of Rocker Switches for Work-Vehicles: An Applications of Kansei Engineering. *Applied Ergonomics* 36: 557-567.

Welch, D., 2002. Luxury Gets More Affordable, *Business Week* April 22: 60-61.

White, G. L., 2001. Car Makers Battle Inferior Interiors in Hopes of Earning Buyers' Respect. *The Wall Street Journal*, December 3, http/online.wsj.com.

You, H., Ryu, T., Oh, K., Yun, M.H. & Kim, K.J. 2006. Development of Customer Satisfaction Models for Automotive Interior Materials. *International Journal of Industrial Ergonomics*, 36(4): 323-330.

Yun, M.H., Han, S.H., Ryu, T., & Yoo, K. 2001. Determination of Critical Design Variables Based on the Characteristics of Product Image/Impression: Case Study of Office Chair Design. *Proceedings of the Human Factors and Ergonomics Society 45th Annual Meeting*: 712-716.

Yun, M.H., Han, S.H., Hong, S.W., & Kim, J. 2003. Incorporating User Satisfaction into the Look-and-feel of Mobile Phone Design. *Ergonomics* 46: 1423-1440.

CHAPTER 21

Measurement of Body Pressures in Double-lane-changing Driving Tests

Youngjin Hyun[a], Nam-Cheol Lee[b] ,Kyung-Sun Lee[c], Chang-Su Kim[a], Seung-Min Mo[c], Min-Tae Seo[d], Yong-Ku Kong[d], Myung-Chul Jung[c], Inseok Lee[b] *

[a] Hyundai·Kia Motors
Hwaseong, Korea

[b] Hankyong National University
Anseong, Korea

[c] Ajou University
Suwon, Korea

[d] Sungkyunkwan University
Suwon, 440-746 Korea

* lis@hknu.ac.kr

ABSTRACT

This study was conducted to measure a driver's dynamic motions responding to the severe maneuver of the car in double-lane changing (DLC) path with varying velocities based on body pressures on the both seat pan and back. The effects of driving velocity as well as the driving stage in the DLC path on the body pressures were analyzed. An expert male driver participated in the experiment. ANOVA showed that driving velocity and driving stage in the DLC path were statistically significant on the mean pressures and contact areas ($p<0.001$). This study showed that the measurement of body pressures during DLC driving test could be used as a quantitative and direct method for evaluating the effect of the performance of the car on the driver's movement.

Keywords: Double Lane Change, Body Pressure, Contact Area, Automobile Driving Test

1 INTRODUCTION

Unprepared vehicle motion in emergency situation is known to make the driver disconcerted (Edward 1987). In order to make their cars safer and more stable in various emergent driving conditions, automobile companies have been testing their newly designed cars in many driving conditions including double lane change (DLC), which is frequently used for testing the stabilizing control of a vehicle in a severe maneuver (Abe et al. 2001, ISO 1999, Smith et al. 1995). While a driver is driving the car in the standardized DLC path with various velocities, various measures, such as wheel angle, steering wheel torque, lateral acceleration, and so on, are measured to represent the dynamic performance of the vehicle.

Even though those measures give very important information on vehicle dynamics, the effect on the driver by the driving conditions should be presumed indirectly. To understand how the severe maneuver of the car driving in DLC path, it is required to measure the dynamics of the driver in some direct ways like motion analysis and body pressure measurement (Chateauroux and Wang 2010, Jurgens 1997, Na et al. 2005, Kyung et al. 2008). Motion measuring system usually has practical difficulties to be installed in a confined space. Body pressure measurement system is sensitive to movement of the driver and relatively simple to be installed in confined space of the car.

This study was conducted to measure the driver's dynamic motion responding to the severe maneuver of the car in DLC path with varying velocities based on body pressures on the both seat pan and back. The effects of driving velocity as well as the driving stage in the DLC path on the body pressures were analyzed.

2 METHODS

2.1 Subject

A male driver, who has been a professional test driver for 22 years, participated in the experiment. He was 45 years old, 173.3 cm in height, 75.2 kg in weight, and right-handed. He had no history of injuries in the last 12 months and was in a good condition of health at the day of the experiment. He gave his consent to participate in the study after being explained the experimental purpose and procedure before the experiment.

2.2 Driving velocities

The driver was required to carry out the driving test through the DLC path at 4 different velocities: 60, 80, 100 and 110 km/h. The highest driving velocity was set to secure the safety of the driver in conducting the test driving with the car.

2.3 Measurement of body pressures and steer angles

Body pressures on both the seat pan and seat back were recorded during the driving tests. Mean pressures and contact areas on the seat pan and the seat back were analyzed respectively.

Steering angle was also measured during the driving tests. It was set to 0 degree when the car was moving forward and increases or decreases as the steering wheel turned right or left.

2.4 Driving stages

The steering angle was used as the criterion of vehicle dynamics for defining the driving stage. Driving stages in the DLC path were defined as 4 sections according to the steering angle. The first section is the first left turning of the car where the steer angle is decreased from the 0 degree and then increased again to the 0 degree. In this way, the other 3 stages were defined as follows: section 2 (1st right turning), section 3 (2nd right turning) and section 4 (2nd left turning). These driving stages are shown in figures 1 and 3.

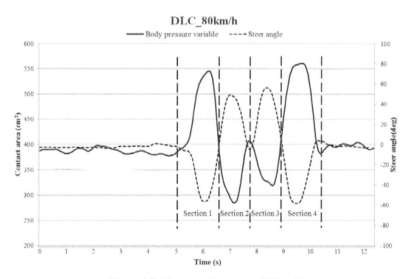

Figure 1 Driving state (example of 80km/h).

2.5 Apparatus and driving path

A midsized, left-handed and hatchback-style car, equipped with a power steering wheel and automatic transmission, was used in the experiment. It was possible to measure the steering angle of the car with 200 Hz of sampling rate.

The body pressures were measured using a body pressure measurement system (X3 PX100, XSensor Inc., Canada) with two pads on the seat pan and the backrest.

The body-pressure measuring pad, of which size is 63.5 cm × 63.5 cm, has 1296 capacitive sensors. The range of pressure measurement was from1.3 kPa to 26.7 kPa, with 20 Hz of sampling rate. The pads were attached on the seat pan and the backrest using Velcro fasteners (Figure 2).

A DLC path was designed according to ISO 1999 and the width of the car (1.79m)(Figure 3). The total length of the DLC path was 125 m.

 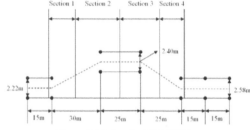

Figure 2 Experimental Set-up.   Figure 3 DLC path (ISO 1999).

2.6   Precedure

The subject was asked to adjust the position of the seat and the steering wheel so that he could keep the most comfortable posture during driving. He leaned against the seat back so that his trunk was totally supported by the seat back, and held the steering wheel at 10 and 2 o'clock positions of the steering wheel. Before starting the experiment, the subject was asked to practice driving on the DLC course. After accustomed to the DLC path, he carried out 20 different driving tests, i.e., 5 repetitions in each driving velocity, in a random order. He was allowed to have a rest at least 1 minute between two consecutive driving tests.

2.7   Data Processing and Analysis

The steer angle and body pressure data were low-pass filtered with 5 Hz of cutoff using 2nd-order and dual-pass Butterworth in order to reduce unwanted noise and phase shift (ISO 2003). Matlab software (The Mathworks Inc., Natick, Massachusetts, USA) was used in processing the signal.

In analyzing dynamic body pressures, 6 mean pressure variables and contact area variables were defined respectively: mean pressure of total seat pan (MP_TP), mean pressure of left seat pan (MP_LP), mean pressure of right seat pan (MP_RP), mean pressure of total seat back (MP_TB), mean pressure of left seat back (MP_LB), mean pressure of right seat back (MP_RB), contact area of total seat pan (CA_TP), contact area of left seat pan (CA_LP), contact area of right seat pan (CA_RP), contact area of total seat back (CA_TB), contact area of left seat back (CA_LB) and contact area of right seat back (CA_RB). These 12 variables were calculated in each driving-state section.

Analysis of variance (ANOVA) was applied in analyzing the effects of driving velocity and driving state on mean pressures and contact areas. The Tukey's Studentized ranged (HSD) tests were performed as post hoc tests following ANOVA at a significance level of 0.05 with SAS 9.1 (SAS Institute).

3 RESULTS

3.1 Mean Pressures

ANOVA showed that driving velocity had statistically significant effects on the all mean pressure variables ($p < 0.001$). The mean pressures of the seat pan decreased as the driving velocity increased except 110 km/h. The mean pressures of the seat back increased as the driving velocity increased except 100 km/h. Overall, mean MP_RP (7.3 kPa) indicated the largest value and mean MP_RB (5.0 kPa) showed the smallest value in all driving velocity conditions (Figure 4).

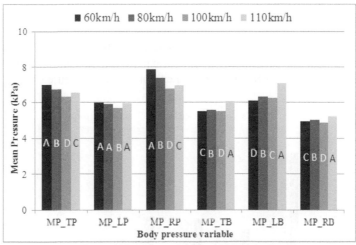

Figure 4 Mean pressures by driving velocity.

Driving stage showed statistically significant effects on all the mean pressure variables ($p < 0.001$). Mean pressures on the left seat pan (MP_LP) showed higher values during right turns (stages 2 and 3) as compared to left turns (stages 1 and 4), while those on the right seat pan (MP_RP) showed higher values during left turns (stages 1 and 4) than right turns (stages 2 and 3). The mean pressures of the seat back according to the driving stage also showed the same results with the seat pan (Figure 5).

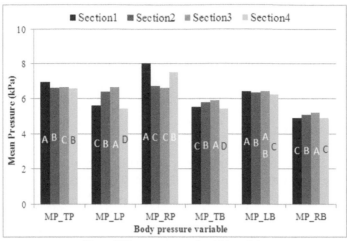

Figure 5 Mean pressures by driving state.

3.2 Contact Areas

Driving velocity showed statistically significant effects on all the contact area variables (p<0.0001). The contact areas of seat pan decreased and those of seat back increased as the driving velocity increased except 100km/h (Figure 6).

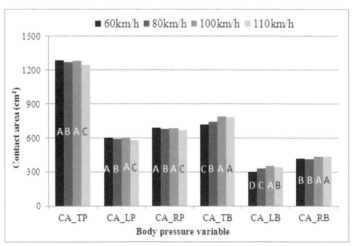

Figure 6 Contact area by driving velocity.

Driving stage showed statistically significant effects on all the contact area variables (p<0.0001). Contact areas in the left seat pan and back showed higher values during right turns (sections 2 and 3) than left turns (section 1, and 4), while those in the right seat pan and back showed higher values during left turns (section 1 and 4) than right turns (sections 2 and 3)(Figure 7).

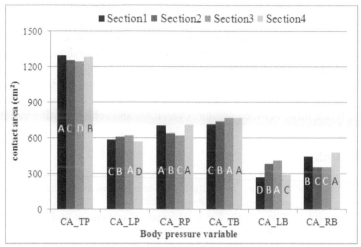

Figure 7 Contact area by driving state.

4   DISCUSSION

It was found that mean pressures and contact areas were influenced by driving velocity and driving stage in DLC driving test. The mean pressures and contact areas of the seat pan decrease as the velocity increased. However, the mean pressures and contact areas of the seat back showed opposite results. These results can be explained by the Newton's third law. To each action there is an equal and opposite reaction (Hall 2006). Therefore, mean pressures and contact areas of seat back increased because the driver should have an equal force in the opposite direction (seat back) as the driving velocity increased in the front direction. Also, the driver seems to lean against the seat back tightly to stretch out his right leg when pushing the accelerating pedal.

MP_RP and CA_RP showed higher values than MP_LP and CA_RP. This result is related to the right leg's role of controlling accelerating pedal. Grujicic et al. (2010) reported that right thigh muscle is needed to exert more force to maintain a consistent driving velocity. Kyung and Nussbaum (2008) and Hanson et al. (2006) reported that pressures in the left side were higher in order to keep balance with the right leg. The difference of this study from previous studies seems to be caused by the different driving conditions such as real driving, DLC, and driving velocities. The previous studies were based on driving in simulators.

The mean pressures and contact areas of the left part of the seat (MP_LP, MP_LB, CA_LP, and CA_LB) showed higher values during right turns than left turns, while those of the right part of the seat (MP_RP, MP_RB, CA_RP, and CA_RB) showed higher values during left turns than right turns. These results are showing how the driver moves according to the vehicle motion. Due to the centrifugal force, the pressures in the opposite side increased during left or right turns. The change of the pressure values and contact areas could be used in explaining stabilizing performance of the car in DLC driving condition.

In conclusion, this study investigated the body pressures and contact areas of a car driver in double lane changing driving tests using body pressures measurement system. This study showed that the measurement of body pressures during DLC driving test could be used as a quantitative and direct method for evaluating the effect of the performance of the car on the driver's movement.

The limitation of this study is that the vehicle motion variables were not analyzed with pressure. Therefore, the future research would be necessary to clearly identify the relationship between pressures variables and vehicle motion variables.

REFERENCES

Abe, M., Y. Kano, and K. Suzuki, et al. 2001. Side-slip control to stabilize vehicle lateral motion by direct yaw moment. *JSAE Review* 22: 413-419.

Chateauroux, E. and X. Wang. 2010. Car egress analysis of younger and older drivers for motion simulation. *Applied Ergonomics* 42: 169-177.

Edwards, M.L. and S. Malone. 1987. Driver crash avoidance behavior. National highway traffic safety administration, Final Report, DOT HS 807 112.

Grujicic., M. B. Pandurangan, and X. Xie, et al. 2010. Musculoskeletal computational analysis of the influence of car-seat design/adjustments on long-distance driving fatigue. *International Journal of Industrial Ergonomics* 40: 345-355.

Gyi, D, and M. Porter. 1999. Interface pressures and the prediction of car seat discomfort. *Applied Ergonomics* 30: 99-107.

Hall, S. J. 2006. Basic Biomechanics, (5th Ed.), The McGraw-Hill., USA.

Hanson, L. M., L. Sperling, and R. Akselsson. 2006. Preferred car driving posture using 3-D information. *International Journal of Vehicle Design* 42: 154-169.

Hegazy, S., H. Rahnejat, and K. Hussain. 2010. Multi-body dynamics in full-vehicle handling analysis under transient manoeuvre. *International Journal of Vehicle Mechanics and Mobility* 34: 1-24.

International Standards Organization. 2011. International Standard ISO 7401: Road vehicles – Lateral transient response test methods – Open-loop test methods, ISO 7401:2011.

International Standards Organization. 1999. International Standard ISO 3888-1: Passenger cars – Test track for a severe lane-change maneuver – part 1: double lane-change, ISO 3888-1:1999.

Jeon, K., H. Hwang, and S, Choi, et al. 2012. Development of an electric active rollcontrol (ARC) algorithm for a SUV. *International Journal of Automotive Technology* 13: 247-253.

Jurgens, H-W. 1997. Seat pressures distribution. *Collegium Antropologicum* 21: 359-366.

Kolich, M, and SM. Taboun. 2004. Ergonomics modeling and evaluation of automobile seat comfort. *Ergonomics* 47: 841-863.

Kyung, GH., M. A. Nussbaum, and K. Babski-Reeves. 2008. Driver sitting comfort and discomfort (part I): Use of subjective ratings in discriminating car seats and correspondence among ratings. *International Journal of Industrial Ergonomics* 38: 516-525.

Na, Sh., Sh. Lim, and H-S, Choi, et al. 2005. Evaluation of driver's discomfort and postural change using dynamic body pressures distribution. *International Journal of Industrial Ergonomics* 35: 1085-1096.

Pick, A. J. and D. J. Cole. 2007. Driver steering and muscle activity during a lane-change manoeuvre. *Vehicle Systems Dynamics* 45: 781-805.

Pick, A. J. and D. J. Cole. 2008. A mathematical model of driver steering control including neuromuscular dynamics. *Journal of Dynamic Systems, Measurement and Control* 130: art no.031004.

Smith, D.E., J. M, Starkey, and R. E. Benton. 1995. Nonlinear-gain-optimized controller development and evaluation for automated emergency vehicle steering, *Proceedings of the American Control Conference,* Seattle, Washington.

Wylie, C. D. and R. R. Mackie. 1988. *Stress and sonar operator performance: enhancing target detection performance by means of signal injection and feedback,* Goleta, CA: Essex Corporation, Human Factors Research Division.

Wylie, C. D., T. Shultz, and J. C. Miller. 1996. *Commercial motor vehicle driver fatigue and alertness study.* Technical Summary (FHWA-MC-97-001), Federal Highway Administration.

CHAPTER 22

Correlation between Muscle Contraction and Vehicle Dynamics in a Real Driving

Seung-Min Mo[a], Youngjin Hyun[b], Chang-Su Kim[b], Dae-Min Kim[c],

Hyun-Sung Kang[c], Yong-Ku Kong[c], Inseok Lee[d], Myung-Chul Jung[a]*

[a] Ajou University
Suwon, Korea

[b] Hyundai·Kia Motors
Hwaseong, Korea

[c] Sungkyunkwan University
Suwon, Korea

[d] Hankyong National University
Anseong, 456-749, Korea

* mcjung@ajou.ac.kr

ABSTRACT

The aim of this study is to evaluate correlation between electromyography (EMG) activities and vehicle dynamics in a real driving condition. One male driver performed a double lane change test at 80 km/h velocity. As dependent variables, EMG signals from *anterior temporalis, sternocleidomastoid, upper trapezius, anterior deltoid, pectoralis major sternal head, serratus anterior, biceps brachii, triceps brachii long head, flexor carpi radialis*, and *extensor carpi radialis* and vehicle dynamics of steering wheel angle, steering wheel torque, and lateral acceleration were collected in this study. The results of correlation analyses demonstrated that *sternocleidomastoid* ($r = 0.65$) and *biceps brachii* ($r = 0.72$) had a significantly high positive correlation with steering wheel angle. A significantly high negative correlation was found between *anterior deltoid* ($r = -0.80$) and steering wheel torque, whereas *triceps brachii long head* ($r \approx -0.78$) showed a significant high negative correlation with all vehicle dynamics variables. Based on this study, driver's muscle activities reflect vehicle dynamics so that muscle activity could be considered as a reasonable physiological response for vehicle dynamics in real driving conditions.

Keywords: electromyography, vehicle dynamics, correlation, double lane change

1 INTRODUCTION

Generally, vehicle performances were evaluated by vehicle motion, stability, and response time. These data could be used as a fundamental data for research and development of vehicles. Vehicle dynamics has been measured by using devices such as steering wheel robot, load cell, and gyroscope censor system (Babala et al. 2002, Ryu and Gerdes 2004). These devices can be obtained steering wheel angle, steering wheel torque, and lateral acceleration (Tseng et al. 1999).

It is important to consider safety, comfort, and convenience of a driver as well as mechanical improvement of a vehicle. Previous studies have been conducted using electromyography (EMG) during a driving condition. Pick and Cole (2006) measured *deltoid, pectoralis major, biceps brachii*, and *triceps brachii* involved in generating steering wheel torque during simulated double lane change conditions. They showed with multiple regression analyses that *anterior* and *middle deltoid, pectoralis major clavicular portion*, and *triceps brachii long head* were significantly high correlations with steering wheel torque around 0.9. Balasubramanian and Adalarasu (2007) analyzed muscle fatigue for *middle deltoid* and *upper trapezius* of professional and non-professional drivers in a 15-minute simulated driving condition. They also found that there was significant difference between professional and non-professional drivers during last one minute. These previous studies showed there was relationship between EMG activity and steering wheel control. Even though facial or neck muscles are irrelevant to steering wheel control, EMG activities of *anterior temporalis* and *sternocleidomastoid* significantly increased as lateral acceleration increased (Kuramori et al. 2004).

However, most aforementioned studies used a driving simulator so that their results may be different in a real driving condition. Therefore, the aim of this study is to evaluate correlation between EMG activity and vehicle dynamics in a real driving condition.

2 METHODS

2.1 Participant

One professional male driver with 22 years of driving experience volunteered to participate in this experiment. His age, height and weight were 45 years, 173.3 cm, and 75.2 kg, respectively.

2.2   Apparatus

A hatchback style vehicle equipped with a power steering wheel and an automatic transmission was used in this study. The vehicle dynamics of steering wheel angle, steering wheel torque, and lateral acceleration were measured by using a DEWE-5000-PM (Dewetron, Austria) with sampling rate of 200 Hz. Steering wheel angle was an angular displacement of steering wheel measured from the straight-ahead position. Steering wheel torque was applied to steering wheel about its rotation axis. Lateral acceleration was defined as the component of vector acceleration in the lateral axis of vehicle (Gillespie 1992). Raw EMG signals were amplified with a gain of 500, a common mode rejection ratio of 100, and sampling rate of 1500 Hz with a filtering bandwidth of 10-500 Hz using a TeleMyo 2400T DTS (NORAXON, USA).

2.3   Procedure

The participant performed a double lane change (DLC) test in five repetitions at 80 km/h velocity. DLC path was followed by ISO 3888-1 track of double lane change recommendation (ISO 1999) in Figure 1. The DLC path could be more effective to identify vehicle motion (Abe et al. 2001).

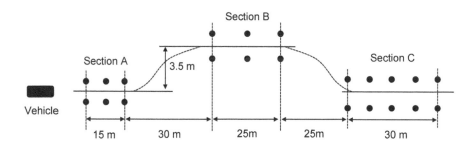

Figure 1 Double lane change path. Lane width in section A: 2.22 m, in section B: 2.40 m, and in section C: 2.58 m where vehicle width is 1.79 m.

2.4   Signal Processing

EMG and vehicle dynamics signals were extracted from before 0.5 seconds of initial maneuver to after 0.5 seconds of final maneuver. The EMG signals were processed with full wave rectification and then, second order Butterworth dual pass filters with a cutoff frequency of 5 Hz (Winter 2005). The vehicle dynamics variables were also filtered by second order dual pass Butterworth low pass filters with cutoff 5 Hz. The signal processing was performed off-line using Matlab software (Mathworks Inc., USA).

2.5   Statistics

This study used muscle and vehicle dynamics as independent variables and Pearson correlation coefficient between EMG and vehicle dynamics as a dependent variable. EMG signals were recorded from 10 muscles of *anterior temporalis* (AT), *sternocleidomastoid* (ST), *upper trapezius* (UT), *anterior deltoid* (AD), *pectoralis major sternal head* (PMS), *serratus anterior* (SA), *biceps brachii* (BB), *triceps brachii long head* (Tlong), *flexor carpi radialis* (FCR), and *extensor carpi radialis* (ECR) in Figure 2. The vehicle dynamics were steering wheel angle, steering wheel torque, and lateral acceleration. SAS software (9.1, SAS Institute) was used for statistical analysis with a significance level of 0.05. This study also considered high correlation when a coefficient was larger than 0.6 (Hopkins 2002).

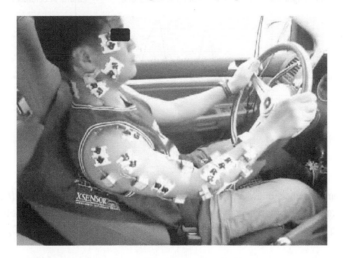

Figure 2 Location of 10 muscles using surface electrodes on the right side of a driver.

3   RESULTS

As shown in Figure 3(a), EMG activities were analyzed for left and right turns during DLC. Although ST, neck muscle, has lower activity than the other muscles, this muscle was activated for right turn. AD, shoulder muscle, was activated for left turn. As an agonist of elbow flexion, BB was activated for right turn, whereas Tlong was activated for left turn as an antagonist. There were no consistent patterns in the forearm muscles of FCR and ECR. The results of vehicle dynamics showed that steering wheel angle had similar patterns to those of steering wheel torque and lateral acceleration in Figure 3-(b).

200

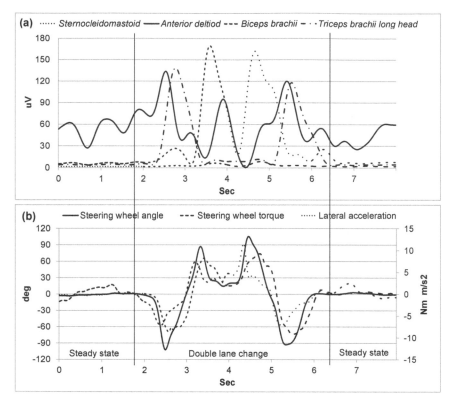

Figure 3 Example signals of EMG (a) for *sternocleidomastoid* (ST), *anterior deltoid* (AD), *biceps brachii* (BB), and *triceps brachii long head* (Tlong) and vehicle dynamics (b) for steering wheel angle, steering wheel torque, and lateral acceleration during double lane change (DLC).

Table 1 shows mean correlation coefficients between EMG signals and vehicle dynamics. The results demonstrated that ST (0.65) and BB (0.72) had a significantly high positive correlation with steering wheel angle, whereas Tlong (-0.82) showed a significantly high negative correlation with steering wheel angle. A significantly high positive correlation was also found between BB (0.62) and steering wheel torque, whereas there was significantly high negative correlation in AD (-0.80) and Tlong (-0.72) with steering wheel torque. Also, there were significantly high positive correlation between ST (0.66) and BB (0.73) and lateral acceleration, whereas Tlong (-0.79) showed a significantly high negative correlation with lateral acceleration. AT, UT, PMS, FCR, and ECR had a significant correlation with vehicle dynamics, but correlation coefficients were relatively low. Especially, AD was analyzed significantly higher correlation coefficient with steering wheel torque than steering wheel angle and lateral acceleration. Tlong had a significantly high negative correlation with all vehicle dynamics above -0.7.

Table 1 Mean correlation coefficient (standard deviation) between EMG signals and vehicle dynamics.

Body part	Muscle	Steering wheel angle	Steering wheel torque	Lateral acceleration
Face	AT	0.29 (0.23)	0.29 (0.16)	0.26 (0.23)
Neck	ST	0.65 (0.11)	0.55 (0.05)	0.66 (0.12)
Shoulder	UT	0.05 (0.18)	-0.26 (0.17)	0.19 (0.15)
	AD	-0.54 (0.11)	-0.80 (0.05)	-0.37 (0.12)
	PMS	0.23 (0.19)	-0.05 (0.13)	0.37 (0.18)
	SA	0.57 (0.07)	0.50 (0.05)	0.57 (0.06)
Upper arm	BB	0.72 (0.05)	0.62 (0.05)	0.73 (0.06)
	Tlong	-0.82 (0.03)	-0.72 (0.08)	-0.79 (0.04)
Forearm	FCR	0.39 (0.09)	0.26 (0.12)	0.47 (0.06)
	ECR	0.35 (0.16)	0.44 (0.14)	0.29 (0.15)

4   DISCUSSION

This study evaluated correlation between EMG activity and vehicle dynamics during a double lane change condition, and revealed that there were significant correlations between muscle activity and vehicle dynamics. Generally, ST, AD, BB, and Tlong were highly reflected with vehicle dynamics.

As shown in Figure 3(a), ST at the neck was not directly associated with steering wheel control so that it had lower EMG activity than the other shoulder and upper arm muscles. However, this muscle showed a significantly high positive correlation with steering wheel angle because the neck rotated before steering wheel maneuver. Doshi and Trivedi (2009) found with a relevance vector machine model that a head motion could be used to predict driver's intention of steering wheel maneuver while driving.

AD had a significantly high correlation for left turn but low correlation for right turn with steering wheel torque because the right shoulder needed to be flexed for left turn (Yu et al. 2011). It could be expected that AD at the left shoulder, which was not measured this time, was used instead during right turn.

This study showed that driver's muscle activities reflected the steering wheel angle, steering wheel torque, and lateral acceleration so that muscle activity might be considered as a reasonable physiological response for vehicle dynamics in driving conditions. The results of this study could be helpful for evaluating of vehicle performance considering with driver's muscle activity as well as vehicle dynamics. Further study is needed to evaluate and compare between muscle activity and vehicle dynamics during driving condition with various vehicle velocities.

REFERENCES

Abe, M., Y. Kano, and K. Suzuki, et al. 2001. Side-slip control to stabilize vehicle lateral motion by direct yaw moment. *JSAE Review* 22: 413-419.

Babala, M., G. Kempen, and P. Zatyko. 2002. Trade-offs for vehicle stability control sensor set. *Society of Automotive Engineers* 34: 2002-01-1587.

Balasubramanian, V. and K. Adalarasu. 2007. EMG-based analysis of change in muscle activity during simulated driving. *Journal of Bodywork and Movement Therapies* 11: 151-158.

Doshi, A. and M. M. Trivedi. 2009. On the roles of eye gaze and head dynamics in predicting driver's intent to change lanes. *IEEE Transactions on Intelligent Trasportation Systems* 10: art. No. 5173535: 453-462.

Gillespie, T. D. 1992. *Fundamental of Vehicle Dynamics*. Warrendale, PA: SAE.

International Standards Organization. 1999. International Standard ISO 3888-1: Passenger cars – Test track for a severe lane-change maneuver – part 1: double lane-change, ISO 3888-1: 1999.

Hopkins, W. G. 2002. *A scale of magnitudes for effect statistics*. Available from: http://sportsci.org/resource/stats/inde.html (Accessed 02.18.2012).

Kuramori, A., N. Koguchi, K. Ishikawa, and M. Kamijo, et al. 2004. Research on evaluation method of vehicle drivability using EMG. *JSAE Annual Congress* 85-04: 19-22.

Pick, A. J. and D. J. Cole. 2006. Measurement of driver steering torque using electromyography. *ASME Journal of Dynamic Systems, Measurement and Control* 128: 960-698.

Pick, A. J. and D. J. Cole. 2008. A mathematical model of driver steering control including neuromuscular dynamics. *Journal of Dynamic Systems, Measurement and Control, Transactions of the ASME* 130: art. No. 031004.

Ryu, J. and J. C. Gerdes. 2004. Integrating inertial sensors with global positioning system (GPS) for vehicle dynamics control. *Journal of Dynamic Systems, Measurement and Control, Transactions of the ASME* 126: 243-254.

Tseng, H. E., B. Ashrafi, and D. Madau, et al. 1999. The development of vehicle stability control at Ford. *IEEE/ASME Transactions on Mechatronics* 4: 223-234.

Winter, D. A. 2005. *Biomechanics and motor control of human movement*. 3rd ed. Newjersey: Jonh Wiley and Sons.

Yu, J., D. C. Ackland, and M. G. Pandy. 2011. Shoulder muscle function depends on elbow joint position: An illustration of dynamic coupling in the upper limb. Journal of Biomechanics 44: 1859-1868.

CHAPTER **23**

Eye-Tracking based Analysis of the Driver's Field of View (FOV) in Real Driving Environment

Sang Min Ko[a], Sun Jung Lee[a], Youngsuk Han[a], Eunae Cho[a], Hoontae Kim[b], Yong Gu Ji[a]

[a]Information and Industrial Engineering
Yonsei University, Seoul, Korea

[b]Industrial and Management Engineering
Daejin University, Pocheon, Korea

ABSTRACT

In recent years In-Vehicle Information System (IVIS) related devices including navigation increased and provided too much information to the driver. This phenomenon caused 'Information Overload' to the driver. To deal with the problems caused by information overload, researchers are studying information expression method and Human-Machine Interface (HMI) operating method of automobile. However, the information provided inside of the automobile has some side effects because each device has different interface. Head-Up Display (HUD) system is suggested as a substitute method in this problematic situation which caused from individual interface. HUD system provides driving related information and variety multi-media device information on the driver's front windshield. The purpose of this method is to minimize the distracting the driver's attention so the driver could achieve the same information faster than before. Not only technical part but also studies related to the driver's Field of View (FOV) takes important part in developing and designing HUD system. In this study, we used eye-tracking method to track the driver's eye direction in actual driving environment. The purpose of this study is to analyze the FOV of the driver by using the result from tracking the driver's eye movement. Thirteen experienced drivers participated in this experiment. The participants whom were in their 20s to 50s used the eye-tracking device and provided information on their habitual driving properties and eye movement in actual driving environment. Through this eye-tracking experiment, we collected some images and analyzed FOV of the participated drivers.

Keywords: Field of View, FOV, Eye-Tracking, Head-Up Display, HUD

1   INTRODUCTION

In the past, drivers only relied on their vision when they judge their driving situation. They operated steering wheel and pedal according to their judgment. Information appliances such as navigation offered more comfortable driving environment to the drivers by providing useful information (Satoshi, Kenichi, Yonosuke, & Takayuki, 2010). Recent automobile provides with integrated information through navigation, multimedia device and control systems for Heating Ventilating and Air Conditioning (HVAC). Furthermore, rapid development of IT technology allowed interworking between smart devices such as smartphone and automobile. However, too much information caused information overload. Researchers are studying on technologies to decrease the drivers' workload and to increase the effectiveness of the drivers considering their levels of situation awareness. Researchers are also actively working on studying Human Machine Interface (HMI). They are studying on solving the problem of the driver's information overload on expressing automobile information and operating automobile in actual driving environment (Lim & Jang, 2011).

As variety high technologies which could support safe driving such as Lane Departure Warning System (LDWS), Tire Pressure Monitoring System (TPMS), and Front Rear Monitoring System (FRMS) applied to automobile, the amount of information provided to the drivers noticeably increased. Recent automobile have 'cluster' and 'center fascia' in its limited space to provide variety driving information to the drivers. However, the information provided inside of the automobile has some side effects because each device has different interface and therefore distracts the driver's attention. For these problems of limited space and distracting the driver's attention, the necessity of practical use of automobile windshield increases. The driver's cognitive overload increases because of limited space and too much information. Therefore Head-Up Display (HUD) system using the automobile windshield is becoming a substitute idea for decreasing the driver's cognitive overload. The HUD technology was applied to aviation field from early 1990s. The technology was then applied to automobile field and made possible of expressing icons, graphics and texts on the automobile windshield. The HUD system has some of its advantages. It provides faster information to the driver and makes the driver easier to make decisions while minimizing the distraction of the driver (Kim, Cho, & Park, 2008).

The purpose of this study is to check on the FOV of the driver on designing and applying the HUD system. We conducted analysis on the driver's eye movement. In this analysis, we used eye-tracking technique and based on the existing studies related to the driver's FOV.

2   RESEARCH ON THE DRIVER'S FIELD OF VIEW

Many studies related to the driver's FOV are based on the study of Andries Sanders in 1963. The main point of his research is that the efficiency of information processing as a function of the visual angle of signals presented declines in stepwise fashion. According to Sanders' studies, the FOV can be divided into three categories. First, a stationary field for which selection required no overt change. Second, an eye field for which only eye movements was needed. Third, a head field in which both head and eyes moved. Within the central 30° this was most often done without any overt change (Sanders, 1963).

In general studies, the reference point of measuring the driver's FOV is 'Eye Point'. In SAE J1050 which is related to the driver FOV of SAE International defines the 'Eye Point' as 'Point representing the location of the eye and from which sight lines may originate. The left and right eye points are 65.0mm apart'. This standard also says 'Eyes may rotate about the eye points (E POINTS) a maximum of 30 degrees left and right, 45 degrees up and 65 degrees downs', and 'The eye may rotate easily 15 degrees lift, 15 degrees right, 15 degrees up, and 15 degrees down from straight ahead' (SAE-J1050, 2009).

Ahn et al. studied on the driver's FOV. They based on the studied of Sanders' and the studies related to the general range of the driver's FOV in driving. Researchers conducted eye-tracking experiment to evaluate the driver's FOV. Regular lanes and alleys were used in this experiment and the twenty five participants were participated. The results of eye-tracking Visual Percentile were divided into four. The results were 62.69% of Effective Visual Field, 14.85% of Optimal Visual Field, 16.80% of Inducible Visual Field and 5.66% of Assistant Visual Field (Ahn et al., 2011).

3   HEAD-UP DISPLAY IN VEHICLE

It is a fact that the most effective information transfer device is a display device. Because it is visual sense which is used the most when driving an automobile device. Recent Hyundai automobiles provide independent display devices such as navigation, multimedia devices. These kinds of display devices provide subordinate information rather than main information. The essential information such as speed, RPM and fuel quantity gage is mostly provided by cluster.

At the present time, most of devices which are used for information transfer are Head-Down Display (HDD). For the driver to get information from the HDD, the driver's eye has to be away from the lane for short time. This property of HDD can seriously effect according to situation. As the information provided to the driver increases in limited space and therefore the cognitive overload increases, the HUD system is becoming a substitute method to replace the problems of HDD system (Liu & Wen, 2004). The HUD system makes it possible to project information directly into the driver's visual field. This principle is based on optical rules. An image is projected onto a glass window and is partially reflected. The reflected fraction is perceived by the observer as a virtual image with the distance of the image source (Ablassmeier, Poitschke, Wallhoff, Bengler, & Rigoll, 2007). The HUD system has an advantage. It minimizes the driver's visual distraction in driving and makes the driver to get faster information because it provides driving related information and multimedia device information by windshield. The HUD system can minimize the risks according to the driver's eye movement because it provides information by windshield in the range of the easy eye rotation. This makes the driver to continuously focus on the front and therefore it decreases the driver distraction (Poitschke et al., 2008; Prinzel & Risser, 2004).

The HUD technology which is used in the field of automobile is not so much different than the technology which is used in the field of aviation. In the case of automobile manufacture companies, they are actively studying the HUD system to apply it to their automobiles. For the Motors companies such as GM, BMW, TOYOTA, CITROëN, PEUGEOT are selling the HUD technology applied automobiles as a part of their premium strategy. For BMW, they set the HUD system development strategy for the next generation to develop the HUD technology and they already applied the HUD system to their automobile from Series 5. For GM, they studied the HUD system with NHTSA, Michigan University, Delphi and etc. from 1999. They applied the HUD system to their resent target automobile (Kim et al., 2008). Although the current HUD system has high degrees of completion, its ability is still remaining at the level of providing driving related basic information or guiding directions by turn-by-turn method. It is expected that the development of HUD related technology could bring insertion of a transparent display system to the windshield to express images on particular region or realizing augmented reality on the windshield.

4   EXPERIMENT

In this study, we collected image data on the driver's eye movement using the Eye-Tracker. The experiment design on the participants and devices are as follow.

PARTICIPANTS

The thirteen participants (7 males and 6 females) were participated in this experiment. The participants were composed of people who are in their age of 21 year-old to 61 year-old. Their average age was 40.8 year-old and average height was 171cm. Their jobs were variety such as undergraduate student, graduate student, employee, house wife and etc. and their average driving experience was 14.4 years. The participants' information according to their age is shown in the table below.

Table 1 The information of participants

Class	Participants		
A	4 persons (2 men, 2 women)		
	A: 27.5 years	H: 174.3 cm	DE: 5.5 years
B	2 persons (1 men, 1 women)		
	A: 34 years	H: 169 cm	DE: 14 years
C	3 persons (2 men, 1 women)		
	A: 41.7 years	H: 171.7 cm	DE: 20.3 years
D	4 persons (2 men, 2 women)		
	A: 56.8 years	H: 167 cm	DE: 19 years

Note: A – age, H – Height, DE – Driving Experience

APPARATUS

In this study, video based image-processing eye-tracker was used. The method which was used in this study is a method of analyzing eye direction by filming the participant's eye using a camera. A subminiature camera films the participant's eye and range of vision. Then the camera measures the participant's eye direction through the location of the participant's eye and location of reflected light by analyzing the images. In this study, we used an eye-tracker from Arrington Research Company. This device is composed of desktop computer, scene camera and analyzing software. The overall composition of the device is as Figure 1 (a). Figure 1 (b) shows the image of the participant wearing Scene Camera in actual driving environment.

(a)   (b)

Figure 1 ArringtonResearch Scene Camera Eye-Tracking System. (a) Desktop computer and Scene Camera (From: ArringtonResearch homepage, www.arrintonresearch.com) and (b) The image of the participant wearing Eye-Tracking Scene Camera in actual driving environment.

There are two methods to track eye movement. One method is binocular method which tracks both eyes movement. The other method is Monocular method which tracks only one eye movement. The method used in this study is Monocular method which tracks only right side eye movement of the participant. Tracking speed was 30Hz or 60Hz, Accuracy was $0.25°\sim1.0°$ visual arc and Resolution was $0.15°$.

EXPERIMENT SETUP

We installed scene camera eye-tracking system to the participant's automobile to conduct this experiment in actual driving environment. After driving the system, we practiced View Point PC-60 program to set Head Mounted Scene Camera and Eye camera. We adjusted Head Mounted Scene Camera on the participant's head to make it easier for the experiment. Individual adjustment was needed according to the participants because each participant had difference in their eye location and angle. We conducted Calibration to get more accurate results after adjusting Eye camera to track the participant's eye pupil more deliberately. Because the calibration point changes according to the participant's automobile and properties, Calibration was practiced before each experiment was conducted. We also conducted a pilot test in actual driving environment to check the accuracy of our eye-tracking device. We set the Calibration point at 6 in the experiment and we could find that a severe error was occurred in this experiment. Therefore we set the Calibration point at 9 to conduct the experiment. The image of actual driving environment and setting of software is as shown below in Figure 2.

(a)   (b)

Figure 2 The Composition of Eye-Tracking Software and Image of Eye-Tracking in Actual Driving Environment (a) The Image of Tracking the Participant's Pupil using Eye-Tracking Software (View Point PC-60) and Setting Calibration Point (9-point) (b) The Image of Tracking the Participant's Eye Movement who is wearing Eye-Tracking Device in Actual Driving Environment.

DRIVING ENVIRONMENTS

In this experiment, we used the participant's own automobile and made the participant to wear our eye-tracking device. We provided the actual lane driving environment to the participant. Then we collected the first and the second image data according to the participant's eye movement. To provide equivalent driving environment to all the participants, we choose one of the three lanes which Road Traffic Authority uses for issuing a driver's license in their Driver's License Examination Office. The total length of the lane was 5.5Km and the experiment was composed of 5 left turns, 2 right turns, 2 U-turns. The average time of filming each participant was 15.4 minutes.

5   RESULTS

It is possible to use eye-tracking Software to track the participant's eye movement in fixed environment such as analyzing the participant's eye movement on a web site. However, our experiment environment was conducted in a dynamic environment. Therefore it was impossible to analyze the participant using our eye-tracking Software in a moving environment. There was another limitation because we performed analysis based on objective figures using the participant's automobile. Because it was impossible for our eye-tracking software package to analyze overall moving images, we used excel program to perform a data analysis sheet on this experiment.

To minimize the observational error between three researchers, we compared the analyzed results after analyzing each participant. When there was a severe difference between these three different results, three researchers were asked to assemble in a same place to discuss and conclude about their results. In the process of analyzing, there was a severe data error of the participant 6 (a male in his 40s) and the participant 11 (a female in her 30s). Therefore we eliminated the two participants from our experiment. The result of the eleven participants' moving image analysis based on the existing studies of driver FOV is shown in Figure 3.

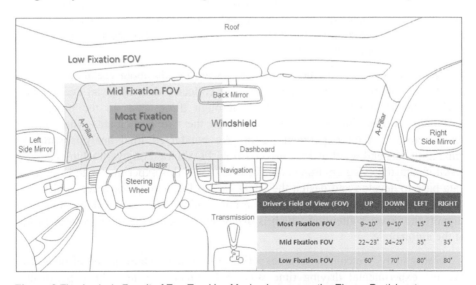

Driver's Field of View (FOV)	UP	DOWN	LEFT	RIGHT
Most Fixation FOV	9~10°	9~10°	15°	15°
Mid Fixation FOV	22~23°	24~25°	35°	35°
Low Fixation FOV	60°	70°	80°	80°

Figure 3 The Analysis Result of Eye-Tracking Moving Images on the Eleven Participants

We studied existing researches on FOV and compared them to our analysis results of eye-tracking moving images and found some differences between the studies and our results. According to SAE J1050 standard 'eyes may rotate about the eye points (E POINTS) a maximum of 30 degrees left and right, 45 degrees up and 65 degrees down' (SAE-J1050, 2009). However, what we have found in this study is that the drivers may rotate their eyes about the eye points (E POINTS) a maximum of 9-10 degrees up and down, 35 degrees left and right in driving their automobile.

6 CONCLUSION AND FUTURE WORK

In this study, we analyzed the driver's eye movement and the driver's FOV in actual driving environment. We used eye-tracking method to conduct this experiment and the driver's FOV analysis was performed through the experiment on the driver's eye movement. The eye-tracking device collects data by using eye-tracking method based on image processing video. This method is analyzing the participant's eye direction by filming the participant's eye using a camera. The subminiature camera films the participant's eye and sight view, analyzes the eye-tracking moving Images, understands the location of the participant's eye or reflected light and measures the participant's eye direction. The eye-Tracker from Arrington Research was used in this study. This device tracks and films the participant's eye pupil in real time and records the participant's pupil movement and overall moving images. Using this eye-tracking device, we provided an actual driving environment to the participants and collected moving images of the participant's driving style and the eye movement. The thirteen driving experienced participants were participated in this experiment (7 male and 6 female). They were in their 20s to 50s and we conducted first and second experiment. In the analysis of collected eye-tracking moving images, we analyzed Area of Interest (AOI) in actual driving environment and draw heatmap.

The existing studies on the driver's FOV were mostly about visual angle based on Sanders' studies (1963). In this study, we collected data using an eye-tracking device and analyzed the driver's FOV through those data. In the process of analyzing the experiment data, we eliminated the two participants who showed severe error on their experiment result. We conducted final analysis with the remaining data and redefined the driver's front FOV based on eye-tracking analysis result in actual driving environment. The result from this experiment can be used as data base to develop and design the HUD system for helping the driver's effective driving. This result also might be used in information expression field of studies to support the driver's attention which is distracted by operating other devices inside of the automobile.

In this study, we conducted an experiment in a rather short period of time. The average experimental driving time was 15.4 minutes because the experimental driving was conducted in local places. There existed a limitation in this experiment that it was impossible to get data of moving images through the eye-tracking software package. We will conduct verification on this rough FOV result by using such programs as CATIA, CAD and etc. Additional analysis will be needed on the driver's FOV. Numerous participants and various kinds of driving environments such as high way driving, night driving and more will be needed in this analysis.

REFERENCES

Ablassmeier, M., Poitschke, T., Wallhoff, F., Bengler, K., & Rigoll, G. (2007, 2-5 July 2007). *Eye Gaze Studies Comparing Head-Up and Head-Down Displays in Vehicles.* Paper presented at the Multimedia and Expo, 2007 IEEE International Conference on.

Ahn, S., Lee, K., Park, P., Jeon, Y., Kim, S., & Choi, J. (2011). *Visual Percentile while Driving on Each Section with an Eye-Tracker.* Paper presented at the Ergonomics Society of Korea 2011 Spring Conference and Workshop.

Kim, K. H., Cho, S. I., & Park, J. H. (2008). Application of Head-Up-Display Technology to Telematics. [Application of Head-Up-Display Technology to Telematics]. *Electronics and Telecommunications Trends, 23*(1), 153-162.

Lim, S.-H., & Jang, C. (2011). *Current Trends and Future Issues of Automotive HMI.* Paper presented at the Ergonomics Society of Korea 2011 Spring Conference and Workshop.

Liu, Y.-C., & Wen, M.-H. (2004). Comparison of head-up display (HUD) vs. head-down display (HDD): driving performance of commercial vehicle operators in Taiwan. *Int. J. Hum.-Comput. Stud., 61*(5), 679-697. doi: 10.1016/j.ijhcs.2004.06.002

Poitschke, T., Ablassmeier, M., Rigoll, G., Bardins, S., Kohlbecher, S., & Schneider, E. (2008). *Contact-analog information representation in an automotive head-up display.* Paper presented at the Proceedings of the 2008 symposium on Eye tracking research; applications, Savannah, Georgia.

Prinzel, L., & Risser, M. (2004). Head-up displays and attention capture. *NASA Technical Memorandum, 213000*, 2004.

SAE-J1050. (2009). Describing and Measuring the Driver's Field of View: SAE International.

Sanders, A. F. (1963). *The selective process in the functional visual field / [by] A. F. Sanders.* Soesterberg, Netherlands :: Institute for Perception RVO-TNO.

Satoshi, K., Kenichi, T., Yonosuke, M., & Takayuki, Y. (2010). The Current State and View of In-vehicle HMI. 自動車技術, *64*(10), 12-17.

CHAPTER 24

Effects of Age and Gender Differences on Automobile Instrument Cluster Design

[1]Sol Hee Yoon, [1]Hwan Hwangbo, [1]Jong Kyu Choi, [1]Yong Gu Ji, [2] Jae Hyeok Ryu, [2]Seung Hwan Lee, [3]Dongsoo Kim

[1]Yonsei University,
Seoul, Korea

[2]Hyundai Motor Company,
Seoul, Korea

[3]Soonsil University,
Seoul, Korea

ABSTRACT

This research aims to study the effect of age and gender on the design of automobile instrumental cluster. Information provided in vehicle has been increasing during the last years with the introduction of smart devices, LCD screen and wireless internet. Hence, automobile instrument cluster is able to provide customization, addition and subtraction of information functionality depending on the driver's preference and driving context. We have conducted an experiment on preference and performance of preselected automobile instrument cluster design factors. The experiment was divided into 3 parts: 1) response time of the different areas of the instrument cluster, 2) satisfaction on warning icons and presentation method, and 3) satisfaction and response time on verification of speedometer with respect to size and location. A total of 28 subjects participated on the experiment, which were divided into 4 groups regarding their age and gender: 1) young male (20~35 years old), 2) young female (20~35 years old), 3) middle male (40~55 years old), and 4) middle female (40~55 years old). A real driving environment was created to proceed with the experiment. The main task consisted on concentrating on the video of real driving situation while performing the secondary task. The secondary task was to answer questions related to the stimuli shown in the LCD cluster in the simulator. The overall results show that there is difference on the performance and preference with respect to the task and design factors of the instrument cluster. There are significant difference with respect to gender and age on the influence of design of automobile instrument cluster such as size, location and presentation method.

Keywords: full LCD Instrument Cluster, design differences, visual perception

1   INTRODUCTION

The rapid growth and development of technology provides the society with wide amount of benefits and comforts in their everyday living. Easy and quick access to information is considered one of the most important improvements, however; this is followed by a decrease on productivity due to information overload. Therefore, the key to this solution is to find a considerable level on the amount of information that humans are capable to process to increase performance.

Driving is one of the most common activities that the human has to perform. Hence, it combines the ability to do correctly a main task, driving, while realizing a secondary task that consists on paying attention and perceiving different kind information provided in the driving context; that is to say, sensory ability such as vision, attention, and memory (Porter & Whitton, 2002). However, the possibility of providing drivers with diverse and large amount of information can provoke serious safety issues as distraction and decrease of main task performance.

Research related with satisfaction and usability of in-vehicle information system (IVIS) it being conducted by many researchers. Most of the drivers considers notoriously comfortable and of preference access to large amount of information provided by IVIS. However, it is remaining as an issue how this information have to be presented to the drivers so as to not interfere with their main task, driving. That is to say, present the desired information without causing big increase on metal workload that can disturb the primary task of the driver.

Previous researched presented that, performance on the divided attention task was affected by age (Wood, 2002). In-vehicle task where the driver have to interact with the entertainment system, this can affect measures of driving performance such as maintaining speed, and preparedness to react to unexpected hazards (Horberry et al., 2006).

Human sensory ability decreases as a result of aging. Consequently, influences on driving performance depending on recognition or divided attention were more affected to older drivers as well as the influence of visual function (Wood, 2002). Moreover, in previous study, older drivers found heavy demands placed on visual attention and information processing while doing driving task (Baldwin, 2002). Different age group show relatively different driving performance with driving distracter in various environment, that is to say, different age group have different reaction and performance (Horberry et al., 2006)

Moreover, the existences of differences on driving performance and sensory ability within drivers of different gender and age, the concept of customization have become relatively important (Rhodes and Pivik, 2011). The introduction of smart devices, LCD screen and wireless internet provides the possibility to customize automobile information system depending on the requirement of each individual. Addition, subtraction and visual modification of information is feasible within automobile information system.

Consequently, in this research, we focus on the differences that exist between gender and age on the perception of information of instrument cluster and their performance. Accordingly, suggesting the effect that age and gender can cause on the design of a vehicle instrument cluster. Therefore, we focus mainly on the overall reaction time on the different areas of the instrument cluster, the preference of warning icon with respect to size and way of presentation, and lastly, preference and performance on speedometer with respect to size and location.

2   METHOD

2.1   PARTICIPANTS

For the experiment, a total of 28 subjects from different age group and gender participated. Subjects group were divided into four: 1) Young Male group (20~35 years old), 2) Young Female group (20~35 years old), 3) Middle Male group (40~55 years old), and 4) Middle Female group (40~55 years old).

We gather information demographic data about gender, age, driving license possession, when the driving license was obtained, real driving experience, automobile possession, and average driving hours per day.

Table 1 Demographic Distribution of participants

	Male Young (n=7)	Male Middle (n=7)	Female Young (n=7)	Female Middle (n=7)
Age Average (Std. Dev.)	27 (3.16)	51 (4.20)	25.5 (1.27)	47 (5.16)
Driving License	100%	100%	100%	100%
Driving Experience (Approx. Avg. in years)	3.3	19.2	2.9	15.1
Avg. Driving Hours per day (min)	107.8	137	92.6	102.6

2.2   APPARATUS

A driving simulator was design for the experiment. The simulator consisted on real car seat with seat control to adjust the body to a comfortable possible for the driver. A steering wheel and a predesign LCD panel that function as a cluster in the experiment. The LCD panel used as instrument cluster had always a black background to make it similar to real instrument cluster. Also, a projector was used to show video of real in road situation.

2.3   PROCEDURE AND EXPERIMENTAL DESIGN

The experiment took place in a quiet and dark room to make it similar to in car real situation. The room was divided into 2 sectors. In one sector, it was located the simulator with the projector and in the other sector the experimenters where located with the necessary computer devices to control and gather data of the experiment.

Figure 1 Image of driving simulator used in the experiment and layout

The experiment consisted on 3 main parts. The first one was conducted to gather information about instrument cluster information recognition time. For that, the previously design full LCD cluster was divided into 45 equal areas (5x9). Then, a number of two digits were presented in each of the area in random order. To reduce learning effect, it was randomly presented so as to avoid increase ability to recognize the number by training. This experiment was conducted three times for each participant. Data about the time taken for participants to recognize the number was gather depending on the 45 area. Number was shown after a sound alarm that indicated the appearance of a number in one of the area. Then, participants press a button to notice that they have recognized the image. After that, they told the experimenter the number that they have seen.

The second part was conducted to know the preferences and satisfaction of driver on instrument cluster icons. Therefore, the experiment was designed with two factors and three levels for each. Factor 1 was Size of icon, which was divided into 12x12 mm, 15x15mm, and 18x18 mm. We found that approximately 15x15mm was the average icon size for instrument cluster. Thus, a smaller and a bigger size from the standard were selected. Factor 2 was Format of presentation, which was divided into icon, icon plus notification, and icon plus command. Most of the current instrument cluster, its only provides icon. However, with the instruction on full LCD cluster, it is possible to customize the information and the way this information is presented. Icon plus notification refers to the presentation of icon followed with a text were it allows the driver know the name and meaning of the icon. That is to say, help drivers to know to what part of the car or the meaning of the icon. On the other hand, icon plus command refers to the presentation of the icon with a simple command of what the driver should do.

A total of 27(3x3x3) different images where presented in random order with 3 times repetition. The repetitions help to be more precise with the preferences of the driver with respect to the design satisfaction. Participants were ask to grade each of the image with a 7 points Likert scale, where 1 represented very low satisfaction and 7 represented very high satisfaction.

Table 2 Table of images used in to evaluate the satisfaction of instrument cluster icons.

Icon	Size of Icon (mm)	Visual representation
	Small 12 x12	Icon
	Medium 15 x 15	Icon & Notification
	Big 18 x 18	Icon & Command

Lastly, an experiment was conducted to test if the cognition and satisfaction of speed information varies depending on the location and size of the speedometer in the instrument cluster. Here, participants had to followed the same task as in experiment one. They have pay attention to the driving video and tell the experimenter the approximated speed that the speedometer showed. A sound alert was used to tell the participants that an image of a speedometer appeared on the instrument cluster. A total of 15 images were presented with three times repetition. Two main factors were taken into account for this experiment: Layout and Size. Layout consisted on the position of the speedometer that was middle, right or left. For the size of the speedometer, it was divided into 5 sizes depending on its diameter: 60mm, 75mm, 90mm, 105mm, and 120cm. After the experiment, participants were asked to grade each of the 15image from 1 to 7 (Likert scale) depending on their satisfaction, where 1 represented very low satisfaction and 7 represented very high satisfaction.

Figure 2 Example of one of the speedometer image used in Experiment 3 to gather data about the satisfaction and reaction time of drivers

The data obtained on the experiments were analyzed by using PASW Statistics 18 software, with Analysis of Variance (ANOVA) with a confidence interval of 0.05. Moreover, to obtained significant difference between groups, it was used Scheffe's method of analysis of variance.

3   RESULT

3.1   PERFORMANCE

The results from the experiment 1 showed that there exist significance difference on the overall performance between each of the group with a p-value <0.05. The Scheffe's method showed that the 4 groups where separated into three: 1) female young with the highest performance, 2) male young and female middle, and 3) male middle with the lowest performance in this experiment. Also, it was found that there exist differences between group and divided areas with a p-value < 0.001. That is to say, for each of the group, performance varied depending on the area the stimuli was shown.

Table 3 Representation of the average performance on each group

	Average reaction time & Distribution of the area									
Male Young	1.9096	1.2330	1.0639	1.0354	0.9359	1.0037	1.1842	1.7374	1.4848	0.8606 ~ 0.9140 se
	1.1574	1.0875	0.8834	1.0218	1.1188	0.9140	0.8897	1.0581	1.8253	0.9141 ~ 1.0218 se
	1.0082	0.8606	0.9907	0.9017	0.9575	0.8747	0.8979	1.0431	1.3542	1.0219 ~ 1.1842 se
	0.9476	0.9607	0.8969	0.9888	0.9721	1.0066	1.0094	0.9716	1.1004	1.1843 ~ 1.3542 se
	1.0450	1.2543	0.9960	0.9359	1.0719	1.0707	1.0484	1.0532	1.0101	> 1.3524 sec
	Average reaction time & Distribution of the area									
Male Middle	2.3738	1.5374	1.4099	1.4106	1.2737	1.4167	1.5635	1.9330	2.0824	1.2674 ~ 1.3194 sec
	1.5237	1.3569	1.5241	1.5415	1.3194	1.3591	1.3872	1.5977	2.3943	1.3195 ~ 1.6087 sec
	1.6024	1.3750	1.4080	1.3382	1.2956	1.5025	1.3002	1.5418	1.8052	1.6087 ~ 2.3943 sec
	1.4492	1.3371	1.4839	1.3633	1.4242	1.2674	1.4370	1.4140	1.4357	2.3944 ~ 3.1211 sec
	1.4414	1.6087	1.4771	1.3541	1.5293	1.5064	1.8805	1.5167	3.1211	> 3.1211 sec
	Average reaction time & Distribution of the area									
Female Young	1.2939	0.8866	0.9695	0.9856	0.8362	0.9081	0.9084	1.3359	1.9005	0.8362 ~ 0.9431 sec
	1.0096	0.8734	0.9024	0.9222	0.8962	0.9259	0.9311	0.9476	1.3791	0.9432 ~ 0.9856 sec
	0.9232	0.9120	0.9659	0.9815	0.9005	0.8797	0.8811	0.9217	0.9352	0.9857 ~ 1.0184 sec
	1.0142	0.8786	0.9052	0.9551	0.9010	0.9331	0.9139	1.0077	1.0782	1.0185 ~ 1.3791 sec
	1.1757	1.0184	1.0043	0.9518	0.9783	0.9431	0.9312	0.9288	0.9164	> 1.3791 sec
Female	**Average reaction time & Distribution of the area**									

Middle											
	1.4286	1.1527	0.9673	1.0704	1.0106	1.0795	1.1351	1.2305	2.2840		0.9589 ~ 1.0261 sec
	1.1312	1.0584	1.0476	1.0210	1.0076	1.0760	1.1232	1.0602	1.5120		1.0262 ~ 1.1527 sec
	1.0451	1.0216	1.1409	0.9589	1.0124	1.0742	1.0879	1.0481	1.0701		1.1528 ~ 1.5120 sec
	1.0351	1.0702	0.9860	1.0919	1.0615	1.0675	1.0490	1.0309	1.1164		1.5121 ~ 2.2840 sec
	1.0763	1.1061	1.0261	1.0067	0.9669	0.9746	1.3141	1.1248	1.1357		> 2.2840 sec

The different color represents the segments form by the analysis of variance with significant difference. Therefore, there exists significant difference in each of the group between the 45 areas.

3.2   SATISFACTION ON WARNING ICONS

Male young drivers showed high satisfaction on warning signs that were represented with notification and between 15mm and 18mm (average of 4.15 & 4.76). Hence, there was significant difference on satisfaction of icon representation style depending on the visual representation and size (p-value<0.001).

The same result was encounter for the female young group that showed significance difference with respect to the visual representation and size (p-value<0.05). Moreover, satisfaction increased as the icon was accompanied with an explanatory text, notification and command (average of 4.38).

For the middle male group there was no significant difference on the size on the icon; however, satisfaction increased depending on the visual representation. That is to say, satisfaction was high for warning sign with notification and ever higher for those with command (average of 5.48).

Lastly, for the middle female group, there was no significance difference between icon & notification and icon & command; however, the score for visual representation of icon alone was very low no matter the size.

Table 4 Differences on preference of warning icons visual representation and size.

	Young	Middle
Male	Icon & Command 15mm & 18 mm (4.15/4.76)	Icon & Command 18 mm (5.48)
Female	Icon & Command 15 mm (4.38)	Icon & Command 18mm (5.00)

3.3   PERFORMANCE AND SATISFACTION OF SPEEDOMETER

The overall performance of the participants was higher when the speedometer was located on the right side of the instrument cluster; however, there was no significant difference on the location of the instrument cluster. That is to say, the location of the speedometer did not influenced on the performance of the drivers. Neither the size of the speedometer had significant difference on performance. However, the size with the highest performance was of 75mm.

Female drivers showed higher performance compared with male drivers regardless of the age. Moreover, there were differences on the performance depending on the location of the speedometer. Young group performance was higher when the speedometer was located on the middle while male middle group had better performance when it was located on the left and female middle group when it was located on the right side.

The preference of the speedometer location and size varied depending on the group of the drivers. Statistic test on the analysis of variance showed that there is significant difference on the location between groups (p-value<0.05). Although all the groups had high satisfaction for speedometer located on the right side of the instrument cluster; middle group also preferred speedometer located on the center. On the other hand, young male drivers had also high satisfaction score for speedometer located on the left side.

Table 5 Differences on performance and preference of speedometer.

		Male		Female	
		Young	Middle	Young	Middle
Performance	Location	Middle	Right	Middle	Right
	Size	90mm	120mm	90mm	75mm
	Avg. reaction time (sec)	1.41	1.78	1.21	1.39
Preference	Location	Left/Right	Center/Right	Right	Center/Right
	Size	105mm	120mm	105~120mm	120mm

4   DISCUSSION

In this research, age and gender influence on the preference and performance of drivers in different aspect. This can be due to the differences on sensory ability, especially to the visual perception that each of the group has.

Table 3 shows the difference on reaction time that drivers had depending on the location that the stimuli. Hence, it is possible to see that depending on the age and gender, the visual field for the drivers on the instrument cluster while driving varies significantly. It can be justify by importance that each group gives to their own car instrument cluster or to the frequency and what type of information they are accustom to verify on the instrument cluster.

With respect to the satisfaction shown to warning icons, it is possible to see that most of the participants preferred icons that where presented with a command of what the driver should do. Consequently this can be a sign that drivers find difficult to remember or have no knowledge on all the features of the car. Therefore, they have high preference on icons presented with command. This was more notorious for female drivers whom frequently find difficulty with some of the car warning signs.

Lastly, preference and performance of speedometer of instrument cluster also varied with respect to the groups. Mostly, participants preferred speedometer located on the right side of the instrument cluster. This is the most common layout presented on car in the market. Therefore, influence on the layout that the participants where more accustom brought this result. On the other hand, young male drivers also show high satisfaction to other types of layout. Hence, desire to personalized and customize the design of the instrument cluster for this group can be justify. In other words, young male group did not consider important where the speedometer was located on the instrument cluster. With respect to the size of the speedometer, the satisfaction of the drivers increased as the size of the speedometer increased. This can be due to the importance that the drivers gave to the speedometer information in driving context. That is say, no matter the gender or age, driver's preference of speedometer was higher as the size increase regardless of their performance.

5   CONCLUSION

The increase amount of information and in vehicle information system that are used in driving context inspire researcher to study about drivers performance and relation. In this study, research on the effect of design factor on drivers performance and preferences was conducted focusing the drivers age and gender. Icon size or speedometer size was mainly affected to middle group drivers while diversity of layout or field of view influence more on younger group drivers. Also, differences on visual recognition varied between group regarding age and gender.

However, this research shows have some limitations. The experiment was conducted in a simulated environment; therefore, there should be some differences to the real driving context. Moreover, it is necessary to research on others design factors that can affect the visual recognition of information presented in automobiles as color, distance, and font.

This research can be applied as basis for automobile instrument cluster design guideline with respect to the division of main areas of the instrument cluster, way of presenting warning sign and it size, and location of speedometer and size. Moreover, this research can be utilized as reference for further HCI studies related to performance of participants in dual task paradigm.

ACKNOWLEDGEMENTS

This research was financially supported by Hyundai-Kia Motors Company in Korea u nder the aim for the Industry-Academy Cooperation research utility.

REFERENCES

Baldwin, C. L. 2002. Designing in-vehicle technologies for older drivers: application of sensory-cognitive interaction theory. *Theory Issue in Ergonomics Science*, Vol. 3, No.4: 307-329.

Harré, N., Field, J., Kirkwood, B. 1996. Gender differences and areas of common concern in the driving behaviors and attitudes of adolescents. *Journal of safety Research*, Vol. 27, Issue 3: 163-173.

Hatfield, J., Chamberlain, T. 2008. The impact of in-car displays on drivers in neighboring cars: Survey and driving simulator experiment. *Transportation Research Part F* 11, 137-146.

Horberry, T., et al. 2005. Driver distraction: The effect of concurrent in-vehicle tasks, road environment complexity and age on driving performance. *Accident Analysis and Prevention* 38: 185-191.

Noy, Y. I. 1997. Human factos in modern traffic systems. *Ergonomics*, Vol. 40, No. 10: 1016-1024.

Porter, M. M., Whitton, M. J. 2002. Assessment of Driving With the Global Positioning System and Video Technology in Young, Middle-Aged, and Older Drivers. *Journal of Gerontology: Medical Sciences*, Vol. 57, No. 9: 578-582.

Rhodes, N., Pivik, K. 2011. Age and gender differences in risky driving: The roles of positive affect and risk perception. *Accident Analysis & Prevention*, Vol. 43, Issue 3: 923-931.

Wood, J. M. 2002. Age and Visual Impairment Decrease Driving Performance as Measured on a Closed-Road Circuit. *Human Factors: The Journal of the Human Factors and Ergonomics Society*, Vol. 44, No. 3: 482-494.

CHAPTER 25

A Study on the Relationship between Pleasures and Design Attributes of Digital Appliances

Sangwoo Bahn[1], Joobong Song[2], Myung Hwan Yun[2], Chang Soo Nam[1]

[1]North Carolina State University, Raleigh, USA 27695
[2]Seoul National University, Seoul, Korea 151-743
panlot@gmail.com, shedtwin@naver.com, mhy@snu.ac.kr, csnam@ncsu.edu

ABSTRACT

As the importance of affective elements of product is becoming more emphasized, increasing number of research on affect and emotion has been conducted for the past decade. Especially, as the pleasure is known to be at the core of affective experience, researchers have proposed various typologies of the pleasure. However, these conceptual typologies and their related elements have not been investigated empirically yet. In this study, an hypothetical model of the relationships between perceived pleasure and design attributes was suggested. The proposed hypothetical model and its elements were then examined by conducting user surveys and assessment experiments on mobile phones. It was found that there are some causal relationships between physiological pleasure and design attributes of the visceral level, and psychological pleasure and design attributes of the behavioral level and reflective level, respectively. Also, a causal relationship was found between sociological and ideological pleasure and design attributes of the reflective level. It is expected that this study lays the foundation for future studies on designing affective and pleasurable product.

Keywords: affect, pleasure, mobile device, structural equation model.

1   INTRODUCTION

The increasing number of research on affect and emotion for the past decade demonstrated that the business had growing inclination toward affective characteristics on customers. Multifaceted approaches have been taken in order to apply these research findings to product development. Research on affect and emotion is referred to as affective design, affective computing or Kansei engineering, comprising of various elements from psychology, ergonomics, design, and cognitive science. These approaches bring about many advances in recognizing, expressing, modeling, communicating, and responding to emotion.

Rusell (1980) introduced a concept of 'core-affect' by combining the affect dimension with physiological arousal in a circular-two-dimensional model. According to Russell, the experience of core-affect is a single integral blend of those two dimensions, describable as a position on the circumflex structure. As its core, a mental representation of emotion is a contentful state of pleasure or displeasure (Russell and Barrett, 1999; Russell, 2003; Barrett, 2006). There are mounting empirical evidences that mental representations of emotion have pleasure or displeasure at their core. People are able to give an explicit account of pleasant and unpleasant feelings using a variety of self-rating scales (Frijda et al., 1989; Russell et al., 1989; Bradley and Lang, 1994; Scherer 1997; Barrett and Russell, 1998; Carroll et al., 1999). Scales that are explicitly built to measure discrete emotions such as fear, anger, or sadness also provide strong evidence of a common core of pleasant and unpleasant feelings (Russell, 1980; Watson and Clark, 1994; Barrett and Russell, 1998).

Researchers have proposed various typologies of pleasure largely based on theoretical arguments. For instance, Duncker (1941) proposed three types of pleasure: sensory pleasures, for which the immediate object of pleasure is the nature of a sensation (e.g., the flavor of the wine, the feel of silk); aesthetic pleasures, derived from sensations expressive of something, offered by nature, or created by man (e.g., sunsets, music); accomplishment pleasures represent the emotional, pleasant consciousness that something valued has come about (e.g., mastery of a skill, sport performance). The latter type of pleasure is conceptually close to Csikszentmihalyi's (1990) concept of flow, although for Csikszentmihalyi pleasure is limited to sensation, as 'beyond that, it becomes enjoyment' (Seligman & Csikszentmihalyi, 2000). Kubovy (1999) defined pleasures of the mind as collections of emotions distributed over time. He also acknowledged the pleasures of nurture and of belonging to a social group as additional varieties of pleasure. Kubovy's typology parallels the sociological account offered by Tiger (1992) who identified four pleasure types: physio-pleasures (sensations or physical impressions obtained from eating, drinking, laying in the sun), socio-pleasures (borne of the company of others), psycho-pleasures (borne of ideas, images, and emotions privately experienced), and ideo-pleasures (borne of ideas, images, and emotions privately experienced). Tiger's typology, like previous ones, has never been tested empirically (Dube and Le Bel, 2003). Considering the importance of affective design is getting more important and the pleasure is at the core of affective experience, the empirical investigation on the relationship between pleasures and design attributes can be a valuable foundation of pleasurable and affective design.

In this study, we investigated the actual effect of the pleasures and attributes related with the pleasures in the context of mobile phone usage. To this end, we proposed a hypothetical model of the relationships between pleasure and design attributes based on the Norman (2003)'s three-level taxonomy and tested the hypotheses using the user's experience on mobile phone.

2. CONCEPTUAL FRAMEWORK

Norman et al. (2003) proposed the theory that characteristics of human are originated from three different levels of brain which is mainly based on biological evolution theory. Brains have evolved into three different levels: 'natural and instinctive level programmed from birth, behavioral level as cognitive function which controls daily actions and contemplative or reflective level'. These three levels interact and control each other. Bottom-up method processes by perceptional function; Top-down method performs by higher level cognitive process. Associated attributes and their definitions are summarized in Table 1.

Table 1. Norman (2003)'s design attributes

Affective design attributes	Definition and description
Visceral design attribute	Design attributes perceived by five senses
Behavioral design attribute	Pleasure and utility during usage.
Reflective design attribute	Meaning or self-image of products formed in memories of individuals or messages desired to be delivered to the others through possessed products.

The visceral level monitors the current state of both the organism and the environment through fast, hardwired detectors that require a minimum of processing (Norman et al., 2003). Visceral level design attributes comprise immediate response to state information coming from the sensory systems. Attributes at this level that are visual, tactile, auditory and olfactory mainly influence the sensing and perception stage of general information process of human (Leder et al., 2004). Also, it is easy to obtain favorable and unfavorable feelings at the visceral level and the reactions at this stage are biological so that almost everyone in the world responds similarly because visceral level attributes are related with superficial looks and forms of the impression using five senses. Taken together, it can be summarized that various stimuli perceived from five senses (look & feel, touch feel, auditory feel, and olfactory feel) influence positively or negatively on affective assessment of a product. Considering that the pleasure is the core element of affect and that these characteristics of visceral level attributes are similar to the physiological pleasure's (Tiger, 1992; Jordan, 2000), the physiological pleasure is thought to have something to do with attributes of visceral level.

- Hypothesis 1: Visceral level design attributes (look & feel, touch feel, auditory feel, and olfactory feel) will influence directly physiological pleasure.

Behavioral level design attributes focus on what a person can do with an object. The behavioral level is the level of skilled and well-learned, largely "routinized" behaviors (Norman et al., 2003). The behavioral level is a quite complex and involving considerable processing to select and guide behavior. It must have access to both working and more permanent memory, as well as evaluative and planning mechanisms (Norman et al., 2003). Interaction elements which have something to do with controlling can be categorized into two major elements: usability and impression (Han et al., 2001). These elements can be further divided into GUI (Graphic User Interface) and PUI (Physical User Interface). These attributes give good affective impressions to users when the product functions as well as is expected to have a good physical feel. These feelings are thought to be related with psychological pleasure.

- Hypothesis 2: Behavioral level design attributes (PUI feel, PUI Usability, Interaction Type, GUI usability) will influence directly psychological pleasure.

Reflection is a meta-process in which the mind deliberates about itself (Norman, 2004). That is, it performs operations upon its own internal representations of its experiences, of its physical embodiment, its current behavior, and the current environment, along with the outputs of planning, reasoning, and problem solving (Damasio, 1994; Norman et al., 2003). This is intellectually driven and is influenced greatly by the knowledge and experience of the designer (user), including the person's culture and idiosyncrasies (Norman et al., 2003). For judgment of taste and fashion, people from different cultures think differently; it all depends on upbringing, traditions, needs, and expectations. Norman mentioned that the meaning of reflective design varies over a broad range. For example, it includes 'things about messages, culture, products and meaning related to the usage of the products' (Norman, 1988). The important field amongst them is the part about meaning of products formed by memories of individual users. Another field of reflective design is about messages or self-image that users send to others through their possessions. These influences of reflective level attributes could be closely related to sociological pleasure and ideological pleasure (Norman, 2004).

- Hypothesis 3: Reflective level design attributes (Experience, Image, Information, and Culture) will influence directly sociological pleasure.
- Hypothesis 4: Reflective level design attributes will influence directly ideological pleasure.

The behavioral level can both inhibit and activate visceral level responses and can pass affective information up to the reflective level when confronted with discrepancies from norms or routine expectations (Norman, 2004). Reflective level only has input from lower levels and neither receives directly sensory input nor is capable of direct control of behavior. However, interrupts from lower-levels can direct and redirect reflection-level processing (Norman et al., 2003).

- Hypothesis 5: Behavioral level design attributes will influence directly visceral level design attributes or visceral level design attributes will influence directly behavioral level design attributes.
- Hypothesis 6 & 7: Reflective level design attributes will be influenced directly by behavioral and visceral level design attributes.

3. METHOD

A survey of 198 (131 males and 67 females) mobile phone users was performed about the relationship amongst pleasure and affective attributes of product. The reasons why mobile phone was selected is because it equally contains the affective design attributes mentioned earlier: visceral level attributes, behavioral level attributes and reflective level attributes. Structural Equation Modeling (SEM) was utilized to understand the relationship amongst each attributes, parent attributes and four types of pleasures. SPSS (ver. 18.0) and AMOSS (ver. 18.0) were utilized to analyze the results statistically. Regression analysis distinguishes variables into dependent and independent variables. On the other hand, in SEM, two dimensions are suggested for discrimination: latent and observed variables; exogenous and endogenous variables. Variables used in this study were a total of 20 (Observed Endogenous Variable: 17, Latent Endogenous Variable: 3) and online survey and user evaluation were conducted considering their ages and genders. The questionnaire was composed of 20 items (7-point likert scale), which asked the satisfaction level of each design attribute and the type of pleasure they felt during the use of their current mobile phones.

In SEM, the match between any particular model and the data is assessed by using several goodness–of-fit indices, because the chi-square test is highly sensitive to sample size, the ratio of χ^2 to its degree of freedom was also computed ($\chi^2 /$ df), with a value of not more than 3.0 being indicative of an acceptable fit between the hypothetical model and the sample data (Carmines and McIver, 1981). In addition, other fit indices are also considered when making comparisons to the baseline model. Following recommendations by Hu and Bentler (1999), and Scheriber et al. (2006), the root mean square residual (RMR) and the root mean square error of approximation (RMSEA) were used as measures of absolute fit and the Normed Fit Index (NFI) and Tucker-Lewis Index (TLI) as indices of incremental fit. From the literature (e.g. Hu and Bentler, 1998), values of 0.9 or more for the NFI and TLI, values of 0.8 or less for RMSEA, and values of 0.5 or less for RMR are reflective of a good fit. At each level of invariance test, if the null model is not rejected, this indicates that the restriction of the parameters did not result in a solution that was worse than the baseline model.

4. RESULTS

4.1. Validation of measurement model

In order to validate a measurement model, CFA (confirmatory factor analysis) of attributes was conducted and the entire factor loading of every observed variable appeared between 0.46~0.87, which makes modification of measurement model unnecessary. Also, observed variables were not added or eliminated because every C.R. value of the measurement model was higher than 1.96 (p < 0.05) and every factor loading was statistically significant. In order to examine how well measurement model fit into sample data, fitness index of measurement model was examined (χ^2 = 298.051(df=105, p=0.000), RMR = 0.078, RMSEA = 0.109, TLI = 0.850, NFI = 0.851). Most of the fitness indices showed the acceptable level so that the measurement model of this study can be considered to be well designed for explanation of data.

In order to examine if measurement is done properly, all of the scales should secure reliability and validity. Construct validity of measured variable can be tested through statistical examination of factor loading. As mentioned earlier, the entire factor loading was statistically meaningful, which secured construct validity of measurement. On the other hand, reliability can be validated through construct reliability and extracted variance that are derived from confirmatory attribute analysis. Theoretically, measurement model is considered to be appropriate when its construct reliability and variance extracted is above 0.7 and 0.5 respectively. According to the result, variance extracted and construct reliability were above 0.5 and 0.7 respectively, which showed the fine level of convergent validity and internal consistency (see Table 2).

Table 2. Construct reliability and variance

Variables	Construct reliability	Variance extracted
Visceral level attributes	0.957	0.853
Behavioral level attributes	0.965	0.875
Reflective level attributes	0.963	0.867
Affective satisfaction	0.971	0.894

4.2. Hypothesis testing

A SEM analysis was conducted to test the hypotheses mentioned in Section 2 and the overall model fit was assessed by five goodness-of-fit indices: χ^2 / df, RMR, RMSEA, TLI, NFI. According to the result, index values of χ^2 / df, RMR, RMSEA, TLI, NFI showed that they all met the recommended guidelines and suggested good model fits (χ^2 / df = 2.31, RMR = 0.047, RMSEA = 0.083, TLI = 0.921, NFI = 0.898), so it is possible to conclude that the proposed model was built properly to account for the data. Also, the result in Table 3 presents each regression coefficient and statistical significance, which showed that all of previously suggested hypotheses are supported together with an additional significant causal effect between reflective attributes and psychological pleasure.

Table 3. Regression coefficient and its significance of structural model

Path			St. Est.	P	Test result
Psychological pleasure	→	Reflective attributes	0.301	0.002	Supported
Sociological pleasure	→	Reflective attributes	0.262	0.008	Supported
Ideological pleasure	→	Reflective attributes	0.199	0.041	Supported
Reflective attributes	→	Behavioral attributes	0.644	<.001	Supported
Psychological pleasure	→	Behavioral attributes	0.271	0.001	Supported
Behavioral attributes	→	Visceral attributes	0.427	0.008	Supported
Reflective attributes	→	Visceral attributes	0.297	0.049	Supported
Physiological pleasure	→	Visceral attributes	0.218	0.002	Supported

5. CONCLUSION

In this study, an hypothetical model of relationship between pleasures and design attributes was suggested based on the theoretical model of affective information processing. The hypothesized relationships model was tested by applying it to mobile phones which contains the affective design attributes equally and diverse use experiences. The results were consistent with the proposed theoretical model, with one exception that psychological pleasure was also influenced by reflective attributes. The exception can be explained from the fact that mobile phone has complex usages which are related to higher level cognitive process. It is expected that this study lays the foundation for future studies on the pleasure and design strategies. Also, the results can contribute the assessment and management of affective quality of products that have been required in academics as well as the industry. In the future research, the proposed model can be generalized and expended by testing in various product contexts.

REFERENCES

Barrett, L. F. 2006. Solving the emotion paradox: categorization and the experience of emotion. *Personality and Social Psychology Review*, 10:20–46.

Barrett, L. F., and Russell J. A. 1998. Independence and bipolarity in the structure of current affect. *Journal of Personality & Social Psychology*, 74:967–984.

Bradley, M. M., and Lang, P. J. 1994. Measuring emotion: The self-assessment manikin and the semantic differential, *Journal of Behavior Therapy and Experimental Psychiatry*, 25(1): 49-59.

Bradley, M. M., and Lang, P. J. 2000. Measuring emotion: behavior, feeling, and physiology. In. *Cognitive Neuroscience of Emotion,* eds. R. D. Lane, L. Nadel, G. L. Ahern, J. J. B. Allen, A. W. Kaszniak, S. Z. Rapcsak, and G. E. Schwartz,

242–76. New York: Oxford Univ. Press.

Carmines, E. G., and McIver, J. P. 1981. Analyzing models with unobserved variables. In *Social measurement: Current issues*, eds. G. W. Bohrnstedt and E. F. Borgattta. Beverly Hills, CA:Sage.

Carroll, J. M., Yik, M. S. M., Russell, J. A., and Barrett, L. F. 1999. On the psychometric principles of affect. *Review of General Psychology*, 3:14–22.

Creusen, M. E. H. 1998. *Product Appearance and Consumer Choice*, Unpublished doctoral dissertation, Delft University of Technology, Delft, The Netherlands.

Csikszentmihalyi, M. 1990. *Flow: The psychology of optimal experience*. New York: Harper Collins.

Damasio, A. R. 1994. *Descarte's error: emotion, reason, and the human brain*. New York: G.P. Putnam.

Dubé, L. and Le Bel, J. 2003. The content and structure of laypeople's concept of pleasure, *Cognition & Emotion*, 17(2): 263-295.

Duncker, K. 1941. On pleasure, emotion, and striving, Philosophy and *Phenomenological Research*, 1(4): 391-430.

Frijda, N. H. 1989. Aesthetic emotion and reality. *American Psychologist*, 44:1546–1547.

Han, S. H., Yun, M. H., Kwahk, J. and Hong, S. W. 2001. Usability of consumer electronic products, *International Journal of Industrial Ergonomics*, 28:143-151.

Helander, M. G. and Khalid, H. M. 2006. Affective and Pleasurable Design, In. *Handbook of Human Factors and Ergonomics* (Third Edition), eds. G. Salvendy. Hoboken, NJ: John Wiley & Sons, Inc.

Hu, P. J., and Bentler, P. 1998. Fit Indices in covariance structure modeling: Sensitivity to underparameterized model misspecification, *Psychological Methods*, 3:424-453.

Jordan, P. W. 1998. Human Factors for Pleasure in Product Use. *Applied Ergonomics*, 29(1):25–33.

Kubovy, M. (1999). On the pleasures of the mind. In *Well-being: the foundations of hedonic psychology* (pp. 134-154), eds. D. Kahneman, E. Diener, and N. Schwarz. New York: Russell Sage Foundation.

Leder, H., Belke, B., Oeberst, A., and Augustin, D. 2004. A model of aesthetic appreciation and aesthetic judgments. *British Journal of Psychology*, 95:489-508.

Norman, D. A. 1988. *The Psychology of Everyday Things*, New York:Basic Books.

Norman, D. A. 2004. *Emotional Design*. New York:Basic Books.

Norman, D. A., Ortony, A. and Russell, D. 2003. Affect and machine design: lessons from the development of autonomous machines. *IBM Systems Journal*, 41(1):9–44.

Picard, R. W. 2003. Affective computing: challenges, *Application of Affective Computing in Human-Computer Interaction*, 59(1-2):55-64.

Ratchford, B. T. 1987. New insights about FCB grid. *Journal of Advertising Research*, 27(4):24-38.

Russell, J. A. 1980. A circumplex model of affect. *Journal of Personality and Social Psychology*, 39:1161–1178.

Russell, J. A. 1989. Affect Grid: A Single-Item Scale of Pleasure and Arousal,

Journal of Personality and Social Psychology, 57(3):493-502.

Russell, J. A. 2003. Core affect and the psychological construction of emotion. *Psychological Review*, 110(1):145–172.

Russell, J. A., and Barrett, L. F. 1999. Core affect, prototypical emotional episodes, and other things called emotion: dissecting the elephant. *Journal of Personality and Social Psychology*, 76:805–819.

Scherer, K. R. 1997. The role of culture in emotion–antecedent appraisal. *Journal of Personality & Social Psychology*, 73:902–922.

Schireiber, J., Nora, A., Stage, F. K. and Barlow, L. 2006. Confirmatory factor analyses and structural equations modeling: An introduction and review, *Journal of Educational Research*, 99(6):323-337.

Snelders, H. M. J. J. 1995. *Subjectivity in the consumer's judgment of product*, Unpublished doctoral dissertation, TU Delft, the Netherlands.

Tiger, L. 1992. *The pursuit of pleasure*. Boston: Little Brown.

Watson, D., and Clark, L. A. 1994. *Manual for the positive and negative affect schedule*. Unpublished manuscript, Univ. Iowa, Iowa City.

Yik, M. S. M., Russell, J. A., and Barrett, L. F. 1999. Structure of self reported current affect: integration and beyond. *Journal of Personality and Social Psychology*, 77:600–619.

CHAPTER 26

Effects of Age, Gender, and Posture on User Behaviors in the Use of Control on Display Interface

Ji Hyoun Lim[1], Yelim Rhie[2] , Ilsun Rhiu[2]

[1]Department of Industrial Engineering, Hongik University
[2]Department of Industrial Engineering, Seoul National University

ABSTRACT

This paper presents an experimental study on behavioral characteristics of old users compared to young users in the use of control on display interface. Thirty two seniors who are over 50 years old and 12 juniors who are in 20s years old were participated in this study. Three basic tasks in touch interface, which are tap, move, and flick, were performed by the users. For the tap task, response time and point of touch were collected and the response bias was calculated for each trial. For the move task, task completion time and the distance of finger movements were recorded for each trial. For the flick task, task completion time and flicking distance were recorded. From the collected data, temporal and spatial differences in interacting behavior between young and old users were analyzed. Although the older users took longer to complete tap, move and flick task, their accuracy of pointing was as good as the younger users in tap task. In the move and flick task, the older users moved their finger less. Gender also effects on the touch behavior. Young female users were slower then young males, however old females were faster than old males in the tap and the move task. There was no statistically significant gender effect on task completion time in the flick task. Using index finger to touch (both handed condition) reduced task completion time and increased accuracy in the tap and the move task.

Keywords: Touch interface, universal design, tap, move, flick

1   INTRODUCTION

Recently, control-on-display interface, which enables users directly control a device on display without separate set of controls, is being spread into mobile communication devices such as smart phones and tablet PCs. As users can operate the device directly on display, control-on-display interface is considered as an easier way of controlling a device than using conventional indirect display control method requiring separate display and set of controls (Kornblum et al., 1990).

Leonardi et al(2010) suggested that touch interface could be delightful to and favored by the elderly. In terms of usability as well, Murata et al(2005) presented the performance gap between the elderly and younger users was less when elderly users use a direct–control touch panel, than using a mouse which is an indirect-control device. In experimental psychology, many studies also have been conducted on elderly users' memory, learnability, and cognitive ability related to their task performance (Salthouse et al., 2002; Neveh-Benjamin, 2000; Echt et al., 1998; Nettelbeck & Rabbitt, 1992), and showed the age effect on behaviors.

Jung(2011) observed elderly users' behaviors in performing basic tasks such as tracking, depth perception, and button manipulation. This study found that elderly users' tracking and depth perception were less accurate, and button manipulation was slower than younger users. According to the existing studies, the typical problems in elderly users' interaction are in interpreting presentation of the machine (understanding displayed information), and then in selecting an appropriate input using structured controller. However, most of the studies were limited to use subjective rating or qualitative methods to investigate the elder usage of IT devices. Although Jung(2011) used objective measures such as task completion time and error rate, the tasks assigned in the study were fundamental level tasks. Therefore, it is difficult to apply the study to the touch interface.

In this study, to understand the behavioral differences by users' age in the use of control-on display, we collected the objective behavior data and analyzed differences on behavioral characteristics between elderly and younger users using basic touch interaction tasks such as 'tap', 'move', and 'flick' interaction.

2   METHOD

2.1. Participants

We observed 44 participants in total recruited in New York area, USA. Among the participants, 12 people were aged 20s (junior group, average age was 23.39, SD=2.80), and nine of them were male(75%) and 3 were female(25%). The remaining Thirty two people were aged over 50(senior group, average age was

53.59, SD=3.10), 14 of them were male(43.75%) and 18 were female(56.25%). Seven of the senior group(58.3%, for 10.9 months on average) and three of the senior group(9.38%, for 9.7 months on average) were experienced of touch screen interfaces at the time the experiment had been conducted. Most of the participants (100% of 20s, 78.25% of over 50s) were using their right hand primarily.

2.2. Apparatus and Tasks

Samsung SGH-F460 which had 320×240 pixels, 2.8 inch screen with capacitive touch panel was used for the experiments. We collected objective data of touch gesture, time and location of finger tip, using software embedded in the device while participants were performing 'tap', 'move', and 'flick' tasks.

For the 'tap' task, participants were asked to touch a visual stimulus appearing on the screen which was divided into 48(6 by 8) districts. We gave circle and rectangle shaped icons as visual stimuli and their diameter were 5mm, 10mm, or 15mm. The location, shape and size of the visual stimulus were controlled by random selection. While a participant performed tap task, the location of finger tip touched and the response time(RT) from the time presenting a visual stimulus to the point participants tap on the screen, were collected. For each of the touched location data, the response bias for vertical and horizontal axes (δx & δy) which is the distance from the center of visual stimulus to the touched location, was calculated.

For the 'move' task, a set of drag and drop tasks were given to participants. When a visual stimulus appeared on one of quadrants, participants should place their finger on the stimuli and drag it to the target location indicated in the screen. There were three possible directions for each quadrant, and 12 directions in total. Two types of stimuli, circle and rectangle, were presented and their size varied among 5mm, 10mm, and 15mm. The shape and size of the stimulus were also controlled. From each of move tasks, task completion time, and path of a finger moved on the screen were collected. From the path data, the moved distance was calculated.

For the 'Flick' task, participants were told to scroll the whole screen upward or downward. When the finger moved more than 3mm in 80msec, the embedded software recognizes the gesture as flicking. Participants flicked on the displayed list according to the direction (up or down) indicated on the top of the screen. For flick task, task completion time and the flicked distance were collected.

2.3. Experiment Procedure

The aim of this research is to compare the behavioral characteristics between two age groups (junior and senior) using the three touch interaction tasks (tap, move, flick). Participants performed tasks in all conditions for a comparative analysis of behavioral differences between the two age groups. In addition to the age factor, gender, posture(one hand- operating with the thumb of their dominant hand versus both hand – holding the device with left hand and operating with index finger of right hand) and task specific factors (location of visual stimulus appearing for tap, moving direction, and flicking direction) were considered as well in the experimental design.

Tasks were arranged in the sequence of tap, move, and flick. The half of the participants performed the first set of experiment with one-handed posture and then the second set with both-handed condition. The rest of the participants performed both-handed posture first. One set of Tap task had 336 trials, and each of move and flick task were composed of 72 and 12 trials. Participants led their own experiment, and took a break time after each task.

3   RESULT

Overall, we collected 21802 tap data(response time, touch location), 5035 move data(response time, and movement path), and 715 flick data(flicking distance, response time) from the senior group. In addition, 8309 tap data, 1746 move data, and 250 flick data are collected from the junior group.

3.1. Analysis on Tap task

For the tap task, age, gender, and posture were tested for their influence on the response time and response bias, δx and δy. There was statistically significant difference in RT ($F=100.994$, $p=0.000$), δx ($F=11.799$, $p=0.001$), and δy ($F=41.756$, $p=0.000$) between the age groups. Gender significantly affected only on the RT ($F=6.655$, $p=0.010$) and δy($F=24.683$, $p=0.000$). The two different postures showed significant difference in RT ($F=5.157$, $p=0.023$) and response bias; δx ($F=42.761$, $p=0.000$), and δy ($F=74.109$, $p=0.000$) as well. The location where the visual stimulus presented also affected significantly on RT ($F=14.979$, $p=0.000$), δx (33.668, $p=0.000$), and δy ($F=53.019$, $p= 0.000$).

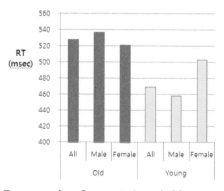

Figure 1. Response time for tap task varied by age and gender

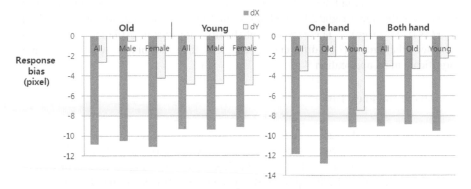

Figure 2. Response bias in tap task varied by age, gender, and posture

On average, the senior group took 58.45 msec longer to response to a visual stimulus, and less accurate than the junior group. The differences in response bias were statistically significant, but the amount of difference was insignificant (1.5~2.2 pixels). On average, females took 12.52 msec longer to respond to the stimulus, but as shown in Figure 1., males in the senior group were slower than the female seniors whereas females in the junior group took almost twice as much time to respond than the male juniors. As shown in the Figure 2., participants made more errors in horizontal (δx) than in vertical (δy). Notable observation was that the junior group was less accurate along with the vertical axis, and females are much less accurate vertically.

There was statistically significant difference in posture of performing the tap task. When participants performed tap task with one hand, it took 17.43 msec longer than when they used both hands. Using one hand decreased the speed of response and also decreased the accuracy. As shown in the Figure 2., using one-hand showed more response errors compared to the both-handed condition.

3.2. Analysis on Move Task

For the move task, age, gender, posture and moving direction were tested with multivariate ANOVA. Age group (F=161.563, p=0.000), posture (F=84.873, p=0.000) and direction (F=2.207, df=11, p=0.12) showed statistically significant effect on task completion time. For the moved distance, all four factors showed significant effects.

On average, the senior group took 372 msec longer to complete a move task. Male participants were 81 msec faster than female participants, and when participants used their both hand, they completed a move task 244 msec faster than when they used only one hand. In the junior group, male participants were faster than females by 105 msec whereas in the senior group the male participants were 33 msec slower than females (Figure 3.).

The moving direction also showed significant influence on task completion time and move-distance. While a move task from the lower right(3rd) quadrant to the upper left(1st) quadrant took the longest time(1653.38 msec) to be completed, a move from the upper right(2nd) quadrant to the lower right(3rd) quadrant took a short time(1410.32 msec). There were statistically significant differences in move-distance, the amount of difference were less than 10 pixels which were not significant in its application.

3.3. Analysis on Flick Task

For the flick task, age, gender, posture, and flicking direction were tested with multivariate ANOVA. Age were the only factor which showed significant influence on task completion time (F=20.556, p=0.000). Flicking distance was significantly varied by age (F=17.239, p=0.000), gender (F=40.158, p=0.000) and their interaction (f=15.627, p=0.000). In flick task, posture and flick direction did not show statistically significant influence on either task completion time or flick-distance.

The senior group took 721.66 msec longer than the junior group to complete a flick task. Flick-distance was also different by the age group. As shown in Figure 4., the senior group made longer flick movements (mean 32.8 pixels) than the junior group. Gender also showed significant effect on flick-distance. Female participants made longer flick-distance (117.05 pixels) than the male participants (90.82 pixels), as shown in Figure 4., the gender difference in flick-distance were bigger in the junior group(49.89 pixels) than in the senior group(11.49 pixels).

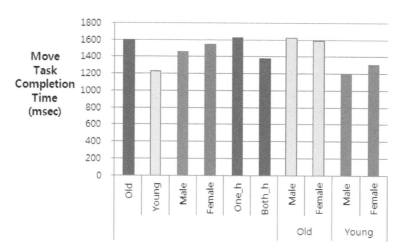

Figure 3. Move task completion time varied by age, gender and posture

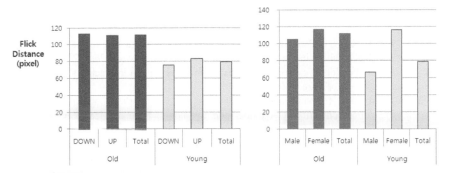

Figure 4. Flick distance varied by flicking direction, age, and gender

4 DISCUSSION

In this study, we identified behavioral characteristics in performing basic interactions such as tap, move, and flick on control-on-display. To perform the tap, move, and flick tasks which require users to detect, interpret, and manage an appropriate finger movement, the senior group needed more time to complete their tasks. The younger male participants were the fastest across all the tasks, and the elder male participants were the slowest. Although the senior group's was slower than the junior in performing tap task on average, the accuracy in tap task was similar or even better on the vertical axis.

In terms of response accuracy, people tend to touch vertically lower and horizontally left point to the center of the target (visual stimulus). Interestingly, the senior group showed more accuracy in vertical axis (δy). In move and flick tasks, which require dynamic control in performance, the elderly showed less movements. Based on these results, it is likely that the elderly took longer time only because they needed longer processing time to identify a stimulus, select response, and control movement.

Use of index finger increased the speed of performance as well as the accuracy in general. To use index finger, a user should hold the device with one hand and operate it with the other hand. This posture improves stability of the operating finger. The index finger also has larger range of motion than the thumb which used in one-hand condition. Therefore, if software designers consider their users to use the device with one hand, they should expect the resulting decreases in both response speed and accuracy.

The significance of this study is that it provides quantitative data for response time and accuracy, which could be applied to design a touch interface system. The data were varied by specific user groups defined by age, gender, and operating gesture. Therefore, this study could help engineers to understand variability in users' response in touch interface, and quantitatively describe the difference.

REFERENCES

Echt, K. V., Morrell, R. W., & Park, D. C., 1998. "Effect of age and training formats on basic computer skill acquisition in older adults". *Educational Gerontology, 24*, 3-25.

Jung, K. T., 2011. "The elderly's error characteristics in some human interactions", *J. of the Ergonomics Society of Korea*, 30(1), 109-115.

Kornblum, S., Hasbroucq, T. and Osman, A., 1990. Dimensional overlap: Cognitive basis of stimulus-response compatibility-A model and taxonomy, Psychological Review, 97, 253-270.

Leonardi, C., Albertini, A., Pianesi, F., & Zancanaro. M., 2010. "An exploratory study of a touch-based gestural interface of elderly", *NordiCHI 2010*, October 12-20.

Murata, A. & Iwase, H., 2005. "Usability of touch-panel interface for old adults". *Human Factors, 47(4)*, 767-776.

Naveh-Benjamin, M., 2000. "Adult age differences in memory performance: test of an associative deficit hypothesis". *J. of Experimental Psychology: Learning, Memory, and Cognition, 26(5)*, 1170-1187.

Nettelbeck, T. & Rabbitt, P. M. A, 1992. "Aging, cognitive performance, and mental speed". *Intelligence, 16*, 189-205.

Salthouse, T. A., Berish, D. E., & Miles, J. D., 2002. "The role of cognitive stimulation on the relations between age and cognitive functioning", *Psychology and Aging, 17(4)*, 548-447.

CHAPTER 27

Subjective Quality Evaluation of Surface Stiffness from Hand Press: Development of an Affective Assessment Strategy

*I. Rhiu, ** T. Ryu,* B. Jin, and * M. H. Yun

*Seoul National University
Seoul, Korea
**Hanbat National University
Daejeon, Korea
1sunr@naver.com , tbryu75@gmail.com, feelbest@gmail.com, mhy@snu.ac.kr

ABSTRACT

The purpose of this study was to analyze customers' feeling of satisfaction on the stiffness of outside panels of passenger cars. Including 'satisfaction', four affective variables were selected for rating the subject assessment of outside panel. Fifty customers evaluated the hood and trunk lid of nine midsize passenger cars in quantitative questionnaire study. Stress-strain curve for the hood and trunk lid for nine vehicles was also produced. It was found that customers were more satisfied as the slope of the stress-strain curve increased, while the decrease at a point in the curve had negative effect on satisfaction. The level of satisfaction on the outside panel stiffness was grouped by stress-strain curves, and it is likely that the affective quality of outside panel stiffness can be controlled by them. With the results of this study, the designers of outside panels are able to know how to make the stress-strain curves of panels for the desired level of satisfaction.

Keywords: Stiffness of outside panel, Stress-strain curve, Passenger cars, Affective design, Affective quality control

1   INTRODUCTION

The affective quality of passenger vehicles such as the look and feel as well as functional performance (e.g. power and fuel consumption efficiency) is becoming an important factor in customers' purchase decisions. Affect is defined as the customer's psychological response to perceptual design details of the product (Demirbilek & Sener, 2003). Product designs that do not consider customers' affection may essentially be weakened (Helander & Tham, 2003).

Stiffness of the outside panel is important for the affective quality of vehicles. Customers usually have a chance to contact with outside panel of vehicles in the events such as car washing, opening door, and repairing panel. In this situation, the unexpected and excessive deformation of outside panel of vehicles, which occurred from customers' contact, can bring out the cheap feel to the whole vehicle. However, making outside panel harder causes much other inefficiency together with cost increase. So, to figure out the optimum level of stiffness of outside panel which could satisfy users is very important in panel configuration.

Few studies have been conducted on the design of outside panel stiffness in terms of customer satisfaction. Many Kansei engineering studies on automobiles have focused on the visual design characteristics of interior and exterior parts (Jindo & Hirasago, 1997; Nagamachi, 2002; Nakada, 1997). Bahn et al. (2006) used visual design characteristics to develop the user's luxuriousness model of crash pad. Also, Tanoue et al. (1997) used the interior images on the affective engineering research. In some studies, tactile feeling was considered as well as visual properties in evaluating the satisfaction of interior materials of passenger cars. But few studies exist focusing on the tactile feeling of the stiffness of outside panels (Yun et al., 2001; Yun et al., 2003; You et al., 2006).

An aspect of mechanical engineering associated to stiffness design was mainly studied in previous studies. Kim (2004) proposed the optimal design of the exterior stiffness which differs throughout the parts of the car using stiffeners. Also, Qian et al (1996) proposed the optimal design of exterior adhesion of cars and its associated exterior stiffness. And there were some studies on the method of measurements of bends on the exterior after exposure to a force (Liu et al., 2000).

This study attempted to analyze customers' satisfaction of outside panel stiffness of passenger cars. Especially, we analyzed the relationship between design variables of outside panel stiffness and users' subjective affection on outside panel stiffness. To do this, this study conducted following tasks: 1) a questionnaire was developed to evaluate customers' affection on the stiffness of vehicle outside panels, 2) design variables related to the stiffness of the panels were selected, 3) an experiment to evaluate customers' affection for the outside panels of various passenger cars was performed, and 4) statistical analysis was conducted based on the experiment data to analyze customers' satisfaction of outside panel stiffness.

2 METHOD

2.1 Design Variables of Outsides Panel Stiffness

The stiffness of an automobile's outside panels was measured using the stress-strain curve (Figure 1 as an example). From a stress-strain curve, two design variables: the slope of curve and type of decrease of the curve at a point (called canning) can be defined. The slope of curve is defined as the slope between the start and end point of a curve. There are infinite types of canning in terms of its range and shapes. It was difficult to collect all kinds of outside panels for the experiment which have various stress-strain curves in the way of factorial design. Thus the stress-strain curve for an outside panel stiffness itself was selected for design variable. The study obtained the stress-strain curve from the weakest point of an outside panel and the value of the two variables related to the stress-strain curve was taken for the further analysis.

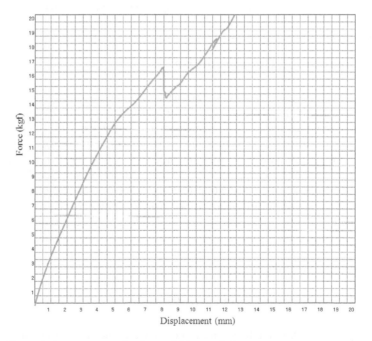

Figure 1 Stress-strain curve example of an outside panel of passenger cars

2.2 Affective Variables

Seven initial affective variables of outside panel's stiffness were collected through Korean adjectives related to touch feel, web survey for customers' experience of contact with vehicle outside panels and expert review. After integrating the initial affective variables, four affective variables were selected based on the result of the pilot-test. Selected variables were 'satisfaction', 'hardness', 'consistency', and 'thickness'. Definitions of selected affective variables are given in Table 1.

Table 1 Selected affective variables for the stiffness of an automobile's outside panel

Affective variables	Definition
Satisfaction	Degree of satisfaction in terms of the automobile outside panel's stiffness when pressing it
Hardness	Degree of how much impact the outside panel can take when pressing it
Consistency	Degree of consistency in deformation of the automobile outside panel when pressing it
Thickness	Degree of how thick the automobile outside panel feels when pressing it

2.3 Evaluation Questionnaire

The developed questionnaire to evaluate customers' affection on the panel stiffness consisted of 1) the questions for basic information of customers, 2) explanation script of the evaluation method and target parts of vehicles, 3) the question to rate customer's affection, and 4) post-test questions. The questions for basic information of customers were included to obtain the demographic data, driving experience and the frequency of contact with the outside panel of vehicles. In the explanation script of evaluation method, a scenario to set up an evaluation context, a detailed task for evaluator to perform (pushing hard the panel with their palm several times), and the explanation of affective s selected in this study were included. In the question to rate customer' affection, the selected affective variables are rated by using 7-points SD scale. The post-test questions were included to clarify and reason customers' rating.

2.4 Outside Panel Parts and Vehicles

Two parts including the hood and trunk lid were selected to examine customer satisfaction with outside panels (Figure 2). These selected parts are most frequently touched with the driver.

The present study used nine midsize passenger cars to measure the design variables of outside panels and customer satisfaction. The vehicles were placed at a yard of an auto manufacturing company; of the vehicles, 2 were domestic and the other foreign, having various characteristics of stiffness.

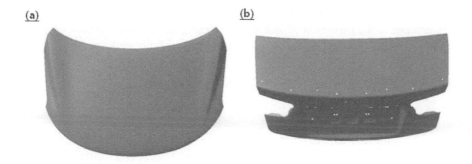

Figure 2 Selected outside panels: (a) Hood, (b) Trunk lid

2.5   Participants and Procedure

A total of 54 males participated in the outside panel affection evaluation for the nine vehicles. Of the participants 25, 17, 8, and 4 were in their 20s, 30s, 40s and 50s respectively.

The evaluation experiment in the study consisted of three sessions: introduction, satisfaction evaluation, and debriefing. At the introduction session, the purpose and method of evaluation were explained to the participants, and the basic questions in the questionnaire were answered. Then, in the evaluation session, each participant visited the 9 vehicles and evaluated the outside panels of the 2 parts in each vehicle by following a predetermined order (the evaluation orders of the vehicles were randomized by the balanced Latin-square design to counterbalance the effects of learning and fatigue). Meanwhile, participant pushed the predefined point of the outside panels, which was marked by the experimenter and on which stress-strain curves were measured. Lastly, at the debriefing session, the post-test questions were answered.

3   RESULTS

3.1   Analysis of Relationship Between Affective Variables and Design Variables

An ANOVA with mixed-factors design was conducted to analyze the effect of outside panels on affective variables. The factors involved in the experiment were type of stress-strain curve, age and their interaction. The variable of stress-strain curve type was within-subject factor and the other variable (age) was between-subjects factor in the experiment. The results are presented in Table 2. All the affective variables were influenced by stress-strain curve type for both the hood and trunk lid. But the effect of age and interaction between the curve and age were not significant on all the affective variables.

Nine stress-strain types were grouped in terms of customer satisfaction by using

SNK (Student Newman-Keuls) method. The nine stress-strain curves of the hood were grouped to the maximum of four groups (Figure 3). The stress-strain curves of the trunk lid were grouped to the maximum of five groups (Figure 4).

Table 2 Summary of ANOVA results(α=0.05)

Part	Independent Variable	Affective variables	df	F	P
Hood	Stress-strain curve	Satisfaction	8	10.46	0.0001
		Hardness	8	19.61	0.0001
		Consistency	8	19.00	0.0001
		Thickness	8	19.68	0.0001
	Age	Satisfaction	3	0.80	0.5012
		Hardness	3	0.43	0.7322
		Consistency	3	0.42	0.7388
		Thickness	3	0.07	0.9769
Trunk Lid	Stress-strain curve	Satisfaction	8	16.27	0.0001
		Hardness	8	16.29	0.0001
		Consistency	8	16.31	0.0001
		Thickness	8	15.22	0.0001
	Age	Satisfaction	3	2.28	0.0910
		Hardness	3	1.67	0.1846
		Consistency	3	1.90	0.1411
		Thickness	3	1.30	0.2832

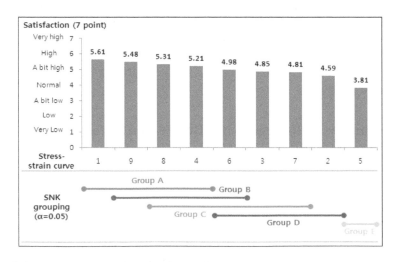

Figure 3 SNK grouping of satisfaction on hood stiffness

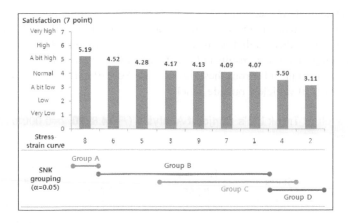

Figure 4 SNK grouping of satisfaction on trunk lid stiffness

3.2 Analysis of Relationship Between Affective 'Satisfaction' and Related Affective Variables

Table 3 Result of Hood's Conjoint Analysis (F=46.9141, p=0.0001, R2=0.6311)

Preference	Attribute	Relative Importance	Attribute Value	Utility
Satisfaction	Hardness	25.7994	1	-0.79392
			2	-0.37392
			3	-0.24092
			4	0.14967
			5	0.38742
			6	0.35038
			7	0.52130
	Consistency	31.7518	1	-0.82487
			2	-0.68116
			3	-0.32824
			4	0.14189
			5	0.11998
			6	0.79380
			7	0.77861
	Thickness	42.4488	1	-1.03201
			2	-0.64922
			3	-0.26271
			4	-0.11734
			5	0.24654
			6	0.68277
			7	1.13198

Conjoint analysis was conducted to analyze the relationship among 'satisfaction' and its related affective variables. The results are presented in Table 3, 4. According to Table 3, results indicated that 'satisfaction' was most influenced by 'thickness' in hood. According to Table 4, results indicated that 'satisfaction' was most influenced by 'hardness' in trunk lid. From these results, the key affective variables could be changed by parts of vehicles' outside panel.

Table 4 Result of Trunk Lid's Conjoint Analysis (F=74.8220, p=0.0001, $R2$=0.7330)

Preference	Attribute	Relative Importance	Attribute Value	Utility
Satisfaction	Hardness	47.0227	1	-1.71442
			2	-0.95498
			3	-0.11814
			4	0.33094
			5	0.59693
			6	0.83033
			7	1.02936
	Consistency	18.2051	1	-0.14561
			2	-0.42257
			3	-0.35755
			4	-0.25838
			5	0.03507
			6	0.51386
			7	0.63970
	Thickness	34.7722	1	-1.10207
			2	-0.43810
			3	-0.34539
			4	0.13849
			5	0.23078
			6	0.58940
			7	0.92689

4   DISCUSSION AND CONCLUSIONS

As expected, it was found that the stiffness was a significant factor on customers' affection on outside panels of passenger cars, and its effect was not different by age. The collected nine stress-strain curves of the hood and those of the trunk lid for the midsize passenger cars could be grouped by the SNK results like Figure 5. Group A was the most satisfied outside panel, and group C was the most unsatisfied outside panel. There was a tendency that the customers rated high as the slope of curve increased in both the hood and trunk lid. In addition, the canning had a negative effect on the satisfaction of the outside panel.

With the results of the study, it is likely that the affective quality of outside panel stiffness can be controlled by the stress-strain curves. Using the result of the study, the level of affective satisfaction for a hood or trunk lid stiffness can be estimated based on its stress-strain curves. And the designers of outside panels are able to know how to make the stress-strain curves of panels for the desired level of satisfaction. Likewise, the method of the experiment and analysis in the study can be extended for the other parts of vehicles' outside panels to control the affective quality of them.

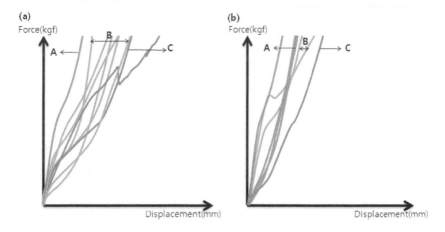

Figure 5 Stress-strain curve of outside panel ((a) : Hood, (b) : Trunk lid)

In addition, the analyzed relationship between 'satisfaction' and its related affective variables can be used to understand the characteristics of satisfaction for vehicles' outside panel. This study specifically explored the quantitative relationship among affective variables and between affective variables and design attributes. Based on these results, it is expected that customers' satisfaction for stiffness of outside panel can be conceptualized more clearly.

ACKNOWLEDGMENTS

This work was funded by grants from Hyundai-NGV.

REFERENCES

Bahn, S., Lee, C., Lee, J., and Yun, M. H. 2006. Development of Luxuriousness Models for Automobile Crash Pad based on Subjective and Objective Material Characteristics. *Journal of the Ergonomics Society of Korea* 24(2): 1-229.

Demirbilek, O., and Sener, B. 2003. Product Design, Semantics and Emotional Response. *Ergonomics*, 46: 1346–1360.

Helander, M. G., and Tham, M. P. 2003. Hedonomics-Affective Human Factors Design. *Ergonomics*, 46: 1269–1272.

Jindo, T., and Hirasago, K. 1997. Application Studies to Car Interior of Kansei Engineering. *International Journal of Industrial Ergonomics*, 19: 105-114.

Kim, J. 2004. Optimization Design of Outer Panel. *Proceeding of the Korean Society of Automotive Engineers*, 754-758.

Liu, L., Sawada, T., and Sakamoto, M. 2000. Evaluation of the Surface Defections in Pressed Automobile Panels by an Optical Refection Method. *Journal of Materials Processing Technology* 103(2): 280-287.

Nagamachi, M. 2002. Kansei Engineering in Consumer Product Design. *Ergonomics in Design* 10: 5-9.

Nakada, K. 1997. Kansei Engineering Research on the Design of Construction Machinery. *International Journal of Industrial Ergonomics* 19: 129-146.

Qian, W., Hsieh, L., and Seliger, G. 1996. On the Optimization of Automobile Panel Fitting. *Proceeding of the IEEE International Conference on Robotics and Automation*, 1268-1274. Minneapolis, USA.

Tanoue, C., Ishizaka, K., and Nagamachi, M. 1997. Kansei engineering: A study on perception of vehicle interior image. *International Journal of Industrial Ergonomics* 19(2): 115-128.

You, H., Ryu, T., Oh, K., Yun, M. H., and Kim, K. J. 2006. Development of Customer Satisfaction Models for Automotive Interior Materials. *International Journal of Industrial Ergonomics*, 36: 323-330.

Yun, M. H., Han, S. H., Ryu, T., and Yoo, K. 2001. Determination of Critical Design Variables Based on the Characteristics of Product Image/Impression: Case Study of Office Chair Design. *Proceedings of the Human Factors and Ergonomics Society 45th Annual Meeting*, 712-716.

Yun, M. H., Han, S. H., Hong, S. W., and Kim, J. 2003. Incorporating User Satisfaction into the Look-and-feel of Mobile Phone Design. *Ergonomics* 46: 1423-1440.

CHAPTER 28

Quantification of a Haptic Control Feedback Using an Affective Scaling Method

S. Kwon, W. Kim, M. H. Yun

Seoul National University
Seoul, South Korea
slowband@naver.com, luckywonjun@naver.com, mhy@snu.ac.kr

ABSTRACT

Haptic feedback device have been widely used and the importance of haptic feedback is becoming more important factor for a control device. Many studies about haptic control device and haptic feedback were researched. Those researches were focused to usability of the device and mechanical approach to haptic feedbacks. However affective approach to haptic feedback of haptic control device has not been studied yet. The aim of this study is to analyze the customers' satisfaction for the in-vehicle haptic control device. Variables related to 'satisfaction' which consists of five affective responses were selected were selected. PC software operating in-vehicle functions was developed for the affective evaluation experiment. Thirty subjects evaluated haptic feedbacks of steering wheel device in vehicle using a questionnaire survey with seven-point semantic differential scale and hundred-point magnitude estimation scale. 16 Haptic feedbacks were developed with computer software in the study. Based on the survey result, 'distinctness' was identified as the most positive affective response factor for satisfaction in a haptic thumbwheel and 'strength' was the most negative affective response. Also, the relationship between 'satisfaction' and its related engineering variables were identified and modeled. The results of the optimal design variables of haptic feedback might help to satisfy customers' affective satisfaction for various kinds of haptic steering devices.

Keywords: Haptic Control Device, Control Knob, Affective Engineering

1 INTRODUCTION

Recently, the load of the major sensory organs - visual and auditory – are on the rise as informatization and digitalization are accelerating further (Brewster and Brown, 2005). This means that the quantity of information became larger when people do single task. And moreover, it means that people would have lack of abilities of recognizing and controlling information. Thus the researches about improving human's recognition and control response by using multiple sensory feedbacks especially tactile feedback are in the limelight at the moment (Lorna et al., 2005). Vilimek and Zimmer(2007) and many researchers argued that multiple sensory feedback using haptic feedback could improve users' recognition and control performance compared to single feedback.

Researches about haptic technology were mainly focused on the physical and mechanical elements that provide tactile stimulus for users. Mark and Ernest(1982) studied that many elements such as location of vibrating equipment, frequency, force, duration determined tactile feeling. Lorna et al. (2005) suggested a term 'Tacton' as a haptic sign compared with visual sign 'icon' or auditory sign 'earcon'. And they argued that rhythm, complex wave type, spatial location could be the factors that differentiate haptic feedbacks.

With these studies about components of haptic feedback, the researches about increasing effectiveness and efficiency of haptic feedback device were done (Cockburn and Brewster, 2005; Penn et al., 2000). Matters(2003) measured the performance with multi feedback device against the performance with single feedback device. And Liu(2001) studied the effectiveness of the multi feedback device in the situation of driving vehicle.

To sum up existing researches, we can divide them into two categories. One is about the physical elements that generate tactile feedback, and the other is about the effectiveness of multiple sensory feedbacks in the driving situation. And the researches from an affective point of view were not conducted yet.

This study aims to evaluate users' affective satisfaction by developing various haptic feedbacks and conducting user survey experiment. Target device was the control knob in the luxury vehicle that generates the haptic feedback.

For this purpose, we collected and selected affective variables related to the haptic control device. And we drew physical elements and converted them to controllable design variables. Then we generated 16 haptic feedback samples and conducted a survey of 28 subjects (male: 18, female: 10). Finally, we analyzed the results and figured out the relationship between affective variables and overall satisfaction, and optimal level of each design variable.

2 METHOD

This research consists of two experiments. The research process is like figure 1. (1) We developed PC software interworking with haptic device to construct real situation of use (figure 2). (2) We collected affective variables and haptic feedback design variables through literature review and experts' evaluation. (3) We reviewed affective variables by pilot study and experts' evaluation. (4) We grasped the relationship between affective variables and haptic feedback design variables by user survey experiments. (5) Based on the results of the experiments, we understood the major affective variables and design variables related to overall satisfaction for the haptic control device. (6) We proposed the combination of haptic feedback design variables for improving user satisfaction for the haptic control device. Figure 3 is the haptic control device used in this research.

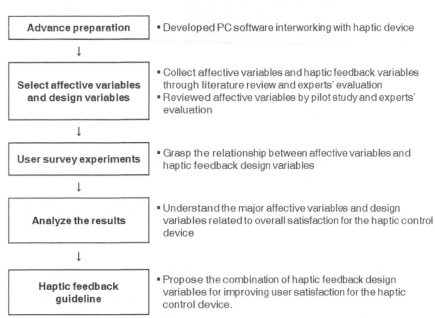

Figure 1 Research process flow chart

Figure 2 PC software developed for this research

252

Figure 3 Haptic control device used in this research

2.1   Affective Variables

We collected affective variables related to tactile impression to grasp affective variables for haptic feedback from the haptic control device by doing literature survey and reflecting experts' opinion. 'Perspicuity', 'softness', 'control power', 'massivity', 'elasticity' were selected among 40 affective adjectives, and 'perspicuity', 'softness', 'control power' were selected for major affective variables by pilot test. With these three affective variables, 'overall satisfaction' was added as a dependent variable (Table 1).

Table 1 Selected affective variables for the control feel of haptic steering device

Affective variables	Definition
Perspicuity	Degree of perspicuity when operating haptic steering device
Control power	Degree of how much power required when operating haptic steering device
Softness	Degree of softness when operating haptic steering device
Overall satisfaction	Degree of satisfaction in terms of the haptic when operating haptic steering device

2.2   Haptic Feedback Design Variables

Haptic feedback design variables were drawn by following process (Figure 4). We collected commercial products that have the control knob. They were collected

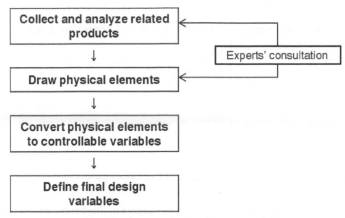

Figure 4 Flow chart for defining haptic feedback design variables

from vehicle control panels, audio equipment, washing machines and so on. After collecting control knobs, we investigated inner parts of the control knob. In this stage, there was a consultation from control knob maker's experts. Control knob is composed of knob and inside switch, and the inside switch consists of detent, ball, and spring. The control feel of the control knob is influenced by the inner parts of the switch. The period of the feedback is determined by the number of detent's sawtooth, and the control force is determined by the spring constant and ball size. The physical elements of control knob are summarized in table 2.

Table 2 Physical elements of the control knob

Classification		Definition
Switch	Spring	Support the ball and determine the control force
	Detent	Determine the unit period of the control knob's force feedback
	Ball	Produce feedback with detent
Knob		Outside part touched by users' hands

In this study, we used Immersion Studio's 'Immersion Rotary Studio' software to add haptic feedback to control device (Grant, 2004). Thus after drawing physical elements, we converted these elements to the haptic feedback design variables that can be controlled in the foregoing software. Finally we selected 4 design variables-magnitude, width, wave type, periodic type-and they were the basic elements that compose the haptic feedback. The physical elements and design variables are summarized in table 3.

Table 3 Physical elements and design variables of haptic feedback

Classification	Physical elements	Design variables of haptic feedback
Switch	Spring	Magnitude
		Periodic effect
	Detent	Wave type
		Width
Knob	Knob weight	Magnitude
	Knob radius	Magnitude

Magnitude is the variable represents the control force for the haptic device and has 3 levels. Width represents the unit angle for the control knob and has 3 levels. Wave type is the type of the detent and has 4 types; 'full sine', 'full triangle', 'half sine', 'half triangle'. Periodic effect is the variable that represents how much additory effect is given. In this study, periodic effect has 3 levels. Using orthogonal design, 16 samples were developed by Immersion Studio's 'Immersion Rotary Studio' software. Design variables of haptic feedback and the levels of the variables are summarized in table 4.

Table 4 Design variables of haptic feedback

Design variables	Definition and levels
Magnitude	Degree of amplitude of haptic feedback : 49.6~79.5 / 89 / 100
Width	Degree of period of haptic feedback : 0.2~0.9 / 1.5~2.3 / 2.7~3.1
Wave type	Type of haptic feedback : Full Sine, Full Triangle, Half Sine, Half Triangle
Periodic effect	Additional effect produced after main feedback : Small, Medium, Large

2.3 Experiment Method

User survey experiment were conducted in the UT room in Seoul National University. The subject was comprised of eight Human factors experts, twenty 20's and 30's (12 men and 8 women). (1) Before evaluating each feedback, subjects have free time to control the device with every haptic feedback for 5 minutes. And they could grasp some relative difference between haptic feedbacks. (2) Subjects also had free time to operate the PC software developed for this study for 3 minutes. And they could get used to the software. (3) Affective variable and overall satisfaction were rated by using 7-points and 100-points likert scale respectively.

Figure 5 Picture of affective evaluation experiment

2.4 Analysis Method

A regression analysis was conducted to understand the relationship between the major affective variables and overall satisfaction when users operate the haptic control device. And conjoint analysis was also conducted to find the optimal level of each design variable of haptic feedbacks.

3 RESULTS

3.1 Relationship Between Affective Variables and Overall Satisfaction

Multiple linear regression analysis was conducted to understand the relationship between the major affective variables and overall satisfaction when users operate the haptic control device. According to the results, the regression model was significant (F=52.249, significant level=0.000). And the statistical power of the regression model was high enough (R^2 was 0.635). The regression equation is as follows.

Overall Satisfaction = 2.8*Softness + 4.8* Perspicuity + 5.7*Control Power

Table 5 Multiple Regression Table

Affective Variable	B	Beta	Sig. level
Softness	2.819	.323	.003
Perspicuity	4.841	.380	.004
Control Power	5.715	.318	.010

According to the results, three affective variables were statistically significant to overall satisfaction (α=0.05). Among three affective variables, perspicuity was the most important factors to overall satisfaction.

3.2 Relationship Between Affective Variables and Design Variables

A conjoint analysis was conducted to understand each design variable's importance to the overall satisfaction and find the optimal level. According to the results, whole conjoint model was statistically significant (Pearson's R=0.925, Kendall's tau=0.725). The importance of design variables ranked in decreasing order is as follows: Width, Wave Type, Periodic Effect, Magnitude (Table 6).

By conjoint analysis, we found the optimal level of each design variable to enhance overall satisfaction. Half Triangle had the highest utility in Wave type variable. In case of Width, 1.56~2.34 level had the highest utility.

Table 6 Results of Conjoint Analysis (Pearson's R=0.925, Kendall's tau=0.725)

Design Variable	Importance	Level	Utility
Magnitude	4.5	49.6~79.5	0.50
		89	0.39
		100	-0.89
Width	40.3	0.2~0.9	-4.58
		1.5~2.3	8.09
		2.7~3.1	-3.50
Wave Type	33.8	Full Sine	0.80
		Full Triangle	0.60
		Half Sine	-6.30
		Half Triangle	4.89
Periodic Effect	21.3	Small	-0.26
		Medium	0.42
		Large	-0.16

According the analysis results, the optimal combination of design variables of haptic feedback are as follows: Half Triangle for Wave type, 1.5~2.3 for Width, 50~80 for Magnitude, and simple Periodic effect.

4   DISCUSSION AND CONCLUSIONS

This study conducted affective evaluation experiment to understand the haptic feedback of control device and find the optimal level for the best haptic feedback. For this purpose, this study follows several steps. (1) We collected and selected affective variables through literature review and applying experts' opinion. They were softness, perspicuity, and control power. (2) We investigated several control knobs and found the physical elements and converted them to controllable design variables. They were Magnitude, Width, Wave Type, Periodic effect. And we developed 16 haptic feedback samples by using orthogonal method. (3) We set up and conducted the experiment and analyze the results.

From the results of the experiment, we could know that softness, perspicuity, control power is the decreasing order of the influence to overall satisfaction. Also we suggest the optimal haptic feedback by figuring out the relationship between design variables and the affective variables. With the results of this study, the guideline to the optimal haptic feedback for the control device could be suggested.

ACKNOWLEDGMENTS

Haptic feedback samples made by Immersion Studio's 'Immersion Rotary Studio' were developed by Daesung electric company.

REFERENCES

Brewster, S. A. and Brown, L. M. 2004. Non-Visual Information Display Using Tactons, Extended Abstracts of ACM CHI 2004, Vienna. Austria, 787-788.

Cockburn, A. and Brewster, S. A. 2005. Multimodal Feedback for the Acquisition of Small Targets, Ergonomics 48: 1129-1150.

Grant, D. 2004. Two new commercial haptic rotary controllers. In Proc. of EuroHaptics 2004, 451–455.

Liu, Y. C. 2001. Comparative Study of the Effect of Auditory, Visual and Multimodality Display on Driver's Performance in Advanced Traveler Information systems, Ergonomics 44: 425-442.

Lorna, M. B. and Stephen, A. B. and Helen, C. P. 2005. A First Investigation into the Effectiveness of Tactons, Eurohatics Conference, 167-176.

Mark, S. S. and Ernest, J. M. 1982. Human factors in engineering and design, McGraw-Hill Companies, 122-177.

Masaya, K. et al. 2002. Analysis of Suture Manipulation Forces for Teleoperation with Force Feedback, Lecture Notes in Computer Science, 2488: 155-162.

Matters, S. 2003. The Lane-Change-Task as a Tool for Driver Distraction Evaluation, Quality of Work and Products in Enterprises of the Future, Ergonomia Verlag, Stuttgart.

Penn, P., et al. 2000 The haptic perception of texture in virtual environments: an investigation with two devices, Lecture Notes in Computer Science, 2058: 25-30.

Vilimek, R. and Zimmer, A. 2007. Development and Evaluation of a Multimodal Touchpad for Advanced In-Vehicle Systems, Engineering Psychology and Cognitive Ergonomics, 842-851.

CHAPTER 29

Effects of Head Movement on Contact Pressure between a N95 Respirator and Headform

Zhipeng Lei and Jingzhou (James) Yang
Department of Mechanical Engineering
Texas Tech University, Lubbock, TX 79409, USA
james.yang@ttu.edu

ABSTRACT

The interaction mechanism between N95 respirators and human faces without head movement has been investigated in our previous work. Whenever humans don respirators, however, their heads move during regular daily work. Does the respirator have the same performance as that without head movement? This paper investigates the effects of head movement on contact pressure between a N95 respirator and headform. A vertebra-joints system is introduced in order to drive human head moving (up and down, left and right). Five headforms and six N95 respirators are utilized. Each headform and N95 respirator pair has five simulations: does not move, moves up, moves down, moves left, and moves right. Results from 180 simulations have shown that the interaction has changed with head movements. Contact pressure distributions have changed before and after head movements. This indicates that respirator has different performances in different work conditions.

Keywords: Respirator; fit and comfort; contact simulation; head movements

1   INTRODUCTION

The OSHA respiratory protection regulation defines the standard eight-exercise procedure in respirator fit tests. During respirator fit tests exercises as (1) Normal breathing without talking; (2) deep breathing; (3) moving the head side to side; (4)

moving the head up and down; (5) talking; (6) grimacing by smiling or frowning; (7) bending at the waist; and (8) normal breathing are performed in sequence (OSHA, 1999). In a modified protocol for respirator fit test developed by Viscusi et al. (2011) the exercises of (1) normal breathing (70 seconds), (2) deep breathing (10 seconds), (3) head side to side (10 seconds), (4) head up and down (10 seconds), (5) talking (10 seconds) and (6) normal breathing (10 seconds) were maintained. The exercises of grimacing and bending at the waist were excluded in this protocol for the purpose of reducing the test duration.

Our previous study investigated respirator fit and comfort by studying the interactions between six N95 filtering facepiece respirators (FFR) and five newly developed digital headforms (small, medium, large, long/narrow and short) (Lei et al., 2011). A few other researchers have developed respirator and headform models and simulated respirator/headform interactions (Bitterman, 1991; Piccione et al., 1997; Zhuang and Viscusi, 2008; Yang et al., 2009; Dai et al., 2011). Their headform models were mostly rigid or deformable single shell, lacking the biofidelity. None of them built a headform model with moving ability.

According to the protocol for respirator fit tests the situations of head side to side and head up and down should also be considered in the contact simulations. How head movements affect the deformation of soft tissue layers (skin, muscle and fatty tissue) is left to be examined. Cervical vertebrae controls head movements, consisting the extension, flexion and rotation (Drake et al., 2005). Most numerical models of the cervical vertebrae include the vertebrae (rigid bone) and the deformable tissue (muscle and ligament) (Yoganandan et al., 2001; Zhang et al., 2005; Lee et al., 2006; Del Palomar et al., 2008; Hedenstierna et al., 2008; Zhuang et al., 2008; Hedenstierna et al., 2009). They were used in impact simulation. Deformation of soft tissue and interaction between cervical vertebrae were studied. While most impact simulations used discrete spring muscles model, Hedenstierna (et al., 2008) incorporated a solid-element muscle model with nonlinear viscoelastic materials into neck model. The passive continuum muscle improved the behavior of the model. The vertebral bones are covered by neck muscle and skin. During head movement, the geometry of the vertebrae droves the muscle and skin to move. Because these available models did not include skin, they cannot be used to simulate their interaction with external objects, such as respirator.

The objective of this work is to study the contact between the headform and the FFR with head movements. The influence of head movements on the contact simulations is to evaluated.

2 HIGH FIDELITY HEADFORM FINITE ELEMENT MODELS

This section describes the method to build high fidelity headform finite element models that can perform the movement of the headform. First, the human head & neck's anatomy that relates to the moving functionality is introduced. Second, a vertebrae-joint system that controls head movements is built. Third, the vertebrae-joint system is incorporated into the finite element headform models.

Head movements are highly related to the cervical region. Cervical vertebrae directly connect the skull. There are 7 cervical vertebrae, CI-CVII, as shown in Figure 1. The vertebrae CI and CII (or atlas and axis) with associated muscles and ligaments have special structures for supporting and positioning the skull. The extension and flexion between the head and the vertebrae CI make the head moving up and down, and relative rotations between the vertebrae CI and CII make the head turning left and right (Drake et al., 2005). Meanwhile, head movements of nodding and rotating are also due to a small amount of extension, flexion and rotation among all the 7 vertebral column. In summary, the cervical vertebrae and related joints make up a complicated multi-body system that controls head movements.

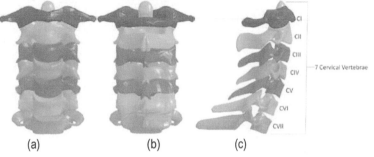

(a)   (b)   (c)

Figure 1 Cervical vertebrae (CI-CVII). (a) Anterior view; (b) Posterior View; (c) Lateral View

Based on the head & neck anatomy the finite element model of a vertebrae-joints system is generated. Our lab has a model of 7 cervical vertebrae without joints. It is imported into LS-DYNA software for further manipulations, such as defining joints.

The connection between the vertebra CI and the occipital bone (of the skull) is the atlanto-occipital joint. It consists of a pair of condyloid joints, which together permits relatively large flexion and extension in the frontal axis that drives the head moving up and down, and small lateral motions that drive the head turning left and right (Drake et al., 2005). As shown in Figure 2, in LS-DYNA software two spherical joints are defined at the center of the superior aritcular surfaces, which are bean shaped and concave. These two joints link the CI vertebra to the occipital bone (not shown in the figure). Each spherical joint has 3 degrees of freedom, but two spherical joints in the same frontal axis give only one degree of freedom. Relative movements between the head and the CI vertebra are constrained to the flexion and extension in the frontal axis.

The dens of the vertebra CII is a distinctive characteristic, which rises superiorly from the body of the vertebra CII and contacts the posterior surface of the CI vertebra, as shown in Figure 3. A closed ring structure is created by a strong transverse ligament and the anterior arch of the vertebra CI, holding the dens in the position. The special structure of the dens, the transverse ligament, and the arc serves as a pivot joint, called Atlanto-axial joint. In LS-DYNA software, we define this joint as a revolute joint that permits the movements of the head rotating side to side.

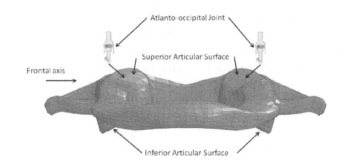

Figure 2 The posterior view of CI vertebra.

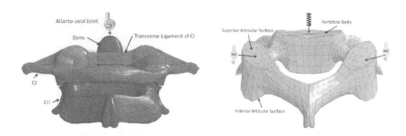

Figure 3 The posterior view of CI-CII vertebrae. Figure 4 The posterior view of CIII-CVII.

Other 5 cervical vertebrae (CIII-CVII) have a typical structure and function, as shown in Figure4. The link between vertebral bodies is a solid joint, so called intervertebral disc. Although the most accurate way to simulate the interaction between vertebral bodies is to build a model of the intervertebral disc, to insert the model between two vertebral bodies and to make a contact simulation between these three objectives, we only care about the movement of the head and this solid joint can be simplified as a spring with high stiffness.

The links between articular surfaces are synovial joints (zygapophysial joints). In LS-DYNA software we use a spherical joint to represent each synovial joint that locates between a superior articular surface of one vertebra and an inferior articular surface of the other vertebra.

After defining all the joints the whole vertebrae – joints system is available, as shown in Figure 5. Generally, it is able to extend and flux along frontier axis, and to rotate along the vertical axis. The joints and springs in this system have stiffness, which will be discussed later. The vertebrae – joints system is ready to be imported into finite element headform models.

In order to add head movement functionality into the headform models, the vertebrae-joint system is incorporated into finite element headform models, which were developed in our previous study (Lei et al., 2011). The assembling is performed in LS-DYNA software. The relative position of the vertebrae-joint system to the headform is based on the anatomy of the head & neck.

The interface between the vertebrae-joint system and the headform is the Atlanto-occipital joint, which connects to the back side of the head and the vertebra CI. By defining the Atlanto-occipital joint in LS-DYNA software, a new finite element headform model that includes the vertebrae-joint system is built with the potential to move side to side, up and down, as shown in Figure 6.

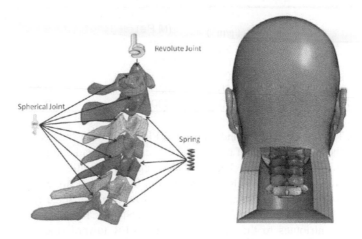

Figure 5 The vertebrae–joints system (lateral view). Figure 6 A complete FEM model.

3   HEAD MOVEMENTS IN CONTACT SIMULATIONS

In our previous study we developed a method to simulate contact between a FRR and a headform without movement. The simulation is divided into two steps: (1) the strap deformation and (2) the FFR and headform interaction (Lei et al., 2011). Considering the new conditions that the head moves, we design an additional step that follows these two steps.

The mechanical properties of each layer are listed in Table I. We give a uniform stiffness, 100 N/radian, to all joints and a uniform stiffness, 100/mm to all springs (solid joints), so that a relative movement of a joint (or a spring) generates a reaction force that equals to the stiffness multiplies the angular displacement.

Six FFR models, including one-size FFR (3M 8210), two-size FFRs (MOLDEX 2200 and MOLDEX 2201), and three-size FFRs (SPERIAN XL, SPERIAN L/M and SPERIAN S) are generated in previous work. From these experiments, a Young's modulus of 10.2 MPa and Poisson's ratio of 0.3 to the straps of the one size FFR were obtained.

Stage I wraps two straps along the back of the headform. At the initial state, the FFR is positioned in front of the headform and about a 3 mm gap is maintained between the FFR and headform. The top strap is located at the top of the back of the headform and the bottom strap is around the neck below the ears. With the FFR in a stationary position, the two ends of each strap are pulled towards the FFR until the straps reach the FFR. As this occurs, the straps are wrapped along the back of the headform. The contact and boundary conditions work together to deform the straps and to generate their internal stresses. In the final state of Stage I, LS-DYNA software automatically exports coordinates and stresses of all nodes as the input of Stage II.

Table I. Mechanical Properties of Headform Layers

Layer	Density (g/m^3)	Young's Modulus (M Pa)	Poisson's Ratio
Skin	1.20	0.6	0.45
Muscle	1.06	0.79	0.42
Fatty Tissue	1.00	0.015	0.48
Bone (Rigid)	4.50	1000	0.30
Back of head (Rigid)	4.50	1000	0.30

Stage II releases the FFR and allows the FFR to contact the headform. We also constrain the displacements of nodes on the back of the head and the bone layers of the forehead, right cheek, left cheek, and chin segments. While the model of the two straps and the model of the FFR are separated in Stage I, they are combined in Stage II. Consider the deformed straps which are extracted from Stage II. At the initial state, the FFR model is in the same position as in Stage I. Potential energy stored in the straps contributes to the movement of the FFR towards the headform. Two contacts exist: the first contact is between the FFR and the headform, and the other one is between the straps and the headform. At the ending state, the FFR rests on the headform. Although the potential energy of the straps now is much less than that at the initial state, it still provides forces on the FFR to generate contact pressure.

In Stage III, the movements of the headform are included in the simulations. Based the respirator fit test procedure there are five conditions, the headform (1) keeps fixed, (2) moves up, (3) moves down, (4) turns left and (5) turns right. The simulation of Stage II achieves the condition that the headform keeps fixed. For each headform-FFR combination, four more simulations are carried out. The contact simulation is terminated at the end of Stage II, when the FFR model contacts the headform model and the whole system is in a steady state. Stage III follows the state of the end of Stage II, and a function, call RESTART, from LS-DYNA software is used for continuing the simulation. The useful feature of the RESTART function is that it permits modifications on a terminated simulation, such as changing boundary conditions, extending calculated time and adding loads.

In the following simulation a load is applied for moving the head. Load of 100 N acting at the back side of the head in the anterior direction makes the head moving down, and load of 100 N acting at the same location in the posterior direction makes the head moving up, as shown in Figure 7(a). The load making the head rotating left is at the bone layer (near nose) in the lateral direction, and the one making the head rotating right is at the bone layer (near nose) in the medial direction, as shown in Figure 7(b). The rigid layers, including the back side of the head and the bone layers of the forehead, the left cheek, the right cheek, and the chin, are rigidly linked as one rigid part. So, when a force is applied at one of them, the five rigid layers would move as a whole. Because a joint links the back side of the head and the vertebrae-joints system, the movements of the head is constrained by the vertebrae-joints system.

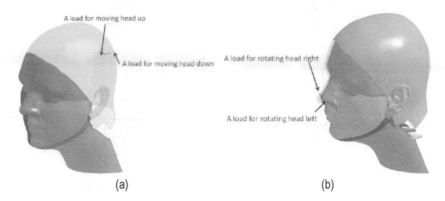

(a)   (b)

Figure 7 Different external loads that can make the head move

Four simulations, including (1) moves up, (2) moves down, (3) turns left and (4) turns right, are carried out separately. At the end of the simulations, the head reaches a stationary state again. Figure 8 compares the final states of the four simulations with the final states of the simulation in Stage II. Figure 9 presents the different pressure distribution on the human face in five simulations.

3   RESULTS

The figures of the pressure distributions on the human face only give us a qualitative way to show simulation results and judge their difference under the condition of four head movements and the condition of no head movement. Figure 10 presents six locations exhibit the highest pressures: the nasal bridge, top of left cheek, bottom of left cheek, top of right cheek, bottom of right cheek, and chin. These six key areas are selected for representing the contact pressure distribution in each simulation.

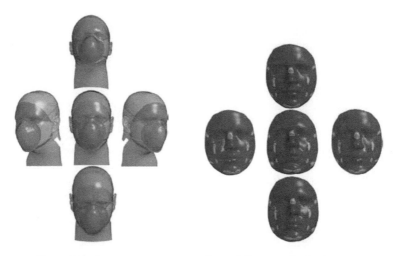

Figure 8 Final states.   Figure 9 Pressure distributions

Figure 10 Six key areas: (1) nasal bridge, (2) top of right cheek, (3) top of left cheek, (4) bottom of right cheek, (5) bottom of left cheek, and (6) chin.

Five size headforms and six FRRs provide totally 30 headform-FFR combinations. For each combination, there are five simulations to calculate the contact under five conditions, such as (1) head does not move, (2) head moves up, (3) head moves down, (4) head rotate left and (5) head rotate right. Tables 2 give the contact pressure values at six key areas from the contact simulation between a large size headform and a one-size FFR. Contact pressure values in other headform-FFR combination are collected in the same manner.

Table 2 Contact pressure values (Unit: MPa) from the contact simulation between a large size headform and a one-size FFR

Location Movements	1	2	3	4	5	6
No move	0.0452	0.0263	0.0295	0.0236	0.0277	0.0235
Head up	0.0383	0.0286	0.0219	0.0188	0.0241	0.0226

Head down	0.0503	0.0292	0.0248	0.0223	0.0333	0.0274
Head left	0.0512	0.0301	0.0201	0.0195	0.0392	0.0404
Head right	0.0518	0.0344	0.0213	0.0300	0.0217	0.0414

After collecting the pressure data in all simulations, we can analyze the effect of head movements on contact pressures between the headform and the FRRs. Tables 3 calculate the pressure difference between simulations with head movements and simulations without head movement for the contact of a large size headform and one-size FFR. Pressure differences in other headform-FFR combinations are recorded in the same way. The first finding is that it is clear that contact pressure distributions change after the movement of the head. Nearly half of the recorded pressure value changes are blew 0.0100 MPa (10 KPa) and the other half are above 0.0100 MPa (10 KPa). We ignore the pressure value differences that are below 0.0100 MPa, and only focus on the differences above 0.0100 MPa. The second finding is that head moving up and down are more likely to change the pressure value at the key area 1 and 6 (the nasal bridge and the chin). The third finding is that the pressure values suffer significant changes when the head rotates left and right.

Table 3 The pressure differences (Unit: MPa) between simulations with head movements and simulations without head movement

Location \ Movements	1	2	3	4	5	6
Head up	-0.0069	0.0023	-0.0076	-0.0048	-0.0036	-0.0009
Head down	0.0051	0.0029	-0.0047	-0.0013	0.0056	0.0039
Head left	0.0060	0.0038	-0.0094	-0.0041	0.0115	0.0169
Head right	0.0066	0.0081	-0.0082	0.0064	-0.0060	0.0179

CONCLUSION

In this study, the effects of head movements on contact pressures between FFRs and headforms were investigated. For each headform-FFR combination, five situations (head does not move, head moves up, head moves down, head turns left and head turns right) were simulated. Results from 180 simulations showed that head movements evidently changed the contact pressure distribution. When head moves up or down, the pressure value at the key area 1 and 6 (the nasal bridge and the chin) are more likely to change. The movements of the head rotating left and right changes pressure values at six key areas significantly.

REFERENCES

Bitterman, B. H.. 1991. Application of Finite Element Modeling and Analysis to the Design of Positive Pressure Oxygen Masks. Air Force Institute of Technology.

Dai, J., J. Yang, and Z. Zhuang. 2011. Sensitivity Analysis of Important Parameters Affecting Contact Pressure between a Respirator and a Headform, International Journal of Industrial Ergonomics, Vol. 41, Issue 3, 268-279.

Del Palomar, A. P., B. Calvo and M. Dolare. 2008. An Accurate Finite Element Model of the Cervical Spine under Quasi-static Loading. Journal of Biomechanics, 41, pp. 523–531.

Drake, R. L., W. Vogl and A. W. M. Mitchell. 2005. Gray's Anatomy for Students. Elsevier Philadelphia

Hedenstierna, S. and P. Halldin. 2008. How Does a Three-dimensional Continuum Muscle Model Affect the Kinematics and Muscle Strains of a Finite Element Neck Model Compared to a Discrete Muscle Model in Rear-end, Frontal, and Lateral Impacts. Spine, Volume 33, Number 9, Page E236-E245.

Hedensitierna, S. and P. Halldin. 2009. Neck Muscle Load Distribution in Lateral, Frontal, and Rear-end Impacts. Spine, Volume 34, Number 24, Page 2626-2633.

Kallemeyn, N., A. Gandhi, S. Kode and K. Shivanna et al. 2010. Validation of a C2–C7 cervical spine finite element model using specimen-specific flexibility data, Medical Engineering & Physics, Volume 32, Issue 5, Page 482-489, Jun.

Lee. S. and D. Terzopoulos. 2006. Heads Up! Biomechanical Modeling and Neuromuscular Control of the Neck. ACM Transaction on Graphics 25, 3 (July), 1188-1198.

Lei, Z., J. Yang, J. and Z. Zhuang. 2010. Contact Pressure Study of N95 Filtering Facepiece Respirators Using Finite Element Method, Computer Aided Design and Applications, 7, Issue 6, 847-861.

Lei, Z., J. Yang, J. and Z. Zhuang. 2012. Headform and N95 Filtering Facepiece Respirator Interation: Contact Pressure Simulation and validation. Journal of Occupational and Environmental Hygiene, Vol. 9, Issue 1, 46-58.

Lei, Z. and J. Yang. Toward High Fidelity Respirator and Headform Models. 1st International Conference on Applied Digital Human Modeling, July 17-20, Miami, Florida, 2010.

Occupational Safety & Health Administration (OSHA). 1999. Respiratory Protection. OSHA Technical Manual (OTM), section VIII: Chapter 2.

Piccione, D., E. T. Moyer Jr. and K. S. Cohen. 1997. Modeling the Interface between a Respirator and the Human Face, Human Engineering and Research Laboratory, Maryland: Army Research Laboratory.

Viscusi, D. J., M. S. Bergman and D. A. Novak, et al. 2011. Impact of Three Biological Decontamination Methods on Filtering Facepiece Respirator Fit, Odor, Comfort, and Donning Ease. Journal of Occupational and Environmental Hygiene, 8: 426-436, July.

Yang, J., J. Dai and Z. Zhang. 2009. Simulating the Interaction between a Respirator and a Headform Using LS-DYNA. Computer Aided Design and Applications, Vol. 6, Issue 4, 539-551.

Yoganandan, N., S. Kumaresan and F. A. Pintar. 2001. Biomechanics of the Cervical Spine Part 2. Cervical Spine Soft Tissue Responses and Biomechanical Modeling. Clinical Biomechanics 16 1-27.

Zhang, Q., E. Teo, P. Lee and K. Tan. "Development and Validation of a Head-neck Finite Element Model for Injury Analysis." XXIII International Symposium on Biomechanics in Sport, Beijing, China, 22-27 Aug 2005.

Zhuang, Q. H., S. H. Tan, and E. C. Teo. 2008. Finite Element Analysis of Head–neck Kinematics under Simulated Rear Impact at Different Accelerations. Journal of Engineering in Medicine, 222: 781.

Zhuang, Z., and D. Viscusi. 2008. A New Approach to Developing Digital 3-D Headforms. SAE Digital Human Modeling for Engineering and Design, Pittsburgh, PA, USA.

Section IV

Product, Service, and System Design

Development of Enhanced Teaching Materials for Skill-based Learning by Using a Smart Phone and Second Life

Akinobu Ando, Toshihiro Takaku,Tomokatsu Saito,
Yasuki Sumikawa, Darold Davis

Miyagi University of Education
Wakuya Junior High School
Replicant AD, LLC
andy@staff.miyakyo-u.ac.jp, toshihiro.takaku@gmail.com,
tomokatsu.saito@gmail.com, liveriver.yasuki@gmail.com,
daroldd@replicant-ad.com

ABSTRACT

In this paper, we present our new learning modules we developed for actions observed in the virtual world called 3Di such as in Linden Lab's Second Life and training by using a smart phone. First, we developed two kinds of learning modules enabling students to understand the differences between a veteran's action and a beginner's. One was "Using a hammer" and another was "Using a saw" as avatar animations in the virtual world. The avatar animations were made using a motion capture system, effectively allowing us to recreate actual actions of both a veteran and a beginner. Second, we developed skill-training material by using a smart phone. In a situation such as using a saw, learners put a smart phone on his/her saw with an attachment, and perform sawing. Then, this application evaluates the learner's motion. Using these types of learning materials helps beginners learn how to use tools without a teacher.

Keywords: teaching materials, skill learning, virtual world, smart phone, sensor

1 INTRODUCTION

Technology education is a skill which students in not only Japanese public education curriculum learn but also students in other countries as well. Particularly in Japan, it is a serious problem for students because the time to learn those skills has been reduced as the public education curriculum changes. Moreover, opportunities of making things for students have been rare as well. It is also important for a nation to nurture people who have creative sensibilities and understanding about industries. We consider that there are two problems about skill learning in Technology Education. The teaching content of technology education includes basic motions: how to use a hammer, a saw, soldering iron and so on. One is that students don't have many opportunities to observe thoroughly someone's motion. Moreover, watching a real situation presents the problem of blind spots because of fixed viewpoints. Therefore they don't understand clearly what the differences are in motion between a model and a beginner. Distinguishing these differences is the first step for learners to grasp the subject matter. Students should get a clear image of their own actions and where their mistakes are before proceeding. Another is that students don't have their teacher out at their home. For example, if a teacher gives skill-based homework to students, few parents will check it for proper form. Without checking and feedback, students cannot develop their own action.

To solve these substantial problems to learn skills, we developed two types of teaching materials for students. For observing motions, we adapted a virtual world: Linden Lab's Second Life and recreated human motions which students need to learn in Technology Education: using a saw and hammer. Moreover, for skill self study and training, we developed application software for smart phones. Current smart phones have acceleration sensors and inclination sensors in them. In this paper, we describe features about two of our two teaching tools we developed.

2 DEVELOPMENT OF TEACHING MATERIALS OF SKILL OBSERVATIONS

In this paper, we focused on one subject area: Technical Education. Junior high school students must take this subject. The students learn to "make things" such as the craft arts (woods, plastics and metals), electricity, metal craft, materials, changing energy and Information Technology. In particular, learning crafts has serious problems. Recent students didn't have much experience in making things using their hands and some tools. Moreover the time required to teach this education is decreasing along with the changes to the Japanese public education curriculum. Thus, students cannot learn and acquire these skills in such a short time frame. They cannot observe an expert's movements using these tools but also beginner's movement.

2.1   Capture and dizitize beginner's motion and model's motion

In order to recreate real human motions in the Second Life virtual world, we needed to overcome two steps. The first was to digitize positions of human joints by using a motion capture system and the other was to adapt captured motion data and the location of human joints into Second Life. We focused on two main tools, which are used to teach these basic skills. These were the hammer and saw. We digitized an expert's motion and a beginner's motion using these tools. Figure 1 shows a scene of capturing a beginner's motion. At the top, two images show views from two cameras. The captured ratio was 30 fps and captured screen size was 640*480. At the end of this step, we made BVH files. These files included coordinates of each joints and simplified human bone structure.

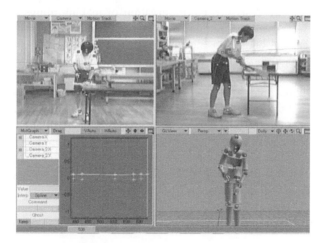

Figure 1 Screen capture of digitizing a beginner's motion

2.2   Recreate motions in the Second Life

To view BVH files as computer animations, a motion viewer is required. In the future we aim to develop a distance-learning application for skill, so the viewer we required must not only have the function of a motion viewer but a function for communication and also be free to get. So we used Linden Lab's virtual world: Second Life. We built a virtual classroom in the Second Life world. To play motion animation, users have to touch a ball object, which has BVH motion data we installed. Figure 2 shows model's sawing animations with changing viewpoints. This function reduces blind spots. When students want to watch these animations, they have to login to Second Life, so we prepared an account for each students group. However, it is expected to be used primarily in a classroom setting.

Figure 2 Screen capture of sawing animation in the Second Life with changing the viewpoints

3 DEVELOPMENT OF TEACHING MATERIALS OF SKILL SELF STUDY AND TRAINING

In chapter 2, we described the observation of making things in craft skills. In this chapter we explain a solution to self-skill learning. Generally speaking, it is difficult to analyze ones own movement. In particular, a beginner doesn't have much Meta-recognition ability. Thus, without a teacher to observe them, students cannot effectively practice how to use tools, e.g. sawing or planing. If they practice and train without the feedback of their results, the will inherently become accustomed to their improper movements. In Japan, the percentage of smart phone users are increasing. It is about 30% according to research done in 2011. All smart phones have three axial accelerate sensors. Using values from sensors we were able to calculate how fast the device is moving and to what degree the device leans and tilts. Our developed application enables a smart phone to be used as a learning device for students' skill simulation.

3.1 The application software for sawing

When a beginner is sawing, incident failure often occurs by a slanted viewpoint. Figure 3 shows the situation. When a beginner moves an elbow backward, it is difficult to move the elbow straight. Our application finds such an improper movement and indicates it. To practice while using our smart phone as an aid, a worker has to fix the smart phone with the installed application to the grip of the saw as an attachment. At the beginning of sawing, the student has to set the saw at an angle until the screen turns blue. This means that it is at the proper angle of sawing. Then, upon touching the start button, a learner simulates the sawing motion. If at any point the movement of sawing becomes improper, the smart phone will indicate it by changing the screen to red, beeping and vibrating (See Figure 4). After finishing the simulation, it shows results in the form of a score and advice to improve. To understand the effects of this application, we conducted a simple test. One group simulated sawing with feedback by using this application and the other group practiced without any feedback. The result of this is shown as figure 5. After four times of simulation or practice, the group with feedback became better significantly than the other group without feedback.

Figure 3 Incident failure of sawing for a beginner

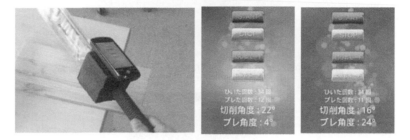

Figure 4 A situation of attached on a saw, and captured screens of good (Blue) and wrong (Red).

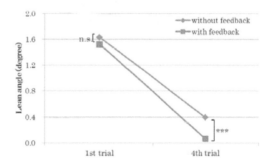

Figure 5. The result of compareing lean angle in our feedback funtion and traditional way.

3.2   The application software for planning

Planing is one of the most important skills of a wood craftsman in the Japanese public education curriculum. A beginner's planing tends to be slow and unstable. In particular, the moving speed at the beginning of planing is an important factor. If there was a teacher near a worker, the teacher would tell the worker concrete advice such as "Move it faster" or "Move it at a breath". Our application is able to indicate this to a worker in place of a teacher.

In order to simulate planing, at first, a leaner fixes a smart phone installed with our application software on a rectangular parallel-piped object like a plan (See

Figure 6). In the application software, proper movement data had already been installed. When the learner finishes the simulation and practice, the application software shows the results of their trial as a graph with proper results that describe movement speed and relative score. Analyzing this graph, the learner can understand how different their own movements are from the proper movement (See Figure. 7).

Figure 6 Fixed smart phone on a rectangular parallel-piped object.

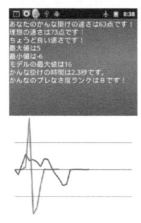

Figure 7 The result of training with a learner's score, speed, time, comment and graph of an ideal motion (Red) and the learner's motion (Blue).

4   CONCLUSION

In this paper, we described our new approaches to observe and simulate how to use some tools without a teacher. We mentioned a hammer, a saw and a plan. When a learner wants to understand how to move their own body and use these tools, it is important to reduce blind spots and watch from various viewing angles repeatedly. So we digitized beginners' motions and a model's motion, and recreated them in a virtual world – Linden Lab's Second Life. By recreating them there, learners can observe movements as frequently as they like and compare proper body motion with improper body motion. The Second Life system also has a voice chat function. For skill distance learning, a learner can ask a teacher via voice chat.

Next, most of the current generations of smart phones have sensors installed. Using values from these sensors, our software application for smart phones we developed can indicate how proper a learner uses a saw and a plane. Most of the learners don't have their own tools at home. Even in such a case, a learner can simulate and practice by using a smart phone without real tools and a real teacher. Their results are shown as a graph with the proper feedback. This application software can also record those results as well. Analyzing the recorded data, the student and the teacher can know the progress of skill improvement.

This application software doesn't have enough precision yet for intermediate level people who want to become better. To improve this, rapid Fourier analysis must be adopted.

ACKNOWLEDGMENTS

This work was supported by KAKENHI (21700783).

REFERENCES

Abiko, H,. and Ando, A, et al. 2005. Basic research on guidance of planing motion: Three-dimensional analyses of planing motion of inexperienced persons, *Bulletin of Miyagi University of Education,* 40: 193-199.

Ando, A,. and Takahashi, H. et al. 2009. Represent of action by using a Second Life viewer, *Proceeding of the 25th Annual Conference of JSET*: 345-346.

Ando, A,. and Abiko, H, et al. 2011. Development and evaluation of a teaching material using a virtual world for watching, *Journal of the Japanese Society of Technology Education (TOUHOKU Branch)*, 4: 34-43.

Ando, A,. and Takaku, T, et al. 2011. Development of a teaching material linked with Twitter for self-instruction to learn skills, *Proceeding of Symposium on Mobile Interactions 2011*: 93-96.

CEC: Center for Educational Computing, 2011. *Application case of educational image resources,* http://kayoo.org/s1_recipe_fr.htm

http://www.impressrd.jp/news/111108/kwp2012

IPA: Information-Technology Promotion Agency Japan. 2011. *Free resources for education,* http://www2.edu.ipa.go.jp/gz/

Keller, J. M. 1987. Development and use of the ARCS model of motivational design, *Journal of Instructional Development,* 10 (3): 2-10.

Sato, K,. and Yoko, A, et al. 2008. The utility of the simple motion capture in the dancing education, *Information Processing Society of Japan SIG Notes,* 100: 9-13.

Shibata, T,. and Yukawa, T, et al. 2008. A Proposal of the CG Contents Production Tool for Explaining the Human Motion, *Information Processing Society of Japan SIG Notes,* 100: 1-8.

CHAPTER 31

Sensor System for Skill Evaluation of Technicians

Noriaki Kuwahara*, Zhifeng Huang**, Ayanori Nagata**,
Kazunari Morimoto*, Jun Ota**, Masako Kanai***, Jukai Maeda***,
Mitsuhiro Nakamura***, Yasuko Kitajima***, Kyoto Aida***

* Kyoto Institute of Technology
Kyoto Japan
nkuwahar@kit.ac.jp, morix@kit.ac.jp
** The University of Tokyo
Tokyo Japan
lnyahzf@gmail.com, nagata@race.u-tokyo.ac.jp, ota@race.u-tokyo.ac.jp
*** Tokyo Ariake University of Medical and Health Sciences
Tokyo Japan
p-kanai@tau.ac.jp, jukai@tau.ac.jp, m-nakamura@tau.ac.jp,
kitajima@tau.ac.jp, k-aida@tau.ac.jp

ABSTRACT

In this paper, we propose self-training system for nursing care students by using the 3D depth sensor and the camera. Our proposed system focuses on the nursing care skills of patient transfer from a bed to a wheelchair and bed making because these skills requires proper use of body mechanics in order to prevent a lower-back injury of nurses and careers. In order to implement such self-training system, the automatic evaluation method of skills is one of the essential functions. Firstly, we discussed the design the checklists for evaluating above-mentioned skills with teachers of the nursing school. Next, we prototyped the system that monitors the trainee's motion during trainee's practice and by using the recognition algorithm, the system automatically evaluates trainee's performance based on above-mentioned checklists. We conducted the preliminary experiment for comparing evaluation results of the system to those of teachers, and examined that the system could evaluate trainee's performance with almost same precision as teachers did. Also, we confirmed the effectiveness of the feedback from our proposed system.

Keywords: Nursing care, Skill, Self-learning system, Kinect

1   INTRODUCTION

In a society with a declining birth rate and increasing percentage of elderly, quality care has become more and more important. Recently, in Japan, while the influx of the people in a wide age group into the care industry, more or less the same amount of the people leave their jobs. Consequently, it causes chronic understaffing problem in the field of the care service. One of the reasons must be insufficient learning of the care skills. Especially, improper conduct of physically intensive cares such as patient transfer to a wheelchair is one of the typical causes of a low back injury of care staffs, for example (Kjellberg, 2004), etc., which leads to their turnover. There are studies in which proper body mechanism is investigated in motions of the nursing care, for example (Collins, 2004), (An, 2010) and so on. The most effective way to learn these care skills is person-to-person guidance, in which a trainer takes a trainee by the hand and teaches a trainee step by step. However, it requires huge numbers of skillful trainers and is not cost-effective. Therefore, we propose the self-training system of care skills that automatically measures the physical movement of a trainee during the care, and reports evaluation results of trainee's care skill.

In order to measure trainee's physical movement, we use Kinect that is the commercial product of the Microsoft. There is the SDK for Kinect that provide the function for extracting the stick model of the human body. However, in the care like patient transfer, it is impossible for this SDK to separate the bodies of the trainee and the patient role, and consequently, is no use. Therefore, we've implemented the technique for separating the physical movement of the patient role and the trainee during the care based on both 3D range data and RGB image data obtained from Kinect. Our proposed system evaluates trainee's performance of the nursing care according to checklists of care skills, and feedback evaluation results to the trainee. Checklists were designed through the discussion with the teachers of the nursing school.

2   METHOD

The purpose of the experiment was to compare our proposed system's evaluation results of trainees with teachers' evaluation results of them, and to confirm the ability of our proposed system. Also, we examined that the effectiveness of our proposed system in terms of self-training by comparing cases with the feedback from our proposed system to those with no feedback.

2.1   Checklist

Table 1 shows the checklist for evaluating the skill of transferring a patient from a bed to a wheelchair, and Table 2 shows the checklist for evaluating the skill of bed making. According to the discussion result with teachers of the nursing school, we identified fourteen steps in transferring a patient from a bed to a wheelchair, and in each step, we defined several important points to check if trainees used body mechanism properly for preventing them from injuries of especially their low-back. Similarly, eight steps were identified in bed making process. Most important point of bed making was to remove the appearance of wrinkles because wrinkles of the surface of the bed might cause the pressure sore of the patient. Additionally, proper use of body mechanism was also important. The grey cells in both Table 1 and 2 represent that these items could not recognized by the system in this experiment for technical reasons.

Table 1 Checklist for evaluating the skill of transferring a patient from a bed to a wheelchair

No.		Check items
1.	a.	Place the wheelchair at the bedside, and orient it at an angle of 20-30 degrees respect to the bedside.
	b.	Place the wheelchair near the patient.
2.		Put on the brake of the wheelchair.
3.	a.	Pull your right foot when making the patient sit on the edge of the bed.
	b.	Put your left foot between the feet of the patient.
4.	a.	Clutch the bottom of the patient.
	b.	Make the patient sit on the edge of the bed by swaying the bottom of the patient.
5.		Make both heels of the patient put near the bed.
6.		Make both arms of the patient lift on your shoulder.
7.		Get your arms around the low back of the patient.
8.	a.	Pull your right foot when making the patient stand up at the bedside.
	b.	Put your left foot between the feet of the patient.
9.	a.	Make your knees bent.
	b.	Make the patient slouch down at the edge of the bed.
10.		Make the patient stand up at the bedside, and turn the patient around your body
11.	a.	Make your knees bent when making the patient sit in the wheelchair.
	b.	Make the patient slouch down in the wheelchair.
12.		Stand at the back of the wheelchair, and grab the forearms of the patient with your hands under the armpits of the patient.
13.		Pull the patient to the seat back of the wheelchair after making the patient slouch down in the wheelchair.
14.	a.	Put the footrest of the wheelchair.
	b.	Put the patient feet on the footrest.

Table 2 Checklist for evaluating the skill of bed making

No.		Check items
1.	a.	Place the mattress pad at the proper position on the bed.
	b.	Place the sheet at the proper position on the bed.
2.		Spread the sheet properly on the bed before putting the edge of the sheet under the mattress
3.		Make the triangle at each corner of the sheet in the proper sequence.
4.		Put the edge of the sheet under the mattress without grabbing both the sheet and mattress.
5.		Turn the back of your hands up when putting the edge of the sheet under the mattress.
6.		Smooth out a wrinkle of the sheet when making the triangle of the last corner of the sheet.
7.		Take a broad-based stance when putting the edge of the sheet under the mattress for using body mechanics properly.
8.		Remove all wrinkles of the sheet on the surface.

2.2 Motion Recognition

Figure 1 shows the experimental setup. Kinect was placed immediately above the bed at a height of about 300cm. Kinect is a motion sensing input device by Microsoft for the Xbox 360 and Windows PCs. It enables users to control and interact with computers with a natural user interface like gestures. Kinect outputs video at a frame rate of 30 Hz. The resolution of the video is 640 × 480 pixels with a Bayer color filter. The monochrome depth sensing video stream is also in VGA resolution (640 × 480 pixels) with 11-bit depth, which provides 2,048 levels of sensitivity. Kinect sensor has a practical ranging limit from 1.2 to 3.5 m distance when used with the Xbox software, although the sensor can maintain tracking through an extended range of approximately from 0.7 to 6 m. The sensor has an angular field of view of 57° horizontally and 43° vertically. There is the SDK for Kinect that provides the function for extracting the stick model of a human body. However, in the care like patient transfer, it is impossible for this SDK to separate bodies of a trainee and a patient role, and consequently, is no use. Therefore, we've come up with the technique for detecting only trainee's physical movement during the care based on both 3D range data and RGB image data obtained from Kinect.

In this prototyping phase, we asked the patient role and the trainee to wear the special markers as shown in Figure 2 in order to reduce complexity of image processing. By using RGB data from Kinect, the system easily determined positions of the head of both the patient role and the trainee by using color information. In the same manner, Wrist, waist, and ankle positions of both of them was also determined. These positions were automatically tracked in the camera coordinate through each session in the training.

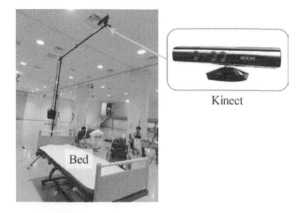

Figure 1　Experimental setup

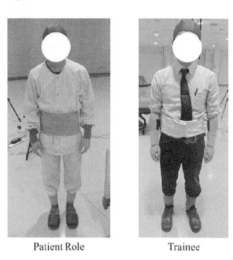

Patient Role　　　　　Trainee

Figure 2　Illustrative example of the experimental setup

2.3　Skill Evaluation

For the skill evaluation of the patient transfer from a bed to a wheelchair, the wheel chair position and orientation were also required. These data were obtained from color data by using some heuristics on its shape.

Then, let's take an example of Figure 3 for introducing the method for determining the step in the nursing care performed by the trainee based on relative positions between each marker, and for evaluating the skill. In Figure 3, the system could easily know that the trainee was in the eighth step as shown in Table 1 by using some heuristics; two wrists' positions of the trainee were behind patient role's head; two wrists' positions of the patient role were behind the trainee's head; trainee's waist and right ankle were observed behind trainee's head.

After the step of the nursing care was determined, the system evaluated the skill of the trainee based on the check items in Table 1. In our previous research, seven cameras traced the markers attached to both the patient role and trainee for evaluating the performance of the trainee (Yonetsuji, 2011); the trainees' performances were compared with the teachers' performances based on the positions of the head, the waist, and the foot, etc. calculated by using marker positions.

In this research, we also evaluated the trainees' performances in the same manner as our previous research did, but we used 3D depth data from Kinect to obtain marker positions in order to make the experiment setup much simpler than that of our previous research.

For evaluation of the skill of bed making, the system observed the size of the sheet region and trainee's position, and inferred the progress of the bed making. In order to evaluate the outcome of the bed making, the system detected edges on the surface of the bed, and counted them for obtaining the number of wrinkles as shown in Figure 4.

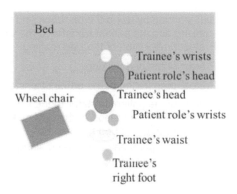

Figure 3   Illustrative example of the experimental setup

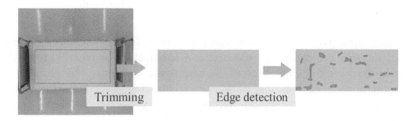

Figure 4   Example of the outcome evaluation of the bed making

2.4   Experimental Protocol

Table 3 shows the experimental protocol. All of the subjects in both the experimental group and the control group could use the textbook and the video for self-learning of bed making and patient transfer from a bed to a wheelchair.

Subjects in experimental group used our proposed self-training system and could get feedback of the evaluation result from the system, although subjects in control group could not get any feedback.

All of the subjects in both groups took pretest and posttest of bed making and patient transfer from a bed to a wheelchair in order to examine the effectiveness of the self-learning. The professional career and the teacher of nursing school also evaluate the trainees' performances in order to confirm the correctness of the system evaluation. Estimated time for the experiment for one subject was on hour and fifteen minutes.

Table 3 Experimental Protocol

Time Line (hh:mm - hh:mm)	To Do
00:00 - 00:06	Understanding the purpose of the experiment, and agreeing in writing.
00:06 - 00:13	Self-learning by using the textbook and the video of the bed making and the patient transfer to the wheelchair.
00:13 - 00:15	Pretest preparation of the bed making.
00:15 - 00:18	Pretest of the bed making.
00:18 - 00:19	Pretest preparation of the patient transfer to the wheelchair.
00:19 - 00:21	Pretest of the patient transfer to the wheelchair.
00:21 - 00:22	Self-practice preparation of the bed making.
00:22 - 00:47	Self-practice of the bed making.
00:47 - 00:48	Posttest preparation of the bed making
00:48 - 00:51	Posttest of the bed making.
00:51 - 00:52	Self-practice preparation of the patient transfer to the wheelchair.
00:52 - 01:12	Self-practice of the patient transfer to the wheelchair.
01:12 - 01:13	Posttest preparation of the patient transfer to the wheelchair.
01:13 - 01:15	Posttest of the patient transfer to the wheelchair.

2.5   Subjects

There were five subjects (two males and three females) in the experimental group, and four subjects (two males and two females) in the control group. All of the subjects were the employees of the care home company. They learned basic skills for nursing care as the workforce training for several months, but their assignments were in the sales, accountant, or personnel divisions after the workforce training.

3 RESULTS

3.1 Precision of the Evaluation of the Bed making

Figure 5 shows the precision of the evaluation of both the patient transfer and the bed making. Horizontal axis in each graph corresponds to check items in Table 1 and Table 2 respectively.

As for the patient transfer, all items achieved more than 50% of the precision rate. On the other hand, as for the bed making, the precision of No.3 was very poor, because the system wrongly recognized trainees' performance inefficient due to improper segmentation of the work area.

Furthermore, as for No.7, the precision rate depended on subjects because the system evaluates proper use of the body mechanism based on the height of the head position and the waist position of the trainee. Therefore the adaptation to subject body size was important.

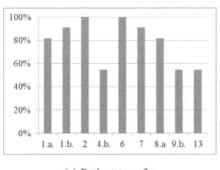

(a) Patient transfer

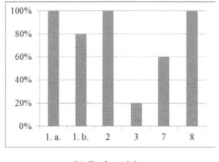

(b) Bed making

Figure 5   Precision of the evaluation of skills

3.2 Performance Improvement of Skills

Figure 6 shows the performance improvement of the patient transfer and the bed making with or without the feedback.

As for the patient transfer, the subjects in the experiment group achieved about 47% at the pretest, and they achieved 80% at the post test. The achievement was improved to 1.7 times. On the other hand, the subjects in the control group achieved about 67% at the pretest, and they achieved about 96% at the post test. The achievement was improved only to 1.4 times.

As for the bed making, the subjects in the experiment group achieved about 32% at the pretest, and they achieved about 64% at the post test. The achievement was improved to double. On the other hand, the subjects in the control group achieved about 19% at the pretest, and they achieved about 33% at the post test. The achievement was improved only to 1.7 times.

Although the number of the subjects was not necessarily sufficient, the experiment showed promising results for our proposed system.

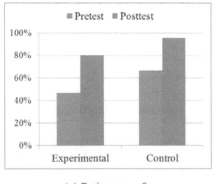
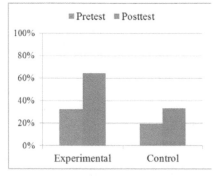

(a) Patient transfer   (b) Bed making

Figure 6   Performance improvements of skills

4   CONCLUSIONS

In this paper, we proposed the system for self-learning of nursing care skills: patient transfer from a bed to a wheelchair: bed making. One of the important features of our proposed system was the automatic evaluation of the skills of the trainee based on checklists that were designed through discussions with teachers of the nursing school.

We confirmed that our proposed system could generally achieve good precision rate of the evaluation of the patient transfer and the bed making, although for some items, further improvement was necessary. As for the patient transfer, only less than half of the items could be automatically evaluated.

The performance improvement of the patient transfer and the bed making by the feedback was also confirmed, although we need further experiment with sufficient number of the subjects in order to investigate the effectiveness statically.

ACKNOWLEDGMENTS

This research was entrusted by the Ministry of Economy, Trade and Industry (METI). A part of this research was supported by a grant-in-aid for scientific research (B) (23300037, 2011-2013) from Japan Society for the Promotion of Science (JSPS).

REFERENCES

Collins, JW., Wolf, L., Bell, J., Evanoff. B., An evaluation of a "best practices" musculoskeletal injury prevention program in nursing homes. Inj Prev, 10, 206-211, 2004.

Kjellberg, K., Lagerstrom, M., Hagberg, M., Patient safety and comfort during transfers in relation to nurses' work technique. J Adv Nurs, 47, 251-259, 2004.

An, Q., Ikemoto, Y., Asama, H., Arai. T., Analysis of Human Standing-up Motion Based on Distributed Muscle Control, 10th International Symposium on Distributed Autonomous Robotic Systems (DARS2010), dars2010_submission_28, 1-12 , 2010.

Yonetsuji, T., Takebe, Y., Kanai, M., et al., A Measurement and Evaluation Method of a Support System to Teach How to Improve Transferring Patients, *Proceedings of the 2011 IEEE International Conference on Robotics and Biomimetics*, 908-913, 2011.

CHAPTER 32

The Design of Adhesive Bandage from the Customer Perspective

John K.L. Ho[1], Steve N.H. Tsang[2], Alan H.S. Chan[2]

[1]Department of Mechanical and Biomedical Engineering
[2]Department of Systems Engineering and Engineering Management
City University of Hong Kong
Hong Kong, China
mejohnho@cityu.edu.hk

ABSTRACT

With increasing demands of more 'personalized' products, the consideration of customer needs in the phase of product development is of great importance. A customer-oriented approach for product development, called Kansei Engineering was developed by Nagamachi for addressing the needs of customers, and has been widely adopted by many designers around the world. In this study, the Kansei Engineering method was applied to study the relationship between the requirements of adhesive bandages and the desirable product features from the customer perspective.

Keywords: Kansei, Kansei Engineering, Design, Adhesive bandage

1   INTRODUCTION

Different from the past, consumers concern not only the price and durability of products nowadays, but also the matching between product design and their personal feelings and expectations for the products. Designers and manufactures can no longer develop a product based only on their own perspectives and concepts of the market needs. On the contrary, the involvement of the customers in the new product development is of the utmost importance for the sake of success.

Kansei Engineering, a consumer-oriented technology for ergonomics product development, was proposed by Nagamachi 40 years ago. Kansei is a Japanese word meaning the consumer's psychological feeling and image to a product, and Kansei Engineering is an approach to convert the consumer's Kansei into the product design elements (Nagamachi, 1995, 2002). The focus of Kansei Engineering is on the relationships between the physical properties of a product and its affective impact on the users. Since its inception, Kansei Engineering has attained recognizable success in many different industries, including automotive, home appliance, construction machine, costume, cosmetic, etc. (Nagamachi, 1986, 1989, 1997, 1999, 2011; Jindo and Hirasago, 1997; Schütte and Eklund, 2005; Hunag, Tsai, and Hunag, 2011).

An adhesive bandage, also called a sticking plaster, is an item used for protecting wounds after injury. In order to optimize the design of adhesive bandages, in this study, Kansei Engineering was applied to design an adhesive bandage from the customer perspective to fulfill the customer needs and requirements.

2   METHOD

2.1   Collection of Kansei words for adhesive bandage design

To understand the customer's psychological feeling towards adhesive bandages, 47 Kansei words describing an adhesive bandage were collected through internet, magazines, product packages and interviews (Table 1).

Table 1 Kansei words collected for adhesive bandage design

No.	Kansei word	No.	Kansei word
1	Waterproof	22	Non-sticky to wound
2	Tough and resistant	23	Easy-open wrapper
3	Long-lasting	24	Stays on in water and shower
4	Soft	25	Allow skin to function normally while providing excellent moist healing environment
5	Flexible	26	Flexes with movement
6	Protect the wound	27	Easily contours to your body, moves and stretches with you
7	Sure-stick adhesive	28	Fitly stick on the skin
8	Air permeable	29	Will not stick to wound for gentle removal
9	Comfortable	30	Protect the wound against germs/bacteria effectively
10	Let your wound breath	31	Prevent dirt from entering the wound
11	Hygienic	32	Enhance healthy skin to grow
12	Sterilized	33	Protect the wound against infection effectively
13	Assorted plasters in a package	34	Advanced healing
14	Low allergy adhesive	35	Stay in place longer
15	Beautiful	36	Reasonable price
16	Not ugly	37	Cheap price
17	Not old fashioned	38	Easy to remove
18	Ideal for face or other visible parts of the body	39	Good reputation of the product
19	Special design	40	Suited for knuckle and fingertip
20	Exaggerated design	41	Suited for moving parts of the body
21	Eye-catching design	42	Suited to protect different kinds of wound

2.2 Identification of major Kansei words as key factors for adhesive bandage design

After collecting the related Kansei words for adhesive bandage design, it was important to understand customers' subjective feelings towards each Kansei word. A questionnaire listing the 42 collected Kansei words was designed and distributed for evaluation. 80 participants (48 males and 32 females) of ages between 17 and 41 years took part in the evaluation. They were instructed to respond to each Kansei word using a 5-point Likert scales (1 for least important, 3 for normal, and 5 for most important). The evaluation was conducted by either face-to-face interview or online survey.

Principal Component Analysis (PCA), a variable reduction procedure for rationalizing the observed variables into a smaller number of principal components (factors) that will account for most of the variance in the observed variables, was then performed on the scores from the evaluations. Kansei words that were correlated would be clustered into a group as a component (factor). The 42 Kansei words were subjected to PCA and the results showed that there were 12 components with eigenvalues greater than 1, and the scree plot (Figure 1) also depicts that the curve start leveling off after component 12. Therefore, the first 12 components were extracted for further investigation. A variable is said to be loaded on a component if the value of factor loading is greater than 0.3 and it was statistically significant if the loading value is greater than 0.6 with the sample size of 80. The 12 extracted components in total explained 74.5% of the variance. The first component, accounted for the largest proportion of the variance, explained 25.5% and 8.7% of the variance before and after rotation, respectively. There were 5 variables loaded to the first component, with 3 variables showing statistically significant values. The first component was labeled as 'Function' to characterize the loading variables (Table 2).

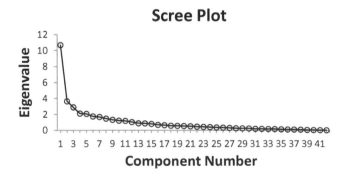

Figure 1 Scree plot of the 42 Kansei words

Table 2 The principal components of the Kansei words (only the first component is shown)

Principal component	Variables	Factor loadings	Percentage of variance explained	The interpretation of the component
1	*12 Hygienic	0.856	8.693	Function
	*13 Sterilized	0.853		
	*9 Air permeable	0.603		
	25 Non-sticky to wound	0.587		
	38 Protect the wound against infection effectively	0.385		

* Statistically significant with factor loading greater than 0.6

2.3 Investigation of the relationship between the selected Kansei words and product design features

Knowing the principal components for adhesive bandage requirements, multiple regression analysis was applied to study the relationships between those components and product design features. To narrow the scope of study, only the Kansei words within the first component were adopted for further study. Within the first component, 3 out of the 5 Kansei words were found to be statistically significant (factor loading > 0.6), and these Kansei words were 'Hygienic', 'Sterilized', and 'Air permeable'.

24 adhesive bandage samples were designed to investigate the relationships of the product features and the selected Kansei words. The 24 samples were the combinations of 8 different colors (red, yellow, blue, black, white, grey, flesh-colored, and transparent and 3 different pore sizes with diameters of 0.187mm, 0.297mm, and 0.456mm for small, middle, and large sizes, respectively. The size of each adhesive bandage sample was 18mm x 64mm, which was the typical size found in the market. A questionnaire showing the real samples of the adhesive bandage designs was constructed and distributed for participant evaluations. 82 participants (49 males and 33 females; age ranged from 17 to 41) were interviewed. They were asked to evaluate the adhesive bandage sample designs in terms of 'Hygienic', 'Sterilized', and 'Air permeable' using a 5-point Semantic Differential Scale (SD-Scale).

The multiple regression equation takes the form $y = c + b_1 x_1 + b_2 x_2 + b_3 x_3 + b_4 x_4$, where $b_1, b_2, ..., b_n$ are the regression coefficients measuring how strongly each independent variable (predictor) influences the dependent variable (criterion), and c is a constant (intercept) of the equation. The multiple regression equations for 'Hygienic', 'Sterilized', and 'Air Permeable' were established ((1) to (3)). For 'Hygienic', the independent variables in regard to the colors of yellow and flesh-colored, and to the sizes of small and large were excluded from the equation because of their insignificant contribution to the dependent variable. Table 3 shows the results of the multiple regression analysis for the three Kansei parameters. Of those significant variables for 'Hygienic', w (0.168), m (0.109), and c (0.063) showed positive values, while b (-0.105), r (-0.129), g (-0.183), and bl (-0.313) exhibited negative values. The results indicated that an adhesive bandage with white or clear color and with middle pore size could lead to the feeling of hygiene, whereas black, grey, blue or red color should be avoided if hygienic image of the adhesive bandage is of great concern. In regard to the dependent variable of 'Sterilized', findings similar to that for the variable of 'Hygienic' were found. Customers would feel sterilized if the adhesive bandage was in middle size in white or transparent, while the opposite feeling would be induced if black, grey, blue, or red color was used. Regarding the feeling of 'Air permeable', w (0.137), c (0.118), and fc (0.047) showed positive values, and g (-0.046), l (-0.125), bl (-0.162), and s (-0.337) showed negative values. The findings indicated that white, clear, or fresh-colored design could provide the feeling of air permeable. On the contrary, grey or black color and large or small size would give an adverse feeling of air permeable for the adhesive bandage design.

$$Hygienic = 3.079 - 1.126(bl) + 0.606(w) - 0.659(g) + 0.228(c) + 0.274(m) - 0.463(r)$$
$$- 0.378(b) \dots\dots\dots\dots\dots\dots\dots\dots\dots\dots\dots\dots\dots\dots\dots\dots\dots(1)$$

$$Sterilized = 3.087 - 1.000(bl) + 0.573(w) - 0.585(g) + 0.268(c) + 0.252(m) -$$
$$0.337(r) - 0.272(b) \dots\dots\dots\dots\dots\dots\dots\dots\dots\dots\dots\dots\dots\dots(2)$$

$$Air\ Permeable = 3.362 - 0.873(s) - 0.598(bl) + 0.504(w) + 0.435(c) - 0.323(l) +$$
$$0.175(fc) - 0.171(g) \dots\dots\dots\dots\dots\dots\dots\dots\dots\dots\dots\dots\dots(3)$$

Where (bl)=black, (w)=white, (g)=grey, (c)=clear, (r)=red, (b)=blue,(fc)=flesh-colored, (s)=small, (m)=middle, and (l)=large

Table 3 Results of multiple regression analysis for 'Hygienic', 'Sterilized', and 'Air permeable'

	Product Feature	Standardized Coefficient (Beta)	Sig.
Hygienic	w	0.168	0.000
	m	0.109	0.000
	c	0.063	0.006
	b	-0.105	0.000
	r	-0.129	0.000
	g	-0.183	0.000
	bl	-0.313	0.000
Sterilized	w	0.170	0.000
	m	0.106	0.000
	c	0.079	0.001
	b	-0.081	0.001
	r	-0.100	0.000
	g	-0.173	0.000
	bl	-0.296	0.000
Air Permeable	w	0.137	0.000
	c	0.118	0.000
	fc	0.047	0.035
	g	-0.046	0.039
	l	-0.125	0.000
	bl	-0.162	0.000
	s	-0.337	0.000

(bl)=black, (w)=white, (g)=grey, (c)=clear, (r)=red, (b)=blue,(fc)=flesh-colored, (s)=small, (m)=middle, and (l)=large

2.4   Establishment of the desirable product features of new design of adhesive bandage

From the results of multiple regression analysis, the design features having significant contributions to the Kansei words were identified. 'Hygienic', 'Sterilized', and 'Air Permeable' were of the greatest concerns for the adhesive bandage design. In order to provide the feelings of 'Hygienic', 'Sterilized', and 'Air Permeable' to the customers, the adhesive bandage should be in white color since white color attained the highest positive contribution to these three Kansei words. Besides the white color, the transparent color also, to a weaker extent, gives customers the feelings of 'Hygienic', 'Sterilized', and 'Air Permeable'. Black, blue, red, or grey color should be avoided as they give customers negative feelings in aspects of 'Hygienic', 'Sterilized', and 'Air Permeable'. For the effect of pore size, middle size is better for the delivery of the feelings of 'Hygienic' and 'Sterilized', whereas small pore size leads to poor 'Air Permeable' feeling. To sum up, taking all 'Hygienic', 'Sterilized', and 'Air Permeable' into consideration, a desirable adhesive bandage should be of middle pore size and in white color.

3.   CONCLUSION

This study demonstrated the procedure of implementing Kansei Engineering for a product design. First, the product, an adhesive bandage, was selected for study, and the Kansei words towards the design of an adhesive bandage were collected by different means such as from advertising materials and interviews from customers. Afterwards, the Principal Component Analysis (PCA) was used to cluster relevant Kansei words as a factor, and multiple regression analysis was applied to identify the relationships between the Kansei words and product features. Finally, a Kansei adhesive bandage fulfilling the customer expectations and needs was successfully designed.

ACKNOWLEDGEMENT

We would like to thank S L Ling for conducting the survey and analyzing the data of the study.

REFERENCES

Huang, M.S., H.C. Tsai, and T.H. Huang. 2011. Applying Kansei engineering to industrial machinery trade show booth design. *International Journal of Industrial Ergonomics* 41: 72-78.

Jindo, T. and K., Hirasago. 1997. Application studies to car interior of Kansei engineering. *International Journal of Industrial Ergonomics* 19: 105-114.

Nagamachi, M. 1986. Image technology and its application. *The Japanese Journal of Ergonomics* 29: 196-197.

Nagamachi, M. 1989. Kansei Engineering. Tokyo: Kaibundo Publishing.

Nagamachi, M. 1995. Kansei Engineering: A new ergonomic consumer-oriented technology for product development. *International Journal of Industrial Ergonomics* 15: 3-11.

Nagamachi, M. 1997. Kansei Engineering: the framework and methods. In. Kansei Engineering, ed. M. Nagamachi. Kaibundo Publishing, Kure, pp1-9.

Nagamachi, M. 1999. Kansei engineering: the implications and applications to product development. *Proceedings of IEEE International Conference on Systems, Man and Cybernetics* 6: 273-278.

Nagamachi, M. 2002. Kansei engineering as a powerful consumer-oriented technology for product development. *Applied Ergonomics* 33: 289-294.

Nagamachi, M. 2011. Kansei/affective engineering., Boca Raton, FL, USA: CRC Press.

Schütte, S. and J. Eklund. 2005. Design of rocker switches for work-vehicles-an application of Kansei Engineering. *Applied Ergonomics* 36: 557-567.

CHAPTER 33

Journeying Toward Female-focused m-Health Applications

Lishan Xue, Ching Chiuan Yen, Leanne Chang, Bee Choo Tai,
Hock Chuan Chan, Henry Been-Lirn Duh, Mahesh Choolani

National University of Singapore
didxl@nus.edu.sg

ABSTRACT

The mobile phone is a commodity tool for personalized health care management and has become a means of health information computing device. A female-dedicative mobile-health (m-health) application was constructed using the Female-focused Design Strategy (FDS), to explore women's reaction and acceptance toward well-being management after interacting with it. Qualitative user feedback was collected and emotional tolerance and user affinity for mobile and health technology may explain one's openness to the services as well as shape their perceptions about the application. In addition, it is possible that acceptance decisions may be influenced by women's health awareness, receptiveness to mobile applications or technological experience which design could help alter. To date, data elements and application features of current m-health applications are often incomplete and not properly secured. It is known fact that emotions are at a basic importance for the success of a system since they influence the evaluation, purchase decision and experience of users significantly. This paper reports an exploration of other non-technical challenges to recommend better design of m-health applications for women and assist in designing optimal informational interventions.

Keywords: user interfaces, mobile health applications, female interactions

1   INTRODUCTION

Mobile phones have become an important part of people's everyday lives. Most people carry their phones with them all the time, everywhere they go, which provides opportunities for timely persuasion and pervasive reminders related to health management (Fogg 2003, Consolvo et al. 2008). Personal health and well-being management seems like a natural extension of a mobile phone's use. A mobile health application (m-Health app) integrates health management to the daily life of the user, and is present in the situations where health-related decisions are made. The m-Health field operates on the premise that technology integration within the health sector has the great potential to promote a better health communication to achieve healthy lifestyles, improve decision-making by health professionals (and patients) and enhance healthcare quality by improving access to medical and health information and facilitating instantaneous communication in places where this was not previously possible (Shields, Chetley, and Davis, 2005). It follows that the increased use of technology can help reduce health care costs by improving efficiencies in the health care system and promoting prevention through behavior change communication. For instance, it enables health management tasks during the idle moments of life, such as when travelling on the bus or waiting for a friend (Srivastava 2005, Fogg 2003).

The motivation to examine the usefulness of m-Health apps for women is three-fold. Firstly, women regardless of ethnicity live longer than men; they have greater morbidity (Rodin and Ickovics, 1990); are more attuned to take care of their health and are also more likely to seek preventive services over the internet (Gijsbers van Wijk et al. 1992; Mustard et al. 1998). Secondly, mobile device technology, as well as mobile content and application development, is increasingly being adopted by women as they now have more economic power and demand that any new treatment options demonstrate cost-effectiveness. The convergence of these factors has resulted in making well-being management for women a viable design opportunity today. There is a real demand of minimally invasive, cost-effective designs needed to address the health needs of women (Waldron 1997). Thirdly, in Singapore where this study is conducted, the mobile network is considered mature; its government is also circumspect about the prospect of the next generation of mobiles and evidence show a penetration rate of 131% among mobile communication usage (Singapore Department of Statistics, 2008).

2   BACKGROUND

A Female-focused Design Strategy (FDS) was developed to encourage the adoption and continued usage of by women incorporating values and attributes they will look for in purchasing and using a self-care information system (Xue and Yen, 2009). A number of quantitative studies which provide the basis for the FDS have been carried out over a period of 5 years in relation to different studies connected to projects concerning female innovations for self-care (Xue, Yen, and Choolani, 2006; Xue and Yen, 2008; Xue et al., 2009; Xue et al. 2012) . The projects aimed at different age groups of women, ranging from participants aged 20 to 82. Different methods ranging from paper surveys to national telephone interviews were employed. Although the studies did not have exact identical research foci, they all focused on collecting user perspectives and acceptance concerning dedicative health IT for women's health. The FDS explains from grounded comparison that women in general are more motivated by systems that benefit them with flexible and responsive interaction and appreciate designs that relate to them almost naturally, with empathy; they like details or features which make the system dependable (see Figure 1).

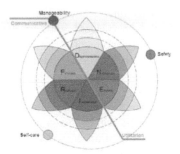

Figure 1 The Female-focused Design Strategy (FDS) is presented in a model using the acronym 'Flexibleness. Responsiveness. Intricateness. Empathy. Naturalness. Dependableness.'. Essentially, FRIEND is an acronym which stands for attributes relating to and forming the FDS.

A test design of a mobile web application specially designed to empower women in prevention and understanding their own conditions in their daily activities was created . It was introduced as *mobileHealthweb*. The layout of content, features and tools were considered based on the attributes from the FDS. The FDS served as a guideline for integrating needs of users with varying abilities into design at an early phase. It was effectively a website designed to be accessed solely from mobile phones to provide women with an easy grasp of health information and visualization of how information at hand can be useful and reliable at all times. It was not narrowly focused on a specific patient type or disease, wellness tracking, or medication tracking. Users could access it online via GPRS (General Packet Radio Service) as long as their phones are equipped with a web browser which all smartphones come with.

The system provided users the option of answering step-by-step questions to cope with uncertainties with advice generated based on their very own replies and it also provides knowledge to increase activities that promote their well-being. A quantitative survey of how women may be engaged with the system seeking health care knowledge in non-clinical environments was conducted (Lim et al. 2011). System logs for each registered user could be obtained and the link between behavioural intention and actual use could thus be established. A total of 164 valid replies were collected with the average age of female participants being 33.6 years, and the range being 21 to 62 years. In brief, the quantitative results show that respondents would consider adopting the application when its use proves effortless and value is properly demonstrated. They would most likely use it if there were different tweets towards their personal preferences and contextual needs. Similar to another research (Boland 2007), participants prefer health information and services to be delivered with relevance to personal characteristics and administered in a method that they prefer. By providing tailored options, users' demands could be met more easily.

On the other hand, qualitative user feedback helped to identify other reasons for the acceptance of the technology and the enthusiasm for it besides perceived usefulness. Content credibility, data security, privacy, the ability to import and export data, upload images and communication with health care providers are all seen as important features for future m-Health apps for women. User interface and ease of finding and entering data are application features mentioned but they can be resolved if the application was solely designed and available via smartphones. We recognize the fact that designers and developers are challenged by the competing and incompatible mobile software platforms, for example, Google Android, Symbian, Java 2 Micro Edition; applications must be customized for each platform and even for different phone models within the same platform. Qualitative results also revealed non-technical factors influencing the acceptance decision, one of which was the personality of the potential user. Emotional tolerance and affinity for health affairs and health technology could explain one's openness to the services as well as shape their perceptions about the application. In addition, it is possible that acceptability could be influenced by women's health awareness, tolerance for health conditions and risks, receptiveness to mobile applications, and demographics.

It is known that emotions are at a basic importance for the success of a system since they influence the evaluation, purchase decision and experience of users significantly. Emotions call forth a coordinated set of behavioural, experiential, and physiological responses that together influence how users respond to perceived challenges and opportunities. Previous studies focusing on the doctor-patient interaction show that when the supplier of information fails to understand the receiver's preferences, the latter will disregard the recommendation completely (Glycopantis and Stravropoulou, 2011). This further strengthens the fact that user considerations begins to shift away from harder, technical and functional performance factors towards softer, more human aspects of emotional engagement, affective, cognitive, and social consequences. Against this background, it is interesting to examine trends from gendered perspectives, linguistics and aesthetics management and cognitive and social science that will impact women's adoption and usage of m-Health apps.

3 THE PRESENT STUDY

It is clear that there are many different ways to regulate emotion for a system design. We seek to establish vocabularies and tactics to create meaningful female-focused user experiences when using m-health apps. User experience is a dynamic, complex and subjective phenomenon. It depends upon reactions to multiple attributes of a design, for example, its behavior, rationale, and appeal that are interpreted through filters relating to individual, social and task significance

(Macdonald, 1998). The experience of virtual interfaces on m-Health apps is no doubt influenced by a dense interplay of technical considerations and contextual factors. We recognize the fact that to achieve a desirable female-focused m-Health app, designers will be confronted with lots of complexity but for the purpose of this paper we focus our discussion on issues such as gendered concerns, choice of aesthetics, cognitive science and social integration. We are not striving to validate or invalidate any factors which may have more significance, but we aim to discuss obvious issues that enable designs to be more female user driven.

3.1   Gendered Tendencies

The gender aspect is a good way of organizing the diverse approaches to information design and system architecture when we take into account that an increasing share of users is women. In general, women are more inclined to be multi-minded and integrated; they believe the best way to absorb information and make decisions is to see the details and understand it comprehensively in order to correctly grasp the information in order to make a good decision with respect to the information. Health messages come from medical, social, scientific, and public sources. To some degree, some of these messages compare with and contradict each other. The range of issues can be examined from political, historical, technological and feminist perspectives. Contradictory information can result in a range of cognitive and emotional effects such as anger, fear, confusion, skepticism, anxiety, and guilt. Learning theory literature states that learning styles may be influenced by gender, especially for adults. Men tend to be autonomous or independent learners. The majority of women, on the other hand, tend to learn in a relational, connected, or interdependent way. Inclusive design practices acknowledge and accommodate these different approaches to learning (Campbell, 1999).

The task of recognizing different sources message executions for different targets even among the female population is not entirely up to the designer, but there are potential gatekeepers who can help shape and reshape health messages. They are administrators, health care providers, consultants, stakeholders, creative and research teams, evaluation and implementation team such as political officers. Communication theories also point to the importance of considering diversity in developing effective health communication interventions. When women communicate, they are concerned with conveying information and building connections. Information provided to women "should be targeted"; women need to perceive information as personally relevant to them in their age group and culture. Boundaries extend to their privacy, relationships, and view of doctors. From previous research, we gathered from the Singaporean women appreciate usefulness of such technologies and will consider adoption base on true benefits. However, this does not reveal how to design messages to convince them otherwise to influence their beliefs or health behavior change. But it may be possible to detect women's preferred choice through linguistics.

3.2   Linguistics

The linguistics applied toward the connections between health messages and a female audience concerns itself with describing and explaining the nature of health management and self-care. Understanding about the different though equally valid conversational frequencies men and women are tuned to can help shape the language to context. According to Bonvillian (2000), men grow up in a world in which a conversation is often a contest, either to achieve the upper hand or to prevent other people from pushing them around. For women, however, talking is often a way to exchange confirmation and support. Likewise, from a health information system and structure, the language used should sound supportive and decrease any possibility of stressfulness. Secondly, since women often think in terms of closeness and support, they struggle to preserve intimacy. Men, concerned with status, tend to focus more on independence. The problem stemmed from a difference in approach. To many men, a complaint is a challenge to come up with a solution. But often women are looking for empathy, not solutions. In general, women accept it better when advice is offered as proposals rather than orders. They are more convinced to act on something when they are won over by agreement first. These traits can lead women and men users to starkly different views of the same situation. Certain types of jargon and clichés are favoured more by females than males. A deeper analysis of how women appreciate linguistics is important to increase positive emotional experience. However, designers should design with universal usability in mind, remembering there are diverse users and understand how to design interfaces which do not require good language abilities. Linguistic used for information systems may have evolved to include emoticons, which are emotion graphics or visual ways to express the way one feel when words alone are not sufficient (MSN. com, n.d.). Icons and other aesthetic representing emotions for users is possibly a new language to motivate users' response which may be effective across the population with different educational levels and cognitive maturation.

3.3   Aesthetics

Our culture is largely visually-orientated and the first encounter with a product or system is usually visual. An inclusive definition of aesthetics is concerned not just with visual form, colour, or texture, but with understanding and predicting the effects of information from all the senses on human perceptions and cognition. For the m-Health app in question, the sensations aroused from sight, touch, hearing a female user experience is highly variable. With reference to existing literature, there is a good overview of work formally attempting to understand women's aesthetics responses to properties of colour, unity, symmetry, regularity, and harmony (Berylne, 1971, 1974; Crozier, 1994). They serve as a guide to consider one part of how users experience with interfaces.

Colour in the m-Health app interface is the basic material of two-dimensional images and visual experiences (see Figure 2). There is much literature about colours women prefer. Eysenck's study in 1941, however, found only one gender difference with yellow being preferred to orange by women. This finding was reinforced later by Birren (1952) who found women placed orange at the bottom of the list. Guilford and Smith (1959) found women were generally less tolerant toward achromatic colours than men. Thus, Guilford and Smith proposed that women might be more colour-conscious and their colour tastes more flexible and diverse. Likewise, McInnis and Shearer (1964) found that blue green was more favored among women than men, and women preferred tints more than shades. Radeloff (1990) has found that women were more likely than men to have a favorite colour. In expressing the preferences for light versus dark colours, there were no significant differences between men and women; however, in expressing the preference for bright and soft colours, there was a difference, with women preferring soft colours and men preferring bright ones. Last but not least worthy of consideration, the colour green is said to be the most restful colour for the human eye. Some claim that green has great healing power and that it can soothe pain. Ideally, it may be used for health-related typography and info graphics.

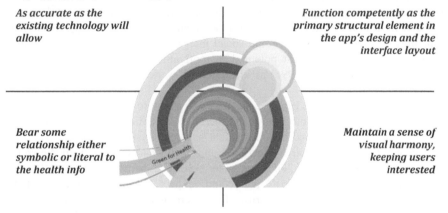

As accurate as the existing technology will allow

Function competently as the primary structural element in the app's design and the interface layout

Bear some relationship either symbolic or literal to the health info

Maintain a sense of visual harmony, keeping users interested

Figure 2 Aesthetics to enhance female-focused user experiences

Aesthetics made up of colour, content layout, icons, graphics and any other display form the comprehensibility of the digital system as far as a m-Health app is concerned and can only function successfully on several levels simultaneously. Firstly, on a technical level, aesthetics must be as accurate as the existing technology will allow, heeding the rules of optics such as providing sufficient contrast. Secondly, it is essential that aesthetics bear some relationship either symbolic or literal to the health message. It may then catch the user's attention in order to convey appropriate information. Thirdly, aesthetics must function competently as the primary structural element in the app's design and the interface layout, creating appropriate spatial and navigational effects on the page and the app as a whole. It is used to direct the eye to the most important areas on the page. Fourthly, aesthetics must create a sense of visual harmony, capturing, sustaining and enhancing the users' interest in the information search and reading experience. She has to feel at ease before discovering the usefulness of the health information.

Moving beyond colour, it is an accepted view women tend to be more spontaneous with appearance and tend to tailor both real and virtual environments and their avatars more frequently and with more embellishments. Certainly, designers should think about gender at a level of sophistication beyond colour and shape. There is a need to be reflective and conscious of the assumptions of use and user being built into the app.

3.4   Cognitive Science and Social Integration

Social integration has recently begun to acknowledge what users, designs and anthropologists have emphasized before that material possessions have a profound symbolic significance for their users, as well as for other people, they influence the ways in which we think about ourselves and about others. Miller (1997) writes of the myriad ways in which 'objects' have significance socially and establishing meaning about our lives and ourselves (Csikszentmihalyi 1991). He suggests that we need to develop a much more conscious psychology of objects which he says might lead to a social ergonomics to parallel the cognitive ergonomics. Likewise, we need to deal with the usage of mobile applications in similar ways. Applying principles from cognitive science and social integration to establish more socially expressive systems is of great potential in motivating users to collaborate with systems (Reeves and Nass, 1996) as illustrated by work on mobile persuasion (de Ruyter et al., 2005).

Social expressive agents has been suggested as a way to build relationships with users; to instill trust, promote liking, and increase perceptions that a system cares about its user (Bickmore and Schulman, 2007). Social system responses that take into account the user's affective experience and circumstances have been shown to lower user frustration (Hone, 2006) and foster perceived caring and support (Brave, Nass, and Hutchinson, 2005). We could expect social, empathic system expressiveness to positively affect trust of a system as a whole and willingness to comply with its requests. However, it is important social expressive behavior has to be adapted to the individual's context, considering the user's social and cultural background, and her personal momentary experience. An overview of experiential and cultural approaches and empathetic design methods that may be helpful in this regard is provided by Wright and McCarthy (2008). Careful consideration of which behavior will match the context, user and system purposes is crucial.

4   CONCLUSIONS

We began with designing and testing a m-Health app for women and then aim to identify more factors needed for researching, designing, using, and understanding m-Health apps for women, present and future. The concept of a m-Health app specially designed for women using female-focused design principles is new, not only in theory but in practice. Our discussion has implications for the theoretical understanding of structuring a model characterizing the attitudes and readiness of women towards m-Health apps as well as for designing m-Health apps for women, and possibly others. Competency in m-Health apps for the genders is likely to grow at a rapid rate in the coming years, due to advances in computing power and the considerable profits at stake. Research and design in m-Health apps for women should progress rapidly as well, because technology will change the way research is conducted. There are still much to investigate because both academia and industry still lack a deep understanding. There is a need to conduct surveys and broader studies to better understand if there are any key issues that we have not identified, or female populations to whom our findings do not apply.

ACKNOWLEDGMENTS

The authors would like to thank the support from the AcRF grant (WBS: R-298-000-001-112) provided by the Ministry of Education, Singapore and the School of Design and Environment, National University of Singapore.

REFERENCES

Berlyne, D.E. 1971. *Aesthetics and psychobiology*. Appleton-Century-Crofts, New York.

Berlyne, D.E.1974. *Studies in the new experimental aesthetics*. Hemisphere, Washington D.C.

Bickmore, T. and Schulman, D. 2007. Practical approaches to comforting users with relational agents. In *Proceedings of CHI'07*, 2291-2296.

Birren, F. 1952. *Your color and yourself*, Sandusky: Prang Company Publishers.

Boland, P. 2007. The emerging role of cell phone technology in ambulatory care. *Journal of Ambulatory Care Management*, 30(2):126-133.

Bonvillian, N. 2000. Language, culture, and communication: the meaning of messages. Upper Saddle River, N.J.: Prentice Hall.

Brave, S., Nass, C. and Hutchinson, K. 2005. Computers that care. *International Journal of Human Computer Studies* 62:161-178.

Campbell, K. (1999). "Learner characteristics and instructional design." Accessed January 10, 2012 www.atl.ualberta.ca/articles/idesign/learnchar.cfm#part7.

Consolvo, S.E., Everitt, K., Smith, I. and Landay, J.A. 2006. Design Requirements for Technologies that Encourage Physical Activity. In: Grinter, R., Rodden, T., Aoki, P., Cutrell, E., Jeffries, R. & Olson, G.M. (eds.).*Proceedings of the SIGCHI conference on Human Factors in computing systems*. Montréal, Québec, Canada, April 22–27 2006. New York, NY, USA: ACM. 1:457–466.

Csikszentmihalyi, M. 1991. Design and order in everyday life. In *Design Issues*, 8:26-34.

Crozier, R. 1994. *Manufactured pleasures: Psychological responses to design*. Manchester University Press, Manchester, UK.

Eysenck, H. J. 1941. A critical and exprimental study of color preferences. *American Journal of Psychology* 54: 385-394.

Fogg, B.J. 2003. *Persuasive technology. Using computers to change what we think and do*. San Francisco, CA, USA: Morgan Kaufmann. 318.

Gijsbers van Wijk, C.M.T., Kolk, A. M., Van den Bosch, W.J.H.M. and Van den Hoogen, H.J.M. 1992. Male and female morbidity in general practice: the nature of sex differences. *Social Science and Medicine* 35: 665–78.

Glycopantis, D. and Stravropoulou, C. 2011. The supply of information in an emotional setting. *CESifo Economic Studies* 57(4): 740-762.

Guilford, J. P. and Smith, P. C. 1959. A system of color-preferences. *The American Journal of Psychology* 73 (4): 487-502.

Hone, K. 2007. Empathic agents to reduce user frustration. *Interacting with Computers* 18:227-245.

Lim, S., Xue, L., Yen, C.C., Chang, L., Chan, H.C., Tai, B.C., Duh, H.B.L. and Choolani, M. 2011. A study on Singaporean women's acceptance of using mobile phones to seek health information, *International Journal of Medical Informatics* 80:e189-e202.

McInnis, J. H. and Shearer, J. K. 1964. Relationship between color choices and selected preferences for the individual. *Journal of Home Economics* 56:181-187.

Miller, H. 1997. The social psychology of objects. In *Proceedings of Understanding the Social World Conference*, The Nottingham Trent University, UK.

MSN.com (n.d.). "Use emoticons in messages". Microsoft Corporation [online]. Accessed January 13, 2012, http://messenger.msn.com/Resource/Emoticons.aspx

Mustard, C.A., Kaufert, P., Kozyrsky, A. and Mayer, T. 1998. Sex differences in the use of health care services. *New England Journal of Medicine* 338:1678–83, 1998.

Radeloff, D. J. 1990. Role of color in perception of attractiveness. *Perceptual and Motor Skills*, 71:151-160.

Reeves, B. and Nass, C. 1996. *The media equation.* Cambridge University Press & CSLI Press.

Rodin, J. and Ickovics, J.R. 1990. Women's health: Review and research agenda as we approach the 21st century. *American Psychologist* 45:1018-1034.

de Ruyter, B. et al. 2005. Assessing the effects of building social intelligence in a robotic interface for the home. *Interacting with Computers* 17:522-541.

Singapore Department of Statistics. "Social Indicators, Mobile phone subscriber, 2008." Accessed August 18, 2011, www.singstat.gov.sg/stats/charts/socind.html#socB

Srivastava, L. 2005. Mobile phones and the evolution of social behaviour. *Behaviour and Information Technology* 24(2): 111–129.

Waldron, E.E. "Tuning into the harmonic convergence in women's health." Medical Device and Diagnostic Industry Magazine, 1997. Accessed August 18, 2011, www.devicelink.com/mddi/archive/97/07/016.html

Wright, P. and McCarthy, J. 2008. Empathy and experience in HCI. In *Proceedings of CHI'08*, 637-646.

Xue, L, Yen, C.C. and Choolani, M. 2006. Framework examining female user response to GUI for e-Health information. *Proceedings of the Design Research Society Conference "Wonderground".* Portugal. (Design Research Society Conference, 1 - 4 Nov 2006, Lisbon, Portugal, Portugal).

Xue, L and C Yen, C.C. 2008. Introducing a female-focused design strategy (FDS) for future healthcare design. *Dare to Desire*, ed. PMA Desmet, SA Tzvetanova, PPM Hekkert & L Justice, comp. Hong Kong Polytechnic University. Hong Kong: The Hong Kong Polytechnic University. (6th Conference on Design & Emotion, 6 - 9 Oct 2008, Hong Kong Polytechnic University, Hong Kong).

Xue, L. and Yen, C.C. 2009. Thinking design for women's health, *Design Connexity Proceeding Book*, ed. Julian Malins, 508-512. Aberdeen: Gray School of Art, The Robert Gordon University. (Design Connexity: 2009 Eighth Conference of the European Academy of Design, 1-3 Apr 2009, The Robert Gordon University, Aberdeen, Scotland.

Xue, L., Yen, C.C., Choolani, M. and Chan, H.C. 2009. The perception and intention to adopt female-focused healthcare applications (FHA): A comparison between healthcare workers and non-healthcare workers. *International Journal of Medical Informatics* 78: 248-258.

Xue, L., Yen, C.C., Chang, L., Chan, H.C., Tai, B.C., Tan, S.B., Duh, H.B.L. and Choolani, M. 2012. An exploratory study of ageing women's perception on access to health informatics via a mobile phone-based intervention, *International Journal of Medical Informatics* In press.

CHAPTER 34

Participatory Design for Green Supply Chain Management Key Elements in the Semiconductor Industry

TA-PING LU[a]*, YI-KUANG CHEN[b], PEI-LUEN RAU[c], and SHU-NING CHANG[d]

[a] Department of Industrial Engineering and Management, National Taipei University of Technology, Taiwan, R.O.C
[b] Taiwan Semiconductor Manufacturing Company, Taiwan, R.O.C
[c] Department of Industrial Engineering and Management, Tsing Hua University, Beijing, P.R. China
[d] Undergraduate Department of Industrial Engineering and Management, National Taipei University of Technology, Taiwan, R.O.C
*Email: robertlu@ntut.edu.tw

ABSTRACT

Green supply chain management (GSCM) has emerged as a major competitive strategy for many globalized companies to be both socially responsible and comply with GSCM policies and regulations such as Waste Electronic and Electrical Equipment (WEEE) and Restriction of Hazardous Substances (RoHS). Semiconductor market exceeded 300 billion US dollars worldwide in 2010. However, studies on the key elements of GSCM in the semiconductor industry are scant. In an attempt to fill this gap, this study adopts the participatory design concept to design GSCM in the semiconductor industry and uses a 2-step approach to propose a hierarchy of key GSCM elements. First, the literature review in Section 1 identifies the initial hierarchy. The second step implements the concept of participatory design using focus group discussions to finalize the proposed key elements hierarchy. These discussions revealed valuable experiences from executives and managers that actively participated in a project that successfully established a seamless SCM implementation between the world's largest semiconductor foundry manufacturing company and the world's largest assembly and testing company. To evaluate the relative importance of these key elements, a second session of focus group discussions was conducted to identify the 3 most important elements. The proposed hierarchy may serve as a checklist for future supply chain managers, ensuring consideration of all key elements when designing the green supply chain in the semiconductor industry. The most important 3 elements were (1) Reducing hazardous substances in the manufacturing process, (2) Design of product for energy and material conservation, and (3) Recycling system in the manufacturing process. The results of this study indicate directions for the continuous improvement and future development of GSCM in the semiconductor industry. The results of this study may serve as a foundation for academic research in fields related to GSCM.

Keywords: participatory design, green supply chain management, semiconductor industry

1 INTRODUCTION

1.1 RESEARCH MOTIVATION

A global innovation in producing technology has begun, especially in the manufacturing industry. High production efficiency and low cost are no longer the only criteria in manufacturing modern products because the ability to comply with greening supply chain requirements is also an important consideration. In the twenty-first century, green supply chain management (GSCM) is not only a business or marketing consideration, but an important competitive element because green products and green production technologies are becoming a competitive advantage. This is particularly true since the European Union (E.U.) passed the Waste Electrical and Electronic Equipment (WEEE) and Restriction of Hazardous Substances (RoHS) directives.

Despite of the importance of the GSCM, a literature review reveals very few GSCM related studies in the semiconductor industry, which had a market value exceeding 300 billion US dollars worldwide in 2010. Most of the GSCM-related studies are aimed at industries such as the automobile industry, electronics industry, and textile and apparel industry. (Hsu and Hu, 2008, Wu et al., Zhu et al., 2007a)

1.1 RESEARCH OBJECTIVE AND CONTRIBUTION

This study adopts the concept of participatory design to identify a hierarchy of key GSCM elements in the semiconductor industry. This study also evaluates the relative importance of these key elements to identify the 3 most important key elements.

The proposed hierarchy serves as a checklist for future supply chain managers, ensuring that they consider all key elements when designing a green supply chain in the semiconductor industry. The results of this study provide directions for the continuous improvement and future development of designing GSCM in the semiconductor industry. The results of this study may also serve as a foundation for academic research in fields related to GSCM.

2   GREEN SUPPLY CHAIN MANAGEMENT

To facilitate environmental performance, minimize waste, and achieve cost savings, many companies have adopted GSCM as an important strategy for improving competitive advantage and profit (Sarkis, 2003;Rao and Holt, 2005; Zhu et al., 2008; Tseng et al., 2009a). As a result, GSCM is attracting increasing interest from researchers and practitioners of operations and SCM (Srivastava, 2007)

Different authors offer different definitions of GSCM. Zsidisin and Siferd(2001)defined GSCM as the set of supply chain management policies held, actions taken, and relationships formed in response to concerns related to the natural environment with regard to the design, acquisition, production, distribution, use, re-use, and disposal of a firm's goods and services. A more specific definition views GSCM as a combination of environmental thinking and supply chain management that includes activities such as 'green design,' 'green sourcing/procurement,' 'green operations' or 'green manufacturing,' 'green distribution/ logistics/marketing,' and 'reverse logistics'(Srivastava, 2007).

3   AN SUCCESSFUL SCM SYSTEM INTEGRATION PROJECT BETWEEN TSMC AND ASE

From 1998 to 2004, Taiwan Semiconductor Manufacturing Company (TSMC), the largest semiconductor foundry and second largest IC manufacturing company in the world and Advanced Semiconductor Engineering Inc. (ASE), the world's largest semiconductor assembly, testing, and packaging service provider, jointly completed an e-Supply Chain Management (e-SCM) project integrating 11 key business processes through the Internet. The result was a seamless interface between TSMC, ASE, and their joint customers. The success of this project allowed them to obtain accurate, timely information on their product status and respond appropriately when needed. Their pioneering experience has evolved from a 2-company project into a potent force, upgrading the efficiency of the entire semiconductor industry through process and data standardization via RosettaNet.

3.1 PROJECT OBJECTIVE

The e-SCM project scope encompassed major business activities between TSMC and ASE in 2 dimensions, as showed in Table 1.

Table 1 project scope

Dimensions	Major business activities
Engineering Collaboration	• Engineering specification • Engineering test data • Yield data
Logistics Collaboration	• e-PO and Order Acknowledge • WIP data • Finished Goods tracking • Shipping notice

Under the 2 dimensions, there were in detailed 11 e-processes established, including yield rates, testing results, order and order acknowledgement, work-in-process, and shipment of finished products in stock; etc. Figure 1 illustrates the identified key processes between these 2 companies.

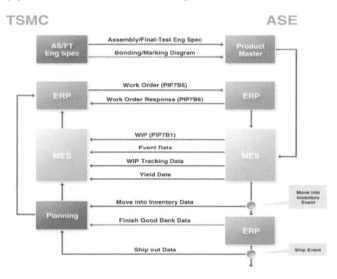

Figure 1 Illustration of TSMC/ASE's key process integration–conceptual overview. (Source: TSMC)

3.2 PROJECT CONTRIBUTION

The benefits of the value chain integration were realized through shorter data transmission time, timely information delivery, and increased data transparency and accuracy due to standardization of data exchange protocols and formats.

For quantitative calculation, take an e-order as an example, the Table 2 illustrates the benefit. The data transmission time and transaction processing time were improved 5 to 6 times and the error rate was significantly decreased 8 times.

Table 2 Before-and-after comparison of e-order.

Items	Before	After	Remarks
Data transmission time	120 minutes/order	20 minutes/order	1.Around 20,000 e-order per year
Order Processing time	100 minutes/order	20 minutes/order	2. The average error recover time is 40 hours
Order Errors (eg, incorrect order line items, mis-matched product id, wrong billing amount	12 orders/month	1.5 order/month	3. 32 man-year cost reduction, equivalent US$ 1M save per year

Over the six-year period from 1998 to 2003, it was estimated that more than US$10 Million value has been created by the 2 companies jointly, compared with the US$ 2 million total investment. The Table 3 illustrated the overall benefits for the eleven e-processes in the 2 collaboration groups.

Table 3 Collective Benefits Summary.

e-operational processes	1998	1999	2000	2001	2002	2003	Total (US$, K.)
Logistics collaboration	7	515	1,481	1,836	1,852	1,955	7,645
Engineering collaboration	0	250	463	535	617	576	2,442
Total	7	765	1,944	2,371	2,468	2,531	10,087

It is worthwhile to note that the above figure is only the "direct" benefits achieved by TSMC and ASE internally. As mentioned in previous sections, more than 30 companies in semiconductor industry have joined the e-Supply chain network ever since. More companies in the high-tech industry adopted the WIP, work order/work order acknowledgement data exchange format after these 3 standards were verified and published by RosettaNet. The "indirect" benefit brought to the whole industry value chain could be in the order of 10 times. (Hwang et al., 2008).

4   PARTICIPATORY DESIGN – A FOCUS GROUP APPROACH

This study adopts the concept of participatory design to propose a hierarchy of GSCM key elements in the semiconductor industry. Focus groups are a powerful research tool for collecting qualitative information across many contexts (Huston and Hobson, 2008). Qualitative information from techniques such as focus groups is particularly vital for design decision-makers (Porter, 1993, Cecil et al., 2006). To define the key elements for designing GSCM in the semiconductor industry, this study uses the focus group method to finalize the proposed key elements hierarchy by extracting valuable experiences from executives and managers that actively joined in the project described in Section 3.

The process of conducting a focus group discussion consists of 3 phases: planning, conducting, and analyzing (Krueger, 1994). These 3 phases are introduced below.

A.   *Planning phase:* clearly define the purpose and topics of the focus group discussion and choose an appropriate moderator with a good understanding of the topics and excellent communication skills. Carefully select the members of the focus group meeting with the following 3 guidelines: homogeneity, heterogeneity, and representativeness (Krueger, 1994). The 4 to 12 participants should possess a similar level of understanding on the topics of the meeting (Morgan, 1988, Lindsay and Hubley, 2006).

B.   *Conducting phase:* To obtain valid information, a questioning route or discussion guide should be prepared in advance. A focus group discussion usually lasts for 90to 150 minutes(Fern, 2001, Easton et al., 2003).There are many different methods of data collection during the discussion; memory, transcripts, notes, and tapes are frequently used. The most common and efficient method of data collection is tape recording, which completely records all ideas generated during the discussion without any judgment made by the note taker.

C.   *Analyzing phases:* This study adopts note-based analysis. An abridged transcript and a brief summary for each focus group discussion were prepared and used to analyze and identify commonalities and patterns. The approaches to conduct a focus group discussion and analyze the results can be very different, depending on topics and participants(Krueger, 1994). Experienced researchers should be consulted to determine the appropriate approaches.

5   DEFINING KEY GSCM ELEMENTS IN THE SEMICONDUCTOR INDUSTRY

This study uses a 2-step approach to construct the hierarchy of key GSCM elements in the semiconductor industry. Sections 5.1 and 5.2 describe these 2 steps.

5.1   STEP 1 - IDENTIFY THE INITIAL HIERARCHY OF KEY GSCM ELEMENTS IN SEMICONDCUTOR INDUSTRY

This section summarizes some of the publications reviewed in the current study. Zhu and Sarkis (2006) defined GSCM practices focusing on 3 typical sectors, the automobile industry, thermal power plants, and the electronic/electrical industry in China. Shang et al.(2010) investigated crucial GSCM capability dimensions and the relationship between these dimensions and firm performance based on electronics-related manufacturing firms. Wu et al. analyzed and reviewed the relationship of GSCM drivers and practices in the textile and apparel industry.

This study collects 44 previously published key GSCM elements from related studies and categorizes them into 5 initial dimensions to construct an initial hierarchy. This initial hierarchy of key GSCM elements serves as a foundation for focus group discussions in Step 2.

5.2 STEP 2 –PARTICIPATORY DESIGN USING FOCUS GROUPS

To finalize the hierarchy of key GSCM elements, focus group discussions were conducted to extract valuable experiences from executives and managers who actively participated in the successful SCM implementation integration project described in Section 3. This process produced a comprehensive hierarchy of GSCM key elements focusing on the semiconductor manufacturing industry. Participants in each of the focus group discussions answered the following questions to ensure that expected results were obtained:

- Which key elements do you think are appropriate for designing GSCM in the semiconductor industry?
- Which dimensions do you think are appropriate for designing GSCM in the semiconductor industry?
- Which key elements and which dimensions can be consolidated?
- Which key element should be put under which dimension?
- Is this key element hierarchy for designing GSCM a comprehensive one?

8 focus group discussions were conducted. All discussions were held with 6 to 8 participants from both industry and academia, and lasted from 40 to 60 minutes. While the same moderator facilitated each discussion, the participants were carefully selected to have different opinions, in compliance with the guidelines of homogeneity and heterogeneity. Table4 summarizes the composition of the participants for the 8 focus group discussions.

Table 4 Participants in focus group discussions

Background \ Field	Engineer	Manager	Senior manager or Director	total
IT	2	8	4	14
Business	3	10	3	16
Total	5	18	7	30

6 RESULTS AND DISCUSSION

6.1 PROPOSED HIERARCHY OF KEY GSCM ELEMENTS IN THE SEMICONDUCTOR INDUSTRY

The results of the focus group discussions were used in the final analysis process to identify the commonalities and differences. The result of this analysis is a hierarchy of key GSCM elements in the semiconductor industry, shown in Table 5. The hierarchy consists of four dimensions: recycling, hazardous substance management, policy and regulation, and energy and material conservation. These four dimensions include 12 key elements. The references cited following the elements denote in which the elements were previously been discussed. The number of references for each dimension or reference does not indicate its relative importance.

Table 5 Key GSCM elements hierarchy

Dimension	Elements	Reference
Recycling	Reverse logistics	(Azevedo et al., 2011, Ru-Jen, Hsu and Hu, 2008, Hu and Hsu, 2006)
	Design of product disassembling, recycling, and reusing	(Azevedo et al., 2011, Zhu and Sarkis, 2004, Ru-Jen, Zhu et al., 2008, Hsu and Hu, 2008, Zhu et al., 2007a, Wu et al., Zhu et al., 2007b, Hu and Hsu, 2006)
	Design of recycling system in the manufacturing process	(Shang et al., 2010, Zhu et al., 2007a, Zhu and Sarkis, 2006)
Hazardous substance management	Design of product for reducing the use of hazardous and toxic substance	(Azevedo et al., 2011, Zhu and Sarkis, 2004, Zhu et al., 2008, Zhu et al., 2007a, Wu et al., Zhu et al., 2007b, Zhu and Sarkis, 2006)
	Hazardous substance control in the manufacturing process	(Shang et al., 2010, Zhu et al., 2007a)
	Hazardous substance disposal procedure and auditing program for third party recycler	(Zhu and Sarkis, 2004, Ru-Jen, Shang et al., 2010, Zhu et al., 2008, Hsu and Hu, 2008, Zhu et al., 2007a, Wu et al., Zhu et al., 2007b, Hu and Hsu, 2006, Zhu and Sarkis, 2006)
Policy and regulation	Waste Electronics and Electrical Equipement (WEEE)	(Shang et al., 2010, Zhu et al., 2007a)
	Restriction of Hazardous Substances(RoHS)	(Shang et al., 2010, Zhu et al., 2007a)
	ISO 14000 certification	(Azevedo et al., 2011, Zhu and Sarkis, 2004, Shang et al., 2010, Zhu et al., 2008, Zhu et al., 2007a, Wu et al., Zhu et al., 2007b, Zhu and Sarkis, 2006)
	Carbon footprint	(Shang et al., 2010, Zhu et al., 2007a, Lee, 2011)
Energy and material conservation	Energy and material conservation in manufacturing and logistics process	(Azevedo et al., 2011, Ru-Jen, Shang et al., 2010, Zhu et al., 2008, Zhu et al., 2007a, Wu et al., Zhu et al., 2007b, Zhu and Sarkis, 2006)
	Designing product to be energy and material conservative	(Shang et al., 2010, Zhu et al., 2008, Zhu et al., 2007a, Wu et al., Zhu et al., 2007b, Zhu and Sarkis, 2006)

6.2 THE TOP 3 ELEMENTS

To evaluate the relative importance of all the elements, a second session of focus group discussions was held with the same directors and senior managers. Experts chose the 3 most significance key elements in each phase. The top 3 elements in semiconductor industry included the following: *(1) Reducing hazardous substances in the manufacturing process, (2) Design of product for energy and material conservation, and (3) Recycling system in the manufacturing process.*

Shang et al.(2010) showed that production planning and control focusing on reducing hazardous substances/waste and optimizing materials' exploitation is the most important factor. This finding agrees with the current study, which indicates that controlling hazardous substances in the manufacturing process is the highest ranking element. The second and third elements, design a product to require little energy and material and designing a recycling system for the manufacturing process, respectively, were also included in 12 the major items for GSCM in a study by Wu et al.

7 CONCLUSION AND PROPOSED FUTURE RESEARCH

This study defines a hierarchy of key GSCM elements in the semiconductor industry. A hierarchy structured under the concept of participatory design was proposed based on a review of GSCM-related publications and especially on the valuable industrial experience of the executives and managers of a successful SCM implementation project. The proposed hierarchy provides a valuable reference for future project managers and ensures that they will consider all key elements when designing GSCM in the semiconductor industry. This study also evaluates the relative importance of the key elements and identifies the 3 most important elements. The results of this study provide directions for the continuous improvement and future development of GSCM. The proposed hierarchy may serve as a foundation for academic research in fields related to GSCM design.

ACKNOWLEDGMENTS

This research is partially supported by National Science Council, Taiwan, R.O.C. Project number NSC 100-2815-C-027 -009-E.

REFERENCES

AZEVEDO, S. G., CARVALHO, H. & CRUZ MACHADO, V. 2011. The influence of green practices on supply chain performance: A case study approach. *Transportation Research Part E: Logistics and Transportation Review,* 47, 850-871.

CECIL, J., DAVIDSON, S. & MUTHAIYAN, A. 2006. A distributed internet-based framework for manufacturing planning. *The International Journal of Advanced Manufacturing Technology,* 27, 619-624.

EASTON, G., EASTON, A. & BELCH, M. 2003. An experimental investigation of electronic focus groups. *Information & Management,* 40, 717-727.

FERN, E. F. 2001. *Advanced focus group research,* Sage.

HSU, C. W. & HU, A. H. 2008. Green supply chain management in the electronic industry. *Environmental Engineering,* 5, 205-216.

HU, A. H. & HSU, C.-W. 2006. Empirical Study in the Critical Factors of Green Supply Chain Management (GSCM) Practice in the Taiwanese Electrical and Electronics Industries. *Management of Innovation and Technology, 2006 IEEE International Conference on.*

HUSTON, S. A. & HOBSON, E. H. 2008. Using focus groups to inform pharmacy research. *Research in Social and Administrative Pharmacy,* 4, 186-205.

HWANG, B.-N., CHANG, S.-C., YU, H.-C. & CHANG, C.-W. 2008. Pioneering e-supply chain integration in semiconductor industry: a case study. *The International Journal of Advanced Manufacturing Technology,* 36, 825-832.

KRUEGER, R. A. 1994. *Focus Groups: Practical Guide for Applied Research,* Sage Pubns.

LEE, K.-H. 2011. Integrating carbon footprint into supply chain management: the case of Hyundai Motor Company (HMC) in the automobile industry. *Journal of Cleaner Production,* 19, 1216-1223.

LINDSAY, A. & HUBLEY, A. 2006. Conceptual Reconstruction through a Modified Focus Group Methodology. *Social Indicators Research,* 79, 437-454.

MORGAN, D. L. 1988. *Focus groups as qualitative research,* Sage Publications.

PORTER, S. L. Year. 14th National Online Meeting proceedings--1993: New York, May 4-6, 1993. *In,* 1993. Learned Information, 265-272

RU-JEN, L. Using fuzzy DEMATEL to evaluate the green supply chain management practices. *Journal of Cleaner Production.*

SHANG, K.-C., LU, C.-S. & LI, S. 2010. A taxonomy of green supply chain management capability among electronics-related manufacturing firms in Taiwan. *Journal of Environmental Management,* 91, 1218-1226.

SRIVASTAVA, S. K. 2007. Green supply-chain management: A state-of-the-art literature review. *International Journal of Management Reviews,* 9, 53-80.

WU, G.-C., DING, J.-H. & CHEN, P.-S. The effects of GSCM drivers and institutional pressures on GSCM practices in Taiwan's textile and apparel industry. *International Journal of Production Economics,* In Press, Corrected Proof.

ZHU, Q. & SARKIS, J. 2004. Relationships between operational practices and performance among early adopters of green supply chain management practices in Chinese manufacturing enterprises. *Journal of Operations Management,* 22, 265-289.

ZHU, Q. & SARKIS, J. 2006. An inter-sectoral comparison of green supply chain management in China: Drivers and practices. *Journal of Cleaner Production,* 14, 472-486.

ZHU, Q., SARKIS, J. & LAI, K.-H. 2007a. Green supply chain management: pressures, practices and performance within the Chinese automobile industry. *Journal of Cleaner Production,* 15, 1041-1052.

316

ZHU, Q., SARKIS, J. & LAI, K.-H. 2007b. Initiatives and outcomes of green supply chain management implementation by Chinese manufacturers. *Journal of Environmental Management,* **85,** 179-189.

ZHU, Q., SARKIS, J. & LAI, K.-H. 2008. Green supply chain management implications for "closing the loop". *Transportation Research Part E: Logistics and Transportation Review,* **44,** 1-18.

ZSIDISIN, G. A. & SIFERD, S. P. 2001. Environmental purchasing: a framework for theory development. *European Journal of Purchasing & Supply Management,* **7,** 61-73.

CHAPTER 35

Comparing the Psychological and Physiological Measurement of Player's Engaging Experience in Computer Game

Wen Cui, P.L. Patrick Rau

Tsinghua University
Beijing, China
cuiwen@umich.edu

ABSTRACT

Despite players' gaming experience is important for game designers and companies, there is few complete theoretical system describing it. Therefore, this study used experimental multi-methods to describe players' immersion experience, and explored the relationship between physiological criteria (blink rate ratio and heart rate ratio) and psychological measurement (immersion level). The research selected Call of Duty 4-Modern Warfare as a gaming platform, which is a 2007 first-person shooter video game. There were totally 30 Tsinghua University male students aging 18 o 24 participated in the experiment. The experiment used a special made helmet with a camera to record each participant's eye blinking movement, and used a chest sensor to capture participant's heart rate changes. Participant's immersion level was measured through a seven-Likert scale questionnaire, including 27 questions. Players' immersion level, blink rate ratio and heart rate ratio were taken as dependent variables and the game's section was taken as independent variable. Through data analysis, there are some conclusions as follows: (1) player's blink rate declined after they getting into the game, and the decline lasted during the whole game; (2) there was no relationship between player's overall blink rate and immersion level; (3) there was no relationship between player's overall heart rate and immersion level; (4) participants' heart rate ratio were correlated with blink rate ratio in some game period. The research combined a variety of measurement methods, and made a connection between physiological and psychological indicators during the game process.

Keywords: computer game, flow, blink rate, heart rate, immersion level

1   INTRODUCTION

Interactive computer game industry is developing rapidly over the world. In order to gain a bigger market share, game companies keep on exploring a better understanding of the audiences' needs. However, it is pretty hard to adequately describe and measure the complicated and intricate game experience, because it would be difficult to take apart to analyze and be characterized in a common, shared vocabulary. Furthermore, different game genres will bring about varieties of experiences to players when they are engaging in playing, for instance, beating a large numbers of enemies for collection treasures and having a simulation car racing game are two distinct different experiences. Besides, even within the same game genre, definition of "fun" may be diverse to different players. This research studied "immersion" basing on the past studies of user's physiological and mental characters. Multi-methods were conducted in the study to capture players' experience in the whole playing process, which includes a questionnaire and two kinds of physiological measurements.

2   LITERATURE REVIEW

2.1   Game Immersion

Computer games can be classified with different genres according to their content, including action, adventure, role-playing, simulation and strategy (Apperley, 2006). Although guided by user-centered design principles, game mainly focuses on fun and entertaining experience, which is different from other electronic applications, such as productivity application software. Malone (1987) characterized three main elements contributing fun in computer games intrinsically motivating players: fantasy, challenge, and curiosity. Fantasy is used to reinforces the instructional goals and stimulate the prior interests of the learner. Varying challenges would make game more appealing. In addition, game should create sensory and cognitive curiosity within the learner. In other research investigating computer games, it was also found that rules, specific goals (Wilson, et al, 2009), high intensity of interaction (Garris, 2002), direct and endogenous feedback (Peixoto, et al, 2010), and game representation (Cooper, 2002) are also important features to make games more attractive.

Immersion, promoted by Brown and Cairns (2004), represents the condition that players become less aware of themselves and their surroundings compared to previous time. Players' attention is totally grabbed and their emotion is directly affected by game. Immersion is described as outcome of a good game experience, which is critical to game enjoyment. There are three levels of immersion: engagement, engrossment and total immersion. To get into the first level the player should spend time, effort and attention in learning how to control and play the game. Then the second level further involves the player to get familiar with the game's construction. When the player is in this level, his emotions will be directly affected by the game and the way of control become not important at all, even "invisible". Further step is to enter the third level "total immersion", which is the highest level of immersion and be described as a sense of presence, which means the player would feel himself being totally in the game.

2.2 Immersion Measurement

2.2.1 Psychological Measurement

Qin (2009) studied narrative game and promoted seven dimensions impacting players' immersion, which include curiosity, concentration, challenge and skills, control, comprehension, empathy and familiarity. Based on the seven dimensions, a seven-level 27-question Likert scale was developed to measure the player's immersion level.

- Curiosity: To stimulate the player's perception and cognition, and to attract them to explore the game world.
- Concentration: Ability to concentrate long-term on the game narrative;
- Challenge and skills: Some relative difficulty in the game narrative for players and corresponding players' skills;
- Control: Ability to exercise a sense of control over game narrative;
- Comprehension: Understanding the structure and content of the storyline;
- Empathy: Mentally entering into the imaginary game world while playing the game;
- Familiarity: Being familiar with the game story. As a key in game acceptance, familiar background will get a quick resonance and new story will be more attractive.

2.2.2 Physiological Measurement

Blinking is the rapid closing and opening of the eyelid. Eye blink rate of a relax person is 15 to 20 times per minute on average, which various between man and woman (Bauer, et al, 1987). Some researches indicated that blink rate would change when people face challenging or stressful situations, like public speech, a formal conversation, etc. (Bentivoglio, et al, 1997). Torkildsen (2009) further pointed out when people facing visual tasks, such like reading and playing games, or focusing the attention at the hearing, the duration and the number of eye blinks are expected to decrease and blink rate will reduce. And he explained the change is caused by increased cognitive demand that directs attention to task-relevant stimuli. Using this theory, an instrument was developed by Tseng (2007) to measure players' eye blink rate and to get an idea of their flow state changes in the process of playing an online game. Through experiment, it was found that when players come to flow state, their blink rate will reduce, yet this phenomenon only happen at the beginning the game but not the whole process of the playing.

Another physiological criteria used in this study is heart rate (HR), which measures the number of heart beats per minute (bpm). The average resting human heart rate is about 70 bpm for adult males and 75 bpm for adult females. Heart rate can be measured by monitoring one's pulse using specialized medical devices. Physiological changes are by-products of emotional states that are generated by brain progress, and early studies indicated that changes in cardiac activity were related to psychological phenomena and emotions (Springer, 1935). In the research of Griffiths and Dancaster (1995), all the subjects experienced significantly higher heart rates in comparison with their own baseline during the game playing period.

3 METHODOLOGY

In this study, participants were asked to play the chosen parts of the game Call for Duty 4. During the game time, participants were required to wear a special adjustable headgear attached with a 500-pixel camera to record their left eye's blink activity, When recording, the camera was placed 8 cm from and 20 degrees above the left eye. After the experiment, the video was analyzed by software "Blink Statistic" to count the blink number, which later helped to calculate the blink rate. At the same time, participants were required to wear a chest sensor to capture their real time heart rate in the whole experiment process, the data of which showed on a special watch and was recorded by a video camera. Future 1 represents the whole setting of the experiment. After the gaming process, each participant was asked to complete an immersion questionnaire based on his feelings from the game. In order to avoid the gender bias on heart rate and blink rate measurement, the participants were only males aging from 18 to 24. Furthermore, they were required to be expert players of First Person Shoot game but novice players of Call for Duty 4. Experiments were individually conducted in Tsinghua University Human Factors and Ergonomics lab.

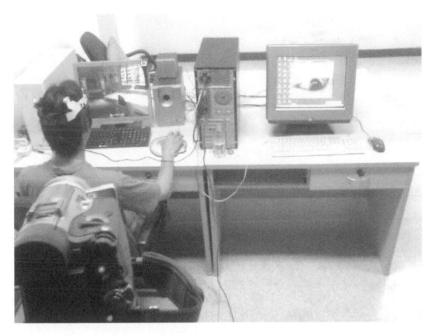

Figure 1 The experiment setting: a participant was wearing the special adjustable headgear with a camera to capture his left eye activities, which was recorded in another computer besides him. At the same time, he also wore a chest sensor to capture his real time heart rate, the data of which showed on a special watch on the right corner of the screen. The camera behind the participant records the data on the watch as well as the compuer screen of the game process.

The independent variable was the game period. Call of duty 4 had a serious of game stories with hurdles. Players need to clear all hurdles to move on to the next story. The difficulty level of each hurdle is different from each other, which requires various level of user involvement. Based on the development of the stories and the difficulties of hurdles, as well as total length of the game time being considered, the experiment game Call for Duty 4 is divided into seven periods: skills training, broking into the enemy warship and killing enemies, finding a way out before the ship sinking, being taken to a execution ground (no tasks and hurdles for plays), sneaking into enemy camps, being found by enemies and going through a hard battle, rescuing a secret agent and leaving.

Figure 2   Scenes in Call of Duty 4: skills training and sneaking into enemy camps

The dependent variables of the study were individual participant's blink rate ratio, heart rate ratio and immersion level. After every participant's left eye blink activities being video recorded and the blink numbers being counted by the software "Blink Statistic" for each game period, the period blink rate could be calculated through blink number divided by the period time. The value then was divided by participant's own normal blink rate, which was to avoid the bias of individual differences. The final value was defined as blink rate ratio. Similarly, every participant's real time heart rate was video recorded, and then the average heart rate for the period could be calculated. Also to avoid the bias of individual differences, the value then was divided by participant's own normal heart rate. The final value was defined as heart rate ratio. The third dependent variable was participant's immersion level, which was measured by a designed questionnaire, including 7 factors and 27 questions using a seven-level Likert scale. This depended variable represented the player's perceptual, cognitive and affective reaction to the game.

4   RESULTS AND DISCUSSION

Descriptive statistics results

There were totally thirty male subjects participating the experiment, who were all Tsinghua University students aging 18-24 (mean= 22.3, Std. Deviation=1.045). Participants had rich experience in computer game: 93.3% participants had began playing computer game since 18 years old, and the average age beginning to play First Person Game was 15 (mean=15.17). 76.7% of them played computer game more than once a week, and their average length of time spending on game was 1.8 hours per time (mean=1.79). For the immersion level, mean=140.33, Std. Deviation=17.36 (totally points is 7*27=189). For the descriptive statistics of heart rate ratio and blink rate ratio, please see Table 1.

Table 1 The descriptive statistics results of participants' heart rate ratio and blink rate ratio

Game section	Heart Rate Ratio		Blink Rate Ratio	
	Mean	Std. Deviation	Mean	Std. Deviation
1	1.1062	.0935	.2858	.14090
2	1.1083	.0991	.1929	.08592
3	1.2054	.1672	.2099	.09310
4	1.0730	.0896	.4469	.21051
5	1.0551	.0777	.2457	.10711
6	1.0819	.0736	.1883	.07491
7	1.1076	.0742	.1779	.07773

When participant's heart rate ratio is higher than 1, his heart rate is faster than normal conditions. From the descriptive statistics results, only in game period three (finding a way out before the ship sinking) and seven (rescuing a secret agent and leaving) all of the participants' heart rate ratio were higher than 1 (min=1.0335 for game period three, min=1.0203 for game period seven), and there was no significant difference between different game period (P=0.69). Similarly, when participant's blink rate is lower than 1, it means his blink rate is slower than normal conditions. From the descriptive statistics results, all of the participants' blink rate ratio were lower than 1 in the whole game period (min 1=0.11, min 2=0.05, min 3=0.05, min 4=0.14, min 5= 0.10, min 6=0.08, min 7=0.06), yet there was still no significant difference between different game period (p=0.24). The descriptive statistic results showed that objects' blink rate would decrease when they become focused on game, and the process lasted through the whole game process.

The relationship between heart rate ratio and blink rate ratio

The result of correlation analysis shows that heart rate ratio was positively correlated with blink rate ratio in game period three, which indicated that in period three of the game when objects' heart rate increased, their blink rate also increased at the same time. Participants in this game period needed to find a way out before the enemies' ship sinking within a limited time, and once they failed, they had to start again of this period of game. The path is intricate with several obstacles, which required the objects highly concentrated on the game and to memorize the path they already tried. As a novice player, most of the participants in this game period were easy to get lost and they had to try several times, which made them nervous. At the same time, when they found the game too difficult for them to go on, they began to be impatient, or even talked to themselves "what can I do?" or "Why I can't do it!" All these situations might lead to the heart rate ratio and blink rate ratio increasing.

From the analysis, it also indicated that heart rate ratio was negative correlated with blink rate ratio in period five of the game, which showed that in this game period when objects' heart rate increased, their blink rate decreased at the same time. This game period required players sneak into enemy camps. The length of game was about 5 minutes, and along the clear path to the enemy camps there are some hurdles to shoot the enemies. During the experiments, most of the participants can successfully found their way and killed the enemies. One possible explanation of the correlation between the changes of heart rate and blink rate is that when participants found the game interesting and difficulty level proper to play, they would easily focus on the game and get involved in the game environment. So their heart rate increased and their blink rate decreased.

Table 2 The correlation analysis results between heart rate ratio and blink rate ratio through the different period of game

	BR1	BR2	BR3	BR4	BR5	BR6	BR7
HR1	.073						
HR 2		.298					
HR 3			.012*				
HR 4				.236			
HR 5					-.020*		
HR 6						.229	
HR 7							.099

*. Correlation is significant at the 0.05 level (2-tailed).

The relationship between heart rate ratio and immersion level & blink rate ratio and immersion level

By correlation analysis of the relationship between the overall heart rate ratio and immersion level, it was found that the overall heart rate ratio was uncorrelated with immersion level ($P=0.442$). And for correlation analysis of the relationship between the overall blink rate ratio and immersion level, it was found that the overall blink rate ratio was uncorrelated with immersion level ($P=0.221$). According to Martos et al (2007), it was indicated that personality characteristics might affect the performance and satisfaction on individuals. The ones who are more aggressive and impetuous want more in less time. When these people participant in a game, they are more anxious and eager to have an achievement, which leads their physical criteria being different from the normal conditions. Yet these physical criteria differences are caused by their nervous and impatient, but cannot indicate a higher level of immersion. Another possibility of the analysis results are the immersion level was measured as an overall involvement experience, even participants had a flow experience during the game, the average heart rate or blink rate of the whole game process may not reflect whether they had a immersion experience.

5   CONCLUSION

This study selected Call of Duty 4-Modern Warfare as the gaming platform and there were 30 Tsinghua University male students aging 18 o 24 participated in the experiment. The objects' heart rate and blink rate were recorded and calculated by special equipment and software. Taken heart rate ratio, blink rate ratio and immersion level as the dependent variables, and game period as the independent variable, the study explored the relationship between heart rate ratio and blink rate ratio, heart rate ratio and immersion level, as well as blink rate ratio and immersion level. Through data analysis, it was found that either player's overall heart rate ratio nor blink rate ratio was correlated with immersion level, yet participants' heart rate ratio were correlated with blink rate ratio in some game period. Especially when the game was too hard for players and they had to try several times to get through, players probably began to get nervous and anxious, then their heart rate and blink rate increased accordingly. And when plays found a game interesting and the game's difficulty level is proper, they might feel exacting and their focus would on the game itself, so their heart rate increased and their blink rate decreased correspondingly. The future study could explore whether these kind of correlation could truly reflect the players immersion level during the game process. Furthermore, players' characteristics could be taken into consideration to explore whether this factor would affect players immersion level.

REFERENCES

Apperley, T. H. 2006. Genre and game studies: Toward a critical approach to video game genres. *Simulation Gaming*, 37 (1): 6-23.

Bauer, L. O., R. Goldstein and J. A. Stern. 1987. Effects of Information-Processing Demands on Physiological Response Patterns. *Human Factors*, 29 (2): 213-234.

Bentivoglio, A. R., S. B. Bressman and E. Cassetta, et al. 1997. Analysis of blink rate patterns in normal subjects. *Movement Disorders*, 12(6): 1028–1034.

Brown, E. and P. Cairns. 2004. A grounded investigation of game immersion. *Proceedings of the CHI '04 extended abstracts on Human factors in computing systems.*

Cooper, D. J. and J. B. V. Huyck. 2002. Evidence on the equivalence of the strategic and extensive form representation of games. *Journal of Economic Theory.* 110(2): 290–308.

Garris, R., R. Ahlers, and J. E. Driskell. 2002. Games, Motivation, and Learning: A Research and Practice Model. *Simulation Gaming*, 33(4): 441-467.

Griffiths, M. D. and, I. Dancaster. 1995. The effect of type A personality on physiological arousal while playing computer games. *Addictive Behaviors*, 20 (4), 543-548

Malone, T. W. and M. R. Lepper, 1987. Making learning fun: A taxonomy of intrinsic motivation for learning. In *Aptitude learning and instruction III: Conactive and affective process analyses*, R. E. Snow & M. J. Farr. Hillsdale, NJ: Erlbaum.

Martos, M. P. B., J. M. A. Garcia-Martinez and E. López-Zafra. 2007. An experimental study about the congruence between Type A behavior pattern and type of task. *Scandinavian Journal of Psychology*, 48, 383–390

Peixoto, D. C. C., R. O. Prates and R. F. Resende. 2010. Semiotic inspection method in the context of educational simulation games. *Proceedings of the 2010 ACM Symposium on Applied Computing.*

Qin, H, P. L. P. Rau, and G. Salvendy. 2007. Measuring Player Immersion in the Computer Game Narrative. *Lecture Notes in Computer Science*, 4740, 458-461.

Springer, N. N. 1935. Cardiac Activity during Emotion. *The American Journal of Psychology*, 47(4): 670-677.

Torkildsen, G. 2009. The effects of lubricant eye drops on visual function as measured by the inter-blink interval Visual Acuity Decay test. *Clinical Ophthalmology*. 3: 501–506.

Wilson, K. A., W. L. Bedwell, and E. H, Lazzara, et al. 2009. Relationships Between Game Attributes and Learning Outcomes. *Simulation Gaming,* 40 (2): 217-266.

CHAPTER 36

Study of the Interface of Information Presented on the Mobile Phones

Hua Qin 1,Songfeng Gao 1, Jun Liu 2

[1] Department of Industrial Engineering, Beijing University of Civil Engineering and Architecture, Beijing, P. R. China
[2] Department of Industrial Engineering, Tsinghua University, Beijing, P. R. China
qinhua@bucea.edu.cn

ABSTRACT

The objective of this study is to explore the impact of the two kinds of mobile phone menu structures for receiving and saving production information on different people, which one is icon-structure with broad structure, the other one is tree-structure with deep structure. Based on the mobile phone simulator with Chinese interface, this study conducts the experiments. The results indicate that between different menu structures there were significant difference impacts on different age or gender persons.

Keywords: Menu structure; Small screen

1   INTRODUCTION

In recent years, mobile phones' users are beginning to need receiving kinds of information, especially product information，which the information such as intro-duction or sales promotion presented by texts or images is tranfferd to the product buyers' mobile phones from desktop computers. However, mobile terminals must have small displays compared with the computers. If people need to look for one piece in the plenty of information with more sophisticated structure on limited size screen, their performance will degrade very severely. So the interface of the infor-mation presented on the mobile phones should be different from on the computers and should be user-friendly and support a business model. Otherwise, users would lost in the plenty of the information (Zhang, Shan, Xu, Yang, and Zhang, 2007).

2.   RELATED WORKS

In the mobile phone menu interface design, structure design is very important. Structure design refers to navigation in the overall structure. Structure design includes depth and breadth design of interface and hierarchy of menu. And navigation is defined as the path and actions needed to find a piece of information on a site and get back when needed. Some studies compared broad to deep structures of information and some indexes indicated that the broad structures were easier to perform for small-screen users (Kaikkonen, and Roto 2003, Tang,2001, Parush, and Yuviler-Gavish, 2004). But for small screen, more actions will be required for users information retrieval in a broad structure than deep one (Tang,2001, Ziefle, 2002). Consequently, more actions will be required for users' information retrieval in a broad structure than deep one. And working memory load does not be alleviated to some degree comparing to the deep structure. Therefore, if lots of information needs to be presented on the websites, it should be organized and classified into optimal levels through breadth-depth tradeoff according to the content (Karkkainen, and Laarni, 2002, Otter, and Johnson, 2001).

In order to avoid disorientation in the plenty of information with sophisitcated structure, two types of context information are needed: spatial context and temporal context. Spatial context tells users where to go from the current location and temporal context solves the question how to get there (Utting, and Yankelovich, 1989, Rau, and Wang, 2004). For navigation, in the mobile phone menu interface design, structure design is very important. Structure design refers to depth and breadth design of interface and hierarchy of menu. According to the previous studies, one of the most commonly used mobile phone menus is matrix (Laarni, 2002, Laarni, Simola, Kojo, and Risto, 2004).

3. METHODOLOGY

Because the production information transferred to mobile phones is diverse and numerous, information structure becomes very important. The well-designed menu will make the cognitive process of operation easily. According to the previous studies, this study considers two kinds of menus for presenting the production information, which one is one of the most commonly used mobile phone menus icon-structure with broad structure, the other one is tree-structure with deep structure. This study was intended to explore the impact of the two kinds of mobile phone menu displays for receiving and saving production information on different people. The two menus used the mobile phone simulator with Chinese interface for experiments. We expected that different type of users' performance would be different when they were given different kind of menu displays. According to the assumption, three independent variables were proposed, which are interface of product information, age, and gender. Dependent variables were reaction time and steps to the target interface.

3.1 Participants

Twenty persons (10 males and 10 females) from Beijing, aged 20 to 60 were recruited for the experiment. These participants were asked to indicate their familiarity with basic mobile phone operation, and with which they had the greatest interests and experience on shopping.

3.2 Apparatus

The trials were completed using a phone-simulator as presented on a 17 inch computer screen. The simulator presented a standard mobile phone keyboard, display, and operation interface (see Fig. 1 and 2). The two kinds of menus patters are presented on figure 1 and 2.

Figure 1 Tree-structure menu pattern   Figure 2 Icon-structure menu pattern

3.3 Procedures

In the experiment, each participant should complete two groups of tasks, which one group of tasks was performed on the one kind of screen menu. After each participant was familiar with the performance of the simulators, they were assigned to conduct the three tasks of the first group. The first was to accept two piece of product information. The second task was to store the information in the sub-menu in terms of types of menu display. The third task was to compare the product information. Then, the participants completed the same three tasks of the second group based on the other kind of screen menu. The order of the two groups was arranged randomly.

4. RESULTS

Table 1 showed the descriptive statistics of reaction time. For tree-structure menu interface, 60.5 seconds were used for participants aged less than 40 to complete the tasks while for aged more than 40 participants, 83.0 seconds were needed. And 61.5 seconds were used for male to complete the tasks while for female 82.0 seconds were needed. For icon-structure menu interface, 71.1 seconds were used for participants aged less than 40 to complete the tasks while for aged more than 40 participants, 80.6 seconds were needed. And 59.2 seconds were used for male to complete the tasks while for female 92.6 seconds were needed.

Table 2 showed the descriptive statistics of steps to the target interface. For tree-structure menu interface, 22.4 steps were needed for participants aged less than 40 to complete the tasks while for aged more than 40 participants, 20.6 steps were used. And 21.8 steps were used for male to complete the tasks while for female 21.2 steps were needed. For icon-structure menu interface, 21.9 seconds were used for participants aged less than 40 to complete the tasks while for aged more than 40 participants, 22.5 seconds were needed. And 21.2 seconds were used for male to complete the tasks while for female 23.2 seconds were needed.

Table 1 Mean time of reaction (second)

Variables	Age				Gender					
	<=40		>40		Male		Female		Sum	
	Mean	SD	Mean	SD	Mean	SD	Mean	SD	Mean	SD
Tree	60.5	25.22	83.0	25.59	61.5	24.76	82.0	13.05	71.8	27.28
Icon	71.1	31.28	80.6	30.04	59.2	24.33	92.6	26.80	92.6	26.80

Table 2 Mean steps to the target interface

Variables	Age				Gender					
	<= 40		>40		Male		Female		Sum	
	Mean	SD	Mean	SD	Mean	SD	Mean	SD	Mean	SD
Tree	22.4	2.84	20.6	2.95	21.8	3.16	21.2	2.90	21.5	2.96
Icon	21.9	2.56	22.5	2.88	21.2	2.89	23.2	2.10	22.2	2.67

Table 3 showed the results of multivariate test of reaction time to check the difference performance while using different menu structures. The repeated measures model was used. The results indicated that the significant was 0.012 less than 0.05 between different genders while participants using different kinds of menu structure. And the significant was also 0.012 less than 0.05 between different range of age while participants using different kinds of menu structure. Table 4 showed the results of multivariate test of steps to the target interface to check the difference performance while using different menu structures. The results indicated that the significant was 0.074 more than 0.05 between different genders while participants using different kinds of menu structure. And the significant was also 0.097 more than 0.05 between different range of age while participants using different kinds of menu structure.

Table 3 Multivariate Test (Time)

Effect		Value	F	df	Error df	Sig.
Time * Gender	Pillai's Trace	.332	7.966	1.000	16.000	.012
	Wilks' Lambda	.668	7.966	1.000	16.000	.012
	Hotelling's Trace	.498	7.966	1.000	16.000	.012
	Roy's Largest Root	.498	7.966	1.000	16.000	.012
Time * Age	Pillai's Trace	.334	8.040	1.000	16.000	.012
	Wilks' Lambda	.666	8.040	1.000	16.000	.012
	Hotelling's Trace	.503	8.040	1.000	16.000	.012
	Roy's Largest Root	.503	8.040	1.000	16.000	.012

Table 4 Multivariate Test (Steps to the target interface)

Effect		Value	F	df	Error df	Sig.
Steps * Gender	Pillai's Trace	.186	3.644	1.000	16.000	.074
	Wilks' Lambda	.814	3.644	1.000	16.000	.074
	Hotelling's Trace	.228	3.644	1.000	16.000	.074
	Roy's Largest Root	.228	3.644	1.000	16.000	.074
Steps * Age	Pillai's Trace	.163	3.105	1.000	16.000	.097
	Wilks' Lambda	.837	3.105	1.000	16.000	.097
	Hotelling's Trace	.194	3.105	1.000	16.000	.097
	Roy's Largest Root	.194	3.105	1.000	16.000	.097

5. DISCUSSION AND CONCLUSION

The repeated measures ANOVA of reaction time in two menu display patterns yielded a significant main effect. The results between different patterns were significant, which tree-structure pattern yielded the highest efficiency. Male users consumed less time with icon-structure menu than with tree-structure menu. But female users spent more time with icon-structure menu than with tree-structure menu. For users more than 40 years old, they consumed less time with tree-structure menu than with icon-structure menu. But users less than 40 years old spent more time with tree-structure menu than with matrix menu. For steps to the target interface, there is no significant difference.

Because product information transferred from computers is complex and the structure is relative deep. The effect of depth design was well-operated than that of the breadth design if the navigation is effective (Laarni, 2002, Laarni, Simola, Kojo, and Risto, 2004). And well-operated logic of the menu content was very helpful for users to operate the menu easily (Shieh, Hsu, and Liu, 2005). However, the operation of icon-structure pattern depends on navigation by memory process if the structure of the information is deep. The users have to switch from one screen to another. If the menu doesn't present the whole structure, users would spend more time on remembering. Therefore, participants more than 40 years old consumed more time with icon-structure than with tree-structure. Because the depth of the tree-structures is deep and icon-structure is broad, steps to the target interface between two kinds of menu structure presenting no significant difference indicate that participants performed much more steps with broad structure than deep structure.

REFERENCES

Kaikkonen, A. and V. Roto: 2003. Navigating in a mobile XHTML application. CHI 2003, Ft. Lauderdale, Florida, USA.

Karkkainen L. and J. Laarn. 2002. Designing for small display screens. NordiCHI, October 19-23.

Laarni, J. 2002. Searching for optimal methods of presenting dynamic text on different types of screens, 2002,10.

Laarni, J., J. Simola, I. Kojo, and N. Risto. 2004. Reading vertical text from a computer screen. *Behavior & information technology*, 23(2), pp. 75-82.

Otter M. and H. Johnso. 2001. Lost in hyperspace: metrics and mental models. *Interacting with Computers*, 13 (1), 40

Parush, A. and N. Yuviler-Gavish. 2004 Web navigation structures in cellular phones: the depth/breadth trade-off issue. *Int. J. human-computer studies*, 60, 753-770.

Rau, P. L. P. and Y. J. Wang. 2004. A study of navigation support tools for mobile devices. *Human-Computer Interaction: Theory and Practice*, Edited by J. A. Jacko, C. Stephanidis.

Shieh, K.K., S. Hsu H, and Y. C. Liu 2005. Dynamic Chinese text on a single-line display: Effects of presentation mode. *Perceptual and motor skills*, 100(3).1021-1035.

Tang, K. E. 2001. Menu design with visual momentum for compact smart products. *Human Factors*, 43(2), 267-277.

Utting, K. and N. Yankelovich. 1989. Context and orientation in hypermedia networks. *ACM Transactions on information systems*, 7, 58-84.

Zhang, X. M., M. W. Shan, Q. Xu, B. Yang, and Y.F. Zhang. 2007. *An ergonomics study of menu-operation on mobile phone interface*. Workshop on Intelligent Information Technology Application.

Ziefle, M. 2002 The influence of user expertise and phone complexity on performance, ease of use and learnability of different mobile phones. *Behavior and Information Technology*, 21 (5), 303-311.

CHAPTER 37

The Design of a Socialized Collaborative Environment for Research Teams

Fan GAO, Ji LI, Yihua ZHENG, Kai NAN

Computer Network Information Center
Chinese Academy of Sciences
Beijing, China
gaofan@cnic.cn

ABSTRACT

Research team collaboration requires supporting for information sharing and relevant communication. The challenge is that affect issues can have significant impact on this form of collaboration. Hence elegant designs are requested to avoid negative effects. This paper describes the attempt to embed affective appearance and social network features into a collaborative environment as a result of affective and pleasurable design.

Keywords: affect, collaboration, social network, product design

1   INTRODUCTION

In the development of a collaborative environment targeting for various research teams, the designers were encouraged to embed affective and pleasurable design. The motivation roots in the collaboration style research teams usually possess, which leads to reasonable solutions with low resistance to emotional disturbances. In this paper, we introduce, discuss and demonstrate a socialized collaborative environment, with affective design in both appearance and function level, attempting to enhance affect and pleasure to cope with collaboration needs and to achieve long-term benefits.

2   COLLABORATION & AFFECT

The high dependency on knowledge, content, and individual work differentiates research teams from other teams. Collaboration in research teams require significant higher weight of data and information sharing than task assignment and progress tracking (Balakrishnan et al., 2010; Johri, 2010). Meanwhile, communication, content sharing and learning are strongly affected by user's psychological status, such as social presence, belonging and other emotional feelings (Culnan & Markus, 1987; Markus, 1994; Tu, 2000; Nonnecke & Preece, 2001; Isen, 1999). Therefore the design of collaborative environment should take into account not only the functionality, but also users' affect and pleasure.

2.1   Collaboration in Research Teams

Classic collaboration involves task assignment and distribution, progress tracking, communication among members about interfaces, individual work and so on. However, since research teams rely heavily on individual work, and usually do not have very strict workflow, the need for collaboration is different. Balakrishnan et al.(2010) identified three types of collaboration in research teams: 1) 50% of co-action; 2) 15% of coordination; and 3) 35% of Integration. This research suggested that for integration, collaboration tools that encourage sharing of intermediate results, merging of tasks, hence facilitate inspirations would overwhelm procedural collaboration tools.

An observation of a 50-member software development team illustrates how this team uses blog and instant messaging as its main collaboration tool. Blogs and corresponding comments and discussions keep members aware of project progress and accelerate their technical growth (Johri, 2010). Although not a typical research team, this team works mainly as integration. Hence the observation supports previous suggestion that sharing of content itself and inspiring works best for such collaborations.

2.2   Affect and Pleasure in Collaboration

There are various definitions and classifications of affect and pleasure, and many theories about how affect and pleasure may impact people's behaviour. Based on Tiger's (1992) work, Helander and Khalid (2006) developed a taxonomy that identifies 5 types of pleasure: 1) physical pleasure, 2) sociopleasures, 3) psychological pleasure, 4) reflective pleasure, and 5) normative pleasure. This taxonomy implies the personal level and social level of pleasure, hence could be used to explain some of the phenomenon found in collaborations.

Isen (1999) found that even mild positive affect improves creative problem solving. On the other hand, LeDoux (1995) and other researchers have been claiming that affect and cognition are conjoint and equal in the control of thoughts and behaviour. Hence affect and pleasure should not be neglected in designs to support research activities.

Little information was found to illustrate the impact of affect and pleasure on collaboration styles, while plenty of researches have shown social status can have significant impacts. Recall the taxonomy of pleasure, which implies that pleasure may come from satisfaction of personal needs (physical and psychological) and social needs. We can therefore assume that social status may have its roots in, or can be altered by affect and pleasure.

At the early stage of the use of computer-assisted collaboration (i.e. Email system and so), researchers have discovered negative situations in working environments due to the lack of information (gestures, facial expressions, tunes etc.) or misuse of such tools on purpose to avoid emotional reactions (Culnan & Markus, 1987; Markus, 1994). Recent studies emphasize people's social status, including perceived existence of groups and communities, relationship among individuals, and interaction styles, can significantly affect communication styles and effects (Tu, 2000; Nonnecke & Preece, 2001).

In summary, affect and pleasure have impacts on people's behaviour and cognition, hence result in changes of communication and collaboration.

3 THE DESIGN

Previous discussions show that to establish a working environment supporting research team collaboration, it is essential to support the sharing of information. Although not directly proven, affect and pleasure issues could leverage the efficiency and effectiveness of such collaboration environment. Hence designers are motivated to carefully embed affect and pleasure into their design. The effort was contributed to both appearance and functionality.

3.1 System Model

The foundation of this collaborative environment (coded as "AI") is a UGC (User Generated Content) system running as a server-browser cloud service. Its content includes resources such as articles written by users, and files uploaded by users as either attachments to articles or individual resources. Figure 1 describes the architecture of the resource-collection system.

Users belonging to the same team share a same resource space, where the resources (articles and files) are sorted into "collections" (similar to folders) manually by team members. Every member has an equal authority to create, modify or delete articles, files, or collections. The articles can be edited by multiple users and keep record of the changes as "versions". This equal and open authority design is based on the assumption that team members can spare trust with each other, respect coworkers' effort, and behave properly, which is usually the case in small-scale research teams.

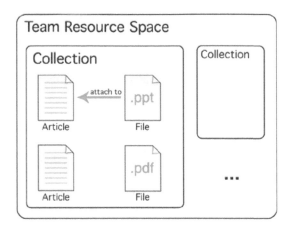

Figure 1. The architecture of resources and collections in A1. Files can exist as attachment to articles (like the .ppt file), or as individual resources (like the .pdf file).

3.2   Building Affective Appearance

Visual appearance leaves with users quick but unconscious short-lasting affect. According to reversal theory, people's arousal state affects their perceived level of pleasure (Apter, 1989). Coping with the context that users are mainly serious-minded and goal-oriented, designers choose to use white, light grey and simple layout to architect the interfaces, using decent grey texture as background. A series of low-saturated blue is used to highlight the hyperlinks, active tabs and other interface objects. The primary buttons were designed to be dark grey with the intention to reduce the association of entertaining and prompt the feeling of professional and decent (See Figure 2). It was later discovered to be a happy coincidence with the keyboard design of Apple Mac Book Pro as shown in Figure 3.

Figure 2. Details of the interface of A1. A snapshot from the "edit article" page, showing background texture, light grey toolbar and dark grey buttons.

Figure 3. The keyboard design of Apple MacBook Pro.
(http://www.apple.com.cn/macbookpro/images/overview_gallery2_20111024.png#gallery3)

3.3   Embedding Social Network Features

There have been various models developed to assist communication and information sharing, but each has its defects. For example, with wiki system, team members can easily edit and contribute to team knowledge library while maintaining perfect structure, yet it's difficult for users to discuss or to track latest updates. Forums use time sequence to highlight updates and use topics to manage discussions, however the weakness is that topics are hardly sorted and not easy to locate.

Social network services, such as Facebook and Twitter, give an inspiration how to design proper features for collaborative environments. Both Facebook and Twitter use news-feed to present recent activities of user's friends or followed subjects. The bond of friendship in Facebook, and the relationship of following in Twitter, each enables the system to present latest updates of selected sources.

Instead of focusing on friends' activities, A1 projected a "friend" to a piece of article or file. The creation and modification of the article or file is treated as an activity, and will appear as a piece of feed, enabling A1 to inform the users with activities of concerned content (See Figure 4).

Another feature learned from social network services is the notification when others reply to the user. This feature makes users aware of messages and helps them to communicate more efficiently.

A1 also developed a "recommendation" mechanism that a user can recommend a certain piece of article or file to selected users and leave a message. These selected users will receive a notification. This mechanism helps users to communicate about certain pieces of content, to spread valuable information accurately, or to inform certain users with important contents.

Figure 4. A snapshot from the "Updates" feed list. The menu on the left shows notifications for messages and followed updates.

The feed, reply notification and recommendation mechanisms together equip the system with both active and passive information flow. They all encourage communication and discussion about the resources. Consequently, a smooth channel for spreading information and filtering valuable resources is formed.

The appearance and interaction style of these mechanisms are designed in the fashion of social network services, so that similar emotional impressions could be reminded, such as relaxing, casual, and socially connected. These emotions help to relieve social and psychological barriers among users, and therefore enhance team members' activity.

3.4   Designing for Advanced Users

Upon basic functions, additional functions are designed for advanced users, to fulfill the basic need for virtuosity (Kubovy, 1999). In the feed mechanism, users can set up a group of "specially followed" resources so that they can track those resources with higher priority. In the organization of resources, advanced users can set shortcuts for important resources so that they can be accessed much easier. Or they can use a grid system to gather related resources for more efficient browsing.

4   DISCUSSION

The team who created A1 has been using it to assist collaboration all the time. Before A1 was developed, the team used a wiki-based system to do the work. Although quantitative analyses are not feasible due to the lack of data from the old system, a brief qualitative analysis is performed. It is discovered that about 1/3 of resource created are of technical experiences, skills, and information from outside sources. Very few of this kind of information could be found in the old system. The amount of work-related documents can hardly be compared, though more visits and a wider range of visitors is observed.

The team has published the beta version of A1 since July, 2011 on http://www.escience.cn, named "Research Online" (the service is currently only in Chinese). This public service keeps evolving and has gained near a thousand users by the end of 2011. Users have shown very different behaviours on this service compared to the old version (identical to the old system used by the team). Most of the users treated the old collaborative environment as a CMS system, and only uploaded static information such as regulations, application forms, annual reports, and data files or documents to be archived. While in the new Research Online, users start to contribute work-related information such as plans, schedules, checklists and intermediate data from experiments. They commented the sharing of data files an effective function. Also, users in the new service are more active than those using the old service.

Defects are also addressed for A1. Novices complain about the complexity of its concept model that they get confused about the relationship between feed, collections and resources, especially when they have little experience of social network services. Some advanced users request for more powerful tools to organize the resources, direct messaging with members in the form of email rather than comment and reply, and so on. All of these feedbacks will be seriously considered in the development of A1.

5   CONCLUSION

The development of A1 as a collaborative environment for research teams learned from the context to provide support for information sharing and communication. Affect and pleasure were taken into account for their potential impact on user behaviour in such systems. As a result, careful appearance design and social network features were embedded into the system. Current tests has shown such design an acceptable one, while more effort is required for further improvement.

REFERENCE

Apter, M.J., 1989. Reversal Theory: Motivation, Emotion and Personality. Routledge, London.

Balakrishnan, A.D., Kiesler, S., Cummings, J.N., Zadeh, R., 2010. Research team integration: What it is and why it matters. Proceedings of the 2010 ACM Conference on Computer-Supported Cooperative Work, 523-532.

Culnan, M., Markus, M.L., 1987. Information technologies. In: Handbook of organizational communication: An interdisciplinary perspective, Jablin, F., et al., Eds. Sage Publications, Newbury Park, Calif., 420-443.

Facebook, 2011. http://www.facebook.com.

Helander, M.G., Khalid, H.M., 2006. Affective and pleasurable design. In: Handbook on Human Factors and Ergonomics, Salvendy, G., Eds. Wiley, New York, 543-572 (Chapter 21).

Isen, A.M., 1999. On the relationship between affect and creative problem solving. In: Affect, Creative Experience, and Psychological Adjustment. Russ, S., Eds. Taylor & Francis, Philadelphia, 3-17.

Johri, A., 2010. Look ma, no Email! Blogs and IRC as primary and preferred communication tools in a distributed firm. Proceedings of the 2010 ACM Conference on Computer-Supported Cooperative Work, 305-308.

Kubovy, M., 1999. On the pleasures of the mind. In: Well-Being: The Foundations of Hedonic Psychology. Kahneman, D., Diener, E., Schwarz, N., Eds. Russell Sage Foundation, New York, 134-154.

LeDoux, J.E., 1995. Emotion: Clues from the brain. Annu, Rev. Psychol, 46, 209-235.

Markus, M.L., 1994. Finding a happy medium: Explaining the negative effects of electronic communication on social life at work. ACM transactions on information systems, Vol.12, No.2, 119-149.

Nonnecke, B., Preece, J., 2001. Why lurkers lurk. Americas Conference on Information Systems.

Research Online, 2012. http://www.escience.cn.

Thu, C.H., 2000. On-line learning migration: From social learning theory to social presence theory in CMC environment. Journal of network and computer applications, Vol.23, No.1, 27-37.

Tiger, L., 1992. The Pursuit of Pleasure. LIttle Brown, Boston.

Twitter, 2011. http://www.twitter.com.

CHAPTER 38

Affective Design and Its Role in Energy Consuming Behavior: Part of the Problem or Part of the Solution?

Kirsten Revell, Neville Stanton

University of Southampton
Southampton, U.K.
kmar1g10@soton.ac.uk

ABSTRACT

To mitigate against the effects of climate change, the UK has legislated to cut greenhouse gas emissions by 80% by 2050 (Climate Change Act 2008). Domestic consumers currently contribute over 25% of total UK carbon emissions (The UK Low Carbon Transition Plan). Significant variations in domestic energy use have been shown to be due to the behavioral differences of householders. The role product design plays in energy consuming behavior was explored with reference to Norman (2004)'s model of the affective system. This papers argues the need for designers to carefully consider the type, magnitude and interaction of affect, at each level of the affective system, when designing energy consuming devices. This paper illustrates through the analogy of a 'pivot scale' how an optimal balance between the benefit offered by the device to the user, and the amount of energy consumed, may be achieved.

Keywords: affective design, behavior, energy consumption, product design

1   INTRODUCTION

To mitigate against the effects of climate change, the UK has legislated to cut greenhouse gas emissions by 80% by 2050 (Climate Change Act 2008). Domestic consumers currently contribute over 25% of total UK carbon emissions (The UK Low Carbon Transition Plan , 2009). Significant variations in domestic energy use have been shown to be due to the behavioral differences of householders. According to Lutzenhiser & Bender (2008), these differences are primarily the result of sociological factors, thwarting attempts at a 'one-size-fits-all' approach to behavior change. The aim of this paper is to 1) consider the role affective design plays in energy consuming behavior in the home, and 2) draw conclusions as to how affective design could be used as a design tool to influence consumption in the home.

Affect is the general term for the judgmental system, of which emotions are part. Norman (2004) distinguishes between emotions and affect, whereby with the former, one is conscious of the cause, object, and type but the latter allows a positive or negative feeling to be experienced without attribution to the cause or object. Norman (2004) proposes that three different levels of the brain make up the affective system; the automatic pre-wired 'visceral level', the 'behavioral level' and the 'reflective level'.

The affective system is an information processing system, like cognition. Norman (2004:p11) positions the role of cognition to interpret and make sense of the world, whilst the affective system "makes judgments and quickly helps you determine which things in the environment are dangerous or safe, good or bad". This idea is reminiscent of Gibson's (1986) theory of affordances, which considers the risks and benefits of the environment as directly perceivable to a person. Affordances are easily incorporated at the behavioral level of the affective system, where these risks and benefits result in negative and positive affect.

Norman (2004) positioned the understanding of the affective system as a tool to prompt designers to think about the effect of their design decisions on the user's pleasure, satisfaction and effective operation of product's, ultimately influencing the products success. Norman (2004) proposes each level plays a different role in the functioning of people, and requires a different style of design. The authors propose a link between how the affective system influences peoples' response to domestic devices, and their resulting energy consuming behavior. To explore this idea, this paper will interpret the levels of the affective system, and their interaction, in terms of energy consuming behavior when using devices. Initial conclusions about how affective design can be applied to optimize energy consuming behavior will then be offered using the analogy of a 'pivot balance'.

1.2   The affective system and energy consuming behaviour

The affective system is composed of three levels; the perceptually based visceral level, the experience based behavioral level and the intellectually based 'reflective level' (Norman, 2004, Norman et al.,2003 and Ortony et al., 2004).

In contrast to the behavioral and reflective levels, where the response to sensory information is influenced by a person's background and experience, Norman (2004;33) defines the visceral level as "automatic (and) pre-wired", where the way sensory information is processed is immediate and universal. Understanding the influence of the visceral level on energy consuming behavior appealed to the authors as design principles applied at this level further the prospect of a 'one-size fits all' approach to behavior change.

A person's experience of affect at the behavioral level, when interacting with man-made devices, was proposed by Norman (2004) to be based on the product's function, performance and usability. These variants are all determined by design so can influence the resulting affect by design strategies. The types of strategies that improve affect, however, may not necessarily have a positive effect on energy consuming behavior, which will be shown in the following section.

The reflective level integrates elements beyond direct interaction with the device, taking into account, for example, social norms, self-image, planning, reasoning and problem solving (Norman et al., 2003). This level could influence energy consuming behavior by considering personal attitudes to waste in general, saving money, genuine concern for the environment or social norms of device and energy use. From a designer's perspective, these elements will vary considerably with socioeconomic factors and as such, the designer will not be able to incorporate a 'one-size-fits-all' approach at this level.

Norman et al. (2003)'s affective system comprises an input, three levels of processing, and an output. The input consists of sensory information, and the output, motor behavior. The authors propose that when using affective design to influence energy consuming behavior, that the input be redefined as 'sensory information that represents energy consumption' and the output as 'motor behavior that impacts energy consumption'.

The visceral (bottom) and behavioral (middle) levels process the sensory information and either directly, or with the interaction with each other or the reflective (top) level, determine the motor behavior. The reflective level does not receive sensory information nor determine motor behavior directly, only via the lower levels (Norman et al., 2003).

Information taken in at the visceral level can either result in an immediate motor reaction, or trigger processing from the higher levels which can serve to strengthen or inhibit the reaction (Norman et al., 2003). The following section illustrates the authors' position on how design choices at the visceral and behavioral level could interact to strengthen or inhibit energy consuming behavior.

Information received at the behavioral level, directly or via the visceral level is likely to be passed up to the reflective level when it does not conform to expectations. At the reflective level, the mind's own internal representations are accessed or amended, and the outputs of planning, reasoning and problem-solving inform the response sent down to the lower levels (Norman et al., 2003). When considering sensory information resulting from device design, the influence of the reflective level on energy consuming behavior therefore relies on first being triggered by the lower levels.

1.3 Using affective design to optimise energy consuming behaviour.

In order to consider pragmatically, the interaction between the levels of affect when designing energy consuming devices, the inputs linked to energy consuming behavior, need to be specified. The authors have provided examples of inputs thought to influence each level of the affective system, to illustrate the consequences of their interaction, but do not claim these criteria as either necessary nor sufficient.

At the visceral level, the type of energy that can be easily perceived when operating a device was considered by the authors to link to energy consuming behavior when using devices (for example energy transmitted via sound, heat, light or motion etc.). The authors suggest that the perception of emitted energy relates, at the unconscious level, to some sort of 'primitive' notion of the feedback associated with consuming fuel or expending personal energy. Norman (2004) provided lists illustrating sensory information which, if perceived, results in positive and negative affect at the visceral level. The authors believe that extensions of these lists would be applicable to the perception of emitted energy by devices.

Given the purpose of this paper was to inform device design, the authors' felt the behavioral level warranted 3 areas of evaluation that reflected design strategies. Norman (2004) described the negative affect that resides from poor usability, and suggested design strategies taken from Norman (2002) such as reducing the gulf between designer and user mental models, natural mappings between controls and function, and pleasure in the sensation of interaction. To encourage energy efficient behavior, the authors propose that the first of these strategies could be translated into 'design provides the user with a mental model of the device that associates 'device benefit' with 'work done' by the device'. The second strategy, could be translated into 'the controls for operation are mapped to consumption levels'. Finally, the idea of pleasure in the experience of interaction, to optimize energy consumption (by making interaction with the device to save energy, more pleasurable).

At the reflective level, the authors have positioned Norman (2004)'s ideas of positive self-image in terms of energy consuming behavior. A simplistic assumption was made by the authors that a positive self-image results from conforming to the 'norms' of society by adhering to government recommended advice available online (at www.direct.gov.org.). How the design of devices help or hinder users' ability to follow this advice, would therefore influence the affect experienced at this level.

From a participant observation study, the authors came to the following conclusions: 1) negative affect at the visceral level may actually promote energy saving behavior, by making the user avoid or consider carefully when and how they use the device; 2) the importance of positive affect at the behavioral level is especially important if the appropriate choice of setting to optimize energy consumption is desired, and 3) whilst design of individual devices (rather than energy monitors) are less well placed to influence behavior change at the reflective level, understanding how this level may interact with the lower levels could contribute to decision decisions aspiring to an optimal balance between device benefit and energy consumption. With the goal of achieving this balance, the authors have depicted in figures 1, 2 and 3 how they consider the levels of the affective system should be approached by a designer, using the discussed inputs.

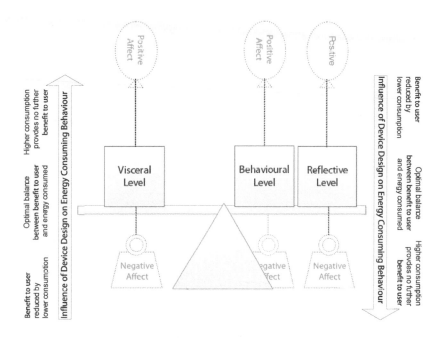

Figure 1 Balancing the 3 levels of affective design to provide an optimal balance between the benefit of a device to the user and energy consumed

Figure 1 depicts the balance between the three levels of the affective system on a traditional pivot scale. The visceral level is positioned on one side of the pivot, and the behavioral and reflective levels on the other side. The authors consider the 'boxes' representing the three levels to be weightless, and the only means of tipping the balance is the attachment of weights (negative affect) or balloons (positive affect). The arrows to the left and right of the balance, show a scale representing the influence of device design on energy consuming behavior.

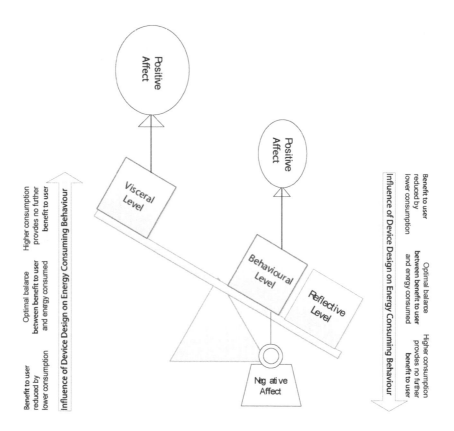

Figure 2   The result of affective design of a gas fire on the balance between energy consumption and user benefit

The top of the scale represents energy consumption exceeding the benefit of the device to the user, and this scenario is presented in Figure 2, using the example of the gas fire from the participant observation study. In this example, the pleasure experienced at the visceral level from the perception of the flicker of flames, encouraged energy consumption to continue for a considerable time after thermal comfort (the intended device benefit) had been achieved. The bottom of the scale represents reduced energy consumption, but at the expense of benefit to the user. This is represented in figure 3 by the example of an extractor fan, whose high noise levels results in the device never being operated. The authors believe the number and size of 'balloons and weights' at each level of affect could be varied in device design to facilitate the desired energy consuming behavior.

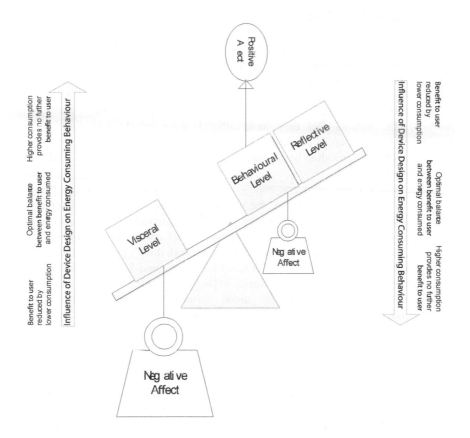

Figure 3 The result of affective design of an extractor fan on the balance between energy consumption

The authors are keen to convey that the pivot scale diagrams are intended to provide a simplistic analogy to illustrate how the levels in the affective system may interact to influence energy consuming behavior. Different inputs identified for each level of the affective system could promote or prevent energy saving behavior in different ways, depending on the class of device, or the stage of operation of the device. The position of the affective levels with respect to the pivot would need to reflect this to provide the appropriate guidance to the designer. Different inputs to the same level of affect, may need to be positioned on different sides of the pivot to provide comprehensive guidance. The questions of how to select categories of input relating to energy consuming behavior via the affective system, measuring the type of affect produced (either 'balloon' or 'weight'), determining magnitude of the affect (the size of the 'balloon' or 'weight') and the relative positions of each level of affect (or their components) on the pivot scale, are beyond the scope of this paper, but could form the basis for further research.

2   CONCLUSIONS

The aim of this paper was to: 1) to consider the role affective design plays in energy consuming behavior in the home and 2) draw conclusions as to how affective design could be used as a design tool to influence consumption in the home.

By interpreting Norman's (2004) affective system in terms of energy consuming behavior, the authors feel they have offered some insight into how energy consuming behavior could be encouraged, discouraged or optimized with reference to device benefit, depending on the design decisions made at each level within the affective system.

Inputs which could influence energy consuming behavior at each level of the affective system were proposed. At the visceral level, the emitted energy perceived when operating the device was proposed. At the behavioral level, three design strategies relating to consumption were recommended, comprising of: 1) providing a mental model associating 'device benefit' with 'work done'; 2) providing controls for operation which are mapped to consumption levels, and 3) making interactions with the device to save energy, a pleasure. At the reflective level, how device design facilitates the user to conform to social norms by adhering to government recommended advice regarding domestic device use, was suggested.

A simplified approach to 'weighing up' the influence of affective design on energy consuming behavior was offered through the analogy of a pivot balance. However, the authors wish to make clear that this model is only illustrative of an interaction between the levels based on the inputs advised, and are not presented as necessary or sufficient. Depending on the chosen inputs, the class of device and the stage within device operation, the positions of each affective level on the pivot scale (and thus guidance to the designer) will vary. These areas, therefore, warrant further investigation if the role affective design plays in energy consuming behavior is to be 'part of the solution'.

REFERENCES

Climate Change Act 2008 [Online]. London: Department of Energy & Climate Change Available:
http://www.decc.gov.uk/en/content/cms/legislation/cc_act_08/cc_act_08.aspx
[Accessed 10th November, 2011 2011].

Top tips on saving energy [Online]. Directgov. Available:
http://www.direct.gov.uk/en/environmentandgreenerliving/energyandwatersaving/d
g_064371 [Accessed January 10th 2012].

The UK Low Carbon Transition Plan [Online]. London: Department of Energy & Climate Change [Accessed November 10th 2011].

Gibson, J. J., 1986. *An Ecological Approach to Visual Perception,* Hillsdale, New Jersey: Lawrence Erlbaum Associates, Inc.

Lutzenhiser, L. a. S. & Bender, S. 2008. The average American unmasked: Social structure and difference in household energy use and carbon
emissions.: ACEEE Summer Study on Energy Efficiency in Buildings.

Norman, D. A., 2002. *The Design of Everyday Things,* New York: Basic Books.

Norman, D. A., 2004. *Emotional Design,* New York: Basic Books.

Norman, D. A., Ortony, A. & Russell, D. M. 2003. Affect and machine design: Lessons for the development of autonomous machines. *IBM Systems Journal,* 42 (1), 38-44.

Ortony, A., Norman, D. A. & Revelle, W., 2004. The role of affect and proto-affect in effective functioning. *In:* Fellous, J. M. & Arbib, M. A. (eds.) *Who needs emotions? The brain meets the machine.* New York: Oxford University Press

Section V

Human Interface In Product Design

CHAPTER 39

Self-Inflating Mask Interface for Noninvasive Positive Pressure Ventilation

Uwe Reischl, Lonny Ashworth, Lutana Haan, Conrad Colby
Boise State University, Boise, Idaho, USA
ureischl@boisestate.edu

ABSTRACT

A prototype facemask interface was developed to help improve patient comfort during non-invasive positive pressure ventilation. The prototype design includes a flexible fabric "skirt" which is attached to the frame of a standard facemask. The interface adapts easily to the contours of the face and provides increased contact area with the skin. The performance of the prototype facemask was compared to the performance of a standard facemask using a manikin. Air pressure and volume airflow through the masks were controlled. The results showed that the flow and pressure characteristics of the prototype facemask including the skirt interface were comparable to those of the standard facemask technology. The prototype reduced the facemask pressure levels on the face by allowing the interface skirt to "float" on the skin.

Keywords: Pressure inflating face mask interface, prototype mask design, mask comfort

1   INTRODUCTION

Positive airway pressure is a mode of respiratory ventilation that is often used in the treatment of sleep apnea and commonly used for those who are critically ill in a hospital with respiratory failure (Garpestad, E., J. Brennan, N. Hill. 2007; Mehta,

S., N. Hill. 2001; Paus-Jenssen, E., J. Reid, D. Cockcroft, K. Laframboise, H. Ward. 2004). In these patients, positive airway pressure ventilation can prevent the need for tracheal intubation or allow earlier extubation. (Meduri, G., R. Turner, N. Abou-Shala, R. Wunderink, E. Tolley, 1996; Liesching, T., H. Kwok, N. Hill. 2003). The mask required to deliver the necessary positive airway pressure must have an effective seal and must be held on the face securely. In general, such respiratory facemasks serve the function of providing patients with supplemental air and oxygen. Figure 1 illustrates such a facemask.

Current facemask designs frequently do not provide an adequate fit and often cause discomfort. Patients may experience pressure sores when facemasks are worn for extended periods. A new self-inflating facemask interface was developed to address these shortcomings. The interface consists of a thin flexible fabric-like "skirt" that surrounds a standard facemask frame. The skirt inflates automatically when the facemask receives external pressurized air. The skirt is flexible, soft, and adapts easily to the topography of the face. This eliminates pressure points and provides comfort for the patient. The self-inflating skirt is simple in design and can be attached to most available facemask frames. Mask performance was evaluated under controlled conditions by using a manikin. Data showed that pressure and airflow characteristics of the prototype were similar to the standard facemask technology while being able to reduce the risk of pressure sores on the face.

Figure 1 Illustration of the use of a standard facemask including support straps. The straps can tightened to increase the seal between the mask and the face.

2 METHODS AND PROCEDURES

Performance characteristics of two facemask designs were compared under controlled laboratory conditions. One design consisted of a standard ResMed Mirage Quattro™ full facemask. The second design consisted of a modified ResMed Mirage Quattro™ full facemask that included a fabric skirt interface which replaced the original facemask cushion assembly. The two facemask design features are illustrated in Figure 2.

A full-size male manikin placed on a gurney in a supine position was used for testing facemask performance. (Figure 4). Airflow was provided by an in-house compressed air supply system. Volume airflow to the facemasks and associated air pressure were measured using the TSI Certifier™ Test System. Facemask strap tension levels were controlled using calibrated weights suspended from the mask strap attachment brackets.

Airflow through the facemasks was measured at the $5cmH_2O$, $6cmH2O$, $8cmH_2O$, and $10cmH_2O$ air pressure levels. Each air pressure condition was evaluated for mask strap tension levels of 500 grams, 1000 grams, 1500 grams, and 1750 grams. All tests were performed three times.

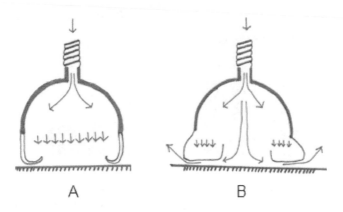

Figure 2 Cross-sectional view of the ResMed Mirage Quattro™ full facemask design (A) and a cross-sectional view of the prototype design (B). The standard design achieves a seal between the mask and the face through pressure applied to the mask pad. The prototype design provides a seal between the mask and face by inducing flotation of the skirt over the skin.

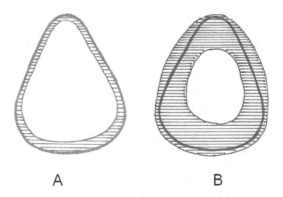

Figure 3 Skin contact patterns for the standard mask design (A) and the prototype design (B). The face contact area for the standard face mask is approximatley 22 cm^2 . The contact area for the pototype design is approximately 70 cm^2.

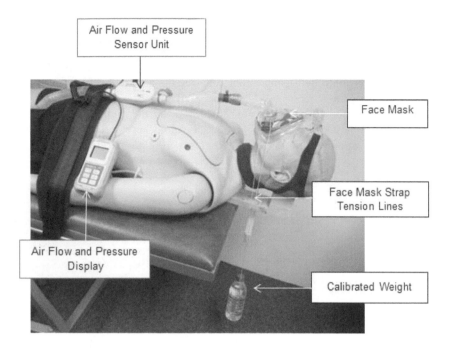

Figure 4 Illustration of manikin equipped with air flow and air pressure sensor unit, data display unit, face mask, mask straps, and calibrated weights for changing contact of mask with the face.

3. RESULTS

Table 1 summarizes the air volume flow rates measured for the "Standard" ResMed[TM] facemask at 5, 6, 8, and 10 cmH_2O pressure levels with facemask strap tension levels of 500, 1000, 1500, and 1750 grams. It can be seen that the volume airflow decreased for all facemask air pressure settings when the facemask strap tension levels were increased. At the 5 cmH_2O pressure level, the volume airflow decreased by 5.1 LPM as the strap tension increased from 500 grams to 1750 grams. At the 6 cmH_2O pressure level, the volume air flow decreased by 6.3 LMP. At the 8 cmH_2O pressure level, the volume air flow decreased by 13.5 LMP. At the 10 cmH_2O pressure level, the volume air flow decreased by 26.6 LMP.

Table 2 summarizes the airflow rates for the Prototype facemask design at 5, 6, 8, and 10cmH_2O pressure levels with facemask strap tension levels of 500, 1000, 1500, and 1750 grams. It can be seen that the volume airflow decreased again for all facemask air pressure settings when the facemask strap tension levels were increased. At the 5 cmH_2O pressure level, the volume airflow decreased by 7.5 LPM as the strap tension increased from 500 grams to 1750 grams. At the 6 cmH_2O pressure level, the volume air flow decreased by 8.5 LMP. At the 8 cmH_2O pressure level, the volume air flow decreased by 11.1 LMP. At the 10 cmH$_2$O pressure level, the volume air flow decreased by 11.3 LMP.

Table 3 summarizes the airflow values obtained for an air pressure level of 5cmH$_2$O in both the "Standard" ResMedTM facemask and the Prototype design with mask strap tension levels of 500, 1000, 1500, and 1750 grams. For the " ResMed^{TM} facemask the volume air flow decreased by 11.5 LPM as the strap tension increased from 500 grams to 1750 grams. For the Prototype facemask, the decrease was 8.6 LPM.

TABLE 1: Summary of airflow values associated with fixed facemask air pressure conditions and facemask strap tension levels for the ResMed™ facemask.

Mask Air Pressure (cmH2O)	Mask Strap Tension (grams)	Mask Volume Air Flow (LPM)			
		Trial 1	Trial 2	Trial 3	Average
5.0	500	38.8	38.9	39.1	**38.9**
5.0	1000	34.3	35.0	35.3	**35.2**
5.0	1500	33.6	33.9	34.2	**33.9**
5.0	1750	33.6	33.9	34.1	**33.8**
6.0	500	43.8	44.1	44.5	**44.1**
6.0	1000	39.8	38.5	40.2	**39.5**
6.0	1500	38.3	38.2	38.1	**38.2**
6.0	1750	37.8	37.6	38.0	**37.8**
8.0	500	58.2	57.8	57.5	**57.8**
8.0	1000	49.4	49.0	49.1	**49.1**
8.0	1500	44.6	44.6	44.5	**44.5**
8.0	1750	44.6	44.1	44.2	**44.3**
10.0	500	75.5	72.6	72.0	**73.3**
10.0	1000	60.4	59.6	59.2	**59.7**
10.0	1500	50.3	51.1	51.1	**50.8**
10.0	1750	49.5	49.8	49.9	**49.7**

TABLE 2: Summary of airflow values associated with fixed facemask air pressure conditions and facemask strap tension levels for the Prototype facemask.

Mask Air Pressure (cmH$_2$O)	Mask Strap Tension (grams)	Mask Volume Air Flow (LPM)			
		Trial 1	Trial 2	Trial 3	Average
5.0	500	48.5	48.9	48.8	**48.7**
5.0	1000	45.8	46.6	46.5	**46.3**
5.0	1500	42.6	43.7	43.3	**43.3**
5.0	1750	41.0	41.5	41.1	**41.2**
6.0	500	54.5	53.8	53.4	**53.9**
6.0	1000	51.4	51.0	50.7	**51.0**
6.0	1500	47.5	47.1	47.4	**47.3**
6.0	1750	45.5	45.3	45.4	**45.4**
8.0	500	64.1	63.8	63.4	**63.7**
8.0	1000	59.7	59.3	58.7	**59.2**
8.0	1500	55.8	55.0	54.6	**55.1**
8.0	1750	53.0	52.4	52.4	**52.6**
10.0	500	71.4	70.4	70.4	**70.7**
10.0	1000	66.2	66.0	65.7	**65.9**
10.0	1500	63.3	63.1	62.6	**63.0**
10.0	1750	59.6	59.6	59.2	**59.4**

TABLE 3: Summary of airflow values associated with a 5 cmH$_2$O mask air pressure level at various strap tension levels for the ResMed™ facemask and the Prototype facemask when an 8mm closed tube was placed between the mask interface and the face.

Mask Air Pressure (cmH$_2$O	Mask Strap Tension (grams)	Mask Volume Air Flow (LPM)			
		Trial 1	Trial 2	Trial 3	Average
ResMed Facemask					
5.0	500	95.6	95.6	95.5	**95.6**
5.0	1000	93.0	92.5	92.7	**92.7**
5.0	1500	89.4	87.2	86.7	**87.8**
5.0	1750	84.6	84.3	83.5	**84.1**
Prototype Facemask					
5.0	500	80.0	80.1	79.4	**79.8**
5.0	1000	78.2	77.5	77.2	**77.6**
5.0	1500	73.6	74.1	73.3	**73.6**
5.0	1750	71.5	71.0	71.1	**71.2**

4 ANALYSIS

The observed facemask airflow conditions are the result of air escaping at the mask-skin interface. The mouth and nose of the manikin were sealed so that no air could escape through these openings. Therefore, the observed flow rates are only due to "leakage" at the mask – skin interface. The tension on the facemask straps determined the seal achieved between the mask and the face. Increasing the strap tension increased the fit and subsequently reduced the leakage. The impact of mask strap tension on the relationship between facemask air pressure and measured facemask airflow is illustrated in Figure 5. As can be seen, both the standard mask design and the prototype design displayed increased airflows as air pressure was increased. However, as tension in the facemask straps were increased, the airflow decreased.

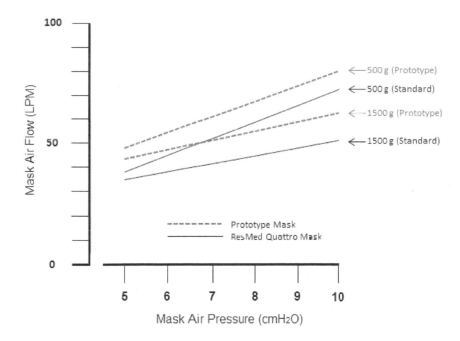

Figure 5 Facemask air flow in relation to mask air pressure and mask strap tension levels.

While the airflows provided by both the standard facemask and the prototype facemask were similar, the functionality of the two designs is substantially different. As seen in Figure 2 and Figure 3, the Prototype facemask presents a larger contact area with the skin of the face than is the case for the standard mask. The Prototype uses the leakage airflow (Venuri suction) to pull the "skirt" down to the contours of the face. This feature not only improves comfort but also offers an advantage when tubes from the nose or mouth must be accommodated such as in a clinical setting. As an example, Table 3 shows the leakage associated with an 8mm (closed) tube placed between the facemask and the skin. The leakage for the Prototype was substantially lower than the leakage observed for the standard facemask. This illustrates that the skirt of the Prototype mask was able to "hug" the tube more closely than the standard facemask pad.

4 CONCLUSIONS

The controlled laboratory tests showed that the Prototype facemask including the pressure-inflating "skirt" feature exhibited performance characteristics comparable to those of a standard high-quality facemask product. However, the study also showed that the Prototype facemask offered a more accommodating placement of the mask on the face. The new "skirt" concept not only has the potential of accommodating external tubing to a patient's mouth, but also has the capacity to improve the comfort of persons requiring positive pressure ventilation for extended periods of time.

REFERENCES

Garpestad, E., J. Brennan, N. Hill. 2007. Noninvasive Ventilation for Critical Care *CHEST*, 132, 711-720
Liesching, T., H. Kwok, N. Hill. 2003. Acute Applications of Noninvasive Positive Pressure Ventilation. *CHEST,* 124, 699-713

Meduri, G., R. Turner, N. Abou-Shala, R. Wunderink, E. Tolley, 1996. Noninvasive Positive Pressure Ventilation Via Face Mask First-Line Intervention in Patients With Acute Hypercapnic and Hypoxemic Respiratory Failure. *CHEST*, 109, 179-193
Mehta, S., N. Hill. 2001. Noninvasive Ventilation. *Am. J. Respir. Crit. Care Med.* 163, 540-577
Paus-Jenssen, E., J. Reid, D. Cockcroft, K. Laframboise, H. Ward. 2004. The Use of Noninvasive Ventilation in Acute Respiratory Failure at a Tertiary Care Center, *CHEST,* 126, 165-172

CHAPTER 40

Product Personality Assignment as a Mediating Technique in Biologically and Culturally Inspired Design

Denis A. Coelho, Carlos A. M. Versos, Ana S. C. Silva

Universidade da Beira Interior
Covilhã, Portugal
denis@ubi.pt, carlos.versos@gmail.com, sofia.silva.design@gmail.com

ABSTRACT

The chapter reviews the product personality assignment technique and proposes its deployment in two kinds of approaches to design, biologically and culturally inspired design, two approaches to design that may contribute to the satisfaction of sustainability goals. While the focus of the first is on efficiency and effectiveness, with decreased resource usage, the promotion of local resource use and local production for local consumption, sought by culturally inspired design, may also be conducive to reduced environmental impacts. Biologically inspired design seeks to inform the process of design with examples and solutions from nature, whether the bionic example is viewed as the trigger for the design process or it is considered in the concept generation phase. The chapter demonstrates, through the report on a design case, the use of the product personality assignment technique within a bionic design process, at the phase of validation of requirements satisfaction. In this case, a set of subjects performed the evaluation directly on the design concepts. This design case consisted of the design of a device to store discs and books, taking inspiration from nature. In another design case, reported in the chapter, seeking transposition of cultural aspects to product design, existing products were initially assigned personality profiles and rated by a set of subjects. The researchers then sought to establish links between the personality assignment made by subjects and by researchers and the features of the products. In parallel, cultural profiles were developed for translation into product personality profiles and from these to product features in order to trigger design processes. The second design case reported led to production of new furniture concepts. Considering the current urgency in achieving sustainability, the two cases presented in the chapter also suggest a systematization of the possible deployments of the product personality assignment technique in a wide array of methodological approaches to design. Taking an even wider perspective, the cases also provide evidence of the interplay between human factors and ergonomics goals in design and sustainability.

Keywords: ergonomics in design, sustainability, user centered design

1   INTRODUCTION

The concept of a product as having a personality was based on the paradigm of the "New Human Factors" developed by Patrick W. Jordan (2000). In this view, the product is anthropomorphized by the user who projects personality traits on the object. This paradigm is in contrast to previous approaches, which tended to look at the product as a mere tool with which the user could perform certain tasks. Jordan, in 2000, used a technique that became known as the "Product Personality Assignment" for the purpose of studying the personality of the product concept and to establish connections between the aesthetic quality of each product and personality. A set of seventeen product personality dimensions were proposed (Table 1). In product design, the transfer from subjective qualities to objective properties represents one of the major challenges for designers (Coelho and Dahlman, 2002), and may benefit from the use of this technique. This chapter reports on two applications of this technique, in two approaches to design. Technology, devised by human ingenuity, can create quality of life and support human well-being, but sustainability needs to be both a limiting factor and a triggering factor for innovation (Coelho, 2012). Biologically and culturally inspired design may contribute to the satisfaction of sustainability goals. Bionic design aims at decreased resource usage, while culturally inspired design assists the promotion of local resource usage and valuation of local production for local consumption. Essential tools in any design process, providing guidelines, goals and technical guidelines for the successful development of products, design methodologies must place emphasis in the validation of the results of the product development process, with respect to initially set specifications and requirements.

The first case reported in this chapter concerns the validation of subjective product qualities, as reported by Versos and Coelho (2012, 2011-a, 2011-b, 2010) and by Coelho and Versos (2011, 2010), within an approach to bionic design. The second study uses the technique to assist in the transfer of cultural traits to product requirements as objective design factors, as reported by Silva and Coelho (2011) and by Coelho, Silva and Simão (2011). While more complete reports on each of the product design studies are available, this chapter focuses on their common steps.

Table 1 Jordan's (2000) 17 product personality dimensions

Product personality dimensions (Jordan 2000).
kind - somewhat kind - neither kind or unkind - somewhat unkind - unkind
honest - somewhat honest - neither honest or dishonest - somewhat dishonest - dishonest
serious minded - somewhat serious minded - neither serious minded or light hearted - somewhat light hearted - light hearted
bright - somewhat bright - neither bright or dim - somewhat dim - dim
stable - somewhat stable - neither stable or unstable - somewhat unstable - unstable
narcissist - somewhat narcissist - neither narcissist or humble - somewhat humble - humble
flexible - somewhat flexible - neither flexible or inflexible- somewhat inflexible - inflexible
authoritarian - somewhat authoritarian - neither authoritarian or liberal - somewhat liberal - liberal
driven by values - somewhat driven by values - neutral - somewhat not driven by values - not driven by values
extrovert - somewhat extrovert - neither extrovert or introvert - somewhat introvert - introvert
naïve - somewhat naïve - neither naïve or cynical - somewhat cynical - cynical
excessive - somewhat excessive - neither excessive or moderate - somewhat moderate - moderate
conforming - somewhat conforming - neither conforming or rebellious - somewhat rebellious - rebellious
energetic - somewhat energetic - neither energetic or non energetic - somewhat non energetic - non energetic
violent - somewhat violent - neither violent or gentle - somewhat gentle - gentle
complex - somewhat complex - neither complex or simple - somewhat simple - simple
optimist - somewhat optimist - somewhat pessimist - pessimist

2   VALIDATION OF SEMIOTIC QUALITIES IN A PRODUCT DEVELOPED USING A BIONIC DESIGN APPROACH

Every product embeds a message, whether its designers consciously controlled for this product aspect or not (Figueiredo and Coelho, 2010). A bionic design project was carried out, following an approach from the design problem to the bionic solution described by Versos and Coelho (2012). The problem considered was the storage and the physical display to enable browsing of personal music collections, focusing on CDs and DVDs. The conduction of the design process led to seek inspiration from nature, having selected the spider web as a natural example that was the basis for the analogy of working principle established. The spider web was one of several natural systems identified as solutions that capture or immobilize certain objects or bodies, as well as natural systems used with the purpose of protecting living beings. Spider webs and cocoons were initially selected as potential matches to the problem requirements. The former were finally chosen given their added features of object grasping with increased lightness and extreme strength (Yahia, 2001). The semiotic requirements established for the project and their corresponding goals are listed in Table 2 (the project was done in an academic setting). Moreover, other requirements were considered seeking the goals of form optimization, organization effectiveness, paradigm innovation for improvement of functional performance and satisfaction of multiple requirements (Versos and Coelho, 2011-a, 2011-b, 2010; Coelho and Versos, 2010).

Table 2 Listing of semiotic requirements set for the bionic design case study and their corresponding goal that was sought.

Semiotic requirements	Goal sought
1. Nice and appealing shape, enabling the user to develop an aesthetic interest in product	Communication effectiveness
2. Sending a message of an avant-garde character, creative and youthful	

The perception by the user of pleasantness and appeal, enabling the development of an aesthetic interest in the product (first requirement in Table 2) was validated through a questionnaire where, among other things, each of the two bionic CD towers was visually compared with a conventional tower (Figure 1). The validation of this requirement is necessarily subjective, because the key issue that arises relates to the taste and sensitivity of each individual questioned. Respondents, gathered through the first author's network of personal contacts, answering by email, accounted to 85, aged between 18 and 60 (mean of 27.8 and standard deviation of 8.5), both male and female, and with diverse professional and knowledge specialties. 116 questionnaires were sent out to valid individual email addresses, with a response rate of 73.3%.

Each respondent indicated which of the towers was personally more aesthetically pleasing and appealing, from 3 paired comparisons presented. The paired comparisons approach applied to this case of three objects enables 8 possibilities of response, two of which are incongruent, since no ranking of preference can be established out of them. Three out of the 85 respondents reported incongruent paired comparisons. Thus, the analysis of results was carried out for 82 responses. The results were analyzed on the basis of the procedure for calculating the Kendall coefficient of concordance (Siegel and Castellan, 1988). The average ranking obtained was bionic tower 1 (first place), bionic tower 2 (second place) and conventional tower (third place). This result is considered significant to represent the overall opinion of respondents to a confidence level of 99%. These results support the validation of the first requirement depicted in Table 2. Both the first and second bionic towers received the preference of respondents over the conventional tower, which proves the validation of the gains in terms of pleasantness and aesthetic appeal, for both versions of the project.

Figure 1   Depiction of a conventional CD tower, and the two bionic CD towers developed: conventional tower(A), bionic tower 1 (B) and bionic tower 2 (C) (designed by the second author).

Considering the second requirement that contributes to the goal of effective communication, validation was sought by means of a technique of anthropomorphizing products through the attribution of personality dimensions. In a first phase, a translation of the requirement into a product personality profile (Jordan, 2002) was proposed. In the second phase of the process, a sample of specialized public (eight undergraduate Industrial Design students) assessed the personality profile of the three objects shown in Figure 1. In such, whether or not the message intended by the designer was transmitted to the public could be verified. The second requirement set in Table 2, was decomposed in a number of concepts to promote the matching process envisaged. This led to considering the attributes of modern, elegant, youthful, joyful, flexible and dynamic. Moreover the attributes consisting of lightweight and stable were also considered from the third and fourth requirements. The correspondence between product attributes intended by the designer to be perceived by the public and product personality dimensions (Jordan, 2002) are shown in Table 3. The outcome of analysis on the respondents' assessment of the personality profiles is also shown based on evaluation of Kendall's coefficient of concordance (Siegel and Castellan, 1988).

For every product personality pair, analysis was performed as exemplified for the pair energetic – non energetic energy, the average ranking of the panel of respondents (with a significance of 99%, given by the assessment of Kendall 's coefficient) resulted in the following rank order: 1st C, 2nd B, 3rd A. As a conclusion to this result, it is understandable that tower C (bionic tower 2) is considered more energetic than the tower B (bionic tower 1), and that tower C (conventional tower) is considered less energetic than tower B. This means that tower C is deemed the least energetic of the three towers and that C is the tower that emerges as the most dynamic and the less dynamic, thus validating this communication requirement.

Table 3 Analysis of the results of the survey on the personality profile of the CD rack and verification of messages perceived.

Designer intent	Personality profile	Average ranking 1st- 2nd – 3rd	Kendall coefficient of concordance	Conclusion
Modern	Bright – Dim	B – A – C	Not significant	Sample did not reveal agreement
Light-weight	Simple – Complex	A – B – C	99%	Tower A is considered most simple (lightweight)
Elegant	Gentle – Violent	B and C – A	Not significant	Sample did not reveal agreement
	Moderate - Excessive	A – B – C	Not significant	Sample did not reveal agreement
Youthful spirit	Liberal – Authoritarian	B and C - A	99%	Towers B and C are the most liberal (youthful)
	Rebel – Conformist	C – B – A	99%	Tower C is the most rebellious (youthful)
	Optimistic – Pessimistic	B – C – A	Not significant	Sample did not reveal agreement
Joyful	Light-hearted – Serious-minded	C – B – A	99%	Tower C is the most light-hearted (joyful)
	Kind – Unkind	B and C - A	95%	Towers B and C are the most kind (joyful)
Flexible	Flexible – Inflexible	C – B – A	Not significant	Sample did not reveal agreement
Dynamic	Energetic – Unenergetic	C – B – A	99%	Tower C is the most energetic (dynamic)
Stable	Stable - Unstable	B – A – C	Not significant	Sample did not reveal agreement

According to the findings obtained, the communication of a message of young spirit, dynamism and joyfulness were validated. Tower C is the one which, according to the survey, more effectively conveys the desired messages, is considered the most dynamic, the most rebellious, most joyful and, together with tower B, most kind and most liberal. Regarding the transmission of the message of lightness, the personality profile related (simple - complex) did not translate so well the associated requirement. This might have led respondents to identify tower A as the simplest, and therefore, according to the tenuous association, the lightest of the three. Interpretative meanings vary from person to person. The absence of actual trial of the towers on the part of respondents, who just exercised visual perception, may have also influenced and contributed to vagueness and lack of agreement among some of the responses. A significant limitation to this study is acknowledged, one that is derived from only showing respondents images of the design via email, rather than having them interact with the real tangible objects.

3   MATCHING SELECTED CULTURAL TRAITS WITH PRODUCT PERSONALITY DIMENSIONS

The study reported in this section, concerning cultural inquiry of the Portuguese and Lusophone countries, was based on literature review to unveil a set of opinions from respected scholars within the humanities disciplines (sociology, anthropology, philosophy) and the relational study of some areas of arts and fine arts. A subset of results was presented by Silva and Coelho (2011) and by Coelho, Silva and Simão (2011). The cultural traits obtained were corresponded to Jordan's (2000) product personality attributes. Each cultural trait was assigned to one or more of the product personality dimensions and a matrix was prepared that translated the cultural traits into personality dimensions. The personality dimensions attained resulted from of subjective transfer of the cultural traits identified.

Fig. 2   Products that were used as a basis for the product personality assignment survey performed

Some examples of objects comprised of four clothes pressing irons and eight coffee machines were chosen (Fig. 2), in order to make an analysis of these objects in respect to the product personality assignment technique by Patrick W. Jordan (2000). The assignment of personality attributes was carried out by a panel of eight third year undergraduate industrial design students (aged from 20 to 23 years old) that rated each object in terms of the personality dimensions in a 5 point Lickert scale ranging from the personality attribute to its opposite (e.g. kind – unkind) and three intermediate ratings (e.g. somewhat kind, neither kind or unkind, somewhat unkind), according to Table 1. The panel analyzed the objects grouped in three sets, one of clothes pressing irons and two of coffee machines. The Kendall coefficient of concordance was used to assess the consistency of ratings among the panel. The rankings obtained and their significance are similar in form to those presented for the bionic design study, with broad consensus among the panel and high incidence of significance. Linking product personalities to characteristics for new products

The 12 objects depicted in Fig. 2 were further characterized, in terms of their product attributes according to a series of dimensions. These included materials, color, shape, graphic markings, archetype, morphology, inferred ease of use, manufacturing process, technological sophistication, multiple functionality and size. As a result, two product attribute lists were attained, one concerning the transfer of Portuguese cultural traits to product properties and the other one concerning the transfer of Lusophone cultural traits (Table 4).

Table 4 Product attributes attained as a result of linking product personalities to characteristics for new products

Product technical dimension	Culturally induced Portuguese product profile	Culturally induced Lusophone product profile
Color	Cold	Cold
Shape	Straight, coherent, contrasting	Straight, coherent, contrasting
Graphical markings	Decorative, instructions	Decorative, instructions
Archetype	Minimalist	Minimalist
Multiple functionality	Single function	-
Size	Small	Small
Ease of use	-	Complex, yet intuitive

Various living room furniture concepts were generated based on two product specifications that took as starting points the results presented on Table 4 and that were enlarged considering anthropometric (Panero and Zelnik, 2002) and other requirements. These initial concept sketches were evaluated by the authors, with respect to criteria derived from the specification and were also subjected to the scrutiny of 21 second year undergraduate industrial design students (aged from 19 to 22 years old). These did not however show significant agreement in terms of their preference among the concepts generated. The authors' evaluation matrix (based on an expanded requirements list developed within the design process) led to the detailed development of the concepts depicted in Figures 3 and 4, respectively, a product line based on the Portuguese cultural traits, named "Vale", and one based on the Lusophone ones, named "Império".

Fig. 3 Render of "Vale" living room furniture line based on the Portuguese cultural traits and their corresponding product technical attributes (designed by the third author)

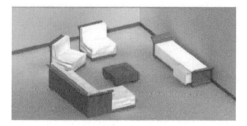

Fig. 4 Render of "Império" living room furniture line based on the Lusophone cultural traits and their corresponding product technical attributes (designed by the third author).

Cultural traits were the starting point to reach at the product profiles that were used as the basis for the design of two furniture lines. The scope of the work reported is not limited to furniture and is deemed applicable in a wider scope, considering its genesis and methodology, based on a literature review of cultural traits, taking into account the personalities of consumer products and consulting industrial design students.

Advancing the knowledge on the transfer of cultural traits to product design features may require further inquiry. The adequateness of the use of the product personality assignment technique in supporting this transfer could not be determined conclusively, as the results of the panel convened to assess the cultural identity of the product concepts produced was not conclusive, lacking agreement among the group.

CONCLUSION

The transfer from subjective qualities to objective properties may benefit from the product personality assignment technique. Validation of results by subjective inquiry may be illusive, as cultural predisposition and proficiency in the visual language of product design, may limit the applicability of the results substantially.

REFERENCES

Coelho, D. A. (2012) Inaugural Editorial: A new human factors and ergonomics journal for the international community is launched, *International Journal of Human Factors and Ergonomics* 1 (1), 1-2.

Coelho, D. A., Dahlman, S. (2002) Comfort and Pleasure, in *Pleasure with Products: Beyond Usability* (edited by William S. Green and Patrick W. Jordan), London: Taylor & Francis, 322-331.

Coelho, D. A., Silva, A. S. C., Simão, C. S. M. (2011) Culturally Inspired Design: Product Personalities to Capture Cultural Aspects, in *Industrial Design - New Frontiers* (edited by Denis A. Coelho), Intech, 55-80.

Coelho, D. A., Versos, C. A. M. (2011) A comparative analysis of six bionic design methods, *International Journal of Design Engineering* 4 (2), 114-131.

Coelho, D. A., Versos, C. A. M. (2010) An approach to validation of technological industrial design concepts with a bionic character, *Proceedings of the International Conference on Design and Product Development (ICDPD'10)* - Athens, Greece, 40-45.

Figueiredo, J. F. D., Coelho, D. A. (2010) Semiotic Analysis in Perspective: A Frame of Reference to Inform Industrial Design Practice, *Design Principles and Practices: an International Journal* 4 (1), 333-346.

Jordan, Patrick W. (2000). Designing *Pleasurable Products: an introduction to the New Human Factors*. London: Taylor & Francis, 216 p.

Jordan, P. W. (2002) The Personalities of Products. In *Pleasure with Products: Beyond Usability* (edited by William S. Green and Patrick W. Jordan), London: Taylor & Francis, pp. 19-48.

Panero, J., Zelnik, M. (2002) *Dimensionamento humano para espaços interiores*. Barcelona: Editorial Gustavo Gili.

Siegel, S., Castellan, N. J. (1988) *Nonparametric Statistics for the Behavioural Sciences*, New York: McGraw-Hill.

Silva, A. S. C., Coelho, D. A. (2011) Transfering Portuguese and Lusophone Cultural Traits to Product Design: A Process Informed with Product Personality Attributes, *Design Principles and Practices: an International Journal* 5 (1), 145-164.

Versos, C. A. M., Coelho, D. A. (2012) Bionic Design: Presentation of a Two Way Methodology, *Design Principles and Practices: an International Journal* (in press).

Versos, C. A. M., Coelho, D. A. (2011-a) Biologically Inspired Design: Methods and Validation, in *Industrial Design - New Frontiers* (edited by Denis A. Coelho), Intech, 101-120.

Versos, C. A. M., Coelho, D. A. (2011-b) An Approach to Validation of Industrial Design Concepts Inspired by Nature, *Design Principles and Practices: an International Journal* 5 (3), 535-552.

Versos, C. A. M., Coelho, D. A. (2010) Iterative Design of a Novel Bionic CD Storage Shelf Demonstrating an Approach to Validation of Bionic Industrial Design Engineering Concepts, *Proceedings of the International Conference on Design and Product Development (ICDPD'10)* – Athens, Greece, 46-51.

Yahia, H. (2001) *The Miracle in the Spider*, New Delhi: Goodword Books.

CHAPTER 41

Why the Optimal Fitting of Footwear is Difficult

Thilina W Weerasinghe, Ravindra S Goonetilleke and Gemma Sanz Signes

Department of Industrial Engineering and Logistics Management
Hong Kong University of Science and Technology
Clear Water Bay, Kowloon, Hong Kong
ravindra@ust.hk

ABSTRACT

Footwear has a long history and has evolved in to a necessary article whether it is fashion, safety or performance. However, fitting footwear is still an art and an ideal fit is predominantly by chance. Poor fitting affects the wearer☐s comfort and posture. The various attributes of a shoe contribute towards the comfort or discomfort of a shoe. For example, the heel inclination can change the wearer☐s spinal shape and body balance, affecting comfort. In this paper we describe the various parameters that affect a person☐s comfort and why an extra effort is necessary to achieve an optimum fit.

Keywords: high heeled shoe, posture, shoe design, COP, spinal shape, fit

1   INTRODUCTION

Drawings, more than 15,000 years old, found in Spanish caves have had people with wrapped furs and animal skins on their feet (Kurup et al., 2011). A recent discovery from a cave in Armenia has proved that customized shoes date back to around 5500 years (Pinhasi et al., 2010). Over the years, the utility of shoes have changed from protecting feet to more fashion. Long time ago, footwear were meant to protect feet from external hazards, but today footwear are claimed to encompass safety, style, and performance. As a result, fit has become one of the most important factors (Chiou et al., 1996, Luximon, and Goonetilleke, 2003, Kuklane, 2009, Lake, 2000, Dahmen, et al. 2001). Surveys have shown that the fit between the shoe and the foot is a prime consideration in the buying decision of a customer

(Chong & Chan 1992). Hence achieving the right fit is quite important to the seller as well as the buyer. Fit can affect thermal comfort too. A person will be comfort-table when the skin temperature of feet is around 33°C at 60% relative humidity (Oakley, 1984). When the skin temperature of the feet drops to 25°C, feet become cold and a further drop to 20°C would make the person quite uncomfortable (Enander et al.,1979). Figure 1 shows the temperature distribution of a foot when wearing differing types of footwear with a room temperature around 23 °C. The images show that the closed shoe provides more insulation to retain the body heat while the open shoe has a thermal image profile similar to the barefoot condition. Thus, closed shoes have thermal benefits when the environmental temperatures are low.

2   TYPES OF FIT

In traditional mechanical engineering, mating parts have three types of fit depending on the application. For example, a hub and shaft will have an inter-ference fit (Norton, 2000); moving parts a loose fit and some others an in-between fit called a transition fit. It seems that the fit between feet and footwear can take any of these types of fit depending on the location. In other words, the differing parts of a foot may and should require a different category of fit depending on subjective preferences and the activity for which the wearer uses it. For example, a ski-boot may require an interference fit all around the foot so that there is minimal movement between the boot and the foot. A casual shoe, on the other hand, may require a loose fit as the shoe may be worn over a long period of time and the looseness can then accommodate the deformation and expansion of the foot over time. The subjective element can be due to the varying properties of the tissue, internal workings of the body primarily related to circulation, and the threshold of discomfort or pain to indicate the potential tissue damage. Many of the past studies have attempted to find the allowances in the different regions and have fallen short of mapping the entire foot. Instead, most researchers have focused on the critical areas of the foot where people have reported greater discomfort and at places where the mismatch could hinder performance. Even the ANSI/ASTM F539-78 (1986) standard concentrates predominantly on two areas when fitting footwear: the toes and the metatarsal region (ball joint).

Figure 1. Thermal images after wearing two types of footwear for one hour

3   SHOE FITTING

In a macro sense, footwear comprise an upper and a bottom. The shoe should have the right fit in the upper as well as the bottom. The upper part in most men's shoes has a fit-adjustment mechanism through the lacing and the stretch characteristics of the material. However, the amount of adjustment and the location of the adjustments are limited. In some shoes the material is reinforced or lined in certain areas to prevent the material from stretching. The bottom of a shoe is called a midsole/outsole combination or just an outsole depending on the type of shoe. The surface on which the foot contacts the shoe is called the footbed. The fit in the different regions within each of these units are very important and the type can range from a loose fit to an interference fit. In other words, different parts of the foot require a different type of fit depending on the structure of the foot and the purpose of the shoe. Witana et al (2004) showed that the foot and shoe should have an interference fit of 8 mm and 15 mm in the forefoot and midfoot regions respectively, for men's dress shoes. In a more recent study, Au et al. (2011) found the interference fit of ladies dress shoes to be 6.4, 12.1 and 10.7 mm for foot breadth, ball girth and waist girth. Tremaine and Awad (1998) proposed an interference fit of 6.35 mm for foot breadth. The numerous bones of varying size and shape require different types of fit and make fitting footwear to feet more difficult.

The footbed contacts the sole of the foot and is the only mechanism to transfer forces. The force distribution will depend on the fit between the foot and footbed. The optimal distribution for performance is not really known even though there are two schools of thought to localize the force in the bony region and distribute the load in the soft tissue region (Goonetilleke, 1998). The load distribution will determine the centre of pressure (COP), which in turn will dictate the stability, posture and the loads on joints and muscles to hold the body in a balanced state. The cause-effect relationships clearly show how fit can affect many variables such as balance, stability, posture and thereby comfort. Should the footbed have the same fit along the surface or should it differ in the differing areas to account for differences in stiffness and resilience of foot tissue? Many past studies have reported COP effects of high-heel shoes without much consideration of the fit at the footbed (Shimizu and Andrew, 1999, Snow, and Williams, 1994, Gefen et al., 2002, McBride, 1991, Han et al., 1999). We have shown (Weerasinghe, and Goonetilleke, 2011) the comfort in inversely proportional to the COP and modeled the relationship as: Comfort = 87.2 − 0.798*COP. So the ability to control the COP through appropriate load distribution can have a positive effect.

It is clear that the right-fit between the footbed and the foot sole is one of the most important issues to achieve the optimal load distribution. The structural integrity and human performance hinge on the proper fit between sole and footbed. Again, the many different regions make the issue complex. The human body can be considered to be in its ideal position when the foot touches the flat floor. Figure 2 shows the various parameters that govern the design of a high-heeled shoe. Most of our shoes have a heel height. Unless the heel is of the platform type, the hindfoot and forefoot will be at different heights. Hence the heel is generally sloping down followed by a curved surface to make a smooth transition to the lower part of the foot. The slope and the curved surface of the shank have their lengths constrained and cannot be more than about 72% of foot length (Xiong et al., 2009).

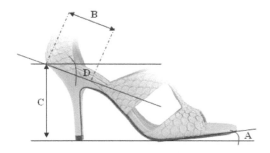

Figure 2. Design parameters of the footbed of a shoe. A= toe spring; B= seat length; C=heel height; D=wedge angle.

The structure of the foot can make these limitations possible. Around 25% of the bones in our body are in our feet. This gives our feet a high level of flexibility even though the flexibility or the range of motion is not the same in all parts of the foot. The forefoot is generally more flexible when compared with the hindfoot. Thus the curvature of the footbed should match the flexibility of the foot and this matching would determine the right fit between the sole and footbed.

A sub-optimal wedge angle and footbed curvature will tilt the body or make the foot slide forward causing discomfort because of looseness or tightness. Then, the foot is squeezed resulting in high-pressures that increase the tissue and joint deformations and hinder movement compromising the foot's performance. Such effects can result in temporary or permanent impairments some of which can be detrimental to the functioning of the feet. Common problems such as callouses and corns are due to undue pressure and relative movement between footbed and foot due to a poor fit of shape and a mismatch of material properties. Hallux valgus is a long-term effect of unwanted pressure in the MPJ area (SATRA, 1993). Figure 3 shows an example of the shift of COP with increasing wedge angle. It can also be seen that a right combination of wedge angle and shank shape can lower the pressures. The negative effects of the present day high-heeled shoes of an anterior shift of COP and an increase in plantar pressure have been shown by many (Gefen et al., 2001; Han, 1999; McBridge, 1991; Shimizu et al., 1999; Snow et al., 1994). The shift in COP results in a feeling of falling forward when wearing high-heeled shoes (Shimizu et al., 1999). Holtom (1995) has shown plantar foot pressure increments of 22%, 57%, and 76% with heel heights of 2 cm, 5 cm and 8.25 cm respectively. Contrary to all such studies, we have shown that COP can be shifted close to a barefoot stance when alterations to footbed geometry are made thereby affecting the fit between foot and footbed (Weerasinghe, and Goonetilleke, 2011).

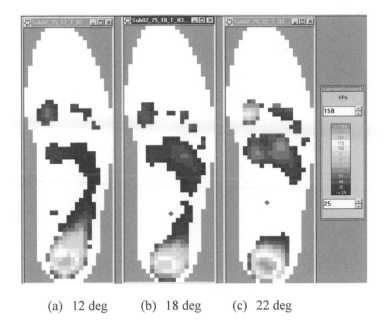

    (a) 12 deg   (b) 18 deg   (c) 22 deg

Figure 3. Effect of wedge angle and shank shape on pressure distribution and COP. (a) 12 deg (b) 18 deg (c) 22 deg

Body tilting and awkward postures as a result of a poorly fitting footbed can propagate up to the spine and beyond since people respond biomechanically like an inverted pendulum. A common belief is that high-heels make the body tilt so that the buttocks and breasts are emphasized (Danesi, 1999). However, some researchers have found opposite effects (Hansen and Childress, 2004). For example, Franklin et al. (1995) used a wooden board 5.1 cm high under the heels to study the standing posture and found that lumbar lordosis actually decreases as a result of a posterior tilt of the pelvis. Decreased lumbar lordosis is one of the common observations in the high heeled shoe wearers (Opila et al., 1987, Franklin et al., 1995, Lee et al., 2001). The inconsistencies among high-heeled posture related studies are possibly due to the use of heel blocks or shoes of a certain height with no control on fit, which is affected by the footbed parameters such as surface geometry (Franklin et al., 1995; Lee et al., 2001).

The footbed fit can affect the spinal shape as well. Figure 4 shows spinal shape data captured using a motion analysis system. The lower comfort ratings are those that are away from the neutral posture. In this case, the neutral posture is when the subject is standing on the ground barefooted. The importance of the footbed fit to minimize injury and increase comfort ought to be clear.

376

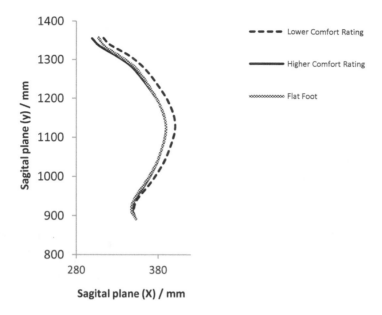

Figure 4. Spinal shape at different comfort rating AT 75mm heel height

4 CONCLUSION

Fit is no doubt an important element in mating parts. With rigid components the tolerances can be defined quite easily. With human tissue, the specification is more complex due to irregular shapes and differing tissue properties. A poor fit between feet and footwear can result in discomfort and injury in the long-term. Even though an optimal fit may be difficult to achieve, the added cost will be a fraction of the value associated with such a condition.

ACKNOWLEDGEMENTS

This study was funded by the General Research Fund of Research Grants Council of Hong Kong under grant HKUST 612711.

REFERENCE

ANSI/ASTM F539-78. 1986. *Standard Practice for Fitting Athletic Footwear*: 229 □235.
Au, E. Y. L., R. S. Goonetilleke, C. P. Witana, and S. Xiong. 2011. A methodology for determining the allowances for fitting footwear. *Int. J. Human Factors Modelling and Simulation* 2(4): 341-366.
Chiaou, S., A. Bhattacharya, P.A. Succop. 1996. Effects of worker☐s shoe wear on objective and subjective assessment of slipperiness. *Am Ind Hyg. Assoc. Journal* 57: 825-831.

Chong, W.K.F and P.P.C. Chan. 1992. Consumer Buying behavior in sports footwear industry. Hong Kong: Business Research Center, Hong Kong Baptist College.

Dahmen, R., R. Haspels, B. Koomen, A.F. Hoeksma. 2001. Therapeutic Footwear for the Neuropathic Foot. *An algorithm Diabetes Care* 24(4):705-709.

Danesi, M. 1999. Of cigarettes, high heels, and other interesting things. St. Martin's Press, NY.

Enander, A., A.S. Ljungberg, I. Holmér. 1979. Effects of work in cold stores on man, *Scand J Work Environ Health* 5: 195⎕204.

Franklin, M.E., T.C. Chenier, L. Brauninger, H. Cook, and S. Harris. 1995. Effect of positive heel inclination on posture. *Journal Orthopedic Sports Physical Therapy* 21(2): 94-99.

Gefen, A., M. Megido-Ravid, Y. Itzchak, M. Arcan. 2002. Analysis of muscular fatigue and foot stability during high-heeled gait. *Gait and Posture* 15:56⎕63.

Goonetilleke, R. S. 1998. Designing to Minimize Discomfort, *Ergonomics in Design* 12-19.

Han, T.R., N.J. Paik, M.S. Im. 1999. Quantification of the path of centre of pressure using an F-scan in shoe transducer. *Gait and Posture* 10: 248-254.

Holtom, P.D. 1995. Necrotizing soft tissue infections, *Western Journal of Medicine* 163(6): 568-569.

Kuklane, K. 2009. Protection of feet in cold exposure. *Industrial Health* 47: 242.253.

Kurup, H.V., C.I.M. Clark, R.K. Dega. 2011. Footwear and orthopedics. *Foot and ankle surgery* (in press).

Lake, M.J. 2000. Determining the protective function of sports footwear. *Ergonomics* 43(10):1610-1621.

Lee, C.M., E.H. Jeong, A. Freivalds. 2001. Biomechanical effects of wearing high-heeled shoes. *International Journal of Industrial Ergonomics* 28:321-326.

Luximon, A., R.S. Goonetilleke. 2003. Critical dimensions for footwear fitting. Proceedings of the IEA 2003 XVth Triennial Congress, Seoul, 2003.

McBridge, I.D., U.P. Wyss, T.D. Cooke, L. Murphy, J. Phillips, S.J. Olney. 1991. First metatarsophalangeal joint reaction forces during high-heel gait. *Foot ankle* 11:282⎕288.

Norton, R. L. 2000. Machine Design. Upper Saddle River, New Jersey: Prentice Hall.

Oakley, E.H.N. 1984. The design and function of military footwear: a review following experiences in the South Atlantic, *Ergonomics* 27: 631⎕637.

Opila, K. A, S.S. Wagner, S. Schiowitz, and J. Chen. 1988. Postural alignment in barefoot and high-heeled stance. *Spine*. 13(5): 542--547.

Pinhasi, R., B. Gasparian, G. Areshian, D. Zardaryan, A. Smith. 2010. First Direct Evidence of Chalcolithic Footwear from the Near Eastern Highlands. *PLoS ONE*, 5(6): e10984. doi:10.1371/journal.pone.0010984.

SATRA. 1993. How to fit footwear. Shoe and Allied Trades Research Association (SATRA) Footwear Technology Center, UK.

Shimizu, M., P.D. Andrew. 1999. Effect of Heel Height on the Foot in Unilateral Standing. *Journal of Physical Therapy Science* 11(2):95-100.

Snow, R.E., K.R. Williams. 1994. High heeled shoes: Their effect on center of mass position, posture, three dimensional kinematics, rearfoot motion, and ground reaction forces. *Archives of Physical Medicine and Rehabilitation* 75:568⎕576.

Tremaine, M.D. and E.M. Awad. 1998. The Foot and Ankle Sourcebook, Lowell House, Los Angeles.

Weerasinghe, T. W., and R. S. Goonetilleke. 2011. Getting to the bottom of footwear customization. *Journal of Systems Science and Systems Engineering* 20(3): 310-322.

Xiong, S., R.S. Goonetilleke, J. Zhao, W. Li, and C.P. Witana. 2009. Foot deformations under different load bearing conditions and their relationships to stature and body weight. *Anthropological Science* 117 (2):77-88.

CHAPTER 42

The Pertinence of CMF between Mobile Phone Design and Fashion Design

Chunlai Qiu, Hua Su

Tsinghua University
Beijing, China
clqiu20112011@gmail.com

ABSTRACT

This paper try to find if there is some pertinence in the color, finishing and the method of their combination between industry design and fashion design, and make the research on how to use them to analysis the trend of CMF in the industry design.

The research bases on the project of luxury mobile phone design for SWAROVOSKI GEMS. During this project there are some research on CMF (color, material, finishing) for luxury mobile phone design. The cell phone is defined as a part of fashion in this project, and there are many key words to describe it like fashion trend, personality, brand DNA and so on. So the CMF analysis for the phone design also needs to connect with fashion trend closely. At the early part of research, material element was leave out, the comparison only between color and finishing and their combination effect. A observation experiment was used to measure 40 participants responses on the 10 picture groups during the research, each group include one part of picture of fashion design and phone design, the purpose is try to find the similar element between those two designs. In the end, the result will be used in the 2012 CMF trend analysis for SWAROVOSKY mobile phone design. Ultimately two results are given, firstly, the pertinence element of CMF between mobile phone design and fashion design, secondly, the CMF trend of luxury phone in 2012 for SWAROVOSKI GEMS.

Keywords: Color, Finishing, Pertinence, Trend analysis, Collocation, Mobile Phone design, Fashion design

1   INTRODUCTION

Looking at the annual Paris, New York, Milan fashion show, both from the style and shape, colors or fabrics, they all let the other faddish industry merchants research and learn to catch up with trend, some fashion collar brands which active in the fashion trend has already become quarter trend symbols. As a mobile phone product which is fashion-oriented and fashion brand developed, besides satisfying the basic functions, highlighting the popular elements of texture, color and other aspects rather need to work hard in. The concept of mobile phones is no longer be defined with industrial products and contact tools, more important things are interpretation of fashion trends and user individual personality, color, material, finishing namely CMF as the primary means to convey the elements of fashion. When we defined mobile phone as a fashion accessory, these three elements are give the effect of concept of the trend in a way, and this trend will inevitably produce a certain relationship with the fashion trend of the main.

How to transform color and fabric trends of fashion design into color and texture trends of the fashion phone design, and appropriate combine and present the CMF trend in the mobile phone and fashion show? It is the main problems to be solved in this study. In mobile phone design, the main trends are more reflected in the texture compare with material which under the finishing, at the same time there are obvious differences in characteristics and lower reference value in material using, so the main consider will not include the trend of the material.

1.2   Literature review

Until this paper is submitted, several kinds of studies about color and finishing are found, when the research results apply in design experience to analysis the trend of color and finishing and the combinations, there always have some problems. Marcel et al. absolutely they found the change of color emotion when it include texture, but the experience which refined color and texture out of products and rebuilt them in computer, make it hard to put the result into design directly. Schloss et which compared the different of color combination of aesthetic, give the psychology guide of color research. Eui showed the CMF research in mobile phone company like Nokia and Motorola, and supplied the CMF marketing and trend research around 2007,but the detail of fashion analysis did not included.

1.3   Methodology

There are two main method, key method will be used for collect and arrange data, pertinence analysis will be used for the CMF analysis of data groups.

Key word method: The key words include time range and other one which could describe emotion, feeling, style characteristic, etc. They all be used to limit the range of data of fashion design and industry design and make sure that these designs and commodity products are appeared nearly at the same time horizon, so the trend of CMF can be seen clearly and accurate. The information is prepared for CMF analysis. Then the product design and cloth design which have the similar finishing effects and their combination and publication time will be put at the same analysis group.

CMF Pertinence experiment: A 7-point List scale was used for visual pertinence experiment, the questions include color purity, lightness and color assortment, and with a series of issues related to texture density , finishing monomer shape , so as to come the trend similar factors that of mobile phone design and fashion design expression in color and texture (Figure1).

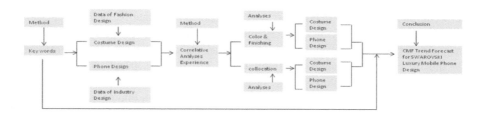

Figure 1   Illustration of methodology

2   SAMPLE SELECTION

2.1   The experimental picture selection

Firstly, the cell phone pictures will be choosed. We selected picture in the range of the past three years, and they all trend-oriented products which including the clothing brand's mobile phone products and professional mobile phone brand's products, their common feature is global brand influence, such as Prada, Versace, Dior, Armani which have mobile phone products, and Moto, Samsung, HTC, Sony Ericsson, Nokia mobile phone brands. Excluding pictures include new materials-oriented phone, for example, with a carbon fiber back shell Moto the RAZR XT910, technology-oriented mobile phone, such as the Sony Ericsson LT26i. Secondly, fashion pictures selecting, there are many fashion brands during each year, but the trends are quite different, so the selection of fashion brands tend to the top ten or top fifteen fashion brands in Paris, London, Milan Fashion Week nearly three years, on behalf of mainstream fashion trends, the study does not include cutting-edge fashion brand and explore the concept of design. The major brands include Hermes, Louis Vuitton, Christian by Dio, Chloé, Armani and so on. (Brand and time detail information is in Table 1.)

2.2 The time difference problem

Releasing time to match the range limit is necessary in order to guarantee the trends correspondence between fashion and mobile phones. Phone release has low amount as compared to the fashion publishing, so the mobile phone launch data will be the time keywords to mainly localize fashion pictures selected. There are nearly one or two season of the time difference between the cell phone and fashion design in new fashion trends releasing, so the time span needs to go ahead one or two season for selecting fashion pictures, for example, a mobile phone CMF trend launched in March 2011 match the fashion trend form the spring and summer to fall and winter of 2011, but the fall and winter of 2010 will not be used.

2.3 Image processing

The pictures were required to be abstracted processing before the experiment, such as remove the background, characters face, foil material, brand information, just retain the main materials and colors. Each image contains a fashion fabric and phone material. Then the image processed into the same proportion. Due to the proportion difference between fashion and cell phone pictures, and cell phone is smaller than clothes, the material texture was used as the standards of phone finishing (Figure 2).

Figure2 Illustration of image processing

2.4 Keywords defined

The keywords include gender, age range, theme or emotion keywords. Theme and emotion describing choosing based on the trend of fashion brands seasonal theme definition and part of style description. The keywords definition pictures from many corresponding filter into a clear one to one relationship. Ultimately we determined ten picture groups, including four groups of male model and six groups of female model, the age range is 25-35 years old. The theme and emotion keywords are City Rhythm, Gentlest and Arbitrary, Dior Lady, Sexy tango, Young Nobleman, Low-pitched romantic, Forthright Geometry, Generations of Individualism, Motorcycle Girl, Hot Rod Sweetness, Capable of Tender and Ably. (Detail information is in Table 1.)

Table 1 Brand & Time Matching Table

Picture Group	Keyword	Fashion Design	Phone Design
Group 1	City rhythm	Hermes 2011Autumn/Winter -Man	Motorola-XT882 03.2011
Group 2	Gentlest arbitrary	Louis Vuitton 2011Autumn/Winter- Man	HTC-Rhyme 10.2011
Group 3	Dior lady	Christian Dior Lady Dior Bag 2011- Woman	Dior Phone 08.2011
Group 4	Sexy tango	Dolce & Gabbana 2010Spring/Summe r Ready-to-wear- Women	Motorola-Aura 12.2009
Group 5	Young aristocracy	Dolce & Gabbana 2012Autumn/Winter -Men	HTC-Evo-3D 10.2011
Group 6	Low-pitched romantic	Valentino 2011Autumn/Winter -Women	Motorola-MOTO- MT870 03.2011
Group 7	Forthright geometry	Versace 2011Spring/Summe r-Women	Versace-Unique Phone 05.2010
Group 8	Individualism generation	Emporio Armani 2010 Spring-Man	Giorgio Armani Phone 10.2009
Group 9	Hot Rod Sweetness	Prada 2012Spring/Summe r-Women	Prada Phone 12.2011
Group 10	Light tender ably	Chloé 2012Spring/Summe r-Women	Nokia-Oro 05.2011

2.5 Questionnaire

Questionnaire is constituted by the humanistic message and 7-point List scale. The meaning of the score is sated as 1 is divided into similar, 4 representatives of uncertainty, 7 represents similar, and each group of pictures are required to answer 10 questions, participants need to answer 100 questions totally in the scale part.

The problem setting is based on the basic properties of color and texture[5], due to the hue difference is not obvious in every photo group, the color hue similarity question is ruled out, because not all pictures have two or more materials, the similarity of material combined is not included. Finally there are 10 questions, they are Color Purity, Color Brightness, Color Combinations, Degree of Texture Concentration, Material Monomer Shape, Material Monomer Combination, Material Overall Gloss, Degree of the Surface Convex, Overall Feeling. (Please see the sample questions in Table 2.)

Table 2 Sample questions

1 Color purity is not similar   1 2 3 4 5 6 7   Color purity is similar
2 Color brightness is not similar   1 2 3 4 5 6 7   Color brightness is similar
3 Color combinations is not similar 1 2 3 4 5 6 7 Color combinations is similar

3   EXPERIMENT

3.1 Analytical methods

Participant sates as the correct posture that face the display under the guidance of staff, there is a plane distance about 60cm between the eyes and display, the line of sight is perpendicular to the plane. Participant needs to look at the reference legend before formal picture group. Participant faces one group of pictures every time, the pictures have a fixed order, and displayed by Microsoft Power Point. Participant is allowed to control answer time themselves, the answer should fill out in questionnaires paper. The pictures themselves have color error which came from photography site, so the experiment on the display and the light source is not being strictly required.

The picture group is independent variable, all 40 participants scores of test question in the corresponding group of pictures were listed, then the columns of score lists from Question 1 to Question 9 were matched respectively, for example, in Picture Group 1 the score list of Question 1 match Question 10, the match method is Paired-Sample T-test in SPSS, in order to calculate the Standard Deviation(correlation coefficient) and Standard Error Mean(significance) of Paired Samples Statistics, the Standard Error Mean value are used as reference for Standard Deviation result, every picture group has 9 calculations, 90 times calculation totally.

One picture group is a unite, the 9 internal calculated correlation coefficient are divided into two groups: the color part of the question group and the finishing part of the question group, in each group, the questions are arrayed from highest to lowest accordance with the Standard Deviation, after that the first three highest scores are received, the same method were used in all 10 Picture Groups. In the case of keeping the Standard Deviation ranking inside the group, the first three highest Standard Deviation were lateral contrasted separately in the color question group and finishing question group, for instance, in the finishing question group, "Degree of texture density" got the maximum frequency and score in the first correlation question, it is the first correlation element without reference to the frequency or score it got in the second or third order of cast of correlation question, it is the most interrelated factor to Overall Feeling in finishing question group.

3.2 Results

From the texture point of view, the greatest impact on the overall similarity is the degree of texture density, followed by the material overall gloss, and the third one is the degree of the surface convex, when the material monomer shape level is apparent, the material monomer shape and monomer combinations will play a larger impact on the overall feeling, for example, the question score in Picture Group, the above two get the correlation coefficient (Standard Deviation) score are 5, 0.580 and 0.574 respectively, the significant (Standard Error Mean)score are all 0.000, they are the second and third score ranked respectively in the finishing group, while in this comparison of Picture Group 1, the above two issues do not appear high correlation, the material monomer shape correlation score is 0.390, significant score is 0.013, material monomer combination score is 0.214, significant score is 0.185. In the picture group that the texture undulations, the degree of surface convex is also one of the main elements for an image similarity, but from the perspective of the pertinence score only appear as the third main impact of the correlation, there is no situation that ranking at the first or second.

From the perspective of color, in purity, brightness and color combination, color combination is the first impact factor to the overall feeling, lightness secondly, the weakest impact element is color purity, Because of the purity have little difference in all selection picture, we could not exclude the results will change when the difference is larger.

3.3 Limitations in the questionnaire and the picture

Some participants who do not have art-related work have many uncertainties in question, they need the guidance from the experiment stuff and legend, the uncertainty focus on compare Degree of Texture Concentration with Material Monomer Shape. Another problem comes from the Question 1, more than 80% participants did not get into the answer state and took a long time on it, maybe next experiment we need to increase a plus question for warm up.

4   TREND ANALYSIS

4.1   Project Background

The research bases on the project of luxury mobile phone design for SWAROVOSKI GEMS, and the phone prize range is 600-800$. The cell phone is defined as a part of fashion in this project, so the CMF analysis for the phone design also needs to connect with fashion trend closely. In this paper, only a part of CMF trend (2/5) of luxury phone in 2012 for SWAROVOSKI GEMS will be opened.

Target group: The male who is in the age range of 25-35, he is the highlight in social activity, free and unruly, always full of vitality and has a lot of friends, chasing for high end consumption. He is pursuing the social belongingness, realizing his unique value. The female who is at the range of 20-30, She clings the fairytale is true and believe in love. She likes sweet color and characteristic adornment. Her personality is very independent. She is the representative personage of new women of time.

4.2   Trend Analysis

The 2012 Spring/Summer trends come from SWAROVOSKI ELEMENT official web, they are all natural theme, two themes were selected for the paper as the trend analysis prototype and reference tone of color scheme, they are Spring of Earth, Soft Romantic. During the analysis of fashion trend in 2012 Spring/Summer, the following clothing material groups were selected for corresponding to the SWAROVOSKI trend, the elements are extracted from them. Ultimately two group of 2012 SWAROVOSKI phone CMF trends are completed with the guidance of experiment results and the office web trend. Every group includes both male and female pattern.

In the male pattern of Group I, the reference picture of fashion design comes from 2012 Spring/Summer Men's trend of Alexander McQueen. The phone of the texture is made up of fine pattern of protruding rounded corners rectangular, another part includes anti-light matte plane and high reflective thread, the reference materials are plastic and brushed metal, the reference color scheme is light gray that reddish and medium brightness, assembles with high light gray; The reference picture of female trends in clothing comes from 2012 Spring/Summer Women show of Channel, the mobile part of the reference material and texture is made up of plastic surface which is engraved with linear knit and paired with ceramic luster metal, the reference colors are rose gold and ceramic white.

In the male pattern of Group II, the reference picture of fashion design comes from 2012 Spring/Summer Men's trend of Adam Kimmel. The mobile part of the reference material and texture is made up of sand crystal plastic face with a brushed metal, the reference colors are low brightness of dark gray and green cast yellow with low lightness; The reference picture of female trends in clothing comes from 2012 Spring/Summer Women show of Armani Prive. The mobile part of the reference material and texture is made up of soft and transparent subsurface with depressions dot textured, paired with silk luster mental, the reference colors are the warm green which is yellow cast and medium brightness, and black (Figure 3).

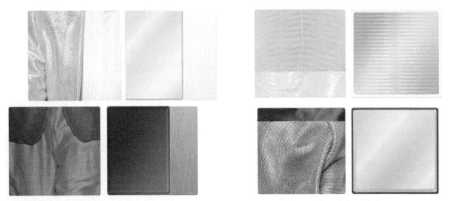

Figure3- Illustration of color and finishing trend picture of Group I and Group II

5   CONCLUSION

This paper provides a method of analysis cell phone CMF trends, and quests a method that put the fashion trends into phone texture. In the CMF design process, the results in this study only apply to the in research fashion trend component (ChongYang Cui), the result also still need to amend by user research and competitive analysis.

ACKNOWLEDGEMENTS

The authors would like to acknowledge to my mentor Professor Hua Su for her careful guidance and strict control to the paper. Thank to Associate Professor Guosheng Wang and Hengfeng Zuo who have given so much guidance and help at the beginning of paper. Thanks to Dr. Jie Yuan for his help in the data analysis phase guidance. Thanks to all participants who have given sincerely support in the experiment.

REFERENCES

Bo Young Kim, Joon Hye Baek, 24 AUG 2011,Leading the Market with Design Thinking and Sensibility.
ChongYang Cui, 2007, Research on using Process of CMF in product design.
Eui-Chul Jung, YongSoon Park, KyuHee Kim. 2011 Framework to Propose a CMF Design Strategy of New Products - Focused on Refrigerators. Journal of Korea Society of Design Forum, http://business.swarovski-elements.com/trend12/ index.html

Jaglarz, A. 2011,Perception and Illusion in Interior Design.

Judy Zzccagnini Flynn & Irene M.Forster, Research Methods for the Fashion Industry, ISBN: 978-1-56367-633-8b, 2009.

Marcel P. Lucassen, Theo Gevers, Arjan Gijsenij, 2010,Adding texture to color: quantitative analysis of color emotions.

Schloss, KB (Schloss, Karen B.)1; Palmer, SE (Palmer, Stephen E.) FEB 2011 ,Aesthetic response to color combinations: preference, harmony, and similarity.

Yong-Keun Lee and John M.powers, 2006,Comparison of the Metrics between the CIELAB and the DIN99 Uniform Color Spaces Using Dental Resin Composite Material Values, Color Research and Application [J].

CHAPTER 43

"Will They Buy My Product?"—Effect of UID and Brand

Manoj K. Agarwal, Alan Hedge and Sajna Ibrahim

Binghamton University, State University of New York, Binghamton, NY, USA
agarwal@binghamton.edu
Cornell University, Ithaca, NY, USA
ah29@cornell.edu
Binghamton University, State University of New York, Binghamton, NY, USA
sibrahi1@binghamton.edu

ABSTRACT

Consumer electronic products have a wide array of user interface designs (UIDs) varying from simple/complex layouts of buttons, and touchpads to touchscreens. This study investigates how product brand, UID, form (appearance) and function (technological capabilities) of a product influence consumer purchase intentions, willingness to pay (WTP) and product choices. Results from a pretest and two studies of reactions to various prototype MP3 players show that UID significantly drives consumer purchase intention, willingness to pay and choice. Also, this effect impacts purchase intention more so for stronger brands and products with more visually appealing form. Strong brands do worse than weak brands when perceived usability of UID decreases. Implications are discussed.

Keywords: user interface design, product design, appearance, technological capabilities, marketing, brand.

1   INTRODUCTION

Consumer electronic products have been designed with novel interfaces that range from innovative layouts of buttons and pointers to new input technologies such as scrollpads (Apple iPOD), touchscreens (Samsung Captivate) and gestural feedback systems (Nintendo Wii). The "user interface" (UI) characterizes how consumers "talk" to the product (Oppenheimer, 2005). The UI also communicates product value because it is the main "sensory" point of contact with the product (Creusen and Schoormans, 2005). For example, Apple's i-Phone relatively high (19%) score on positive interactivity serves as the one of its main "unique selling propositions" (Koca, Karapanos, and Brombacher, 2009). For consumers, the user interface design (UID) is a key element which drives perceptions of usability (ease of use) of the product (Norman, 1983). For marketers, UID as a component of product design presents a strategic opportunity for differentiation of products to gain advantage (Reibstein, Day, and Wind, 2009).

Literature in marketing on consumer response to product design has mainly conceptualized product design as form (package shape (Raghubir and Greenleaf, 2006), aesthetic appearance (Page and Herr, 2002)) or function (technical specifications or product performance such as long battery life, or picture quality (Hsee et al., 2009)). But the concept of the UID is missing in the conceptualization of product design in research in marketing. To address this gap, we adopt the concept of "user interface" from human computer interaction (HCI) research as the "point of interaction between humans and machines" (Booth, 1989) and conceptualize that product design should include the design elements of form, function *and* user interface. We define UID as "the design element which facilitates interactions between users and products to enable the use of functions." The UID can be comprised of both physical and logical elements (Han et al., 2000).

With the wide range of UIDs available for the consumer, UID is emerging as a critical influencer on consumer evaluation, preferences and purchase decisions. Our research examines the role of a product's UID and its effect on consumer purchase intentions, willingness to pay and choice. Since products are evaluated in a holistic manner (Veryzer and Hutchinson, 1998) with interactions between various product features (Holbrook and Moore, 1981), we study the interactions of UID with the design elements of form and function. In response to a call to understand the synergy between marketing signals and design (Bloch, 2011), we also include the brand in our investigation. Brands are important marketing signals that influence customer's relative preference for product design characteristics (Chitturi, Chitturi, and Raghavarao, 2010); we explore consumer response to UID for strong and weak brands.

The UID of a product is instrumental in creating perceptions of both how the product can be used and how easy it will be to use (Creusen and Schoormans, 2005). We propose that the perceived usability of UID of a product influences consumer purchase intentions, willingness to pay (WTP) and choice over and above the influence of other product elements as form, function and brand. We also propose that for stronger brands, purchase intentions and WTP increase with higher perceived usability of the UID compared to weaker brands. Alternatively, purchase intentions and WTP are expected to reduce more for stronger brands than weaker ones when the perceived UID is less. Furthermore, we propose that UID influences consumer preferences for products with varying form (appearance) and function (capabilities) differentially. Keeping in mind the fundamental importance of physical interactions when using consumer electronic products (MacDonald, 2001), we vary the UID by manipulating the design of the physical user interface (specifically controls).

Research studies were undertaken to test hypotheses that higher the perceived usability of the UID, higher the consumer purchase intentions and willingness to pay. Also, this effect is higher for higher form (high visual appeal, high ease of use)/ higher function of the product and stronger brand. Also, higher the perceived usability of the UID , higher the consumer's choice of the product.

2   RESEARCH STUDIES

2.1   Pretest

Prior to our main study, a pre-test was done to help choose the appropriate product design elements of form, function and UID. CAD (computer aided design) product models were constructed to better control the multiple design elements. At the same time, we used design elements as close as possible to the real products to avoid being perceived as too innovative. To understand elements of form, we varied the shape and thickness of MP3 players. To comprehend elements of UID, we varied the number and type of buttons on a digital camera.

Product Model construction for form & UID stimuli: First, pictures of four MP3 Players were created using software (Siemens PLM Solid Edge), varying the "form" design element. Overall aesthetic attractiveness of a product is influenced by roundedness, size and color (Creusen and Schoormans, 2005). Consequently, two basic elements of form were varied: shape and thickness - (i) Rectangle with sharp edges (2(length)X3(width) inches) and (ii) Rectangle with rounded edges ((3(length)X2(width) inches) with thicknesses of 0.5 and 1 inch each. Other design elements as display, control buttons and stereo jack were kept constant across the models. The products were presented as gray-scale pictures and all other details of the product were kept constant. The basic Gestalt principles of unity, symmetry, proportion and balance were followed for creating the pictures (Veryzer and Hutchinson, 1998). Second, for varying the UID for the product, pictures of four digital camera models (selected due to the higher range of controls that could be shown compared to MP3 players) were created with varying UIDs, viz. 2 pushbuttons, 14 pushbuttons, 4 arrow buttons + 2 pushbuttons and 4 arrow buttons + 8 pushbuttons. The number and placement of buttons results in perceptions of ergonomic value or usability of the product (Norman, 1988). Pushbuttons were square in shape while arrow buttons were triangular. The shape and all other design characteristics of the camera were kept constant across different pictures.

Pretest Results: Twenty undergraduate students from a North American University participated for course credit. The pretest was administered as an online survey. The students were asked to rate the four MP3 player models on visual attractiveness, look, style and shape on a 7 point scale ranging from 1 (Very bad) to 7 (Very good). The two models, rectangle with round edges with 0.5 inch thickness (M=5.63) and rectangle with sharp edges with 1 inch thickness (M=3.13) were selected as the best "high" and "low" form design element stimuli. Paired t-test between the means of these 2 models were significant (t=6.953, p<0.0001). We also asked about the importance of functional features considered for purchasing MP3 players. The features with the highest ratings were Storage space (M=6.0) and WiFi Capability (M=5.71).

Next, the participants were asked to rate the four digital camera models (with varying UIDs) on "ease of use" on a 7 point scale ranging from 1 (Very difficult) to 7 (Very easy). The two UIDs, 4 arrow buttons+ 2 push buttons (M=5.25) and 4 arrowbuttons+8 pushbuttons (M=3.8) were selected as the "high" and "low" levels of usability of UID stimuli. Paired t-test between the means of these 2 models were significant (t=3.621, p<0.002). After the pretest, participant's comments on the UIDs shown in the models were collected. Some of their comments included "It may be easier with arrow buttons... ideally, the combination of arrow and pushbuttons would be best", "Lot of square pushbuttons that do not have obvious functions make it difficult to use", "arrow buttons are more intuitive and interactive" and "gives a sense of direction". These comments corresponded with the buttons shown on the "high" and "low" UID products. Note that we used cameras to understand the relationships between physical buttons and their usability, and we assumed that similar associations would hold for MP3 players.

Brand strength as perceived by consumers in relation to specific product characteristics was also elicited. We presented them with 10 MP3 player brands currently available in the marketplace (Microsoft, Coby, Cowon America, Creative Labs, Apple, Sony, Samsung, Sandisk, Archos & Philips). They were asked to rate the brands on their opinion on how "Visually Attractive" they find their MP3 Players and on how much their MP3 Players have "Innovative Features/Functions". From these results, for the main study, Sony (M=5.5) and Coby (M=2.8) were chosen as exemplars of "high" and "low" levels of brand strength. Apple (M=6.9) was excluded due to its high positive bias. This pretest also showed that the importance of UID in consumer purchasing decisions for MP3 player was quite high (M=6.1), further justifying the choice of MP3 player in the study.

2.2 STUDY 1

2.2.1 Method

Participants and Design: One hundred and forty undergraduate students from a North American University participated in the experiment for course credit. The study used a 2 (form- high versus low on visual appearance) X 2 (brand - high strength versus low strength) X 2 (UID- high versus low on perceived usability) X 2

(Function – high versus low functional features) mixed design. Form and brand were manipulated as between subject variables. UID and function were manipulated as within subject variables. Based on pretest data, the form, function, UID and brand stimuli were combined to generate pictures of 16 MP3 player product models. For high and low function, three features as storage (8GB or 64GB), WiFi (Yes or No) and Song Management (Advanced or Standard) were provided. (See Figure 1 for product design stimuli). Participants rated four randomly selected products. The presentation order of the products was also completely randomized.

Procedure: The experiment was administered as an online survey. Participants were told that the questionnaire was part of an ongoing market research study on high technology consumer products. They were informed that we would like to know their opinion on 4 new MP3 players that are going to be introduced by a consumer electronics manufacturer. Each participant was presented with four product models. The participants were asked to provide their purchase intentions ("How likely are you to buy this product?") on a 7 point scale (1- very unlikely to 7- very likely). For eliciting the WTP, the participants were asked to provide the money (in dollars) they were willing to pay for the product. A range of $40-$200 was given based on current market prices to choose from. For each MP3 model, the participant were also asked to rate the products on visual appearance, product features, ease of use and overall product quality on a 7 point scale ranging from 1 (Very bad) to 7 (Very good). A 'Not sure' option was provided.

Consumer expertise with using products may influence the usability perceptions of the product (Langdon, Lewis, and Clarkson, 2007). To control for this alternate explanation, participant expertise ratings (MP3Expertise) on their using an MP3 player (5 point scale-novice to skilled) were collected. Previous experience with using products of the same brand could also influence overall usage perceptions about the product. To account for this, the participants were asked to provide information on whether they were using or have used MP3 players (Yes/No response) from the respective brands (MP3EarlyUse). These measures were collected after the main product evaluations and preference elicitation.

2.2.2 Results

Manipulation check: The participant ratings on visual appearance, product features and ease of use were used as manipulation checks for the independent variables form, function and UID. Mixed analyses of variance revealed significant main effects for all manipulations. Consistent with pretest findings, products high on form (rectangle with rounded edges with 0.5inches thickness) were rated higher on visual appearance (M=4.63) than products low on form (rectangle with sharp edges with 1 inch thickness) (M =3.54) ($F_{(1,135)}$ = 24.36, p < 0.0001). Across all factors, the products that had high function ratings were rated higher in "Product features" (M=5.97) than those with low function (M =3.45) ($F_{(1,134)}$ = 296.03, p < 0.0001). Also, the products that have high UID (i.e. high perceived usability of UID) were rated higher in "Ease of Use" (M= 5.05) than the products with low UID (M = 4.01), ($F_{(1,128)}$ = 53.48, p <0.0001). As a proxy for checking the manipulation for brand strength, the participant rating for overall product quality rating was analyzed. Across all factors, the products that have high brand strength were rated higher in "Overall Quality" (M=4.94) than the products having low brand strength (M = 4.44; $F_{(1,123)}$ = 6.15, p < 0.01).

Data Analysis: A mixed ANCOVA model was used to test consumer purchase intentions and WTP separately for the product, with form (high vs. low on visual appearance) and brand (high vs. low strength) as between-subject factors and UID (high vs. low perceived usability) and function (high vs. low features) as within-subjects factors. 'MP3Expertise' and 'MP3Earlyuse' measures were added as covariates in the model. The 'not sure' responses were coded as missing values.

Results indicated that participant purchase intention ($F_{(1,134)}$=25.17, p<0.0001) differed with respect to UID. Main effects of the other three design elements of form ($F_{(1,134)}$=6.87, p<0.01), function ($F_{(1,134)}$=229.88, p<0.0001) and brand ($F_{(1,134)}$=6.83,p<0.01) were also significant. Participant WTP ($F_{(1,134)}$=15.21, p<0.00) differed with respect to UID. The mean WTP varied from \$73.95 to \$81.05 for low to high perceived usability of UID. Main effects of the two design elements of form ($F_{(1,134)}$=5.857, p<0.02) and function ($F_{(1,134)}$=301.71, p<0.00) were significant but the main effect of brand was not significant.

For consumer purchase intention, UID interacted with brand strength ($F_{(1,134)}$=4.65,p<0.03). When perceived usability of UID was high, purchase intention for stronger brand was high (M=3.91) compared to weaker brand (M=3.13). At low UID, stronger brand had higher purchase intention (M=3.19) than weaker brand (M=2.831). But in comparison to the high UID condition, the purchase intention decreased more for stronger brand (Mean difference = 0.72) than for weaker brand (Mean difference=0.29). Further post-hoc tests showed that the influence of UID on purchase intention was significant for both strong ($F_{(1,67)}$=24.88,p<0.0001) and weak brands ($F_{(1,67)}$=4.41,p<0.04), but an examination of the effect sizes show that the effect of UID was stronger for higher brands (η^2=0.28) compared to weaker brands (η^2=0.06). For WTP, the interaction of UID with brand strength was only marginally significant ($F_{(1,134)}$=3.44,p<0.06).

For consumer purchase intention, there was also a significant interaction of UID with form ($F_{(1,134)}$=11.54,p<0.0001). Post-hoc tests showed an effect of the UID for the high-rated form ($F_{(1,67)}$=48.89,p<0.0001) but no significant effect for the low-rated form. When perceived usability of the UID was high, purchase intention for product with the high-rated form was high (M=3.94) compared to the low-rated form (M=3.10). Alternately, for a product with low UID, purchase intention decreased more for the high-rated form (Mean difference = 0.84) than for the low-rated form (Mean difference=0.17). For WTP also, the interaction of UID with form was significant ($F_{(1,134)}$=10.96,p<0.00). For consumer purchase intention and for WTP, the interactions of UID and function were not significant. Overall, one of the covariates 'MP3Expertise' remained significant ($F_{(1,134)}$=4.25,p<0.04) whereas 'MP3Early' was not significant.

2.2.3   Discussion of Study 1

Results of this study showed that UID is a critical driver of consumer purchase intentions and willingness to pay, over and above the influence of other product design elements of form and function. An examinition of the effect sizes shows function to be most important, followed by UID, form and brand. Importantly, when consumers evaluate the UID in presence of brand, the strength of the brand influences the consumer response to perceived usability of UID. For stronger brands, purchase intentions and willingness to pay are higher for products with high perceived usability of UID than for weaker brands. Lower perceived usability of UID reduces consumer purchase intentions and WTP significantly for both brands. It is seen that purchase intentions reduces more for the stronger brand than the weaker brand. Also, results reveal that UID influences consumer preferences more for products with higher form than lower form. For higher form products, purchase intentions and WTP are higher with high perceived usability of the UID than lower form products. At low UID, purchase intentions and WTP for products with higher form is more negatively influenced than products with lower form. These results hold even after controlling for user expertise.

2.3   STUDY 2

2.3.1   Methods

Participants and Design: One hundred and forty undergraduate students from a North American University participated in the experiment for course credit.

Materials and Procedure: A discrete choice experiment (DCE) survey was designed. DCE is an established technique to model choices and estimate utilities for each attribute/level comprising the alternatives (Hensher, Rose, and Greene, 2005). Respondents are typically shown choice sets depicting various choice alternatives described on a number of attributes. Respondents choose one of the shown choices. We used MP3 players and the attributes and their levels were selected based on pretest results discussed earlier. The attributes and levels are displayed in Table 1 below.

Table 1 Attributes and Levels for DCE

Attributes	Levels	1(LOW)	2(HIGH)
UID	2	4 arrowbuttons+8 pushbuttons	4 arrowbuttons+ 2 push buttons
Form	2	Rectangle with sharp edges with 1 inch thickness	Rectangle with round edges with 0.5 inch thickness
Function	2	Storage: 8GB WiFi : No Song Management : Standard	Storage : 64GB WiFi : Yes Song Management : Advanced
Brand	2	COBY	SONY

The form, UID and brand attributes were shown visually on the choice cards. The functions were listed in text along with the visual picture of the MP3 players. A factorial design using SAS (Kuhfeld, 2010) was created using 16 choice sets, with two choices on each set. All main effects and two factor interactions could be estimated with the design. The 16 sets were blocked into four subsets, with each participant randomly getting one of the blocks. A sample choice card is shown in Figure 1.

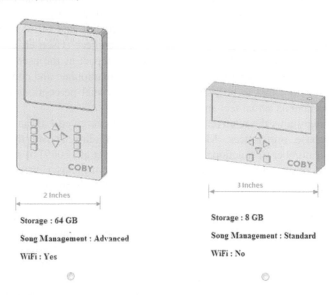

Please make your choice from the options given below.The prices and other product features are similar.
(Please wait till the picture loads)

Storage : 64 GB

Song Management : Advanced

WiFi : Yes

Storage : 8 GB

Song Management : Standard

WiFi : No

Figure 1 Sample Choice Card with product stimuli

Data Analysis: We employed LIMDEP to fit a multinomial logit model to the data (Greene, 2007) . The model used incorporates the main effects of UID, Form, Function and Brand, and all 2-way interactions.

2.3.2 Results

Results show that the model had a loglikelihood value = -221.14 (p < 0.0001) and and pseudo R^2=0.408. The values of the coefficients were found to be Form = 0.63 (p<0.0001), Function=1.43 (p<0.0001), Brand=0.30 (p<0.0001), UID=0.40 (p<0.0001) and there was a significant interaction between UID and Brand = -0.26 (p<0.04). The remaining variables were non-significant. The positive coefficient of the UID indicates that higher the perceived usability of the UID, the higher the utility and hence the choice probability. Results show brand, function and form are also good predictors of consumer choice. It is also seen that choice probability for high brand and low brand reduces for low perceived usability of UID. These results are in line with those reported from Study 1.

3   DISCUSSION

Going beyond the form and function aspects of product design (Luchs and Swan, 2011), our research shows the critical importance of product UID in shaping consumer preferences. An examination of the estimates of the effect sizes in study 1 show that after function, the perceived usability of the UID is the main driver of consumer purchase intentions and WTP, followed by brand and form. For choice data, function is most important, followed by form, UID and brand. In choice making, respondents seem to be focusing more on form rather than the UID when compared to preference data. This could be due to the tendency of the consumer to focus on easier to evaluate objective attributes (like function and form) rather than softer experiential attributes like UID while making choices (Hsee et. al., 2003). Also, in these studies form could be having a higher impact as the users only had visual images of the products rather than prototypes to manipulate and consequently the ease of interaction had to be judged based upon visual perceptions of design elements rather than its actual feel and use.

This work highlights the relative importance of the perceived UID in impacting consumer purchase preferences. Managerially, this finding shows that simply adding a greater number of functional capabilities (Hsee et al., 2009) to a product is questionable. As observed by Norman (1988) the design of the user interface with a simple approach (lesser number of control buttons) makes a product look easy to use and gives perceptions of high usability, and perceptions of ease of use are important for consumer preference judgments (Creusen, Veryzer, and Schoormans, 2010). Adding too many functions is likely to make the UID much more complex, which decreases the perceived usability and utility of the product. We also established the critical synergy between marketing signals such as brand and design elements, which is currently lacking (Bloch, 2011). Our research shows that for stronger brands, the focus on good UID during product design is critical as they stand to gain much more and if they ignore this they will lose more. Even for weaker brands, the perceived usability of the UID can add to the value proposition. When products are designed with high form and function, the UID is a critical factor which should be kept in mind during product design, especially when the product is having high aesthetic appearance and high technical capabilities.

Future research in can look into the influence of UID when consumers fully interact and use the products and its subsequent effect on consumer experience and repurchase intentions. Researchers should explore a richer holistic perspective on consumer response to product design that includes form, function and UID elements.

REFERENCES

Bloch, P. H. 2011. Product Design and Marketing: Reflections After Fifteen Years. *Journal of Product Innovation Management,* 28(3):378–380.

Booth, P. A. 1989. *An introduction to human-computer interaction.* Psychology Press.

Chitturi, R., P. Chitturi, and D. Raghavarao. 2010. Design for synergy with brand or price information. *Psychology and Marketing,* 27(7):679–697.

Creusen, M.E.H, and J. P.L Schoormans. 2005. The Different Roles of Product Appearance in Consumer Choice. *Journal of Product Innovation Management,* 22(1):63–81.

Creusen, M. E.H, R. W. Veryzer, and J. P.L Schoormans.2010. Product value importance and consumer preference for visual complexity and symmetry. *European Journal of Marketing,* 44(9/10),1437–1452.

Greene, William H. 2007. *LIMDEP Version 9.0. Econometric Modeling Guide,* New York: Econometric Software Inc.

Han, S. H, M. Hwan Yun, K. J Kim, and J. Kwahk. 2000. Evaluation of product usability: development and validation of usability dimensions and design elements based on empirical models. *International Journal of Industrial Ergonomics,* 26(4):477–488.

Hensher, David A., John M. Rose, and W. H. Greene. 2005. *Applied Choice Analysis: A Primer.* Cambridge University Press.

Holbrook, M. B, and W. L Moore. 1981. Feature interactions in consumer judgments of verbal versus pictorial presentations. *Journal of Consumer Research.* 103–113.

Hsee, C. K, Y. Yang, Y. Gu, and J. Chen. 2009. Specification Seeking: How Product Specifications Influence Consumer Preference. *Journal of Consumer Research,* 35(6):952–966.

Hsee, C.K., J. Zhang, F. Yu, and Y. Xi. 2003. Lay rationalism and inconsistency between predicted experience and decision. *Journal of Behavioral Decision Making,* 16(4):257–272.

Koca, A., E. Karapanos, and A. Brombacher. 2009. Broken Expectations' from a global business perspective. *Proceedings of the 27th international conference extended abstracts on Human factors in computing systems,* 4267–4272.

Kuhfeld, W.F. 2010. Marketing research methods in SAS,. http://support.sas.com/techsup/tnote/tnote_stat.html#market.

Langdon, P., T. Lewis, and J. Clarkson. 2007. The effects of prior experience on the use of consumer products. *Universal Access in the Information Society,* 6(2):179–191.

Luchs, M., and K. S Swan. 2011. Perspective: The Emergence of Product Design as a Field of Marketing Inquiry. *Journal of Product Innovation Management,* 28(3):327–345.

Macdonald, AS. 2001. Aesthetic intelligence: optimizing user-centred design. *Journal of Engineering Design,* 12(1).37-45.

Norman, D. 1983. Some Observations on Mental Models. *Mental models 7.*

Norman, D. A. 1988. The psychology of everyday things. *Basic books.*

Oppenheimer, A. 2005. From experience: Products talking to people—conversation closes the gap between products and consumers. *Journal of Product Innovation Management,* 22(1):82–91.

Page, C., and P. M Herr. 2002. An investigation of the processes by which product design and brand strength interact to determine initial affect and quality judgments. *Journal of Consumer Psychology,* 133–147.

Raghubir, P., and E. A Greenleaf. 2006. Ratios in proportion: what should the shape of the package be? *Journal of Marketing,* 70(2):95–107.

Reibstein, D. J, G. Day, and J. Wind. 2009. Guest editorial: is marketing academia losing its way? *Journal of Marketing,* 73(4):1–3.

Veryzer Jr, R. W, and J. W Hutchinson. 1998. The influence of unity and prototypicality on aesthetic responses to new product designs. *Journal of Consumer Research,* 374–394.

CHAPTER 44

Evaluating the Usability of Futuristic Mobile Phones in Advance

Huhn Kim, Jonghyeok Lim

Department of Mechanical System Design Engineering
Seoul National University of Science and Technology
Seoul, Korea
huhnkim@seoultech.ac.kr, lvzerovl@naver.com

ABSTRACT

Recently, the development cost for mobile phones has been increasing dramatically, i.e., the competition is getting more intense. In this situation, many companies have been trying to develop the mobile phones with innovative shapes beyond usual square-shaped designs. Thus, many designers have been presented diverse concepts on futuristic mobile phones. According to the survey in this study, the future concepts could be classified into wearable and flexible. And their physical user interfaces (UIs) were characterized by screen-flexible, foldable, displayed-by-projector, wrist-attached, wearable-ring and wearable-in-body.

However, the usability of physical UI (PUI) of the future concepts is yet to be known, in spite of being innovative. To reduce unnecessary development costs, their usability should be investigated in advance. In this study, six representative future concepts were chosen and their rapid prototypes were developed with low fidelity materials. Then an experiment was performed to evaluate their preference and usability in typical mobile tasks such as making a call, writing a message, checking the time, and so on. The results showed that the screen-flexible, wearable-in-body and wrist-attached type were more preferable and usable than the other concepts. On the contrary, foldable or displayed-by-projector design had low preference but high usability. The wearable-ring type had low preference and usability. From the experimental results, a series of design guidelines for PUI of future mobile phones was proposed.

Keywords: futuristic mobile phones, usability evaluation, low-fidelity prototype, physical user interfaces

1   INTRODUCTION

In the near future, the shapes of mobile phones will be getting diverse due to the advances of new technologies like flexible display, voice/gesture recognition, and so on. Many companies have been trying to develop the futuristic mobile phones with innovative shapes beyond usual square-shaped designs. Many designers have been presented various futuristic concepts as well. The concepts were fresh and innovative, and thus almost all users might prefer them. However, both preference and usability of physical UI (PUI) of the future concepts is yet to be known, in spite of being important. In order for the companies not to spend unnecessary costs for developing the concepts, a method for investigating the preference and usability of their PUIs in advance is imperative.

The user preference is one of the important factors that affect the success of a product (Baxter, 1995). Most of all, profitable and high-quality products should be designed dependent on a detailed understanding of consumer preferences (Swift, 1997). However, the user's preference is not simple but "a multidimensional psychological construct that might be composed of perceptive, affective, and behavioral dimensions" (Chuang et al., 2001). Usability is also a key factor for a product's success (Thomas and Thimbleby, 2002). The usability is a behavioral aspect of the user preference. On the contrary, the user preference is one of important performance criteria for measuring the usability (Nielsen, 1993).

With only product design images, the perceptive and affective aspects of the preference could be measured but the behavioral dimension or usability might not be measured. According to Artacho et al. (2008), dynamic or interactive representations of products are more effective than static ones when transmitting the products' semantic concepts into users. Buxton (2008) also emphasized that it is important for designers to experience interactive systems in advance before developing them completely. One of representative methods for gaining the prior experience for the interactive systems is paper prototyping. Generally, the paper prototyping has been usually used to rapidly evaluate usability and functionality of products in initial development phase, without realizing their software or hardware (Synder, 2003). In diverse studies, found usability problems between a paper prototype and a computerized prototype were not different significantly (Virzi et al., 1996; Sefelin et al., 2003; Lim et al., 2006). Besides, the paper prototyping is used to verify new product concepts or observe users' behaviors in using the new ones. For instance, Lee et al. (2010) observed deformation-based user gestures by observing users interacting with a paper or plastic sheet as a substitute of flexible displays.

In this study, six representative future concepts were chosen and their rapid prototypes as the interactive representations for the products were developed with low fidelity materials. Then an experiment was performed to evaluate their preference and usability in typical mobile tasks such as making a call, writing a message, checking the time, and so on. From the experimental results, a series of design guidelines for PUI of future mobile phones was proposed.

2   EXTRACTING TYPICAL FUTURISTIC MOBILE PHONES

In this study, diverse design concepts for futuristic mobile phones were investigated through Internet (http://www.concept-phone.com, etc.). The concepts were mainly designed from 2008 to 2010. After leaving out redundant concepts out of them, sixty-two ones were chosen and were classified by the levels of both wearable and flexible (Figure 1). And then similar concepts were categorized into five groups like Figure 2. The five groups were screen-flexible, foldable, displayed-by-projector, wrist-attached and wearable-in-body.

The wrist-attached designs were necessary to be separately handled because there were so many relevant concepts, although they were included in the wearable-in-body. In addition, the types of ring out of the wearable-in-body ones were also handled as a separate group because of their design specialty. Taken together, total six concepts could be extracted as typical futuristic mobile phones: screen-flexible, foldable, displayed-by-projector, wrist-attached, wearable-ring and wearable-in-body.

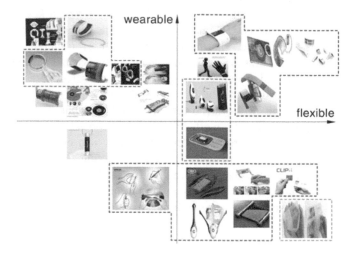

Figure 1 Image mapping of chosen design concepts for futuristic mobile phones

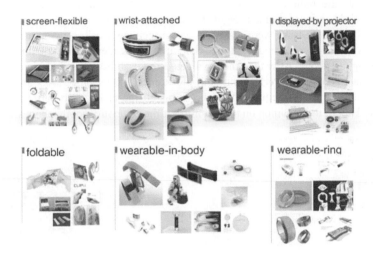

Figure 2 Extracting typical futuristic design concepts by image grouping

3 THE USABILITY EVALUATION METHOD

3.1 Developing Low-fidelity Prototypes

In order to investigate usability of the six futuristic design concepts, low-fidelity prototypes were developed by using papers, plastics, form boards, and so on. Figure 3 shows the six design concepts and their corresponding low-fidelity prototypes. The developed prototypes in this study were not fully interactive but good enough to be communicating with users with regard to core mobile activities.

3.2 Experimental Design

In this experiment, sixty participants (male 30, female 30), who were undergraduate students at Seoul National University of Science and Technology, were employed. Each participant was equally assigned into the two of six concepts in order to diminish the learning and order effects. In other words, this experimental design was an incomplete block design which the participant was a block factor and two design concepts per the block were allocated.

3.3 Tasks and Procedures

The participants performed three kinds of tasks. First, they checked their preference (9-point Likert scale) on each design concept from the five viewpoints such as convenience of making a call and sending a message, product design, mobility, and overall usability. At the time, they could see just design images. Second, the participants were asked to evaluate usability of each design concept from the same viewpoints as the preference check. In this time, they could also see just design images. Third, they performed eight basic mobile tasks with the low-fidelity prototypes. The eight tasks were making a call, receiving a call, sending a message, receiving a message, checking the time, checking the sender, taking a calling pose, and carrying the phone. After that, they evaluated the usability of each design concept with 9-point Likert scale. Performing both preference and usability tasks took about twenty minutes per a participant. After finishing all the tasks, each participant was interviewed about their opinion on the usability of the futuristic mobile phones.

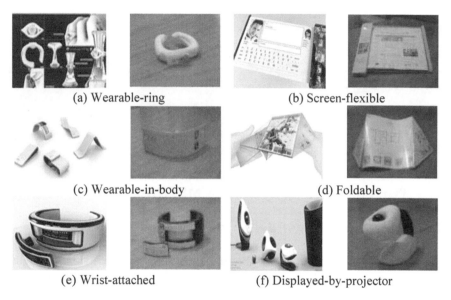

(a) Wearable-ring   (b) Screen-flexible

(c) Wearable-in-body   (d) Foldable

(e) Wrist-attached   (f) Displayed-by-projector

Figure 3 Futuristic design concepts (Left) and their corresponding low-fidelity prototypes (Right)

4   RESULTS

4.1   Design Preferences

The Kruskal-wallis non-parametric test was employed because the preference data could not meet the normality assumption for ANOVA. The preference differences among six design concepts were statistically significant (p=0.000). However, the preference was not different between male and female (p=0.138). Figure 4 shows the average preferences of each design concept. The screen-flexible, wearable-in-body and wrist-attached design concepts showed higher preferences. This result might be induced by the participants' thought that the three best concepts were most likely to be similar to the usage scene of general mobile phones. On the contrary, the foldable and displayed-by-projector design concepts were the worst because their design images might be unfamiliar with the participants. The preference of the wearable-ring design was also not high (5.05 on the average).

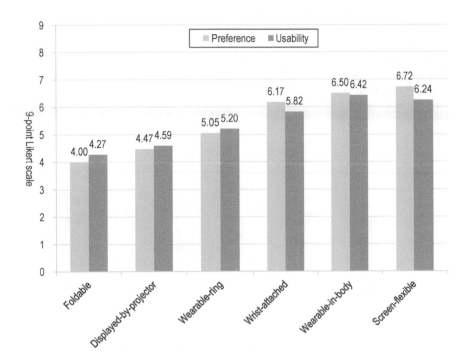

Figure 4 Result of preference and usability evaluations performed with futuristic design images

4.2   Usability Evaluations with only Design Images

The Kruskal-wallis non-parametric test was also employed due to the same reason as the preference data. The usability among six design concepts and among eight tasks were significantly different (p=0.000, p=0.000, respectively). Figure 4 shows the result of usability evaluations performed with only design images. The usability result was very similar to the preference result because the participants did not understand how they could use the products with only design images.

4.3   Usability Evaluations with Low-fidelity Prototypes

The Kruskal-wallis non-parametric test was also used. The usability among six design concepts and among eight tasks were significantly different (p=0.000, p=0.000, respectively). However, the usability was not different between the genders (p=0.267).

Figure 5 shows the result of two usability evaluations performed with design images (this was redundantly shown in Figure 4) or low-fidelity prototypes. On the whole, the usability values performed with the prototype were higher than those with the design images. The screen-flexible concept showed the highest usability value (7.04 on the average). The screen-flexible, wearable-in-body and wrist-attached design concepts were better than the others in both preference and usability. In special, the foldable and displayed-by-projector concepts made a big difference between before and after using low-fidelity prototypes. After performing several tasks with the prototypes, they took a little bit higher usability values, in spite of taking low preference and usability values with only design images. Performing the tasks with low-fidelity prototypes was likely to contribute to their thought's change on the usability of the two concepts.

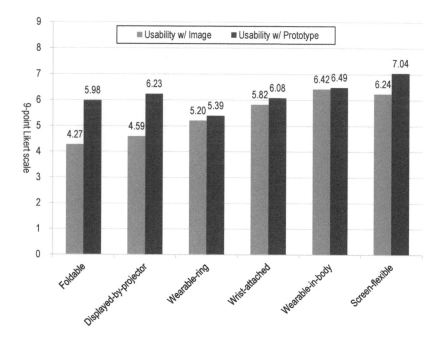

Figure 5 Result of two usability evaluations performed with design images or low-fidelity prototypes

Figure 6 shows the average of usability on performing each task with the prototypes. The screen-flexible concept showed high usability points in almost all tasks. On the contrary, the usability points of the wrist-attached and wearable-in-body concepts were not high in both sending and receiving a message. This was due to a small size of display screen in the wrist-attached concept. Relatively, the foldable and displayed-by-projector recorded low usability values in both taking a calling pose and carrying the phone. However, the wearable-ring design concept showed the lowest usability values in almost all tasks except for carrying the phone.

Figure 7 shows the differences between two usability values on performing three same tasks with design images or prototypes. The figure indicates that evaluating task usability with only design images is very limited because participants have never been experienced such kinds of phone as the futuristic shapes are different with those of general mobile phones.

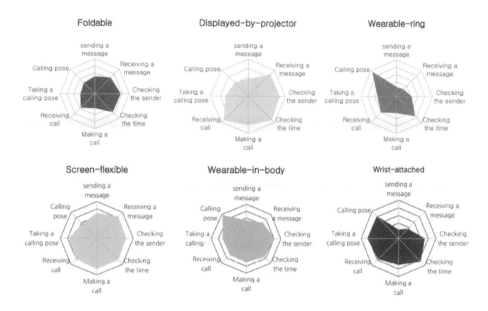

Figure 6 The average of usability on performing each task with each design concept

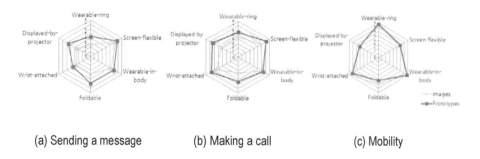

(a) Sending a message   (b) Making a call   (c) Mobility

Figure 7 Differences between two usability values on performing each task with design images or prototypes

5   CONCLUSIONS

Through a thorough design survey in this study, the futuristic mobile phones were classified into screen-flexible, foldable, displayed-by-projector, wrist-attached, wearable-ring and wearable-in-body. Then, the low-fidelity prototypes for the six representative design concepts were developed and an experiment was performed to evaluate their preference and usability in typical mobile tasks.

The experimental results showed that the screen-flexible, wearable-in-body and wrist-attached types were more preferable and usable than the other concepts. On the contrary, foldable or displayed-by-projector concept had low preferences but high usability points. Specially, the usability results of the two design concepts between before and after using low-fidelity prototypes were highly different. And the wearable-ring type was worse in both preference and usability. From the experimental results, a series of design guidelines for PUI of future mobile phones was proposed. The design guidelines are shown in Table 1. These guidelines were drawn from comparing the characteristics of the best and worst design concepts for each task. However, these guidelines are not complete and this study has the limitation that employs only low-fidelity prototypes for evaluating their usability. Despite of such low-fidelity, these results may be meaningful as basic information because the tasks performed with the prototypes were very simple and basic in a mobile context.

Table 1 PUI guidelines for designing futuristic mobile phones

Tasks	Best concept and its reason		Worst concept and its reason	
Carrying a phone	Wearable-in-body	Wearable in any part of the body	Displayed-by-projector	Swollen shape
	Design shapes should not have bended or swollen parts			
Taking a calling pose	Wrist-attached	Bluetooth	Foldable	Uncomfortable grip
	Design shapes must consider users' grip feelings or substitutive methods such as the Bluetooth function should be employed			
Receiving / Making a call	Screen-flexible	Familiar shapes with general phones	Foldable	Uneasy to grip & press buttons due to their thin shape
	Design shapes should not be deviated from familiar phone shapes and should be properly thick enough to grip the phone and press buttons.			
Checking the time	Wrist-attached	Similar shape with a watch	Displayed-by-projector	Uncomfortable hologram
	LCD screen should be activated with a simple and familiar way			
Checking the sender	Wearable-in-body	-	Wearable-ring	Too small screen
	If the screen size is small, the complementary techniques like TTS (Text-to-Speech) or hologram should be employed to cover the size issue			
Receiving a message	Screen-flexible	Familiar shapes with general phones	Wearable-ring	Too small screen
	If the screen size is small, the complementary techniques like TTS (Text-to-Speech) or hologram should be employed to cover the size issue			
Sending a message	Screen-flexible	Familiar shapes with general phones	Wrist-attached	Too small buttons and uneasy layout
	The buttons for writing a message should be large enough to press them easily. And it is recommended that the layout of the buttons is familiar with users. If the button sizes have to become smaller, other complementary techniques like STT (Speech-to-Text) should be employed to cover the small sizes			

REFERENCES

Artacho, M.A., Diego, J.A., and Alcaide, J. 2008. Influence of the mode of graphical representation on the perception of product aesthetic and emotional features: An exploratory study. *International Journal of Industrial Ergonomics* 38: 942-952.

Baxter, M. 1995. *Product design, practical methods for the systematic development of new products*. Chapman & Hall, London.

Buxton, B. 2007. *Sketching user experiences: getting the design right and the right design.* Morgan Kaufmann Publishers.

Chuang, M.C., Chang, C.C. and Hsu, S.H. 2001. Perceptual factors underlying user preferences toward product form of mobile phones. *International Journal of Industrial Ergonomics* 27: 247-258.

Lee, S., Kim, S., Jin, B., Choi, E., Kim, B., Jia, X., Kim, D., and Lee, K. 2010. How users manipulate deformable displays as input devices, *CHI 2010*, April 10-15, 2010, Atlanta, Georgia, USA.

Lim, Y., Stolterman, E., and Tenenberg, J. 2008. The anatomy of prototypes: prototypes as filters, prototypes as manifestations of design ideas. *ACM Transactions on Computer-Human Interaction*, Vol. 15, No. 2, Article 7.

Nielsen, J. 1993. *Usability Engineering*. Morgan Kaufmann Publishers.

Sefelin, R., Tscheligi, M., and Giller, V. 2003. Paper prototyping - what is it good for?: a comparison of paper- and computer-based low-fidelity prototyping. *Conference on Human Factors in Computing Systems CHI '03*: p.778-779.

Snyder, C. 2003. *Paper Prototyping: The fast and easy way to design and refine user interfaces*. Morgan Kaufmann Publishers.

Swift, P.W. 1997. Science drives creativity: a methodology for quantifying perceptions. *Design Management Journal*, Spring 1997, 51-57.

Thomas, P. and Thimbleby, H. 2002. The new usability: the challenge of designing for pervasive computing. *ACM Transactions on Computer Human Interaction*: 1-3.

Virzi, R., Sokolov, J, and Karis, D. 1996. Usability problem identification using both low- and high-fidelity prototypes. *Proceedings of the SIGCHI conference on Human factions in computing systems*: 236-243.

CHAPTER 45

Towards a More Effective Graphical Password Design for Touch Screen Devices

Xiaoyuan Suo, Janet L. Kourik

Webster University
Saint Louis, MO, USA
Xiaoyuansuo51@webster.edu, kourikJL@webster.edu

ABSTRACT

The objective of this project is to conduct exploratory research into the use of graphical passwords on touch screen devices. Graphical password schemes have been proposed as a possible alternative to text-based password schemes. Humans can remember pictures better than text, thus graphical passwords may contribute to a more positive user experience. We also propose a novel graphical password scheme designed specifically for touch screen devices. Since touch screen devices dominate the mobile world, this project has the potential to influence design of mobile user experiences.

Keywords: Graphical password, iPad/iPhone app, Mobile app design, Mobile

1   INTRODUCTION

Touch screen interface designs have attracted rising attention in recent years; devices such as ATM (automated teller machines), ticket machine, PDA (personal digital assistant), have been widely used in various occasions. Lately, touch screen devices are technologically becoming more accurate, usable and popular in many sizes; such as smart phones, or Apple's iPad, iPhone, and iPod touch (Apple) etc. Media reports estimate that touch screen devices will account for more than 80 percent of mobile sales in North America by 2013 (Gartner, 2010, Fleming, 2010).

Graphical passwords have been proposed as a possible alternative to text-based schemes, motivated partially by psychological studies (Shepard, 1967) that show human can remember pictures better than text. Graphical password techniques include recall-based click password (e.g. imposing background image so user can click on various locations on the image), and recognize-based selection password (e.g. selecting images or icons from an image pool). In particular, recall-based click passwords have received much more attention in recent years (Wiedenbeck et al., 2005). However, relatively little study of usability has been done for graphical passwords. In addition, as Cranor, et al. (Cranor and Garfinkel, 2004) noted, little work has been done to study the security of graphical passwords as well as possible attacking methods. In fact, some recent studies have shown that there are in fact unintentional patterns in user created graphical passwords. In reality, security and usability of graphical passwords are often at odds with each other, (Suo et al., 2005) but both factors are critical to all authentication systems.

In this work, we propose a novel graphical password scheme designed specifically for touch screen devices. Since touch screen devices dominate the mobile world, this project has the potential to influence design of mobile user experiences and implications for the future growth in mobile devices. The project is built on iOS platform using Objective C, and deployed as an iPad/iPhone application. To our knowledge, this is the first attempt in studying graphical password design scheme on touch screen devices. This study directly concerns user experiences and effective designs of touch screen applications.

Further, expert reviews and usability studies were used to explore user interactions in order to gain a more complete understanding of the following:

a. *Approaches to overcome limitations of a touch screen computer for graphical password designs.*
Touch screens have special limitations such as: user's finger, hand and arm can obscure part of the screen. Human finger as a pointing device has very low "resolution". It is also difficult to point at targets that are smaller than the users' finger width.

b. *The relationship between user password choices and the complexity of the background image.*
The complexity of an image is defined as a combinational quantitative measure of the number of objects presented, the number of major colors, and the familiarity of the image to users as well as other factors. Carefully selected background images can enhance effective graphical password design. The study highlights the vulnerability of click based graphical passwords. As a result, we discuss several graphical password attacking methods based on the choice of pictures. This study aims to achieve better balance between security and usability of click-based graphical passwords through better choices of background images.

c. *The relationship between background image choice and successful authentication rate.*

Success rate, in this case, is defined as the number of successful authentications versus the number of trials within a critical time frame.

d. *The relationship between tolerance rate and successful rate.*

The tolerance rate is defined as a number of pixels permitted to be selected around the original selection. A properly selected tolerance rate has direct impact on user experiences.

The goal of our work is to introduce a graphical password design for touch-screen devices and provide a study on the usability of such an application design. We hope to provide fellow researchers and practitioners in the field with more complete guidance to achieve more usable touch screen device designs.

2   A NOVEL SCHEME OF GRAPHICAL PASSWORD ON TOUCH SCREEN MOBILE DEVICE

The architecture of our novel graphical password scheme is shown in Figure 1. The flow chart is based on two main logic paths, namely, registration and authentication. Users are given a database of images to select from; the images are categorized into different groups to improve user experiences.

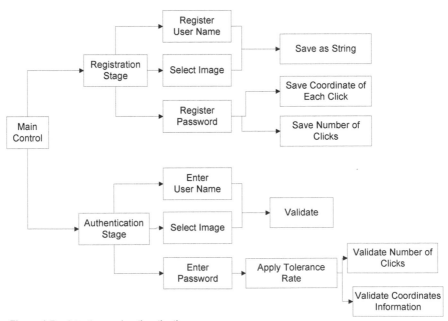

Figure 1 Registration and authentication process.

In the screenshot shown below (figure 2), two categories are shown near top of the screen: personal images and natural images. During registration, user is also given the choice of importing images from the personal collection.

Figure 2 Screen shot of graphical password on iPad.

During the registration stage, user is required to enter a user name and select an image first. The size of each image is adjusted to fit into the screen. After the image is loaded onto the screen, user will be asked to click on various places on the image. The sequence of clicks, with an X and Y coordinates, is recorded and saved along with the user name and choice of image into our database.

During the authentication stage, the user will first be asked to enter a user name and select their registered image. Unless the user name and the choice of images match those in the database, the authentication will fail. When the image is loaded, the user will be allowed to tap the image to authenticate. During this process, user will be allowed to select a tolerance rate, which means they can tap within a circle of pixels rather than the specific pixel they pre-selected.

3   FINDINGS FROM USER STUDIES

In usability experiments conducted with a group of experts recruited from information technology or IT education related fields, the successful authentication rate of our scheme is more than 80%.

3.1 Touch Screen Device Design Limitations

One difficulty for interface design on mobile devices is lack of screen space caused by their small size (Brewster, 2002). Small displays and multiple inputs, especially with the presence of a figure, require users to register click-based password with pinpoint accuracy.

When using a touch screen device, the user's finger, hand and arm can obscure part of the screen. Also, the human finger as a pointing device has very low "resolution". These limitations have been observed and tackled before, mostly notably by Sears, Shneiderman and colleagues (Sears, 1991, Sears et al., 1992). Their basic technique, called *Take-Off*, provides a cursor above the user's finger tip with a fixed offset when touching the screen. The user drags the cursor to a desired target and lifts the finger (takes off) to select the target objects. They achieved considerable success with this technique for targets between finger size and 4 pixels. Instead of using a bare finger, in some cases the user may use a stylus (pen) to interact with touch screens. A stylus is a much "sharper" pointer than a finger tip, but its resolution may still not be as good as a mouse cursor. Recent work by Diller (Diller) also studied various techniques to improve input areas of touch screen mobile devices.

3.2 The Choice of Tolerance Rate

The bigger the tolerance rate, the easier the user task is. At the same time, the bigger tolerance rate permits a bigger chance of educated guess and limits the number of click points for graphical password. To balance security and usability, it is noted that a better tolerance rate is essential. After a brief user study, we found a tolerance rate bigger than 30 pixels is preferred among all users. When the tolerance rate is set below 30 pixels, the failure rate significantly increases. The following figure demonstrates how different tolerance rates affect the authentication failure rate.

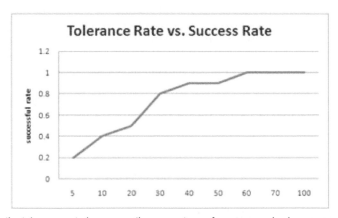

Figure 3 As the tolerance rate increases, the percentage of successes also increases.

3.3   Effective Password Space Analysis

In this section, we introduce a term "Effective Password Space". In touch pads or similar electronic devices that rely on finger touches or a hand-held tool for input, effective password space should be smaller than the touch pad screen size. It is generally assumed that touch input cannot be accurate because of the fat finger problem, i.e., the softness of the fingertip combined with the occlusion of the target by the finger (Diller 2010). In the average case of user selecting five different regions as their graphical password; it is required none of the five regions should have overlap.

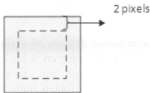

Figure4 Total effective password space for a touch screen, assuming the pressing area is 4 pixels square.

Studies have shown using a stylus pen or computer aided input device would potentially achieve an ideal accuracy of 1 pixels square (Go and Endo, 2008), but the usability of such an approach remains a question. In particular, selection time and error rate data were not reported under this scheme and this mechanism was reported to be impractical to carry out in regular users' daily life.

It also should be noted that "tolerance rate" is also a key factor in determination of effective password space. Typically, a tolerance area should be set to accommodate the usability of the system. When tolerance rates are too high, users end up with less effective password space, and vice versa. In the equation below, password space or the number of possible clicks can be defined in two possibilities, where n and m are the length and width of a particular screen in number of pixels:

$$password\ space = \begin{cases} 1, & if\ n \leq 4\ or\ m \leq 4 \\ (n-4)*(m-4), & if\ n > 4\ or\ m > 4 \end{cases}$$

In the case of iPAD, which became a popular hand-held touch pad device since April 2010, the screen size is $1024 \times 768\ pixels$. In a short experiment we conducted with users, we chose a simply designed screen of 16 rows and 12 columns, representing a total of 192 pixels for the experiments:

1. The first figure press can happen in almost anywhere on the screen, except the edge areas. Therefore, the possibility is 140—minus the edge pixels.
2. To calculate the second finger press possibility, we wrote a very simple program to statistically calculate the overall-possibility based on the first press's location. As we mentioned before, the first press has 140 possibilities. After measuring 140 different possibilities, the possibility of second press is in average close to 131 possible locations.
3. The third finger press is relatively harder to calculate. There should be four different types finger presses in general. The likelihood of each finger press is similar. After a brief calculation, we found122 possible areas for the third finger press in average.

4. When similar rules were applied, 113 possible areas are found for the fourth finger press. And 104 possible areas are found for the fifth finger press.

5. In total, we had a possibility of 140*131*122*113*104 password space—fairly large, but not sufficiently large enough to prevent brutal force search. After all, $2.6*10^{10}$ is still a finite number that can be cracked by brutal force attack.

In fact, applications for touch screen devices all suffer the "fat-finger" problem. We performed an experiment with our users on a 850×650 screen with simple white background, and no assistance from any commercial software. We then measured the pixel size of 41 sparsely distributed finger presses made by our users. The average size was 67.9 pixels; the smallest pixel size was 6 and the largest was 213; with most sizes found between 10 and 100.

With all factors discussed, effective graphical password space for touch screen devices is indeed very small.

3.4 Background Image Selection vs. User Experiences

Users were given the option of using their personal pictures for background. In the presence of a background image, it is arguable if the password space will remain as large as possible. Users do have significant preferences when it comes to different background images. In other words, background images reduce password space; yet regular users cannot live without background images (Suo et al., 2009).

Complexity of the background image directly affects the usability of the graphical password. Some of the factors that define the image complexity are listed below:

1. Colors: We cannot always provide the user with meaningful pictures. When the graphical password is generated in a semi-automatic fashion, color can play a practical role.

2. Objects: Objects in the image are other contributing factors. Face recognition (Davis et al., 2004) is one type of graphical password that uses objects as its main theme. Depending on the size of the object and the proportions the object occupies compared to the entire image, the user may only able to focus on one or a very limited number of objects at a time.

3. Location and Shapes: There can be two types of shapes in a graphical password images: the shape of the objects in an image, or the shape formed by patterns of clicks.

4 CONCLUSION

Expert review of the system revealed the scheme to be effective and promising. During the debriefing following the experiment, experts proposed improving graphical targets through a variety of design innovations that may be avenues for future research. Expert reviews and usability experiments indicate graphical passwords have potentials to work well with touch screen devices.

Through experimental examination of touch sizes, tolerance areas and recall based passwords, our scheme produced successful authentication rates of greater than 80%. This study contributes additional data to advance the field of graphical password research.

REFERENCES

APPLE http://www.apple.com.

BREWSTER, S. (2002) Overcoming the Lack of Screen Space on Mobile Computers. *Personal and Ubiquitous Computing,* 6.

CRANOR, L. & GARFINKEL, S. (2004) Secure or Usable? . *Security & Privacy,* 2, 2.

DAVIS, D., MONROSE, F. & REITER, M. K. (2004) On user choice in graphical password schemes. *13th conference on USENIX Security Symposium.*

DILLER, F. (2010) Target Practice: Current Efforts to Improve Input Areas on Touchscreen Mobile Devices.

FLEMING, K. (2010) Report: Touch Screen Mobile Device Sales Booming. *CRN.*

GARTNER (2010) Gartner Says Touchscreen Mobile Device Sales Will Grow 97 Percent in 2010. *http://www.gartner.com/it/page.jsp?id=1313415.*

GO, K. & ENDO, Y. (2008) Touchscreen Software Keyboard for Finger Typing. *Advances in human-computer interaction.*

SEARS, A. (1991) Improving Touchscreen Keyboards: Design issues and a comparison with other devices. *Interacting with computers,* 3, 253.

SEARS, A., REVIS, D., SWATSKI, J., CRITTENDEN, R. & SHNEIDERMAN, B. (1992) Investigating Touchscreen Typing: The effect of keyboard size on typing speed *Behaviour & information technology,* 12, 17.

SHEPARD, R. N. (1967) *Recognition memory for words, sentences, and pictures.* Journal of Verbal Learning and Verbal Behavior. *Journal of Verbal Learning and Verbal Behavior,* 156-163.

SUO, X., ZHU, Y. & OWEN, G. S. (2005) Graphical Password: A Survey. *Proceedings of Annual Computer Security Applications Conference (ACSAC).* Tucson, Arizona, IEEE.

SUO, X., ZHU, Y. & OWEN, G. S. (2009) A Study of the Vulnerability of Click Based Graphical Password. IN SCIENCE, L. N. I. C. (Ed.) *5th International Symposium on Visual Computing, Lecture Notes in Computer Science.*

WIEDENBECK, S., WATERS, J., BIRGET, J. C., BRODSKIY, A. & MEMON, N. (2005) Authentication using graphical passwords: Effects of tolerance and image choice. *Symposium on Usable Privacy and Security (SOUPS).* Carnegie-Mellon University, Pittsburgh.

CHAPTER 46

Material Sensibility Comparison between Glass and Plastic Used in Mobile Phone Window Panel

Jae Hee Park*, Eun Ha Kim*, Min Uk Kim* , Seung Hee Kim* ,

Sang Bom Ha**, Jae In Lee**

*Hankyong National University, Anseong, Rep. of Korea, maro@hknu.ac.kr
**LG electronics Inc., Pyeongtaek, Rep of Korea

ABSTRACT

The aim of this study is to investigate the sensibilities of various materials including glass and plastic. Material sensibility is critical in shaping one's image for a product, then may affect on the decision of purchasing the product. However, the basic studies on the sensibility on various materials are rare. Five different materials including glass and plastic were presented to 50 subjects each. They evaluated the materials in terms of transparency, plainness, hardness and sleekness, after experiencing the materials in three different ways; tactile sensing only, visual sensing only and whole sensing allowed. Although it was difficult to identify the materials, the subjects could differentiate relatively the properties of the materials. Also, the result shows subject depend on more the tactile sense than visual sense in deciding what the material is. Especially hardness is most dominant in the sensibility evaluation. People may depend on tactile sense, especially on hardness, in characterizing a glass-like material.

Keywords: material, sensibility, glass, plastic

1 INTRODUCTION

Glass and plastic are commonly used materials in daily necessaries. Plastic have been substituted for glass in eyeglasses by reason of light weight and safety. In contrast, glass has substituted for plastic in tableware for reserving foods by reason of sanitation. Generally, plastic is cheaper, lighter and easier to be processed than glass. Glass is harder and more transparent than plastic. Also, glass is more preferred in touch than plastic. Therefore, glass has been substituting for plastic after rapidly introducing smart phones which need window touches for operations (Weiss, 2002; Burdea, 1996). However, there are few studies that compare the materials and find why glass is preferred in touch sense.

This study aims to compare transparent materials including glass and plastic which can be used in window pane of smart phones in terms of adjective words representing material sensibilities and preference. If one plastic material is perceived as glass, it can be substituted for the expensive glass. To compare alternative window pane materials for smart phones, an experiment for subjective evaluation was conducted.

2 METHOD

For an experiment, five different materials were prepared; (1) phone glass currently used in mobile phones, (2) plastic currently used in mobile phones, (3) polycarbonate, (4)acryl, (5) building glass. Two materials, (1) and (5) are glasses. The others, (2), (3) and (4) are plastics. The five materials were framed with a window (110x57mm). A picture sheet with smart phone icons was positioned under the materials to give subjects feel of smart phone (Figure 1). Twenty nine subjects, 26 males and 3 females, participated in the experiment. Their average age was 23.4 years. They had no impairment in visual and tactile senses. They all have owned smart phones.

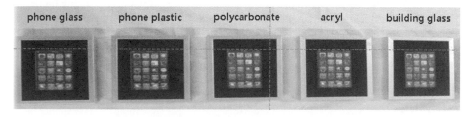

Figure 1 Five materials used in the experiment

The experiment was performed in three modes; visual sense only, tactile sense only, and whole senses used. In visual sensing mode, the subjects were not allowed to touch the materials, but could only see the materials. In tactile sensing mode, the subjects were blinded with an eye patch. They could only touch the materials with fingers. In whole sensing mode, the subjects could use visual and tactile senses.

Visual sensing mode   Tactile sensing mode   Whole sensing mode

Figure 2 Three modes of the experiment

For each material, after seeing or/and touching it, the subjects were asked to answer whether the material is glass or plastic. Then they rated the certainty on their decisions in the 0-100 scales. After experiencing a series of five materials, they were asked to rank the materials in terms of four adjectives which describe the materials; plain, transparent, hard, and sleek. They also ranked the materials in terms of preference. Due to the high similarity among the materials, it was very difficult to rate the words on Likert scales in the preliminary experiment. Therefore, we adopted rank scale in the main experiment (Sinclair, 1990; Meister, 1986). Finally, the subjects ranked the adjectives and rated them in 100% in how much each adjective attributes to their decisions (Table 1).

Table 1 Questionnaire and scales

No.	questionnaire	scale	experiment
Q1.1	Is it glass or plastic?	0/1	All experiments
Q1.2	How much is the certainty of your decision?	%	All experiments
Q2.1	Rank the transparency of a material	rank	In visual mode
Q2.2	Rank the plainness of a material	rank	In visual mode
Q2.3	Rank the hardness of a material	rank	In tactile mode
Q2.4	Rank the sleekness of a material	rank	In tactile mode
Q3.1	Rank the preference of a material	rank	All experiments
Q4.1	Rank 4 importance of adjectives in decisions	rank	Each subjects
Q4.2	How much is the attribution ratio of 4 adjectives	%	Each subjects

3 RESULTS

3.1 Identifying glass and plastic

Figure 3, 4, and 5 show the results of the subjects' decisions on what a material is glass or plastic. To find the statistical significant difference of correct and incorrect answers, we conducted chi-square test. The significant probabilities of the test are summarized in Table 1. For phone plastic, the subjects significantly could identify it as plastic in any modes. For building glass, if the subjects use their tactile sensing, they could identify it as glass. However, for phone glass and acryl, the subjects significantly could not identify them even though the correct answer ratios are slightly greater than incorrect answer (Figure 3, 4, 5). For phone glass, it was very similar with plastics in physical properties. It is softer and more flexible than conventional building glass. It made the subjects confused in their decisions. Comparing sensing modes, there was no significant difference but correct answer ratios were increased for phone glass, phone plastic, polycarbonate, and building glass if whole sensing was allowed.

Table 1 Significant probabilities of identifying glass or plastic correctly

Evaluation Adjectives	Phone glass	Phone plastic	Poly-carbonate	Acryl	Building glass
Visual mode	0.564	0.016*	0.178	0.059	0.353
Tactile mode	1.000	0.000*	0.083	0.450	0.000*
Whole sensing mode	0.131	0.000*	0.016*	0.847	0.002*

3.2 Material sensibilities

For the four adjectives, transparent, plain, hard, and sleek, the subjects ranked five materials. The same rank was allowed in ranking if they could not differentiate materials. The results of the average ranking for four adjectives are shown at Figure 6 and 7. For all four adjectives, there were significant differences among the five materials from Friedman rank test (p=0.002 for transparent, p=0.000 for the others). Generally, glass materials, phone glass and building glass, were better than plastic materials for all adjectives. For the only plainness of carbonate showed low rank approaching glass materials (Figure 6). There was no significant difference among four adjectives for glass materials; phone glass (p=0.202) and building glass (p=0.075). However there was significant difference among four adjectives for plastic materials; phone plastic (p=0.000), polycarbonate (p=0.003), and acryl (p=0.021).

. The subjects finally were asked to rank the materials in terms of preference. They preferred glass materials, phone glass and building glass, rather than plastic materials (p=0.000)(Figure 9). Preference was mainly related with the hardness. (Compare Figure 7 and 8). We also asked the subjects which material sensibility attribute to their decisions between 'glass' and 'plastic'. Figure 8 shows that hardness of material mainly gives the subjects clue for their decision. Therefore, first above all, if someone wants the plastic to be regarded as a glass material, it is needed to harden a plastic material to the level of glass.

420

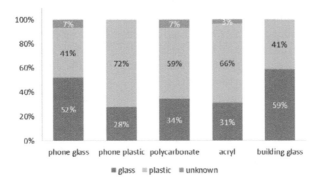

Figure 3 Correct and incorrect answer ratios in visual sensing mode

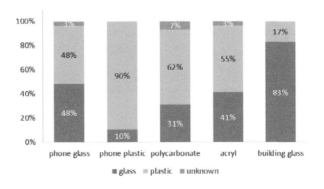

Figure 4 Correct and incorrect answer ratios in tactile sensing mode

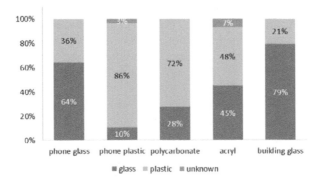

Figure 5 Correct and incorrect answer ratios in whole sensing mode

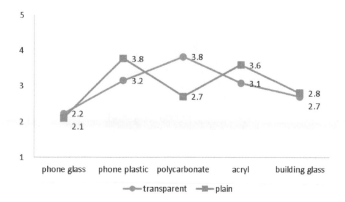

Figure 6 Ranks of five materials in transparency and plainness

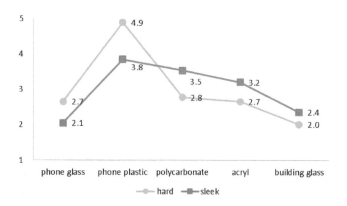

Figure 7 Ranks of five materials in hardness and sleekness

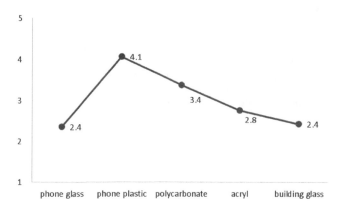

Figure 8 Ranks of five materials in preference

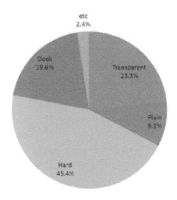

Figure 9 Ratio of adjectives which are referred in the subjects' judgment

4 CONCLUSIONS

The importance of emotion and sensibility in products are increasing in these days. Especially, in mobile phone industry with short product life cycle, consumers' sensibilities and preference are critical. In selecting the materials for mobile phone window panel, the material sensibilities should be considered at first. We conducted a material sensibility evaluation experiment. As a result, we found that the subjects could identify the plastic from glass. Although only visual sensing could not identify materials between glass and plastic, but tactile sensing greatly attribute to identifying glass. Especially, the hardness is main characteristics of material. Therefore, it is not easy to substitute glass with plastic in this moment. It is difficult to harden plastic at the level of glass. To develop new plastic materials which can fake glass, the sensibilities on glass and plastic should be explored further more.

REFERENCES

Burdea, G., 1996, Force and touch feedback for virtual reality, Jhon Wiley & Sons.
Meister, D., 1986, Human Factors testing and evaluation, Elsevier.
Sinclair, M.A, 1990, Subjective assessment, in Evaluation of human work (edited by J.R. Wilson and E.N. Corlett), Taylor and Francis.
Weiss, S., 2002, Handheld usability, Jhon Wiley & Sons.

Section VI

Emotion and UX Design

Developing New Emotional Evaluation Methods for Measuring Users' Subjective Experiences in the Virtual Environments

Nermin Elokla, Yasuyuki Hirai

Kyushu University, Japan
nelokla@hotmail.com

ABSTRACT

Nowadays, users' emotions have been recognized as valuable parameters for product design development. We all know from our personal experiences that products can elicit strong emotional responses. These product emotions influence both the decisions to purchase a product and the pleasure of owing and using it after the purchase. For designers it is essential to create products that fit the emotions of users, that is, products that elicit the emotions that the user would like to experience. In the present study, the authors focused on user experience and how it can be captured during the use of product. This work aimed to introduce new types of emotional evaluation methods that can be used to examine and understand user experiences while they are interacting with the products in the virtual environment. The two novel methods are called; *Kansei* sheets, and Read Body Language sheet. The results of this study will be of interest for design educators, and decision makers.

Keywords: emotional evaluation methods, user experience, pleasant product design

1   INTRODUCTION

Emotions pervade our daily life. They can help guide our choices, avoid a danger and they also play a key role in non-verbal communication. Assessing emotions is thus essential to the understanding of human behavior and need. Emotion assessment is a rapidly growing research field, especially in the area of human product interaction (HPI) (Guillaume, et al., 2006). Nowadays, there are few techniques for gathering affective data (Eva & Muriel, 2007). In this paper, we introduced and discussed new methods to interpret and assess emotional and physical responses of users while they are interacting with the products. Our measurement methods provide interactive designers and researchers with valuable information about a user easily. To demonstrate how to use our new emotional evaluation methods, a workshop was carried out at Kyushu University.

2.   METHODS

2.1   Kansei Sheets

They are visual measurement tools (Figure 1). Sheet #1 aims to measure the emotional responses of a user while he or she is interacting with a product. Sheet #2 aims to measure physical responses of users. On sheet #1, the user's emotional responses are measured according to a Likert-type scale: I feel this to some extent (10%); I feel this (40%); I feel this very much (70%); and I strongly feel this (100%). On sheet #2, the user's physical responses are measured according to a similar Likert-type scale: I suffer somewhat from this (10%); I suffer from this (40%); I suffer from this very much (70%); and I strongly suffer from this (100%). Through using these sheets, participants will be able to recognize all emotional and physical responses in two differentiated components; qualitative and quantitative.

The two sheets can be used cross-culturally because they do not ask respondents to verbalize their emotions.

2.2   Read Body Language (RBL) Sheet

This tool was proposed to be used by a designer to understand the users and their non-verbal communications/behaviors (Figure 2).It aims to interpret users' dynamic expressions by observing them while they are interacting with a product. Based on the ten emotion heuristics (TEH) (Eva & Muriel, 2007) and several studies regarding human emotions (Tingfan, 2009) and (Cynthia & Martinovich, 2002), 20 distinct emotional responses were interpreted and analyzed in one sheet which is called "read body language RBL sheet".

2.3   Workshop Details

The workshop (WS) aimed to introduce new methods for designing pleasant product and user experience. It was carried out in a laboratory in Kyushu University in June 2010.

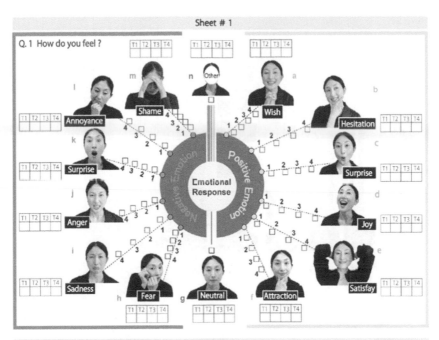

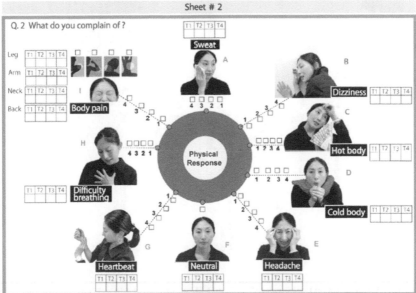

☐ 1 (10%), 2 (30%), 3 (70%), and 4 (100%) "Emotion Degree"
T1, T2, T3, and T4 "Tasks" ▬The users were asked to do different tasks for evaluating a product

Figure 1   The Product Emotion Measurement Method (Kansei Sheets)

428

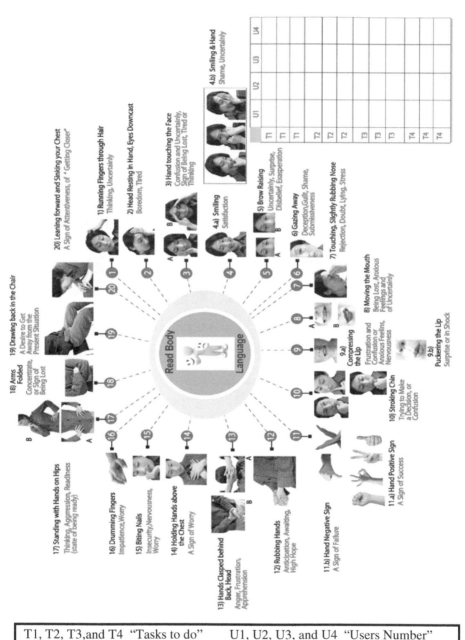

T1, T2, T3, and T4 "Tasks to do"   U1, U2, U3, and U4 "Users Number"

Figure 2   the Product Emotion Measurement Method (Read Body Language Sheet)

A blood pressure (BP) device was selected as a case study. The workshop was done in three stages (Figure 3); capture, develop, and present.

Four users (20 to 30 years old) were asked to evaluate the device's usability, and its appearance (including size, color and shape).

In the first stage "Capture", each user was asked five questions: First one was regarding aesthetics appeal of a BP device — "Q.1 Is this design of a BP device attractive for you?" Second, third and fourth questions were about product usability— "Q.2 Can you set up this device?", "Q.3 Can you measure your blood pressure?", and "Q. 4 Can you dismantle the device and pack it back into the original state? The fifth question was about the device value — "Q.5 What do you think about the value of this device?" Three different elements were measured for understanding and identifying user's Kansei: a) internal emotional and conscious physical responses, b) body language, and c) spoken words. The following different methods were used to measure each element respectively; Kansei sheets, RBL sheet, and user interview.

First and second methods were firstly used together "at the same time". Then, interviews were conducted to identify the reasons behind the positive and negative emotional responses of each user. The collected data of the above different methods were compared and analyzed.

In the second stage of the workshop "Develop", based on the findings of the former step, unique and pleasant ideas were suggested for the new design of the BP device. As for the third stage of the workshop "Present", the best ideas were presented and discussed.

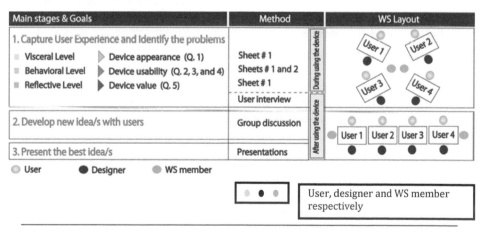

Figure 3   Workshop Stages

3.   RESULTS AND DISCUSSIONS

Using emotions is tricky. As Pieter Desmet points out, "there is no one-to-one relationship between the design of a product and the emotion it elicits. An emotion is not elicited by a product as such, but by the appraised significance of this product for our concerns". The basic fact is well-known, for many products induce strong, but contradictory emotions in different people – some loving it, some intensely disliking it. This means that different products will satisfy different classes of people, or different setting and usages. A colorfully decorated lunch pail would work just as well for children as for distinguished business executives, but the executives might very well judge the pail to be emotionally pleasing and fun for the children while simultaneously viewing it with contempt for themselves. Here, the same product receives different emotional assessment even by the same person when the intended role of the product is changed (Pieter, 2006 & Donald, 2003).

Recently, in Japan, the results of several studies regarding healthcare sector found that some medical devices are difficult to use and have unattractive design. This may lie in the fact that some companies haven't gained sufficient insight into users' needs (Japan External trade Organization, 2006). Therefore, the WS stages focused on redesigning a BP device that is currently used in Japan. In the WS, three stages were carried out as follows:

3.1   Stage 1 of WS: capture users' experiences

In this stage, five questions were asked to each user throughout three evaluation levels; visceral, behavioral, and reflective (Norman, 2003). These levels were created by Donald A. Norman.

First: Visceral Level (superficial first impression):
We asked users one question in order to evaluate the device appearance and aesthetics "aesthetics experience"; "Q.1 is this design of a BP device attractive for you?" In other words, how the product (whether new design or current design) looks, feels and sounds. Each user was asked to use *Kansei* sheet #1 to perform this task.

Second: Behavioral Level (usage of product):
We asked users three questions in order to evaluate product usability and functionality "experience of usability". The questions are; Can you set up this device?", "Q.3 Can you measure your blood pressure?", and "Q. 4 Can you dismantle the device and pack it back into the original state? In this level, each user was asked to use *Kansei* sheets # 1 and 2 to carry out the tasks.

Third: Reflective Level (overall impression / the meaning of things):
Emotional product experience is as dynamic as the human-product interaction itself. Products do not elicit a single emotion but complex emotional episodes. The emotional impact of a product 'as such' differs from the emotional episode elicited by product usage. For designers it is relevant to understand the holistic nature of this emotional episode. We cannot simply break down the interaction into a sequence of actions because emotion in a particular action is determined by all previous actions.

You can compare it to reading a book. The emotions experienced at any point of time are not just elicited by the particular page one is reading, but by all previous pages that have already been read (Pieter, 2006).

Therefore, in this level, we asked users to evaluate a product as a whole, including its value and importance for him or her "experience of meaning". In addition, what does using this product say about you (or user)? This level of evaluation is essential for creating things we want to show off to our friends.
Sheet # 1 was used by each user to achieve this evaluation level.

In first stage of WS, the designers observed the users during their interaction with the device. Designers used the RBL sheet in order to understand the meaning of users' emotional responses. The main results of this stage revealed the following (Figures 4 and 5);

● In terms of observable body language (including outward facial expressions), there are two different types of users' emotional responses; visible and no visible. In case of visible responses, the RBL sheet enabled us to interpret outward facial and body expressions of many participants. However, we found some difficulty in interpreting emotional responses of some participants.

● The comparison that was done between the collected data of *Kansei* sheets and the RBL sheet revealed that the inner emotional responses (heart / mind) and the outer emotional responses (outward facial expressions) of some participants were different. In other words, visible reactions of some participants were not correctly interpreted and were thus misunderstood. The reason is that obvious facial and body expressions of those participants did not precisely reflect their true internal emotional responses.

It can be said that the collected data of *Kansei* sheets helped us to understand the true emotional responses of all participants, especially people who misread their emotional responses and others whose emotions were not visible to us. We could recognize all emotional and physical responses in two differentiated components; qualitative and quantitative. The first component is regarding the identification of the emotional (happiness, anger, etc.) and physical (sweat, headache, etc.) signatures. On the other hand, the second component is regarding the degree of each emotional (such as; I feel this to some extent, I feel this, I feel this very much, and I strongly feel this) and physical signature (such as; I suffer from this to some extent, I suffer from this, I suffer from this very much, and I strongly suffer from this).

According to the above finding, designers need to use several methods together for capturing comprehensive information of user experience. Using only one method or relying on designers' observation may lead to defective results of user experience in user-product interaction.

● Based on the interviews that were carried out with each user, different opinions and comments were collected about the device appearance, usability and value. In addition, the most important problems were identified.

Regarding the device appearance, it was obvious that the appearance of a device was more acceptable among males than females. The results of the interviews with the participants showed that the device appearance has many problems, such as; the belt color is unpleasant, the form of the device is not attractive, and the size of the device is big. As for the problems of the device usability, female participants had

432

more negative feelings than male participants during their use of a device. The results of the behavioral evaluation level revealed that there are several problems related to the device usability, such as; a) the belt material is not soft, b) the color of the belt is dark, so that it gets dirty very easy and is difficult to clean, c) the start button is not recognized easily, and d) the users should do many steps in order to measure their blood pressure.

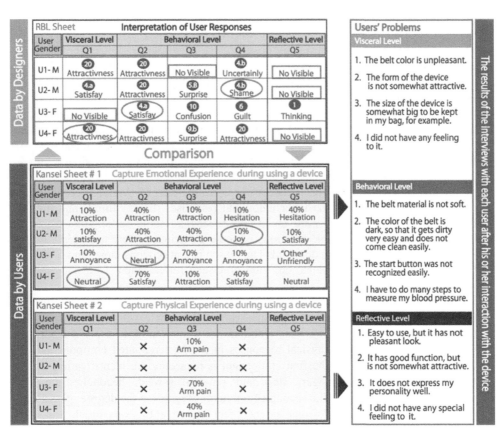

Figure 4   The results of Kansei sheets and RBL sheet

Regarding the users' opinions on the device value, four points we got from the users, include: the device is easy to use, but it does not have pleasant appearance. It has good function, but it is not somewhat attractive. It does not express the user personality well. The user did not have any special feeling about it.

In the end of the first stage, we found that most of the users believe that *Kansei* sheets are easy to use and understand. In addition, the new methods are assistance, not a restriction, and therefore allow rich and amiable discussions.

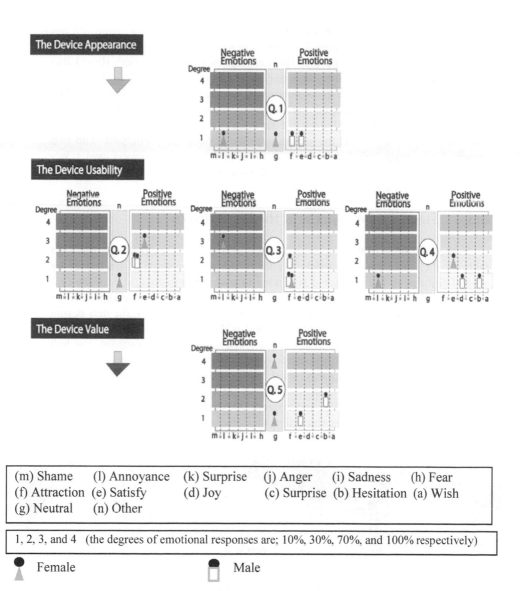

Figure 5   Analysis of the users' answers

3.2   Stage 2 of WS: develop new idea/s with users

The consideration of people who are going to be benefit from design during design processes is not a new concept in design practice - many humanitarian designers have emphasized this relationship. However, these relationships between designers and design users seem restricted to a quantitative approach based on measuring people's bodies and analyzing the usability of designs in relationship

434

with people's abilities or disabilities. Gradually, this 'designing for' approach has been challenged. Recently, there is a new democratic design development, which encourages designing 'with' people. This indicates that design practices should also consider people's emotions rather than only their capabilities to use design.

In the workshop, we tried to consider users' emotions and physical abilities. The results of the first stage detected that two issues should be regarded for designing appropriate and pleasant a BP device. First issue is related to the belt material and color, and the device size is the second one. We divided the participants into two groups. Each group included 5 people as follows; 2 users, 2 designers, and 1 WS member. Different ideas were proposed with users in order to develop the design of a BP, including its size and belt (Figure 6).

3.3 Stage 3 of WS: present the best idea/s

In the third stage of the workshop, the best idea/ s were presented and discussed with all the workshop participants. Basically the best proposal characterizes by several features as follows; small size, easy to use, and attractive. As for the first and second features, the size of the new design is small enough to be kept it in the users' bag, so that it can be used easily in everywhere. In addition, the new design can be operated easily because it is provided with clear and enough instructions. Regarding the third feature, the new design of the BP device considers the belt quality and color. It was selected carefully in order to match the users' tastes and preferences. Overall, this study tried to emphasis on the importance of understanding users' emotions and experiences throughout the design process. The intention is to design pleasant and appropriate design that fits their emotional, physical needs and life styles (including their life, homes, and the context within which they operate).

Figure 6   Developing and Presenting Different Ideas

4   CONCLUSIONS

The present study aimed to introduce two emotional evaluation methods. The two methods are called; Kansei sheets and Read Body Language (RBL) sheet. These novel methods were proposed to examine and interpret users' emotions, while their interaction with the products. The results of this study revealed the following; 1) both "Kansei Sheets and RBL sheet" are effective ways to understand users' experiences (UX) and needs. The users find expressing their emotions using Kansei sheets pleasant tasks, in contrast to the difficulty of expressing their emotional reactions in words. Furthermore, we found that using the two methods are convenient to create an informal atmosphere in which the users feel free to discuss experiential aspects of a product.

2) For capturing "UX" successfully, "triangulation" of methods is important to be used. In other words, designers should pay a close attention to collect information in different ways and comparing the results. The results of this study reveal that however "direct observation" is an appropriate method to identify the user's problems, and also to break the ice between user and design, it is not often an effective way to understand user emotions and experiences.

REFERENCES

Cynthia, M., and Martinovich, L. (2002). What you Speaks volumes- How Body Language can be used to understand others, Michigan Bar Journal, 36-39.

Guillaume, C., Julien, K., Didier G., and Thierry, P. (2006). Emotion Assessment: Arousal Evaluation Using EEG's and Peripheral Physiological Signals. Springer - Verlag Berlin Heidelberg, MRCS 2006, 530-537.

Japan External trade Organization JETRO (2009). Attractive Sectors. Medical Care, Invest Japan Division, Invest Japan Department.

Eva, L., Muriel, G. (2007). Ten Emotion Heuristics: Guidelines for Assessing the User's Affective Dimension Easily and Cost-effectively. Proceeding of the 21st BCS HCI Group Conference, Lancaster University, the British Computer Society, Vol. 2.

Norman, D.A. (2003). Measuring Emotion. *The Design Journal*, Vol. 6, issue 2.

Norman, D.A. (2003). Attractive Things Work Better. New York, NY: Basic Books.

Pieter, D. (2006). Design & Emotion: The Emotional Experience of Product, Services, and Brands: available at:

http://www.design-emotion.com/2006/11/05/getting-emotional-with-dr-pieter-desmet/

Tingfan Wu, Nicholas J. Butko, Paul Ruvulo, Marian S. Bartlett, and Javier R. Movellan (2009). Learning to make facial expressions, IEEE 8TH International Conference on Development and Learning.

CHAPTER 48

Hemispheric Asymmetries in the Perception of Emotions

Sol I. Lim, Sangwoo Bahn, Jin C. Woo, Chang S. Nam

North Carolina State University
Raleigh, NC, USA
silim@ncsu.edu, panlot@gmail.com, jwoo3@ncsu.edu, csnam@ncsu.edu

ABSTRACT

The purpose of this study is investigating differential hemispheric activations in the perception of emotion. Electroencephalography (EEG) data were recorded while subjects viewed pictures of three types of emotional contents (negatively arousing, calm, and positively arousing) chosen from the International Affective Picture System (IAPS). The results showed that in the upper alpha (10~12 Hz) frequency band subjects showed greater activation of the left hemisphere than the right hemisphere in response to the positively arousing pictures, whereas negative emotional pictures resulted in greater activation of the right hemisphere relative to the left. The other frequency bands, including lower alpha (8~10 Hz) and gamma (30~45 Hz), did not show any significant interaction effects between emotion type and hemispheric activation. These findings are of practical importance in neuroadaptive system design whose functional characteristics of the system change in response to meaningful variations in the user's emotional states measured by brain signals.

Keywords: Emotion, EEG, Hemisphere asymmetry, IAPS, SAM

1   INTRODUCTION

Emotion is a complex set of interactions among subjective and objective factors, mediated by neural systems, which can give rise to affective experiences such as feeling of arousal, pleasure and displeasure (Kleinginna and Kleinginna, 1981). The investigation of brain asymmetries for cognitive functioning has shown that there are hemispheric differences in the perception of emotion. In general, the right hemisphere is more involved in negative affective processing while the left hemisphere is mainly responsible for positive emotional processing (Tucker, 1981; Silberman and Weingartner, 1986). Waldstein et al. (2000) showed that happiness-inducing tasks evoked more prominent left frontal EEG activation than the right, while greater right frontal EEG activation was induced by anger-inducing tasks. Davidson et al. (1990) also showed that disgust is associated with less alpha power (i.e., more activation) in the right frontal areas than is happiness, while happiness is associated with less alpha power in the left frontal areas than is disgust.

A guiding assumption underlying the interpretation offindings involving frontal EEG alpha asymmetry is that greater alpha power is indicative of less cortical activity in broad underlying regions. The activity within the alpha range is inversely related to underlying cortical processing, since decreased in alpha tend to be observed when underlying cortical systems engage in active processing (Allen, Coan & Nazarian, 2004). Cortical alpha power is strongly and inversely related to glucose metabolism in the thalamus (Larson et al., 1998), so the thalamic activity in response to sensory or cortical input will disrupt alpha rhythmicity. These results support the hypothesis that the right hemisphere is specialized for the experience of certain negative affects, whereas the left hemisphere is responsible for certain positive emotions. Davidson (1992) explained these findings with the role of the cerebral hemispheres in emotional processing. Approach and withdrawal are funda-mental motivational dimensions which may be found at any level of phylogeny where behavior itself is present. Davidson related this theory to emotion processing and suggested that the anterior regions of the left and right hemispheres are specialized for approach and withdrawal process, respectively. During the experi-mental arousal of withdrawal-related emotional states (e.g., fear and disgust), the right frontal and anterior temporal regions are selectively activated. However, Muller et al. (1999) showed that 30±50 Hz gamma band had relatively more power for negative valence over the left temporal region as compared to the right and a laterality shift towards the right hemisphere for positive valence, which is contrary to the previous research. Muller indicated that cortical activity in the gamma band is related to visual and language information processing, and possibly to feature integration.

Taking these findings together, it is of significance to note that brain asymmetries for emotional processing still remain unclear. The main goal of this study is to systematically investigate hemispheric asymmetries between different emotional types.

2   METHODS

2.1   Participants

Eight healthy right-handed male participants (mean age 24.6 ± 4.3 years) were re-cruited from the student population of a local university in return for extra course credit.

2.2 Data Acquisition

Brain signal was recorded using an EEG cap (g.tec Medical Engineering) embedded with 16 electrodes covering Fp1, Fp2, F3, F4, F7, F8, C3, C4, T7, T8, T9, T10, P5, P6, P7, and P8, based on the modified 10 -20 system of the International Federation and using Fpz as reference. Figure 1 depicts the electrode montage used in the study. Each electrode placement was chosen carefully according to their related position of Brodmann Area (BA). BA is a region of the cerebral cortex defined based on its cytoarchitectonics, or structure and organization of cells. Many of the areas Brodmann defined based solely on their neuronal organization have since been correlated closely to diverse cortical functions. In this research, BAs with emotion-related functions were chosen and then its corresponding electrodes were matched. The signal was amplified with a g.USBamp amplifier (g.tec Medical Engineering). The EEG was bandpassfiltered 0.1 –60 Hz and digitized at a rate of 256 Hz. Data collection and signal processing were all conducted by the WolkPack BCI system, developed at the North Carolina State University.

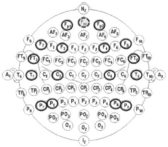

Figure 1 The electrode montage used in the present study.

2.3 Emotional Stimulation

A total of 150 colored pictures were selected from the International Affective Picture System (Lang, Bradley & Cuthbert, 1999) and divided into three groups according to their valence and arousal levels. All pictures in the IAPS excepting particular images (for example erotic images) were clustered into three groups and each 50 pictures were selected based on their proximity to the center of each cluster: 50 pictures of negatively arousing images (valence: $M = 2.43$, $SD = 0.47$, arousal: $M = 6.21$, $SD = 0.48$), 50 pictures of calm images (valence: $M = 5.00$, $SD = 0.35$, arousal: $M = 2.80$, $SD = 0.41$), and 50 pictures of positively arousing images (valence: $M = 7.00$, $SD = 0.60$, arousal: $M = 6.22$, $SD = 0.59$). Figure 2 shows the data points of three groups of images and their sample images. Their ratings were significantly different from each other, $F(2,49) = 2565.63$, $p < .0001$ and $F(2,49)=1307.45$, $p < .0001$ for valence and arousal, respectively. The IAPS numbers of the pictures in the three groups are given in Appendix. All pictures were presented in the center of a 19-inch computer screen with a frame rate of 75 Hz. The screen was placed 0.5 m in front of the viewer, resulting in a picture presentation with a visual angle of 15 degrees horizontally and 11 degrees vertically. Each picture was presented for 3000 ms.

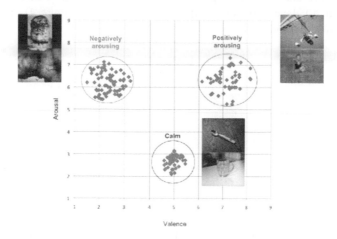

Figure 2 Groups of negatively arousing, calm, and positively arousing images based on their arousal and valence level from IAPS

2.4   Procedure

To systematically investigate hemispheric asymmetries in the perception of emotions, this study used two types of measures: (1) two dimensions of emotion, valence and arousal, using Self Assessment Manikin (SAM), and (2) electroencephalographic (EEG) signals from the brain. The procedure of experiment is described in Figure 3 in detail. Each subject was asked to complete Edinburgh Handedness Inventory (Oldfield, 1971) for handedness qualification before starting the experiment. Subjects were asked not to drink coffee, tea or other caffeine-containing beverage on the day of the experiment. For brain signal recording, the subjects wore an EEG cap, which contains 16 electrodes. After attaching the electrodes, brain signals were recorded for 2 minutes in eyes-open resting conditions. Only last 1 minute of the signals were used to calculate the subjects' baseline for each of the 16 electrodes. In each trial, the rest period differed based on the subjects' brain signals. Our assumption was that the subject's brain signals would be affected by the previous trial's emotional contents. From pilot test, we found that over 95% of the subjects' brain signals returned to their baseline after watching the emotional stimulus within 10 seconds. If their brain signals keep falling into these baseline ranges for 10 seconds, next trial automatically begins. Otherwise, the system waits for the subjects' signals to become stable for up to 20 seconds. To set up a range of baselines, we used the amplitude interval of $mean \pm 2\sigma$.

During the experiment, subjects viewed 150 pictures, resulting in 2 sessions with 15 trials each. Each picture was presented for 3 seconds. In all trials, a set of five images from the same affective category was presented. After one trial, subjects were asked to rate the respective picture on three categories, affective valence and arousal through the SAM (Lang, Bradley & Cuthbert, 2008). The SAM consists of 2 scales, valence and arousal, ranging from 1 to 9 with higher values representing more positive valence (pleasureness) and higher arousal (Figure 4). The order of trials was randomized.

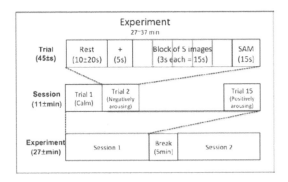

Figure 3 Procedure of experiment (Rest period: 10 seconds ± 20 seconds based on the subject's brain signals, cross fixation: 5 seconds for concentration, Block of 5 images: 3 seconds each for 5 group of images, SAM: 15 seconds for ratings)

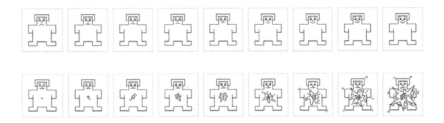

Figure 4 Self Assessment Manikin (SAM) – Above scale for valence, below scale for arousal

2.5　Data Reduction and Analysis

For EEG analysis, single epochs of 3,000 ms length were extracted. These epochs were submitted to preprocessing state for artifact rejection using methods of Common Average Reference (CAR) and detrending in sequence. CAR subtracts the common activity in the brain to the position of interest (Bertrand, Perrin & Pernier, 1985). The main idea is to remove the averaged brain activity which can be seen as EEG noise. The formula to calculate CAR is

$$V_i^{CAR} = V_i^{ER} - \frac{1}{n}\sum_{j=1}^{n} V_i^{ER}.$$

Detrending is a mathematical tool for removing trends from the data. It is very efficiently used in preprocessing to prepare data for analysis by methods that assume stationarity. Trends can be defined as slow and gradual change in the statistical property of the process under the whole interval under investigation (Garg and Binderup, 2007). Detrending has the same effect on the frequency spectrum as that of a high pass filter, which means the variance at the low frequencies is diminished in comparison to the variance at high frequencies.

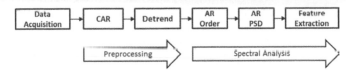

Figure 5 Procedure of signal processing

After the preprocessing stage, spectral analysis was done in the sequence of Auto Regression (AR) order, Auto Regression Power Spectral Density (AR PSD), and feature extraction. For spectral analysis, AR is preferred over Fast Fourier Transform (FFT) for the following reasons: 1) AR modeling provides a better frequency resolution (Polak and Kostov, 1998) and 2) We can obtain good spectral even with short EEG segments. Signals were then extracted in the lower alpha (8~10 Hz), the upper alpha (10~12 Hz), and gamma (30~45 Hz) frequency ranges.

3   RESULTS

3.1   SAM-ratings

The valence and arousal SAM-ratings for all three groups of images across subjects were summarized in Figure 6. The statistical analysis showed a highly significant effect for valence ($F(2,49) = 519.41$, $p < .0001$) and arousal ($F(2,49) = 80.89$, $p < .0001$).

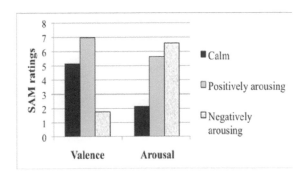

Figure 6 Mean SAM ratings of arousal and valence across subjects

3.2 Electrophysiological recordings

There were three within-subjects factors manipulated in the study: emotion type (3 levels), hemisphere (2 levels), and frequency (3 levels). Analyses of variance (ANOVAs) were performed on mean baseline corrected Power Spectral Density (PSD). The analysis showed significant main effects of emotion type, $F(2, 38390) = 34.73$, $p < .0001$, hemisphere, $F(1, 38390) = 863.24$, $p < .0001$, and frequency, $F(1, 38390) = 2476.68$, $p < .0001$. All two-way interactions were significant, emotion type x hemisphere [$F(2, 38390) = 4.61$, $p < .0099$], emotion type x frequency [$F(2, 38390) = 10.58$, $p < .0001$], and hemisphere x frequency [$F(1, 38390) = 252.09$, $p < .0001$]. Post-hoc ANOVAs for each frequency band showed that significant main effects for emotion type and hemisphere in both lower and upper alpha bands, and interaction effect only in the upper alpha band (Table 1).

Table 1 A summary of significant effects for emotional response

Effect	DF	Lower Alpha		Upper Alpha		Gamma	
		F value	p value	F value	p value	F value	p value
Emotion type	2	42.59	<.0001	23.81	<.0001	0.97	0.378
Hemisphere	1	519.68	<.0001	1103.4	<.0001	0.09	0.764
Emotion type x Hemisphere	2	2.84	0.391	3.82	0.022	0.25	0.778

Consistent with previous literature (Tucker, 1981; Silberman and Weingartner, 1986; Allen, Coan & Nazarian, 2004; Davidson, 1988), positively arousing picture elicited less relative left upper alpha EEG, which is indicative of higher cortical activity in broad underlying regions, whereas negatively arousing pictures elicited less relative right upper alpha EEG (Figure 7).

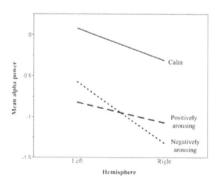

Figure 7 Mean baseline corrected upper alpha power (µV²/Hz) for the left and right hemisphere during calm, negatively arousing, and positively arousing picture conditions (More negative numbers indicate less alpha power. Lower numbers are associated with increased activation).

4   DISCUSSION AND CONCLUSION

We found that the upper alpha band measured at each hemisphere distinguished the emotion types. Subjects exhibited greater relative left EEG activation during the presentation of positively arousing pictures and greater relative right EEG activation during the presentation of negatively arousing pictures. These findings are consistent with the emotion valence models discussed by Tucker, Silberman and Weingartner, and Davidson et al. (see Tucker, 1981; Silberman and Weingartner, 1986; Davidson et al., 1990). We also noted that the upper alpha is the only frequency band that showed emotion type by hemisphere interaction effects. Doppelmayr et al. (1998) pointed out that conventionally subdivided frequency bands (i.e. alpha (8~13 Hz)) blur specific relationships between cognitive performance and power measurements. We subdivided the alpha frequency range into the lower alpha (8~10 Hz) and upper alpha (10~12 Hz), and found the result that the upper alpha is a better indicator for emotion type by hemisphere relationship. This result is supported by Klimesch (1999)'s findings with respect to the alpha frequency range. Klimesch found that desynchronization in the lower alpha (6~10 Hz) band reflects attention and desynchronization in the upper alpha band (10~12 Hz) reflects semantic memory performance. When emotionally aroused, we form semantic and episodic memories about such situations (LeDoux, 2007). Since the upper alpha waves are present during semantic memory performance, which is formed by emotional arousal, we could conclude that the upper alpha is a better indicator for emotion type by hemisphere interaction.

Findings from this research are of practical importance in neuroadaptive system design whose functional characteristics of the system change in response to meaningful variations in the user's emotional states measured by brain signals. To build such adaptive systems, identification of biological or psychophysiological signals that reliably index the emotional state of the user is needed (Hettinger et al, 2010). Further studies will also be needed to highlight individual differences in processing emotions that are congruent in emotional tones with either current mood states or stable personality traits.

APPENDIX

The number of IAPS images used in the present study.

Negatively arousing pictures: 9414, 9635.1, 9413, 9183, 3110, 3059, 3068, 3150, 9410, 9252, 9910, 6540, 9902, 3001, 6520, 3266, 2352.2, 3071, 9570, 3120, 2703, 3062, 6560, 6022, 3100, 3130, 3350, 9412, 3069, 3016, 3140, 3030, 9920, 3010, 6570, 2811, 9903, 3000, 6415, 9254, 9560, 9520, 3530, 3170, 9500, 3000, 9810, 3261, 3060, 3015

Calm pictures: 2038, 2200, 2214, 2320, 2393, 2411, 2495, 2570, 2580, 2840, 2850, 2870, 2880, 5120, 5390, 5520, 5530, 5731, 5740, 7000, 7002, 7003, 7004, 7009, 7020, 7026, 7030, 7032, 7034, 7035, 7038, 7041, 7050, 7055, 7059, 7060, 7090, 7161, 7179, 7185, 7187, 7205, 7217, 7233, 7235, 7491, 7705, 7950, 8465, 9700

Positively arousing pictures: 8080, 1650, 8341, 8490, 2300, 5460, 5470, 5600, 7405, 8178, 8180, 7570, 5621, 8260, 8163, 5450, 8251, 8191, 8186, 8499, 8370, 5629, 8179, 8040, 1720, 8090, 8380, 8206, 8158, 8185, 7499, 8170, 8501, 8190, 8470, 8400, 5700, 5260, 8492, 8340, 8200, 2030, 5660, 5910, 8300, 8030, 5982, 5833, 2034, 6910

REFERENCES

Allen, J.J.B., Coan, J.A. and Nazarian, M. 2004. Issues and assumptions on the road from raw signals to metrics of frontal EEG asymmetry in emotion. *Biological psychology* 67: 183-218.

Bertrand, O., Perrin, F. and Pernier, J. 1985. A theoretical justification of the average reference in topographic evoked potential studies. *Electroencephalography and Clinical Neurophysiology/Evoked Potentials Section* 62: 462-464.

Davidson, R.J. 1992. Anterior cerebral asymmetry and the nature of emotion. *Brain and cognition* 20: 125-151.

Davidson, R.J. 1988. EEG measures of cerebral asymmetry: Conceptual and methodological issues. *International Journal of Neuroscience* 39: 71-89.

Davidson, R.J., Chapman, J.P., Chapman, L.J. and Henriques, J.B. 1990. Asymmetrical Brain Electrical Activity Discriminates Between Psychometrically-Matched Verbal and Spatial Cognitive Tasks. *Psychophysiology* 27: 528-543.

Davidson, R.J. and Fox, N.A. 1982. Asymmetrical brain activity discriminates between positive and negative affective stimuli in human infants. *Science* 218: 1235.

Doppelmayr, M., Klimesch, W., Pachinger, T. and Ripper, B. 1998. Individual differences in brain dynamics: important implications for the calculation of event-related band power. *Biological cybernetics* 79: 49-57.

Garg, A. and Binderup, A. 2007. Implementation of Online Brain Computer Interface Using Motor Imagery In LabVIEW.

Goldstein, J.M., Seidman, L.J., Horton, N.J., Makris, N., Kennedy, D.N., Caviness, V.S. Jr, Faraone, S.V., and Tsuang, M.T. 2001. Normal sexual dimorphism of the adult human brain assessed by 571 in vivo magnetic resonance imaging. *Cereb Cortex* 11: 490-497.

Hettinger, L.J., Branco, P., Encarnacao, L.M. and Bonato, P. 2003. Neuroadaptive technologies: applying neuroergonomics to the design of advanced interfaces. *Theoretical Issues in Ergonomics Science* 4: 220-237.

Jones, N.A. and Fox, N.A. 1992. Electroencephalogram asymmetry during emotionally evocative films and its relation to positive and negative affectivity. *Brain and cognition* 20: 280-299.

Kleinginna, P.R. and Kleinginna, A.M. 1981. A categorized list of emotion definitions, with suggestions for a consensual definition. *Motivation and Emotion* 5: 345-379.

Klimesch, W. 1999. EEG alpha and theta oscillations reflect cognitive and memory performance: a review and analysis. *Brain Research Reviews*, 29: 169-195.

Lang, P., Bradley, M. and Cuthbert, B. 1999. International affective picture system (IAPS): Technical manual and affective ratings.

Lang, P., Bradley, M. and Cuthbert, B. Gainesville, FL: University of Florida; 2008. International affective picture system (IAPS): affective ratings of pictures and instruction manual.

Larson, C.L., Davidson, R.J., Abercrombie, H.C., Ward, R.T., Schaefer, S.M., Jackson, D.C., Holden, J.E. and Perlman, S.B. 1998. Relations between PET-derived measures of thalamic glucose metabolism and EEG alpha power. *Psychophysiology*, 35: 162-169.

Müller, M.M., Keil, A., Gruber, T. and Elbert, T. 1999. Processing of affective pictures modulates right-hemispheric gamma band EEG activity. *Clinical Neurophysiology*, 110: 1913-1920.

Nishizawa, S., Benkelfat, C., Young, S.N., Leyton, M., Mzengeza, S., de Montigny, C., Blier, P., and Diksic, M. 1997. Differences between males and females in rates of serotonin synthesis in human brain. *Proc Natl Acad Sci U S A*, 94: 5308-5313.

Oldfield, R.C. 1971. The assessment and analysis of handedness: the Edinburgh inventory. *Neuropsychologia*, 9: 97-113.

Polak, M. and Kostov, A. 1998. Feature extraction in development of brain-computer interface: a case study. *Proceedings of the 20th Annual International Conference of the IEEE*, 2058.

Silberman, E.K. and Weingartner, H. 1986. Hemispheric lateralization of functions related to emotion. *Brain and cognition*, 5: 322-353.

Tucker, D.M. 1981. Lateral brain function, emotion, and conceptualization. *Psychological bulletin*, 89: 19.

Waldstein, S.R., Kop, W.J., Schmidt, L.A., Haufler, A.J., Krantz, D.S., and Fox, N.A. 2000. Frontal electrocortical and cardiovascular reactivity during happiness and anger. *Biological psychology*, 55: 3-23.

Wrase, J., Klein, S., Gruesser, S.M., Hermann, D., Flor, H., Mann, K., Braus, D.F. and Heinz, A. 2003. Gender differences in the processing of standardized emotional visual stimuli in humans: a functional magnetic resonance imaging study. *Neuroscience letters*, 348: 41-45.

CHAPTER 49

Understanding Differences in Enjoyment: Playing Games with Human or AI Team-mates

Kevin McGee, Tim Merritt, & Christopher Ong

National University of Singapore
Singapore
mckevin@nus.edu.sg

ABSTRACT

Increasingly, humans are interacting not only with other humans, but also with artificial agents – and there is a growing effort to develop artificial team-mates for different team-based activities. Related work comparing human-human and human-computer conversational interactions, competitive interactions, and cooperative interactions shows that there are significant differences between the way humans feel about, treat, and behave when interacting with humans – compared to the way they feel about and act towards/with artificial agents. One issue that has not been sufficiently studied in the related work is whether there are differences in player *enjoyment* when playing a team-mate game as a member of a human-human team versus a human-agent team – and if so, how to explain such differences. A number of empirical comparative studies were conducted in which participants played team-mate games with either an artificial team-mate – or with a human team-mate who was connected over the network from another location. In fact, unknown to the participants, the "human" team-mate was the same as the artificial team-mate, so gameplay behavior was constant under all conditions. Participants consistently made claims about increased enjoyment when the team-mates was perceived to be human. This paper summarizes these results and proposes an explanation in terms of players' inability to imagine artificial agents are capable of certain states or behaviors.

Keywords: Team-mate games, multiplayer games, team-mates, enjoyment

1   INTRODUCTION

There is growing effort to develop artificial team-mates for non-game purposes (Van Diggelen, Muller, & Van Den Bosch, 2010, Doherty, 2003, Babu et al.., 2009), for team-based games, e.g., sports games, team-based shooters, *RoboCup* (McGee & Abraham, 2010), and for arts/performance ensembles (Baird, Blevins, & Zahler, 1993, Mizutani, Igarashi, Suzuki, Ikeda, & Shio, 2010). As there is increasing potential for teams that consist of a mixture of human and artificial team-mates, one question that arises is whether (and how) human team-members will treat their human team-mates compared to how they will treat their artificial team-mates under the same conditions.

Although there is a considerable body of work in the *CASA* ("Computers Are Social Actors") tradition (Reeves & Nass, 1996) that suggests that "the social rules and dynamics guiding human-human interaction apply to human-computer interaction" (Nass, Fogg, & Moon, 1996), most of this research involved showing that when people interact with computers under different conditions they will treat each computer in a different way that parallels the way they would treat different people.

The work that has examined similarities and differences between the way humans feel about, treat, and behave when interacting with humans – compared to the way they feel about and act towards/with artificial agents – falls broadly into studies about interactions with conversational agents (Cassell, 2000), with competitive game agents (Gajadhar, Kort, & IJsselsteijn, 2008), and with cooperative agents (Kiesler, Sproull, & Waters, 1996). Overall, this research shows that there are significant differences between the way humans feel about, treat, and behave when interacting with humans – compared to the way they feel about and act towards/with artificial agents.

There has been some research that has examined differences in enjoyment when competing against computer players or against human players, showing that human competitors are preferred (Ravaja ct al.., 2006); evoke higher levels of enjoyment due to a greater sense of social presence (Weibel, Wissmath, Habegger, Steiner, & Groner, 2008) or increased possibilities for communication (Gajadhar et al.., 2008); and lead to greater emotional response (Mandryk, Inkpen, & Calvert, 2006). In terms of work on enjoyment in the context of team-mates and cooperative games, one study in which participants engaged in inventory exchanges in a modified version of *World of Warcraft* measured physiological arousal during the session and then asked the participants to rate their emotional valence, but the findings did not show any significant differences of enjoyment (Lim & Reeves, 2010).

2. STUDYING DIFFERENCES IN ENJOYMENT

One issue that has not been sufficiently studied in the related work is whether there are differences in player enjoyment when playing a cooperative team-mate game as a member of a human-human team – compared to playing as a member of a

448

human-agent team. The one study that looked at enjoyment in a cooperative game made use of a fairly structured scenario in which participants merely negotiated the exchange of inventory items.

In order to explore this issue in a real-time context, we made use of the cooperative, real-time two-player game Capture the Gunner (CTG) (Abraham & McGee, 2010) to do a quantitative study with 40 participants (for full details of the study, see (Merritt, McGee, Chuah, & Ong, 2011)).

Figure 1: Capture the Gunner game elements: a) human-controlled avatar b) computer-controlled agent c) gunner d) gunner's field of view (FOV)

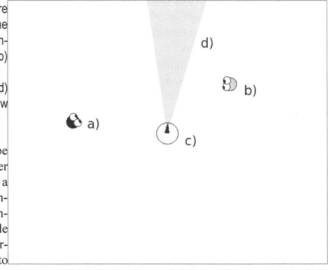

CTG can be played with either a human or with a computer team-mate. Team members must evade bullets and cooperate in order to "capture" (touch) the gunner which is rotating and firing within its "field of view" from the middle of the game space. At each level, both players must touch the gunner (though, not necessarily at the same time); once this occurs, the game proceeds to the next level, with the gunner rotating faster.

The game allows (but does not require) players to perform risky actions that benefit their team-mates – specifically, player's can "draw gunfire" towards themselves (and away from their team-mates).

During the study, all participants were familiarized with the game and then played the game twice: once with an AI team-mate and once with a "presumed" human team-mate (i.e., an AI team-mate that they believed was a human team-mate). Thus, the team-mate performance and behaviors were identical for both cases – and in both cases, the team-mate "drew gunfire" an equal amount of the time.

The participant response to the game was measured in two different ways: self-reporting and game logs. They were asked to rate their subjective experience through simple Likert scale questionnaires (e.g., "How much did you enjoy the game session? " or "Considering both sessions, which team-mate would you choose? (Select one: Computer or Human)").

The main result of this study is that participants overwhelmingly chose the PH ("presumed human") team-mate over the AI team-mate even though the team-mates were in fact the same. Specifically, the participants reported significantly higher levels of enjoyment and cooperation during the game sessions with the PH team-mate.

3   DISCUSSION

The results indicate that, even though the team-mate behaviors were equivalent (i.e., the identical AI team-mate for both conditions), the *belief* in the team-mate's identity influenced the subjective experience of the game.

It is important to emphasize that these differences in perception and subjective experience are based on cooperative game sessions in which the participants *did not communicate with their team-mates*. This is similar to the results of another study we conducted in which participants played the real-time, cooperative game of *Dearth* (Singapore-MIT GAMBIT Game Lab, 2009) and we explored communication between players (McGee, Merritt, & Ong, 2011).

In that study, 40 participants played over a network (i.e., they were not in the same location) with either no communication or with microphone/headsets. The significant finding of that study was that although participants consistently claimed that the version with the communication channel made the game more fun, on average, players did not actually say more to each other when they had the means to do so. In fact, partly because of the real-time nature of the game, players did not say very much under either condition.

In that research, we hypothesized that the desire for (and use of) communication channels – *even in purely goal-oriented games of coordinated skill* – might actually be a desire for (and satisfaction derived from) social companionship of a very specific kind: experiencing the *presence* of (or "being with") fellow players.

Further evidence about the desire for companionship from team-mates comes indirectly from the results of a thought experiment the participants were asked to perform. They were given 12 different options for playing variations of the game and asked to order them from favorite to least favorite. One of the parameters was whether there was a spectator present. The game option that involved playing with a computer team-mate using full communication was the second to least favorite; this is not surprising as most AI team-mate did not respond to commands and was otherwise mute. However, by adding a silent human *spectator* to the configuration of human and computer team-mate with full communication was the 3rd most popular option. In other words, the mere presence of a silent spectator shifted a configuration from near the bottom to near the top.

These results are similar in some ways to the related work that has shown differences in enjoyment of participants when interacting with humans versus interacting with computer agents.

4 CONCLUSION

There is mounting evidence that people treat computer agents in ways that are quite different from the ways they treat other people. This in itself is not surprising.

When we think of the richness of human interaction it is difficult to imagine anything like that richness in our interactions with software or hardware (robotic) agents.

But stopping there is a mistake on two counts.

First, some of the research constrains users such that all their interactions are impoverished. In other words, the "richness" that is evident to participants when interacting with human team-mates seems at least partly a richness that the *participants* bring to the experience.

Second, there is something telling in the phrase that "it is difficult to imagine" such richness being present when interacting with computer agents. In fact, in many of our studies participants consistently indicate that they simply cannot imagine that an AI team-mate could be flexible, could pro-actively move to protect them or "draw fire" on their behalf, and so on. And many of these participants are experienced gamers. In other words, this inability to imagine is not due to limited exposure to AI in computer games.

In some ways, these two points are two aspects of the same issue: partcipants are able to project richness onto some interactions and unable to project it onto others. One area of future research will be to explore whether there are design changes that can increase the richness of experience with both human and artificial team-mates.

ACKNOWLEDGMENTS

Portions of this work were funded under a Singapore-MIT GAMBIT Game Lab research grant, "Designing Adaptive Team-mates for Games" as well as a Singapore Ministry of Education Academic Research Fund grant, "Understanding Interactivity", NUS AcRF Grant R-124-000-024-112.

REFERENCES

Abraham, A. T., & McGee, K. (2010, Aug). AI for dynamic team-mate adaptation in games. In *Proceedings of the 2010 IEEE Conference on Computational Intelligence and Games* (pp. 419–426).

Babu, S., Grechkin, T., Chihak, B., Ziemer, C., Kearney, J., Cremer, J., . (2009, Mar). A virtual peer for investigating social influences on children's bicycling. In *Virtual Reality Conference, 2009. VR 2009* (pp. 91–98).

Baird, B., Blevins, D., & Zahler, N. (1993). Artificial intelligence and music: Implementing an interactive computer performer. *Computer Music Journal, 17*(2), 73–79.

Cassell, J. (2000, Apr). Embodied conversational interface agents. *Commun. ACM, 43*(4), 70–78.

Doherty, S. M. (2003). Human-centered design in synthetic teammates for aviation: The challenge for artificial intelligence. In I. Russell & S. M. Haller (Eds.), *Proceedings of the Sixteenth International Florida Artificial Intelligence Research Society Conference, May 12-14, 2003, St. Augustine, Florida, USA* (p. 54-56). AAAI Press.

Gajadhar, B., Kort, Y. de, & IJsselsteijn, W. (2008). Influence of social setting on player experience of digital games. In *CHI '08 Extended Abstracts on Human factors in Computing Systems* (pp. 3099–3104). ACM.

Kiesler, S., Sproull, L., & Waters, K. (1996, Jan). A prisoner's dilemma experiment on cooperation with people and human-like computers. *Journal of personality and social psychology*, *70*(1), 47–65.

Lim, S., & Reeves, B. (2010, January). Computer agents versus avatars: Responses to interactive game characters controlled by a computer or other player. *International Journal of Human-Computer Studies*, *68*(1-2), 57–68.

Mandryk, R. L., Inkpen, K. M., & Calvert, T. W. (2006, April). Using psychophysiological techniques to measure user experience with entertainment technologies. *Behaviour & Information Technology*, *25*(2), 141–158.

McGee, K., & Abraham, A. T. (2010). Team-mate AI in games: A definition, survey & critique. In *FDG 2010: The 5th International ACM Conference on the Foundation of Digital Games. 19-21 June, 2010, Monterey, USA*.

McGee, K., Merritt, T., & Ong, C. (2011, 28 Nov – 2 Dec). What we have here is a failure of companionship: communication in goal-oriented team-mate games. In *Proceedings of the 2011 Annual Conference of the Australian Computer-Human Interaction Special Interest Group (CHISIG) of the Human Factors and Ergonomics Society* (pp. 198–201).

Merritt, T. R., McGee, K., Chuah, T. L., & Ong, C. (2011). Choosing human team-mates: perceived identity as a moderator of player preference and enjoyment. In *Proceedings of the 2011 Foundations of Digital Games Conference*.

Mizutani, T., Igarashi, S., Suzuki, T., Ikeda, Y., & Shio, M. (2010). A realtime human-computer ensemble system: Formal representation and experiments for expressive performance. In F. Wang, H. Deng, Y. Gao, & J. Lei (Eds.), *Artificial intelligence and computational intelligence* (Vol. 6319, p. 256-265). Springer Berlin / Heidelberg.

Nass, C., Fogg, B. J., & Moon, Y. (1996, December). Can computers be teammates? *International Journal of Human-Computer Studies*, *45*(6), 669–678.

Ravaja, N., Saari, T., Turpeinen, M., Laarni, J., Salminen, M., & Kivikangas, M. (2006, August). Spatial Presence and Emotions during Video Game Playing: Does It Matter with Whom You Play? *Presence: Teleoper. Virtual Environ.*, *15*, 381–392.

Reeves, B., & Nass, C. (1996). *The Media Equation: How People Treat Computers, Television, and New Media Like Real People and Places*. Cambridge University Press.

Singapore-MIT GAMBIT Game Lab. (2009). *Dearth*.

Van Diggelen, J., Muller, T., & Van Den Bosch, K. (2010). Using artificial team members for team training in virtual environments. In *Proceedings of the 10th International Conference on Intelligent Virtual Agents* (pp. 28–34). Berlin, Heidelberg: Springer-Verlag.

Weibel, D., Wissmath, B., Habegger, S., Steiner, Y., & Groner, R. (2008, Sep). Playing online games against computer- vs. human-controlled opponents: Effects on presence, flow, and enjoyment. *Computers in Human Behavior*, *24*(5), 2274–2291.

Williams, R. (2002, Sep). Aggression, competition and computer games: computer and human opponents. *Computers in Human Behavior*, *18*(5), 495–506.

CHAPTER 50

Does User Frustration Really Decrease Task Performance?

Gloria Washington, Ph.D.

The MITRE Corporation
Washington, DC USA
gwashington@mitre.org

1   ABSTRACT

It is a natural assumption in research devoted to human-computer interactions that user frustration has a negative impact on task performance. This assumption stems from the very nature of the emotion frustration. Frustration occurs when a user is blocked or impeded from completing a goal or task. The intensity or amount of frustration varies according to how important the goal is to the user. However, is this assumption true for all amounts of user frustration experienced by a user in human-computer interactions?

The objective of this research is to determine if all user frustration amounts (or intensities) decrease task performance in human-computer interactions. This research is significant to usability practitioners because it goes beyond asking what the user feels and utilizes signals from human-body based measures to signify frustration or other performance-decreasing emotions. It also examines the unique impact of user frustration intensities on task performance metrics. Affective computing has utilized human-body based measures like skin temperature, respiration, and heart rate to identify user frustration. This research also uses these various measures to identify user frustration. In this research, users were asked to interact with a modified program that included frustration-inducing events to understand how task performance measures suffered. Data from human-body measures such as heart rate, skin temperature, respiration were then transformed into frustration intensities or amounts using an adaptation of the OCC Theory of Emotions.

The OCC Theory of emotions was used in this research because it is a cognitive model that says emotions have various intensities or amounts. Within the OCC model the emotion frustration does not exist, however only the emotion type anger is represented. Since user frustration is an emotion that occurs when a user is blocked from completing a task or goal, it was natural to link the real-world emotion frustration to the OCC emotion type anger because it is also a goal-based emotion. OCC says that the intensity of the emotion type various according to the importance of the goal and how sure the user believes he can complete the goal. After the emotion user frustration was transformed into intensities, it was necessary to determine from task performance measures how each calculated intensity impacted user productivity metrics. For each person in the study, the amount of intensities were calculated that decreased, increased, or did not change task performance metrics. This was important in usability studies because it is necessary to understand what level of frustration starts to decrease user productivity. Results from the study showed that most users in the study has user frustration intensities that allowed their task performance metrics to remain unchanged. However, about 13% of the users experienced frustration intensities that allowed their task performance to increase. Future work of this research seeks to study if certain amounts of frustration or intensities can be maximized across the time spent by a user in usability testing.

Keywords: user productivity, human-body based measures, OCC Theory of Emotions, user frustration

1   HISTORY OF USER FRUSTRATION IN HUMAN-COMPUTER INTERACTIONS

Ever since the creation of the computer, humans have been studying the effects of man-machine interactions on humans. One notable study performed by Reeves and Nass introduced The Media Equation, which theorizes humans communicate with computers like they are human . Although computers do not contain the cognitive abilities to correctly respond to emotions, this study showed that people will talk to, yell at, hit, or even reason with computers to try and get them to understand our frustrations. All types of computer users (beginner to novice) experience some type of frustration when a computer produces an unexpected outcome. Frustration occurs when a situation hinders or impedes a person from completing a goal; user frustration is experienced while interacting with a computer that interferes with task completion. Research in human-computer interaction assumes that annoyances or frustration-inducing events will negatively affect user productivity. However, before we can make this assumption, we must first the basic emotion frustration and how the human body reacts to stress caused by task-blocking events.

Freud believed that frustration occurs when a situation hinders or stops someone from reaching their goal. Frustration occurs when two aspects are not met: 1) an expected outcome is not produced and 2) goal-blocking interference or occurrence prevents the expected outcome. Therefore in order for user frustration to take place

1) a person must expect a certain outcome to occur after providing input or a series of input to the computer and 2) the computer, through some form or fashion, interferes with task-completion by producing an unexpected outcome. Although humans can without great difficulty categorize most frustrating events, it is very hard to convert these categorizes into language that the computer can understand. One reason is duality of the frustration emotion; frustration is more than just an emotion. It is a concept containing the external frustrating event and emotional response experienced by the user. The Stanford Encyclopedia of Philosophy says that frustration, like all emotions, causes four major reactions in the human: cognitive reactions (perception and memory involved), affective reactions (i.e. either positive or negative state), physiological reactions (i.e. changes to the human autonomic system), and a behavioral reactions (i.e. expression of feelings by yelling or changing facial expressions). All of these reactions vary according to the user and might negatively affect task performance.

Frustration dimensions relevant to this study are emotional arousal and intensity of the emotion. Arousal and intensity relate to the amount or how strongly a person has experienced an emotion. Cognitive psychologists have various explanations on why humans react certain ways to various events. Goal theory, Locke and Latham 1990, suggests that goal formation and achievement is related to task performance. In this theory Lock and Latham state that a users frustration will increase depending on how important the goal is and if the person perceives they can accomplish the task.

Since people utilize the computer to complete tasks, HCI research has focused on ways to identify the causes of user frustration to reduce their occurrence. The following lists outlines various causes of user frustration, Bessiere et al 2006.

- Users' lack of knowledge or training with a piece of software
- Users' unwillingness to read instructions
- Slow response time between communications,
- Poorly built user interfaces,
- Unclear error messages,
- Computer jargon used in help menus,
- Interruptions,
- Computer hardware or software failure, and
- Users' unfamiliarity with computers

As a result of studying and identifying the causes of user frustration, prevention techniques have been created in hopes of reducing the occurrence frustrating events. Prevention techniques such as interactive help systems, automated helper tools, "undoable" edits for changes made by automated helper tools, and automated menu placement and selection have helped to make the life of the computer user much easier. However, these methods, although useful, have not been totally successful at eliminating frustration experienced by computer users. Often times prevention technologies can annoy the user far worse than any frustrating event. Take for example, the Microsoft PaperClip used to help users as they compose documents in MicroSoft Word. This feature although very helpful to some users, was phased out of subsequent releases of the word processing tool. A segment of HCI research is devoted primarily in understanding the differences in computer users.

Computer users are not all the same as demonstrated in the example above. Although some may view a technology as helpful others may find the technology to lead more frustrating events. Computer users often fall into three categories: the beginner, the novice, and the advanced user. The beginning user has no experience with a computer, the novice has some experience, and advanced users interact with computers on a daily basis. Today, there are not many new computer users because of the ubiquitous nature of computers in our society. Humans interact with computers constantly, either through surfing the web, their cell-phones, or even watching downloaded shows on television. However, there are individuals that ignore or refuse to interact with computer devices; this type of user is usually older and accustomed to life without computers. Unfortunately, user frustration is experienced more by this group because of their unfamiliarity with the computer and unwillingness to learn. Users who are unfamiliar with the computer are less likely to want to use a device again if they have had a bad experience, Bessiere 2006.

Knowing more about the types of computer users and they way they might perceives technologies designed to assist them is one other research area within HCI. Researchers in HCI have developed computing models to try and understand user frustration and how it is experienced by different types of computer users. One such model is the Computer Frustration Model developed by Bessiere. The Computer Frustration Model was developed from analyzing the psychological and social effects of user frustration in workplace. In the model, Bessiere outlines various causes impacting the amount of frustration experienced by a user including a users mood, cultural and societal influences, and the user's experience while interacting with the computer. This model was used by developers of business applications to understand the psychological and social characteristics of the user and to assist in determining ways to reduce frustration from knowing these characteristics. Although this model identified the common causes of user frustration and different level of frustration; it did not affective computing techniques to detect for this emotion.

2 APPLICATION OF PHYSIOLOGICAL INDICATORS OF FRUSTRATION IN HCI

HCI research has also focused on ways to identify when frustration is experienced by computer users by analyzing human biological signals. Pickard was the instrumental in using heart-rate, blood-pressure to identify frustration in computer users using a Hidden Markov Model to learn attributes of frustrating events. Research in identifying user frustration through the use of human body-based measures has exploded since the introduction of affective computing by Picard in 1997 Since this study, researchers have been leveraging human-body measures as input to adaptive user interfaces and other tools to gain more information about what the user is experiencing during frustrating events. Although these studies were transformative, they did not examine the dimension of the emotion frustration that may help to understand the true impact on user productivity: intensity or arousal level.

Signals of frustration can be monitored with electronic sensors that collect data about changes to the human body. The sensors that measure the physiological indicator can be categorized by its method of gathering data. Sensors gather data intrusively and/or passively. Intrusive sensors are usually attached to human and may get in the way of a user and the computer. However researchers have found ways to overcome this by embedding intrusive sensors into everyday devices used along with the computer, Anttonen and Surakka, 2005.

Table 1: Categorized and grouping of physiological indicators of frustration.

Bio-signal	Type	Sensor	Description/Explanation
Pressure	Intrusive	Pressure sensitive touchpad	Squeezing objects, or pressure increases as users experience frustrating incidents in word processing tasks
Facial expressions	Musculatory, Intrusive and Passive	Video Camera	Muscles in the face called zygomaticus major and corrugator supercilli expand contract to produce frustration expressions
		EMG	
Heart rate	Circulatory, Intrusive	ECG	Heart rate increases as frustration is experienced. Heart rate increases significantly as anger is experienced
		EMFi	
		earlobe PPG	
Heart rate variability	Circulatory, Intrusive	ECG	Variations in heart rate occur when frustration or anger is experienced
Blood pressure	Circulatory, Intrusive	Stethoscope and sphygomanometer	Blood pressure increases as frustration is experienced.
Breathing	Passive	Rubber belt	Rapid breathing is associated with some sort of stress
Voice	Intrusive	microphone	Frequency, pitch, loudness and speech-rate are all affected by frustration.
Skin conductivity	Intrusive	galvanic skin response	Skin's ability to conduct electricity changes as stress is experienced
Body temperature	Intrusive and Passive	Thermal imaging	Body temperature increases as frustration is experienced
		Gas masks	

So now that we have proven methods to identify user frustration through human bio-signals, wouldn't it be helpful to analyze the arousal level or intensity of this emotion to understand their impacts on task performance in HCIs? The rest of this paper explains the study used to test and how analyzing this dimension of user frustration may benefit HCI research.

3   OCC THEORY OF EMOTIONS

The OCC Theory of emotions was created in 1988 by Ortony, Clore, and Collins to model human emotions for use in artificial intelligence applications. It was not meant for use in emotion generation, but to interpret human emotions and understand circumstances that affect emotion intensity. In this theory, the authors impose a clear structure on emotion-eliciting situations and OCC says that emotions are not really experienced until they have reached and surpassed a certain threshold. Emotions are grouped together into 22 groups called emotion types. Emotion types have different factors affecting the intensity of an emotion. Factors that tend to increase emotion intensity often increase the potential for other emotions to occur in that emotion type. For instance, a student may have a deadline to submit his/her homework assignment and his/her computer crashes. Also, One differentiating characteristic from other emotional theories is the characterization of emotions by situations that cause them to occur.

Figure 1: Original OCC Model

```
IF (EMOTION-POTENTIAL) > (EMOTION-THRESHOLD)
THEN
SET (EMOTION-INTENSITY) = (EMOTION-POTENTIAL) –
            (EMOTION-THRESHOLD)
ELSE
SET (EMOTION-INTENSITY) = 0;
```

In OCC, The amount of emotion experienced by an individual pertains to:
*   the congruence of an event's consequences with one's goals (i.e. the user is pleased when his computer "helps" him by automatically typing a word into a report, but displeased if it inserts an incorrect word),
*   the consequences of actions of agents(one's self, people, or inanimate objects such as computers) according to some standard (i.e. a person is displeased when he realizes he has lost his report due to his failure in saving the document), and
*   the consequences of people's attitudes or disposition to like or dislike certain objects or aspects of objects (i.e. people's attitudes about root canals causes the idea of going to the dentist to be unappealing.)

For this study, we were most concerned with the first and second bullets since real-world human-computer interactions often contain competing and conflicting goals a user must accomplish and the human's expectation of a successful outcome by the computer that may impact the intensity of an emotion.

4   APPLICATION OF OCC

The original OCC Model was used to calculate the arousal or intensity of frustration experienced by a human in man-machine. Since OCC Theory does not have an emotion type called frustration. For this study, the emotion type *anger* was used for frustration because it is a goal-based and standards-based emotion that directly relates to the task participants will be asked to complete. In the study to understand frustration intensities, users were asked to complete several tasks (accomplish goals) within a certain amount of time and each user may make judgments about the blame in response to system delays added to the interface.

OCC Theory theorizes that an emotion has not occurred until reached or surpassed a certain threshold. The theory states that if an intensity variable is lower than this threshold then the emotion has not been experienced. This is a wonderful axiom to apply to the emotion frustration because its human physical response is similar to other emotions such as fear and excitement. Therefore for this research, the threshold value is to be calculated based off of a user's normal or usual behavior when reacting to frustrating incidents.

5   APPLICATON OF FRUSTRATION INTENSITIES ON TASK PERFORMANCE

Data from human bio-signals was collected along with task performance measures such as number of errors, numbers of typos, etc. to determine intensity effect on these measures. Task performance was measured by:
- number of completed tasks
- number of times a subject skips a task
- number of typos
- consecutive number of typos
- number of formatting errors
- number of uncompleted tasks
- number of times student did not follow directions

6   RESULTS

43 subjects participated in the study. Two of the individuals were excluded because task performance data was missing or incomplete. There were 20 females and 23 males. The age of the participants ranged from 18-45. The study was conducted in the computer laboratories of the George Washington University. Users were given the option to skip a task and go on to the next tab, if they did not want to

complete the task. When asked why participants did not skip a task and go on to the next, some subjects said they wanted to follow directions and felt that the task was easy enough without skipping a task. A few users opted to skip a task, but when on to complete the skipped task before the end of the study. The only participant to skip a task and stay with his decision was the oldest participant in the study.

Results from the study showed that on average users were able to complete the task assigned to them without going over the allotted time of 15 minutes. The duration of the study with added delays was faster. This was surprising since the study included system delays that created a barrier to task completion. Perhaps, this is due to the user being familiar with system delays and knowing how to recover quickly from such events.

Most users in the study did not experience frustration intensities that dramatically decreased task performance measures. Some users even experienced an increased task performance during the actual experiment than the control part of the study. An explanation of this, could be that the users were did not find the frustrating inducing event to be actually that frustrating or users have become very familiar with these types of events and are able to recover quickly from them after their occurrence. In fact, 55% of users experienced an amount of frustration that allowed user productivity to remain constant and not decrease. In sports, stress from frustration is often a motivating factor to continue and often out-perform some hidden goal. Maybe the same can be said in beating frustrating events experienced during human-computer interactions?

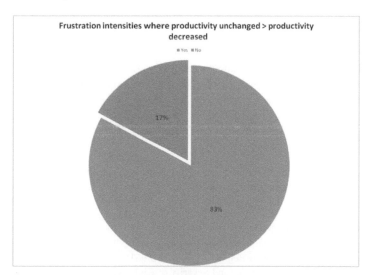

Figure 2   Pie chart showing relationship between frustration intensities that left productivity unchanged and intensities that decreased task performance.

460

7   DISCUSSION

Users are more familiar with frustrating incidents than ever before due to the ubiquity of computers. There is a natural assumption within HCI research that user frustration negatively impacts user performance. However, this is not true in all cases. More research is needed to study what intensities or arousal levels of frustration that do not decrease task performance. Perhaps, identifying these situations that indicate high stress, yet unchanged or even increased productivity will help us to create technologies that can optimize these user frustration levels.

8   CONCLUSIONS & FUTURE WORK

The key to understanding user frustration in human-computer interactions is to understand that user frustration is not one continuous emotional state experienced by a human in one session. Past research has examined it as one continuous state, however psychology shows this emotion is a culmination of various arousal levels, amounts or intensities. This research examines the effect of user frustration intensities on task performance and takes one step forward in analyzing user frustration and its various sub-components.

Future work will explore how to identify or optimize these intensities that help the user stay productive. Also future work will explore when frustration has the potential to turn into anger; thereby reducing aggressive behaviors against devices.

9   ACKNOWLEDGMENTS

The authors would like to acknowledge Booz Allen Hamilton for providing a grant through their Virginia Center of Excellence to purchase equipment related with the study.

10   REFERENCES

Anttonen, J and V Surakka. "Emotions and Heart Rate While Sitting on a Chair." Human Factors in Computing Systems. Portland, 2005.

"Autonomic Nervous System". Accessed 28 April 2007. www.britannica.com/eb/article-9011379.

Bessiere, K, et al. "A Model for Computer Frustration: The Role of Instrumental and Dispositional Factors on Incident, Session, and Post-Session Frustration and Mood." Computer in Human Behavior. 2006. 941-961.

Ceaparu, I, et al. "Determining Causes and Severity of End-User Frustration." Human Factors in Computing Systems. 2004. 333-356.

Dollar, J, et al. Frustration and Aggression. London: Kegan Paul, Trench, Trubner and Co Ltd., 1944.

Freud, S. Beyond the Pleasure Principle. London: Hogarth Press, 1922.

Freud, S. "Types and Onset and Neurosis." Freud, S. The Standard Edition of Complete Psychological Works of Sigmund Freud. Vol. 12. 1921. 227-230.

Katsionis, George and Maria Virvou. "Adapting OCC theory for affect perception in educational software." 2005.

Locke, A E and G P Latham. A Theory of Goal Setting and Task Performance. Englewood Cliffs, NJ: Prentice-Hall, 1990.

Ortony, A, G L Clore and A Collins. A Cognitive Structure of Emotions. Cambridge: Cambridge University Press, 1988.

Picard, R W. Affective Computing. Cambridge: MIT Press, 1997.

Reeves, B and C I Nass. The Media Equation: How People Treat Computers and New Media Like Real People and Places. New York: Cambridge University Press, 1996.

CHAPTER 51

Comfortable Information Amount Model for Motion Graphics

Masato SEKINE[†], Katsuhiko OGAWA[††]

[†, ††]University of Keio
[†]JSPS Research Fellow
Fujisawa, Japan
sekine@sfc.keio.ac.jp

ABSTRACT

This paper describes the development of a computational model for the estimation of a comfortable amount of information in motion graphic videos; image structure analysis is used in the development. A time series of frame differences (TSFD) is used as a computable index of the amount of motion in order to investigate the correlation between the spatiotemporal feature of motion and subjective impressions, including "activity," "complexity," and "comfort." By performing correlation analyses between impression scores based on the results of a subjective impression evaluation and the power ratios of the time series of frame differences, it is suggested that each impression has a correlation with a particular band of spectrum of TSFD. These bands correspond to human information processing characteristics. The retention time of VIS and Visual STM and the property of pattern chunking have a strong relation with the impression evaluation of a structural feature of motion graphics.

Keywords: motion graphics, human information processor model, comfort, impression, pattern chunking

1   INTRODUCTION TO MOTION GRAPHICS AND THEIR INTERPRETATION USING COGNITIVE SCIENCE APPROACH

"Motion graphics" are graphics designed as non-narrative, non-figurative visuals that change over time (Matt, 2003). The beginning of this art form is disputed, but it was derived from experimental films and animation by the 1950s at the latest. Saul Bass, who produced the title roll graphics for popular films such as *The Man with the Golden Arm* (1955) and *Psycho* (1960), is commonly referred to as the pioneer of motion graphics design. Nowadays, motion graphics are rapidly evolving with computer graphics and are widely applied to visual content such as music videos, television commercials, screen savers, and of course title rolls of films.

As indicated by the name, the motion of or the change in the structural elements that include the form, color, and arrangement of objects is the most appealing feature of this art form, rather than the semantic content of the graphics. Suppose, for instance, that a designer uses only plain quadrangles as the graphic elements. Whether his work can attract a viewer's interest for long or bores them depends on the spatiotemporal composition of the elements. For the viewers, viewing a motion graphic movie might be expressed as some kind of visual pattern finding game in which they pick up patterns and orders of visual data sequences without conscious volition and follow or continuously predict the changes in the sequence. It can be assumed that the viewer's objective evaluation or impression of a motion graphic work has a significant relationship with human information processing characteristics.

By referring to the Model Human Processor (Card et al., 1986), which involves the use of a cognitive modeling method and cognitive, perceptual, and motor processors along with the visual image, working memory, and long-term memory storage, we can describe the process of information processing in viewing motion graphics as follows:

1) The perceptual processor transfers sensory information from the retina to the visual information storage (VIS).
2) The cognitive Processor analyzes the local features of the sensory information, such as line segments, brightness, and angles, and then encodes these features and transfers them to the working memory (Visual STM).
3) In the working memory, the information encoded as the local features is further encoded in a relatively abstract and global representation by the cognitive processor by utilizing the information stored in the long-term memory.

In order to impose a limitation of the human information processing ability, each processor has a cycle time and each memory has a decay time in the MHP model. Lang (Lang et al., 1999) also adopted the idea that people are information processors and their ability to process information is limited; he suggested that in a television viewing situation, mental resources are selectively allocated to important parts of information processing at each stage. Lang hypothesized that a fast-paced production of the information presented to a viewer increases the allocation of processing resources to message encoding and elicits both self-reported and physiological arousal. On the basis of these ideas and the hypothesis, it is possible to develop a computational model for the estimation of a comfortable amount of information in motion graphic videos; the development involves image structure analysis. However, there are few researches that examine the relation between the impression of comfort and the structure of moving images with the computable indexes of image engineering. As the initial step in the development of the model, it would be meaningful to examine the relation by using some basic image features as analysis variables.

2   FRAME DIFFERENCE AS A COMPUTABLE INDEX OF MOTION

In the realm of computer vision, the frame difference is frequently adopted as a useful index to simulate the human perception of an object's motion. Its effectiveness is supported by a physiological theory that the first signal processing in the human visual system is the detection of the brightness change at the receptive fields. The frame difference is also used for estimating the subjective judgment of the activity of a television image in the field of video compression research. A study conducted by Candy (Candy, 1971) presents statistics that indicate a relation between the amount of frame difference and the subjective judgment of activity. The amount of frame difference, however, gives the number of changed pixels only between two successive frames. Therefore, this subjective judgment implies only an instantaneous emotion. In order to evaluate the impression of a motion graphic video whose duration ranges from several tens of seconds to several minutes, the analysis of only instantaneous change is inadequate. Hence, the use of a time series of frame differences as an analysis variable is more suitable for the study of motion graphic videos.

3   METHOD

The basic design for this study required subjects to watch motion graphic videos and to evaluate impressions of "complexity," "activity," and "comfort." The following step is to calculate the time series of frame differences (hereafter referred to as TSFD) and to analyze the correlation between the TSFD and the result of the impression evaluation. Subjects consisted of 23 undergraduate and graduate students, 10 males and 13 females.

3.1   Material

Stimulus motion graphic video clips are composed of moving particles as the graphical element. For focusing attention on the effects of motion on the viewer's impression, and in order to exclude the effects of color, the stimulus videos are monotone and have no color depth image (binary format image with white particles and a black background). The duration per stimulus is 45 s, which is sufficient for evaluating the impression. The resolution is 720 × 480 pixels, and the frame rate is 30 frames per second. The video clips are generated with the following two algorithms:

A.   Boids algorithm (Reynolds, 1987), which simulates the flocking behavior of birds to generate graphics characterized by irregular motion and asymmetric form

B.   Vector field and sinusoidal parameter control to generate graphics characterized by cyclic motion and a symmetric form

The number of type A video clips is 16 and that of type B is 5. Each video was generated with a unique parameter set such as motion speed, diffusivity, and arrangement of particles. Figure 1 shows sample images of stimulus videos.

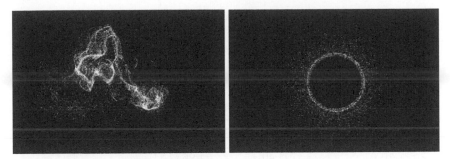

Figure 1 Sample images of stimulus videos. The left one is generated by algorithm A, and the right one is generated by algorithm B.

3.2 Apparatus and procedure

Stimulus videos were played for the subjects with EIZO Flex Scan S2411W color LCD monitor having a screen size of 518.4 × 324.0 mm. The distance between the display and a subject was around 50 cm. For canceling the stimulus order effects, the playing order of the video clips was shuffled for every subject. The evaluation items were "complex - simple," "active - gentle," and "comfort - exhausting." As indicated in figure 2, the subjects selected the most suitable grade in the seven-grade evaluation.

	Very much	Quite	Slightly	Neutral	Slightly	Quite	Very much	
Simple	├	┼	┼	┼	┼	┼	┤	**Complex**
	0	1	2	3	4	5	6	

Figure 2 Example of evaluation items.

The following list describes the procedure cycle of the impression evaluation.

1) Play a stimulus video clip for 45 s.
2) Take a 45-s interval.
3) A subject evaluates each item during the interval.
4) Play an alert sound for preparing the subject for the next stimulus.
5) Play the next stimulus video clip.

By repeating this cycle, the subjects evaluated all 21 video clips.

4 RESULT OF IMPRESSION EVALUATION

The impression score of each video clip was calculated on the basis of the evaluation of each subject. For instance, in a complexity section, the items ranging from "very much simple" to "very much complex" were assigned 0 to 6 points, and the mean score and the variance were calculated. Table 1 presents the score for each video clip.

Of particular note is that the complexity scores of type B video clips are obviously lower than those of type A ones. Further, through a correlation analysis among impressions, a weak negative correlation between "activity" and "comfort" was found when the coefficient of determination $R2 = 0.339$ ($p = 0.0078$).

Table 1 Impression evaluation scores and variance

No	Type	Complexity	Activity	Comfort
1	A	2.96 (1.26)	1.13 (0.87)	3.74 (1.42)
2	A	3.17 (1.40)	1.96 (1.11)	3.26 (1.45)
3	A	3.48 (1.59)	5.0 (1.0)	2.13 (1.55)
4	A	3.91 (1.12)	4.04 (1.07)	2.87 (1.74)
5	A	3.87 (1.22)	4.40 (1.12)	2.87 (1.52)
6	A	3.43 (1.07)	0.78 (0.85)	3.21 (1.59)
7	A	3.87 (1.10)	2.22 (1.41)	3.67 (1.40)
8	A	4.35 (1.11)	4.96 (0.77)	2.70 (1.55)
9	A	3.35 (1.40)	2.26 (1.48)	3.60 (1.59)
10	A	4.09 (1.38)	1.91 (1.24)	4.17 (1.07)
11	A	4.04 (1.36)	4.65 (0.93)	2.70 (1.29)
12	A	2.30 (1.29)	0.83 (0.72)	3.09 (1.62)
13	A	3.65 (1.58)	3.0 (1.76)	1.83 (1.70)
14	A	4.17 (1.30)	4.65 (1.43)	1.61 (1.78)
15	A	1.87 (1.42)	3.48 (1.50)	1.91 (1.70)
16	A	3.39 (1.34)	2.52 (1.34)	3.09 (1.16)
17	B	1.35 (0.71)	3.61 (0.89)	3.17 (1.44)
18	B	2.70 (1.52)	3.13 (1.22)	4.09 (1.50)
19	B	1.78 (1.52)	3.26 (1.21)	2.48 (1.62)
20	B	1.43 (1.27)	3.61 (1.03)	2.74 (1.67)
21	B	2.04 (1.33)	2.30 (1.29)	3.91 (1.68)

4   CALCULATION OF TIME SERIES OF FRAME DIFFERENCES (TSDF)

From the 1350-frame (30 fps × 45 s) bitmap data of each stimulus video clip, we measured the number of changed pixels between successive two frames; all quantity data within each clip were integrated as a time series data array having a data length of 1349. Figure 3 shows the TSFD of stimulus video no. 7 and no. 8 plotted on a 2D graph. The vertical axis denotes the number of different pixels, and the horizontal axis represents the frame number. In 2D graphs of TSFD, a steep slope indicates a rapid change in the frame difference rate. For instance, when a considerable number of particles drastically change their motion speed in a short period of time, the phenomenon is depicted as a steep slope in a 2D graph. In contrast, a small change in the frame difference rate is expressed as a smooth slope irrespective of the absolute number of changed pixels.

Figure 3 Sample graph of TSFD plotted in 2D. The left one is for TSFD of video clip no. 7 and the right one for video clip no. 8.

The following tendencies were identified by the observation of the waveforms of TSFD and impression scores.

1) Amplitude scale of TSFD seems to have a weak correlation with the scores of "activity" and "comfort."
2) Video clips whose waveforms have a considerable number of sharp peaks and troughs tend to have a high score of "activity."

In order to quantify these observable tendencies, it would be reasonable to compute the frequency spectrum of TSFD and to calculate the power ratios of the classified spectrum and the total power as variables for the correlation analysis with the impression scores.

The frequency spectrum of TSFD was calculated by using FFT analysis (Nyquist frequency: 11.1 Hz). Then, the spectrum band was split evenly into six bands in the logarithmic scale, and the power ratio was calculated (Table 2).

Table 2 Power ratios [%] and total power (T.P). Cycle time: Low-1 [20.23–60.7 s], Low-2 [6–20.23 s], Mid-1 [2–6 s], Mid-2 [0.74–2 s], High-1 [250–740 ms], and High-2 [90–250 ms]

No	Low-1	Low-2	Mid-1	Mid-2	High-1	High-2	T.P
1	74.50	20.61	4.65	0.16	0.02	0.07	12.33
2	30.37	28.37	36.15	4.61	0.39	0.11	2.01
3	38.31	10.66	22.50	22.45	5.45	0.62	13.90
4	37.63	35.90	20.01	5.34	0.90	0.22	84.76
5	26.75	37.58	26.86	7.13	1.58	0.10	40.05
6	88.18	6.86	2.92	1.10	0.45	0.49	10.43
7	65.01	31.87	2.55	0.38	0.07	0.12	18.65
8	30.94	29.59	26.20	10.35	2.67	0.25	120.26
9	76.18	19.46	3.56	0.51	0.14	0.14	37.00
10	74.09	21.54	3.10	0.83	0.30	0.15	126.06
11	48.32	24.35	21.63	5.12	0.55	0.03	144.61
12	87.75	6.59	2.53	1.11	0.59	1.44	2.93
13	47.00	49.95	2.46	0.28	0.14	0.17	10.02
14	21.34	51.78	24.37	2.12	0.27	0.12	112.13
15	23.43	52.70	18.57	3.53	0.62	1.14	2.97
16	61.01	35.13	3.31	0.20	0.13	0.22	16.44
17	0.92	43.96	21.26	24.0	9.09	0.73	1.17
18	1.08	34.24	37.60	18.07	7.71	1.30	0.05
19	0.61	5.06	66.99	18.95	6.17	2.21	0.05
20	0.06	0.23	61.29	32.27	5.16	0.99	1.54
21	0.74	72.50	24.72	1.52	0.21	0.31	0.89

5   CORRELATION ANALYSIS BETWEEN IMPRESSION SCORE AND SPECTRUM DATA OF TSFD

The correlation between the impression score and the spectrum data of TSFD was analyzed by using multiple linear regression analysis (MLRA), taking the impression scores (3 impressions × 21 stimuli) as the dependent variables and the spectrum data as the independent variables. The spectrum data were composed of the power ratios of the classified spectrum (6 bands × 21 stimuli) and the total power of TSFD (1 × 21 stimuli). The independent variables were sifted through a stepwise regression for excluding the multi-colinearity.

5.1   Result

"Activity"
The significant variables were the Low-2 (coefficient value = 0.03, t = 3.15, p = 0.0059), the Mid-2 (CV = 0.103, t = 5.32, p < 0.0001), and the total power (CV = 0.015, t = 4.43, p = 0.0004). The adjusted R^2 was 0.659 (p < 0.0001). In the analysis carried out with only the data of type A videos, the adjusted R^2 was 0.911 (p < 0.0001). This result suggested the possibility of a strong effect of the power ratio of the Mid-2 band (0.74–2 s) on the impression of activity; however, this effect seemed to depend on the certain characteristics of the graphics.

"Complexity"
The significant variables were the High-2 (CV = -0.774, t = -3.08, p = 0.0062) and the total power (CV = 0.009, t = 3.14, p = 0.005). The adjusted R^2 was 0.633 (p < 0.0001). In the analysis carried out with only the data of the type A videos, the adjusted R^2 was 0.730 (p < 0.0001). This result suggested that the power ratio of the High-2 band (90–250 ms) had a weak negative correlation with the impression of complexity.

"Comfort"
There were no significant variable in the type-combined analysis. In the analysis carried out with only the data of the type A videos, the significant variable was the Low-1 (CV = 0.021, t = 3.48, p = 0.0037) and the adjusted R^2 was 0.425 (p = 0.003).

5.2   Discussion

"Activity"
There was a high correlation between the power ratio of Mid-2 (cycle length: 740–2000 ms) and the impression score of activity, particularly in the case of type A videos. Considering that the memory holding time of Visual STM is around 600 ms (Phillips, 1974), the Mid-2 was just the area over the limit of Visual STM. By referring to the study by Inui and Miyamoto(1979), we concluded that 600 ms was also the limit time of the pattern chunking of the sequentially displayed parts. Therefore, it could be concluded that the impression of activity had a strong relation with the human ability of pattern chunking and the limitation of Visual STM. Further, Phillips (1974) suggested that the retention rate of a random pattern in Visual STM is based on its

complexity. As mentioned in the previous section, the impression scores of the complexity of type B videos (regularity and symmetry) were obviously lower than those of the type A videos (irregularity and asymmetry), and the adjusted R2 of MLRA without the consideration of the type B videos was 0.25 higher than that obtained by considering the type B videos.

It was speculated that the order of repeated motion and symmetric form would be easier to encode and would require less storage space in Visual STM than that of irregular motion and asymmetric form. In contrast, the identification of the order in the case of irregular motion and asymmetric form would require a considerable amount of mental resource and a large storage space in Visual STM. Therefore, it was reasonable to speculate that the limit time of Visual STM was critical for encoding irregular, asymmetric patterns and that a large amount of motion in this time range (740–2000 ms) would require a considerable amount of mental resource for the encoding process.

"Complexity"

With regard to the impression of complexity, a weak negative correlation between the impression of complexity and the power ratio of High-2 (cycle length: 90—250 ms) was suggested by the result of MLRA. The cycle length corresponded to the range of the shortest time required to recognize visual information from the retina (1 cycle of perceptive processor and 1 cycle of cognitive processor, around 200 ms). Further, the average retention time of VIS was 200–250 ms, and the successive sensory information in this time range tended to merge in VIS. Hence, when the power ratio of High-2 became dominant in a video, it was difficult to perceive the scene as a sequence of discrete motions. For instance, with respect to the snow noise of analog TV that contains equal power within a fixed bandwidth and has a higher power ratio of High-2 than ordinary video, we would not perceive the scene as discrete motions. Perhaps, similar to the case when the high-frequency parts of a still image tend to be recognized as texture, a motion in this cycle might be recognized as the "texture of motion." Hence, details pertaining to the information would be compressed and they would recede into the background through the encoding process. Although this consideration explains how the power ratio in the cycle reduces the impression of complexity, it does not explain the causality of the impression. In order to identify the main causality of the complexity in motion graphics, we need to analyze other spatiotemporal features such as the optical flow.

"Comfort"

In the analysis with type A videos, Low-1 (cycle length: 20—60 s) had a medium correlation with the impression of comfort. However, considering that there was no significant variable in the type-combined analysis, we concluded that the evaluation of subjective comfort was affected by the degree of regularity and symmetry. As in the case of the impression of activity, further analyses are required to investigate the mechanism of the comfort impression.

6   CONCLUSION

In this paper, the development of a computational model for the estimation of the comfortable amount of information in motion graphic videos was discussed; image structure analysis was used in the development. The result of the correlation analysis between the subjective impression score of the motion graphic videos and their power ratio of TSFD as the computable index of motion indicated the effects of the temporal characteristics of human information processing on the impressions, including "complexity," "activity," and "comfort." In order to develop the model for estimating the comfortable amount of information, it was essential to take into account the retention time of VIS and Visual STM and the property of pattern chunking.

This study is the first step in the development of the model. TSFD would be the simplest feature of motion that can provide a considerable amount of information on temporal structure. However, TSFD contains less information on spatial structure such as form and arrangement. Therefore, further analysis involving the consideration of the spatial features of motion graphics is required.

REFERENCES

Candy, J. C., et al. 1971. Transmitting Television as Clusters of Frame-to-Frame Differences, *The Bell System Technical Journal*, 50(6): 1889–1917.

Inui, T. and K. Miyamoto 1979. Spatio-temporal Properties of Face Recognition, *Bulletin, Osaka University*, 5: 191–221.

Lang, A., et al. 1999. The Effects of Production Pacing and Arousing Content on the Information Processing of Television Messages, *Journal of Broadcasting & Electronic Media*, 43(4): 451–457.

Matt, F. 2003. Changing over Time, The Future of Motion Graphics: http://www.mattfrantz.com/thesisandresearch/ motiongraphics.html.

Phillips, W. A. 1994. On the Distinction between Sensory Storage and Short-Term Visual Memory, *Perception and Psychophysics*, 16: 283–290.

Reynolds, C. W. 1987. Flocks, Herbs, and Schools: A Distributed Behavioral Model, *Computer Graphics*, 21(4): 25–34.

CHAPTER 52

The Kansei Research on the Price Labels of Shoes

Saromporn Charoenpit[1], Michiko Ohkura[2]

[1]Thai-Nichi Institute of Technology
[2]College of Engineering, Shibaura Institute of Technology

ABSTRACT

The Kansei values of industrial products are considered very important. In this study, we focused on the factors affecting the price-labels of shoes for sale both in Japan and Thai from the Kansei engineering aspect. Experiments were planned for 160 participants in Japan and Thai that separated 8 groups related to the conditions of gender, age and nationality. We designed an experiment of answering questionnaire via internet. The experiment was divided into two parts. As for the former part, we designed questionnaires in web page in which participants evaluates each item comparing two figures of the price labels of shoes that affect decision to buy shoe. The number of the repetition of the comparison is 78 in Japan and 66 in Thai. As for the latter part, we designed the questionnaires in web page having 5 questions related to the reasons for the selection of the price labels of shoes in the former part. The candidates of shoe-labels some screen shots of the questionnaire system. From the results of the analysis of variance, the results of (F.4), (F.5), (F.6) and (F.12) of the price labels of shoes in Japan have statistically significant difference from age groups between $20 - 25$ years old and $40 - 50$ years old at the $p < 0.05$ level.

Keywords: Kansei Engineering

INTRODUCTION

At present, customers easy to buy shoes because that cheap, As results their decisions depend increasingly and greatly on subjective factors, such as feelings, images, fashion, price, quality, impressions and demands of the product. Most people have shoes to wear for every state of affairs that life may offer them. A woman's and man's shoes really go a long way towards telling you who they are, what they are like, and what they does with them life. There are many cases when they really needs a lot of different colors and styles, such as when they works everyday and needs a variety of shoes as well as some that are comfortable.

The labels are tools used for relaying to customers the cost of your items, and you can make your own personalized tags. Price labels made in different shapes and ink colors are attractive and eye-catching. All goods in shops and in display windows shall be clearly labeled, either on the products themselves.

Kansei engineering is a technology for translating human feelings into product design. Several multivariate analyses are used for analyzing human feelings and building rules Using Kansei engineering (Ishihara S., Nagamachi M., and Ishihara K., 2011) it is possible to incorporate consumer emotion into the product design process, creating products that appeal to customers on a subjective level.

We have researched the factors affecting the price labels of buying shoes in Japan and Thai with Kansei engineering. We designed an experiment of answering questionnaire by internet. The experiment was divided into two parts. As for the former part, we designed questionnaires in web page in which participants evaluates each item comparing two figures of the price labels of shoes that affect decision to buy shoe. The number of the repetition of the comparison is 78 in Japan and 66 in Thai. As for the latter part, we designed the questionnaires in web page having 5 questions related to the reason for the selection of the price labels of shoes in Thai and Japan.

2. METHOD

2.1 Construction of web questionnaire system

We conducted with sample from Japan and Thai that cannot create a standalone program, we designed questionnaire survey on web base and link to the internet system, used PHP and MySql then samples to test through the Internet by using web browser such as Internet explorer, chrome, firefox (Andrew S. Tanenbaum.,1996) shown in figure 1.

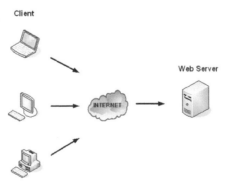

Figure 1 Experimental System

2.2 Labels and figures of sues

We surveyed the price labels of shoes at Ginza, Harajuku, Shibuya, UENO, Yurakucho, Odiba, Ikebukuro, Toyosu and Nishi Kawaguchi in Tokyo, Japan. Based on our survey mentioned above, the labels in Japan and Thai difference in text number price and symbol ¥ yen shown in figure 2. We classified 13 labels in Japan and 12 labels in Thai because the labels (F.13) have only in Japan shown in figure 3.

(Japan)   (Thai)

Figure 2 Difference the price labels in Japan and Thai

Figure 3 The price labels of shoes in Japan

2.3 Outline Web Questionnaire System

The outline of the web questionnaire system have 3 steps as follow:
(1) Selection of participant's attributes
Selection of gender, age group, and nationality.
(2) Selection of labels of shoes
In Japan, Each participant was randomly shown 78 pairs of price labels. In Thai, Each participant was randomly shown 66 pairs of price labels the cumulative data are shown in figure 4.
(3) Questionnaire for selection reasons
We designed the questionnaires in web page having 5 questions related to the reason for the selection of the price labels of shoes in Thai and Japan [2]. Participants evaluate each item on a scale of 5-points Likert's scale (Izumiya A., Ohukura M., Tsuchiya F., 2009) as follow:
Q.1 : The characters are easy to read?
Q.2 : The colors are easy to view?
Q.3 : The display is easy to understand?
Q.4 : Familiarity?
Q.5 : Complexity?
as shown in figure 5.

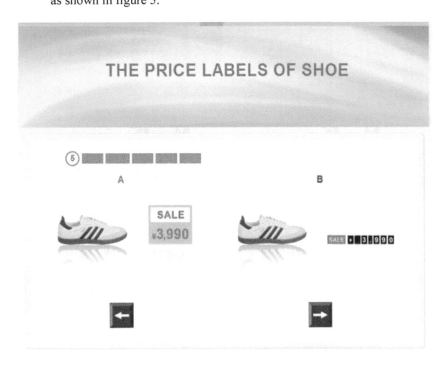

Figure 4 Example of comparison

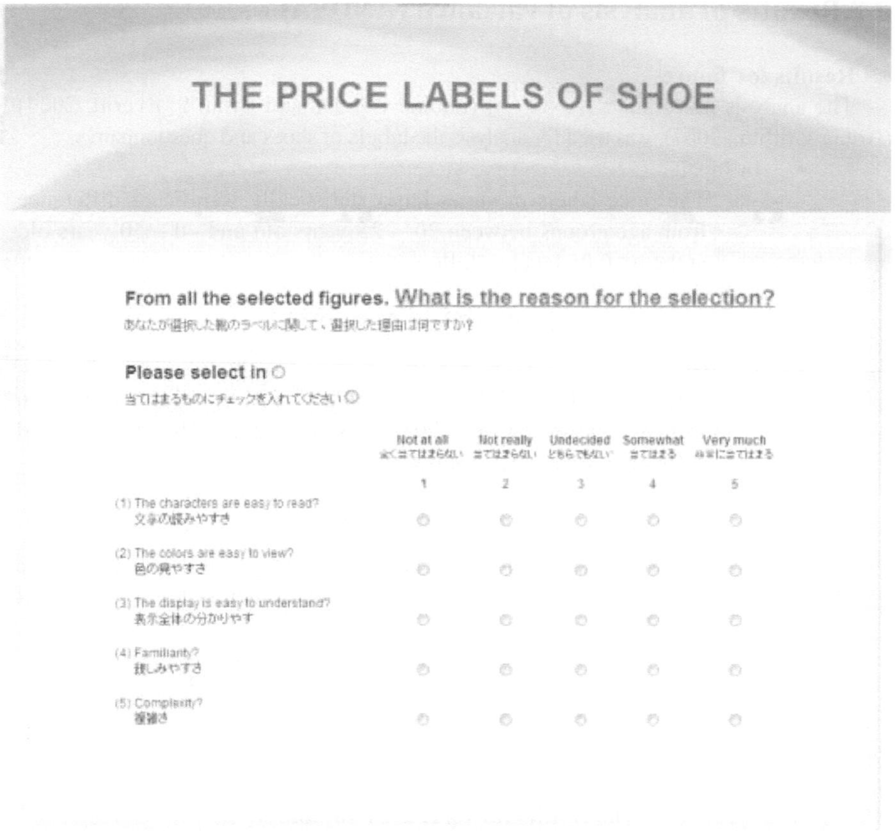

Figure 5 Example of questionnaire

3. RESULTS

3.1 Participants

Experiments were planned for 160 participants in Japan and Thai that separated 8 groups related to the condition of gender, age and country:

Group 1 : Men20 – 25 years old in Japan
Group 2 : Men 40 – 50 years old in Japan
Group 3 : Women 20 – 25 years old in Japan
Group 4 : Women 40 – 50 years old in Japan
Group 5 : Men20 – 25 years old in Thai
Group 6 : Men 40 – 50 years old in Thai
Group 7 : Women 20 – 25 years old in Thai
Group 8 : Women 40 – 50 years old in Thai

3.2 Results of analysis of variance (ANOVA)

Results for figures

The analysis of variance (ANOVA) (Sabine Landau and Brian S. Everitt. 2004) (Arthur Griffith., 2007). was used for analysis the labels of shoes and questionnaires.

- In Japan

    The price labels of shoes have statistically significant difference from age groups between 20 – 25 years old and 40 – 50 years old at the $p< 0.05$ level as follow:

    (F.4), (F.5), (F.6) and (F.12)

- In Thai

    The price labels of shoes have statistically significant difference from age groups between 20 – 25 years old and 40 – 50 years old at the $p< 0.05$ level as follow:

    (F.2), (F.3), (F.8) and (F.12)

    The price labels of shoes have statistically significant difference from gender groups between man and woman at the $p<0.05$ level as follow:

    (F.3), (F.4), (F.5), (F.10) and (F.11)

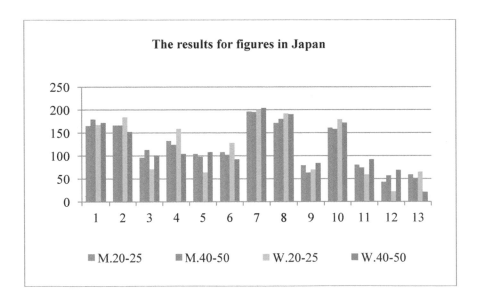

Figure 6 Frequency of the price labels of shoes in Japan

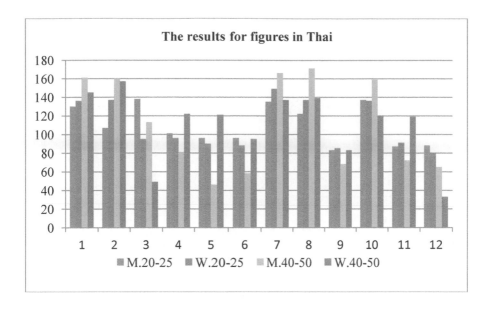

Figure 7 Frequency of the price labels of shoes in Thai

Results for questionnaires

From the results of the analysis of variance (ANOVA), the results of question 1 to question 5 of the price labels of shoes in Japan and Thai.

- In Japan

    The results of (Q.1) to (Q.5) of the price labels of shoes have no statistically significant difference from gender groups between man and woman, and age groups between 20 – 25 years old and 40 – 50 years old.

- In Thai

    The results of (Q.4) of the price labels of shoes have statistically significant difference from gender groups between man and woman at the $p < 0.05$ level and the results of (Q.1), (Q.2), (Q.3) and (Q.4) of the price labels of shoes have statistically significant difference from age groups between 20 – 25 years old and 40 – 50 years old at the $p < 0.05$ level.

478

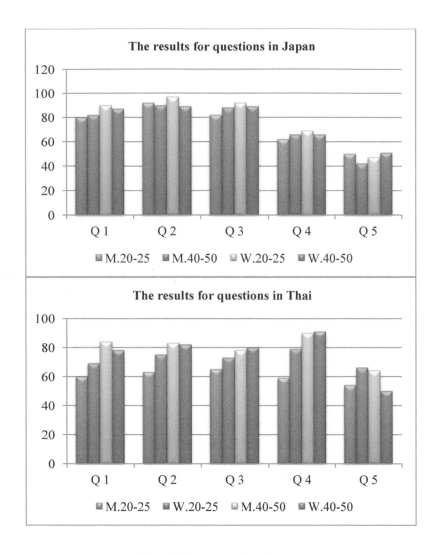

Figure 8 Frequency of questionnaire

4. DISCUSSION AND CONCLUSION

In Japan
- The best of the price labels of shoes for every the price labels of shoes (F.7) for all participants.
- The results compare from age groups between 20 – 25 years old and 40 – 50 years old that effective for a sale labels in Japan are (F.4), (F.5), (F.6) and (F.12).

In Thai
- The best of the price labels of shoes for every the price labels of shoes as follow:
    (F.7) for man groups and 20 – 25 groups
    (F.2) for woman groups and 40 – 50 groups
- The results compare from age groups between 20 – 25 years old and 40 – 50 years old that effective for a sale labels in Japan are (F.2), (F.3), (F.8) and (F.12)

All participants
- The results compare from gender groups between man and woman that effective for a sale labels in Thai are (F.3), (F.4), (F.5), (F.10) and (F.11)
- These results of all participants chose the price labels of shoes show the importance of red background color as candidates for the price labels of shoes.
- The price labels of shoes (F.13) is a symbol labels of shoes in Japan that not effective for a sale labels in Japan which the character is not easy to read, color is not easy to view and display is not easy to understand .
- The shops in Japan and Thai should keep the price labels of shoes (F.7).

ACKNOWLEDGEMENT

This research was party supported by Shibaura Institute of Technology, Japan. We also thank all the participants in our experiments.

REFERENCES

Andrew S. Tanenbaum., "Computer Networks", Prentice-Hall, Inc., 1996, pp.10. 17.

Arthur Griffith., "SPSS for Dummies", Wiley Publishing, Inc., 2007, pp. 229-242.

Ishihara S., Nagamachi M., and Ishihara K., "Neural Networks Kansei Expert System for Wrist Watch Design", HCII 1995-07-09 Vol. I. Human and Future Computing, Kansei Engineering, p.167-172,

http://www.hcibib.org/bibtoc.cgi?file=bibdata/HCII95* [retrieved 23 May 2011].

Izumiya A., Ohukura M., Tsuchiya F., "The Evaluation of Pharmaceutical Package Designs for the Elderly People", Proceedings of the Symposium on Human Interface 2009 on Human Interface and the Management of Information. Information and Interaction. Part II: Held as part of HCI International, 2009.

Sabine Landau and Brian S. Everitt., "A Handbook of Statistical Analyses using SPSS", Chapman & Hall/CRC Press LLC, 2004, pp. 135-177.

CHAPTER 53

Toward Emotional Design: An Exploratory Study of iPhone 4

Bahador Saket*, Lim Tek Yong*, Farnaz Behrang⁺

* Multimedia University
Cyberjaya, Malaysia
Bahador.saket@gmail.com, tylim@mmu.edu.my

⁺ Amirkabir University of Technology
Tehran, Iran
fbehrang@aut.ac.ir

ABSTRACT

Emotions are essential part of how people experience the world. This suggests that emotions are important for the success of a product and influence the purchase decision of users. In other words, a successful design is linked to the emotions of user. Web reviews are important sources of information for analyzing user experiences. In this paper, using iPhone 4 as a case study, an analysis of web reviews was conducted to elicit the emotional responses of reviewers and gain insights into how different design specification impact on these emotions. Then, some significant points of results were accentuated to help mobile designers in the design process.

Keywords: user emotional response, emotional design model, iPhone 4

1   INTRODUCTION

With the invention of the first series of mobile phones, a massive and impressive revolution occurred in human life. Early mobile phones, in comparison with today's ones, had a simpler technology. Gradually, with progress of mobile technology, more and different capabilities were release. Nowadays, mobile phones are the most essential tools in human life. Up to the middle of 2010, there were more than five billion connections worldwide. India with the population of 1,210,193,422 people had around 851,695,668 million mobile phones in use in the middle of 2011 (Telecom Regulatory Authority of India, 2011). It means the proportion number of mobile phones in use to population in India is around 0.71. That makes the difference between mobile phones is virtually every design specifications such as color, size, functions, and material (Han, Kim, Yun, Hong, and Kim, 2004). These specifications will be evaluated by users.

In fact, designers and researchers evaluated these design specifications by gathering user emotions. Garreta-Domingo, Almirall-Hill, and Mor (2007) researched on the suitable methods for eliciting, analyzing users emotional responses and evaluating the mobile design. Identifying user's requirements and applying them into design process is a big problem for designers (Guo and Tian, 2010). Researchers in Human Computer Interaction (HCI) acknowledged the importance of user emotion in "user-centered design". This motion has helped researchers to explore the nature of human emotions and behaviors that emerges in interactions with technologies (Garreta-Domingo, Almirall-Hill, and Mor, 2007).

Norman and Ortony (2003) proposed an emotional design model that consists of three different levels of the brain. The first level is called visceral, where some emotions are responded automatically by the brain after seeing or feeling a mobile device. Behavioral is the second level, where emotions are generated by the brain after performing a specific task (such as sending SMS to a friend). The highest level is known as reflective. This level involves conscious thought, where human stop from current routine and start to think whether owning the mobile is worthwhile or suit to their lifestyle. Whether it is negative or positive emotional state, it changes the way human thinks (Norman, 2002). The goal of this paper is to explore the following two questions.

1) What is the most important design specification for users while they are interacting with those mobile phones?

2) Is this design specification related to visceral, behavioral or reflective level of emotional design model?

Thus, it is the significant point for mobile phone designers to know the design specifications which they must consider carefully during designing a new mobile phone.

2   RELATED STUDIES

Lim, Donaldson, Jung, Kunz, Royer, Ramalingam, Thirumaran, and Stolterman (2008) conducted a study to elicit and analyze users' emotions on interactive products that have owned or used before. They defined affect, emotion and experience as different qualities. These qualities could help designers and developers to understand what should be considered in design, especially concerning user's emotional responses. They used a post-use interview called "disposable camera study". 12 participants volunteered for this study, where half of the participants had a background in HCI. Participants were asked to take 5 to 10 pictures of products that they frequently used. This method (taking picture) encouraged participants to choose the ones in which they have strong emotional responses. Then, participants were asked to explain their memories or experiences they have with the product. Participants were also asked to pick any 16 semantic values to summarize their emotional responses. Four semantic values were related to visceral level, four for behavioral level and the remaining eight concerning reflective level. Two types of information were extracted from this study, namely the qualities of interactive products and qualities of user experiences.

However, it is obvious that users' emotional response to a specific product could vary during a period of time (Folkman and Lazarus, 1985). For example, when a user learns to make a new folder using iPhone for the first time, the user may express his/her emotion as "It is very hard to do". If the user is very familiar with iPhone features, he/she may say "That is simple and easy". Thus, the result of their study tend to focus more on reflective level rather than visceral and behavioral. When people discussed their emotional experiences with products, they usually refer to functional and interactive qualities of the products in comparison with other design specifications, such as tactile and visual. For interactive products like mobile phones, the design of functionality and interactivity is influential. Their study found out that functionality is important for users.

2.1   Relationship Between Functionality of Mobile Phone And Behavioral Level

Functionality is refers as the set of actions and services that a device provides for users (Karray, Alemzadeh, Saleh, and Arab, 2008). The main function of a mobile phone is voice communication, which allows a user to call another and talk from a distance. Besides voice functions, most mobile phones offer text or data transfer as well. Users can send written messages, share pictures and videos, or use specific internet applications. Other functions include using contact lists, calendars, clocks, calculators, word processing, spreadsheets, document viewing, taking pictures, listening to music and playing games. Norman and Ortony (2003) suggested that when a user feels pleased using a tool (or a function), this emotional response is derives from behavioral level. Emotions that user associated with after using a specific function in mobile phone, can be divided into positive and negative states. A positive emotional state is normally associated when a user completed a task successfully, while the negative emotional state is triggered when the user failed to complete the task.

2.2   Using Web Reviews For Data Collection

Wenger (2008) conducted a study to analyze blog entries on trip to Austria (which posted to www.travelblog.org). The study established similarities and differences between those posting blog entries and Austria's tourism markets. The researcher also analyzed the content of blog entries to identify negative and positive perceptions of Austria as a tourism destination. Wenger (2008) went through 114 blogs to collect the demographic information about each author of these weblogs (such as age, gender and country of origin), modes of transport, services used and problems encountered in their journeys.

On the other hand, Bosangit, McCabe, and Hibbert (2009) researched travel blogs as textual artifacts to gain insights into how tourists construct order and make meaning from their experiences as part of the process of identity management. They used three websites (www.travelpod.com, www.travelblog.com and travelbuddy.com) in their research. The 10 most recent travel blogs entries on each of the three websites were selected (on February 21, 2008), downloaded and printed for their analysis. They proved the importance and reliability of web reviews, and then suggested that web reviews can be considered as a new tool for consumer narrative analysis.

Figure 1 Sample of gathering emotional responses.

3   METHODOLOGY

Previous studies have highlighted the potential of web reviews as sources of information regarding user experiences. For this paper, iPhone 4 was selected as a case study. Many professional and personal reviews related to iPhone 4 were posted on the internet. Google search engine generated a list of websites with the keyword "iPhone 4 Reviews". Only the first ten results (such as www.engadget.com, www.gizmodo.com) in English language were selected. The use of search engine is an accepted technique for accessing websites (Wenger, 2008; Carson, 2008).

While reading each web review (between April 1, 2011 to April 14, 2011), two important aspects were identified, namely emotional response and related design specification. In this paper, emotional response is refers to adjectives used by reviewer to describe a design, while design specification is refers to detailed of iPhone feature or part. For instance; "iPhone improved optics for its camera", *improved* is an adjective and *camera* is the design specification. Then, the identified emotional responses were categorized into three levels of emotional design model, namely visceral, behavioral and reflective.

Some categorization activity was ambiguous and difficult, so the original sentence or paragraph is used to determine the suitable levels of emotional design. For instance, the word "*good*" can be used in both behavioral and reflective levels. When a reviewer expressed his/her emotion about retina display and said "There's no denying that it looks good. We haven't seen the yellow areas that users have complained about", then the word "*good*" is referred to reflective level. On the other hand, when a reviewer expressed emotion about face time and said "we tried it and it looked good", the word "*good*" can be associated to behavioral level. Or when a reviewer stated that "The iPhone's display is beautiful" or "Downloaded videos look beautiful on the retina display", the word "*beautiful*" is related to visceral level. The same adjective word "*beautiful*" can be associated to behavioral level when a reviewer describes the photo taking task "the iPhone 4 takes beautiful photos". A further analysis was carried out to break down the emotional responses into positive and negative states.

4   RESULT

The design specifications of iPhone 4 were extracted from the Apple official website (see Table 1). Other factors were highlighted in the web reviews such as CPU, material, face of iPhone and price.

Table 1 Design specifications of iPhone 4

Features	Features	Features	Features
Face Time	Game Center	Capacity	Accessibility
Retina display	Phone	Size	Search
Multitasking	Mail	Weight	Voice Control
iPod	Safari	Cellular	Air Print
Video Recording	Photos	Power	Air play
Camera	CPU	Headphones	App Store
Folders for Apps	Home Screen	Messages	iMovie
Game Center	OS	Keyboard	iBooks
Keynotes	Color	Numbers	Pages
Price	Face of iPhone	Material	Button

Total 108 emotional responses were found in the 10 web reviews (see Table 2). 14 emotional responses were considered as repetitive words (such as good and great) and the remaining were unique.

Table 2 List of emotional responses found in web reviews

Emotion	Level	Emotion	Level
Resistant	Visceral	Large	Behavioral
extremely fragile	Visceral	slightly poor	Behavioral
Familiar	Visceral	Compact	Behavioral
Nice	Visceral	Dense	Behavioral
Expensive	Visceral	Happy	Behavioral
Shocked	Visceral	Bright	Behavioral
Flat	Visceral	Natural	Behavioral
Stainless	Visceral	Flexible	Behavioral
Shiny	Visceral	Fluid	Behavioral
Varied	Visceral	Annoying	Behavioral
slightly chunky	Visceral	Speedy	Behavioral
Sleek	Visceral	Ultrahigh	Behavioral
Breakable	Visceral	Fantastic	Behavioral
Crisp	Visceral/ Behavioral	Stunning	Behavioral/ Reflective
Sharp	Visceral/ Behavioral	Impressive	Behavioral/ Reflective
Angular	Visceral/ Behavioral	Great	Behavioral/ Reflective
Beautiful	Visceral/ Behavioral	Limited	Behavioral/ Reflective
Nice	Behavioral	very well	Behavioral/ Reflective
Strong	Visceral/ Reflective	Good	Behavioral/ Reflective
Hard	Visceral/ Reflective	Excellent	Behavioral/ Reflective
Thin	Visceral/ Reflective	New	Reflective
Favorable	Behavioral	Eager	Reflective
Visible	Behavioral	Integrated	Reflective
Exciting	Behavioral	Faulty	Reflective
Pleased	Behavioral	Tall	Reflective
Fun	Behavioral	not good	Reflective
Useful	Behavioral	Outrageous	Reflective
Futuristic	Behavioral	Small	Reflective
Easy	Behavioral	Gorgeous	Reflective
Readable	Behavioral	Dazzling	Reflective
increasing accuracy	Behavioral	Narrow	Reflective
Full	Behavioral	not true	Reflective
Clear	Behavioral	Same	Reflective
Admirable	Behavioral	Different	Reflective
Smooth	Behavioral	very responsive	Reflective
high resolution	Behavioral	not noisy	Reflective
Super	Behavioral	very comparative	Reflective
Simple	Behavioral	Lucrative	Reflective
Classic	Behavioral	Improved	Reflective
Glad	Behavioral	Incomplete	Reflective
convenient	Behavioral	Tremendous	Reflective
Usable	Behavioral	high quality	Reflective
irretrievable	Behavioral	Competitive	Reflective
not great	Behavioral	Beefed-up	Reflective
Decent	Behavioral	Quick	Reflective
Solid	Behavioral	Fast	Reflective
Awesome	Behavioral	Super slick	Reflective

Besides that, 85% were positive emotional states and 15% were negative emotional states. Based on three levels of emotional design (see Figure 2), most of the emotional responses were categorized under behavioral level (47%), followed by reflective level (34%) and visceral level (19%) has the smallest share.

486

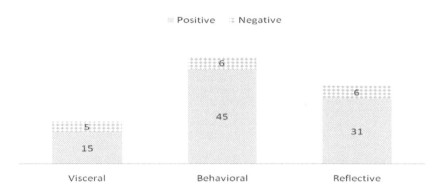

Figure 2 Total number of positive and negative emotional states for each emotional design level.

Design specifications like retina display, photos, video recording and editing, face time, CPU, home screen, and material have the largest share in the number of positive emotional states. Multi-tasking was the only design specification that elicited negative emotional states from all web reviewers. In other words, this design specification failed to meet web reviewers' expectation. The reviews highlighted only positive emotional states for 15 design specifications. It showed that these design specifications were successful in attracting the reviewer's attention.

Retina display, video recording and editing, photos, face time, material, voice control, and price were design specifications that received both positive and negative emotional states. 14 of design specifications did not receive any emotional response (see Figure 3). It can be inferred that these design specifications were not consider by web reviewers in comparison with other specifications. Besides that, the reviewers gave emotional responses for material, size, and photos in all three emotional design levels (see Figure 4). These three design specifications were noteworthy for web reviewers.

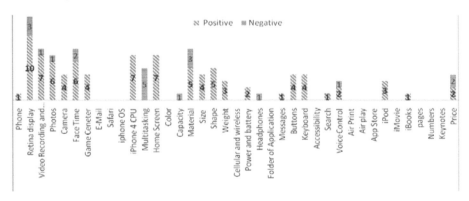

Figure 3 Total number of positive and negative emotional states for each design specification.

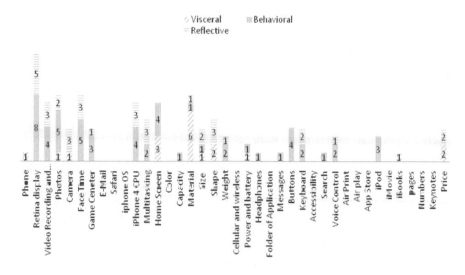

Figure 4 Total number of emotional design levels for each design specification.

5   DISCUSSION

Web reviewers were more concerned on behavioral level compare to visceral and reflective levels. In other words, when evaluating a mobile phone like iPhone 4, web reviewers would perform certain tasks such as taking photos, sending SMS, and making calls. They would describe or summarize their emotional responses as they complete a task. Thus, usability of iPhone 4 is important to ensure that users can complete their tasks effectively and efficiently. This finding confirmed that behavioral level is important for iPhone 4, as proposed by Lim, Donaldson, Jung, Kunz, Royer, Ramalingam, Thirumaran, and Stolterman (2008) and Karray, Alemzadeh, Saleh, and Arab (2008).

5.1   Positive and Negative Emotional States

It is worthwhile to notice that all emotional responses related to CPU and home screen were positive states. This is an important point for mobile designers to learn that all web reviewers satisfied with these design specifications. However, there were also negative emotional states found in web reviews. The mobile designers and developers could improvise the related design specifications instead of removing it completely. Thus, iPhone 4's designers need to know which design specifications can elicit positive or negative emotional state.

Another important point was the relative symmetry in emotional responses about price. Half of emotional responses were considered as negative states (such as expensive and shocked) and the other half were categorized as positive states (such as lucrative and comparative). It was interesting to see that in these web reviews, reviewers complained about iPhone 4's price, but they concluded that iPhone 4's price is also lucrative and comparative after a detailed comparison was completed.

6   CONCLUSIONS

In recognition of the rising popularity of mobile phones among users, this paper has demonstrated the importance of emotional experience in mobile design. The users' emotional experiences were elicited using web reviews as sources of information. Then, it was determined each emotional response relates to which design specification. The results show that behavioral level has the greatest impact on users because this level is closely related to functionality of mobile phone. These emotional responses were also classified into negative and positive states. The need of designers to know which design specifications can elicit positive or negative emotional state was also accentuated. Besides web reviews, future studies should explore alternative methodology to elicit user emotional responses. An in-depth examination may help to find other important results for mobile designers.

ACKNOWLEDGMENTS

This research work was supported by Ministry of Higher Education, Malaysia under Fundamental Research Grant Scheme (FRGS/2/2010/SG/MMU/03/4).

REFERENCES

Bosangit, C., McCabe, S., and Hibbert, S. 2009. What is told in travel blogs? Exploring travel blogs for consumer narrative analysis, Information and Communication Technologies in Tourism, 61-71, Wien, Springer-Verlag.

Carson, D. 2008. The "blogosphere" as a market research tool for tourism destinations: A case study of Australia's Northern Territory. Journal of Vacation Marketing, 14:111-119.

Folkman, S., and Lazarus, R.S. 1985. If it changes it must be a process: Study of emotion and coping during three stages of a college examination, Journal of Personality and Social Psychology, 48(1):150-170.

Garreta-Domingo, M., Almirall-Hill, M., and Mor, E. 2007. User-centered design gymkhana, Proceedings of ACM CHI, 1741-1746.

Guo, F., and Tian, T. 2010. Consumer demand oriented study on mobile phones' form perception design method, *Proceedings of the International Conference on Management and Service Science*, 1-4.

Han, S.H., Kim, K.J., Yun, M.H., Hong, S.W., and Kim, J. 2004. Identifying mobile phone design features critical to user satisfaction, Human Factors and Ergonomics in Manufacturing & Service Industries, 14(1): 15–29.

Karray, F., Alemzadeh, M., Saleh, J.A., and Arab, M.N. 2008. Human-computer interaction: Overview on state of the art, International Journal on Smart Sensing and Intelligent Systems, 1(1):137-159.

Lim, Y., Donaldson, J., Jung, H., Kunz, B., Royer, D., Ramalingam, S., Thirumaran, S., and Stolterman, E. 2008. Emotional experience and interaction design, Affect and Emotion in Human-Computer Interaction, 4868:116–129.

Norman, D.A. 2002. Emotion & design: attractive things work better, Interactions, 9(4): 36-42.

Norman, D.A., 2005. *Emotional Design: Why We Love (or Hate) Everyday Things*, Basic Books.

Norman, D.A., and Ortony, A., 2003. Designers and users: Two perspectives on emotion and design, *Proceedings of the* Symposium on Foundations of Interaction Design at the Interaction Design Institute.

Telecom Regulatory Authority of India. 2011. "Highlights of Telecom Subscription Data as on 30th June, 2011" Accessed December 1, 2011, http://www.telecomindiaonline.com/highlights-of-telecom-subscription-data-as-on-30th-june-2011.pdf

Wenger, A. 2008. Analysis of travel bloggers' characteristics and their communication about Austria as a tourism destination, Journal of Vacation Marketing, 14:169-176.

CHAPTER 54

Invariant Comparisons in Affective Design

Fabio R Camargo, Brian Henson

University of Leeds
Leeds, UK
mnfrc@leeds.ac.uk
B.Henson@leeds.ac.uk

ABSTRACT

The principle of invariant comparisons is a key concept for verifying whether the measurement of affective responses has been achieved. Invariant comparisons show a valid evidence for the relationship between the mathematical properties on numbers and the properties of the affective characteristic that is measured. An empirical investigation demonstrated that the Rasch probabilistic model can implement invariant comparisons across items with free sample distribution, yielding a solid basis for inference. The research aimed to establish a common metric for comparing affective responses to different packaging characteristics through tactile perception when squeezing containers of everyday products. The results point to a range of compliance from which consumers can infer what kind of product they might obtain from the containers.

Keywords: affective design, invariant comparisons, Rasch model, packaging.

1 INTRODUCTION

Integrating established engineering topics with subjective factors has been a challenge. A current problem is that some methods for eliciting affective responses to products have presented shortcomings for violating measurement assumptions. One of the reasons is the difficulty to relate the diverse adjectives that characterize the semantic space to the affective attribute a manufacturer wishes to know about. Another reason is that in current affective approaches frequently the variables are not measured in an interval scale, so that it is not possible to infer reliable meaning to the numbers obtained. The consequence is that various sources of inaccuracy are introduced in the process. This can motivate excessive low predictability for product requirements and difficulties when comparing results.

The principle of invariant comparisons is a key concept for verifying whether the measurement of affective responses has been achieved. Invariant comparisons show a valid evidence for the relationship between the mathematical properties on numbers and the properties of the affective characteristic that is measured. As a result, comparisons can be made by the difference between the numbers associated to responses, where a particular difference has the same interpretation across the measurement scale. Thus, comparisons between individuals can be generalized beyond the particular conditions under which they were observed (Andrich, 1988).

The Rasch model embodies the principle of invariant comparisons (Rasch, 1960, 1980). The formal structure of the model allows independence between persons parameter and items (i.e., variables) parameter, and therefore, their mathematical separation. Accordingly, the Rasch model can overcome many of the problems when measuring responses in affective design (Camargo and Henson, 2011).

This research aimed to establish a common metric for comparing affective responses to different packaging characteristics through tactile perception when squeezing containers of everyday products. An empirical investigation was designed to examine whether after calibration using the Rasch model the scale would yield invariant comparisons. The results indicate there is a range of compliance where consumers can infer on what kind of product they might obtain from the containers when squeezing them.

2   INVARIANT COMPARISONS USING THE RASCH MODEL

In affective design there is clearly an effort to measure whether a variable is more or less endorsable as compared against different design characteristics (i.e., stimuli) or against different individuals. Nevertheless, if a measurement instrument is valid, it should be more than a description of the analysts who create the instrument (Thurstone, 1959). Typically in affective design, analysts have used prevalent models that can characterize the data. A different approach is to identify a model that not only can characterize the data but can meet determined assumptions of measurement. This is the case of the Rasch model.

Georg Rasch was a Danish mathematician who developed a class of models for measurement in education and in the social sciences. The distinctive perspective of the Rasch model, if comparing with traditional approaches, is that any samples from a group of relevant persons, called population in traditional terms, should present invariance of comparisons of independent variables, called items. If a stable relation between items i and j holds, then it is expected the ratio of their relative endorsement rates to remain statistically equivalent irrespective of the people who respond to them. As a significant variation of the ratio is found between different groups of people, then the sources of variation should be investigated to detect anomalies between item content and persons' characteristics. Estimating the ratio between items i and j requires a probabilistic model, which can implement invariant comparisons across items with free sample distribution.

The Rasch model predicts responses' probabilities from two independent parameters; item difficulty and person ability. These terms were derived from the first model's approach in education and are adapted in this paper for a better understanding in affective design. Thus, the term 'difficulty' expresses the proportion of respondents that endorse an item. That is, an item endorsed by comparatively few respondents is more difficult than an item endorsed by many respondents. The term 'ability' represents the readiness of respondents to endorse an item, and therefore, it corresponds to their affective responses.

The model is originally represented by the probability that a person will endorse an item with two-category responses (e.g., yes or no) as a logistic function of the difference between the person's location and the item's location on a linear scale. The model was extended to accommodate scales with more than two categories (e.g., disagree, neutral and agree) (Rasch, 1961; Andrich, 1978; Masters, 2002). Linacre (1989) introduced the concept of facets to the original Rasch model. A facet can be defined as a component or variable of the measurement condition that is assumed to affect the scores in a systematic fashion (Linacre and Wright, 2002). In the particular case of affective design this facet is called stimuli.

The Rasch model incorporates two key assumptions. One is that items count independently in the analysis. That is, the response of a person to an item *i* in the scale should not interfere with his or her response to any other item within the same scale. Another assumption is each and every item must contribute to the measure of a single attribute; i.e., the structure must be unidimensional (Andrich, 1988; Wilson, 2005).

If data fit the model, the requirement of invariance of comparisons will be formalized within a frame of reference, which comprises the group of specified items and the group of targeted persons. Therefore, the comparison between two persons will be independent of the items in the measurement structure and the comparison between any two items will be independent of the persons, yielding a solid basis for inference.

3   METHOD

The scope of the experiment was to measure the relative importance of the packaging material for obtaining an intuitive perception of a delicate cream as a product feature, denoted 'perceptiveness' henceforth. Five products with different characteristics related to compliance of their containers were used to provide a variety of values of physical properties that were of interest to the study. Participants were neither able to see the containers nor able to make contact with the product inside them. After squeezing each container (Figure 1) participants rated their endorsement on a five-point Likert-style scale to statements related to 'perceptiveness' of the product using computer-base self-report questionnaires.

Figure 1 Experiment lay-out for touching a stimulus.

The experiment was split in two stages. The first stage was concerned with the collection of words and statements. The second stage aimed to obtain affective responses to a preliminary set of items and, subsequently, to calibrate the measurement structure.

In the first stage, a preliminary pool of statements was established to capture participants' responses to a set of stimuli related to the context of packaging of everyday products. These statements, such as 'I feel the product in this packaging could be sticky', emerged from words and statements collected by qualitative research using a focus group with six participants, which documented verbatim statements to express affective requirements. Other statements stemmed from publicly available online consumers' reviews, manufacturer catalogues and advertisements for everyday products such as personal care, food, healthcare and household.

3.1   Pool of Statements

Analysis of the transcription and collection of statements and words consisted of clustering them according to their characteristics. These characteristics were arbitrarily elected as those of functionality and usability, graphics and form, and tactile perception. Sixteen clustered statements that represented affective requirements related to or stemmed from characteristics of tactile perception were included in a preliminary pool of items (Table 1).

3.2   Stimuli

Five everyday products with different characteristics related to compliance of their containers were used as stimuli and presented to the participants. Commercially available products were selected according to capacity, dimensions, proportions, packaging material and characteristic of the content (Table 2). The containers were classified as: cylinder tube, oval bottle, downward taper tube and gusseted pouch.

Table 1 Original pool of items and codification of Items

Code	Items
I1	The product in this container would give me a heavy, greasy film on my skin.
I2	The product in this container is likely to look and smell delightful.
I3	I might get a bit watery product in this container.
I4	I feel the product in this container would hydrate my skin.
I5	The product in this packaging might be pricey.
I6	The container feels only half filled when squeezing it.
I7	The container makes me feel like I would be buying a great product.
I8	The product inside the container would spread easily.
I9	There is a lightweight cream in this container.
I10	It is easy to know how much is left in the packaging.
I11	The product inside this container could be sticky.
I12	The product in this packaging is likely to flow easily.
I13	The product in this packaging might seem more medicinal than anything else.
I14	It is quite hard to explain the product when touching its packaging.
I15	The product in this container could give me a refreshing sensation.
I16	The product in this packaging could be a bit boring.

For the compliance measurement a testing system was used that consisted of a force platform (MiniDyn: multi-component dynamometer Type 9256C2, Kistler), an X–Z motion table (Series 1000 Cross Roller, Motion link), a steel ball of radius 7mm, a controller and a PC. The containers were positioned between the steel ball and the force platform. The ball was pressed against the surface of each stimuli and the ball's displacement Dy with increasing load Fy was recorded. The measure of compliance was empirically taken to be the value of Dy (mm) when Fy was 3N (Chen et al., 2009).

3.3 Data Collection and Rasch Analysis of Affective Responses

One hundred and twenty volunteers, 65% male and 35% female, 51.6% between the age of 18 and 25, 31.7% between the age of 26 and 35, and 16.7% over 35, took part in the second stage of the experiment.

Data from participants' affective responses were obtained from ratings collected by computer-based self-report questionnaires. The order in which participants were required to consider the containers was determined using a counterbalanced design. The order of the statements on the questionnaires was automatically randomized.

Table 2 Packaging material and compliance

Packaging	Product	Material	Compliance (3N)
Stimulus 1	Baby food	Polypropylene/aluminum/polyethylene laminated, gusseted pouch.	6.08 mm
Stimulus 2	Toothpaste	Polyethylene/aluminum/polyethylene laminated tube.	5.70 mm
Stimulus 3	Hair conditioner	Low density polyethylene squeeze tube.	4.74 mm
Stimulus 4	Moisturizer	Multi-layer low density polyethylene.	4.11 mm
Stimulus 5	Baby bath lotion	Oval, flat based, multi-layer, high density polyethylene.	1.02 mm

Items were calibrated for each stimulus independently using RUMM2030®, licensed version (2011). Subsequently, items and stimuli were placed on the same continuum through the faceted Rasch approach.

Two test-of-fit were performed in the analysis. Items' fit residual statistics assessed the degree of divergence between the expected value and the actual value for each person-item as summed over all items for a given person, taking as an indicator of fit a SD$\leq$1.4 (Pallant and Tennant, 2007). The second test of fit was a formal test of invariance across the trait. A significant chi-square probability; i.e., p>0.05 indicated invariance across the measurement scale (Tennant et al., 2004).

The item-person interaction score indicated the degree of discrepancy of each person from the model. Residuals between ±2.5 were assumed as random errors (Andrich, 1988). Differential item functioning (DIF) or item bias was also tested through an analysis of variance (ANOVA), which was conducted for each item comparing scores across different respondent groups and across each level of the person factor, in this study sex and age group.

Response dependency was identified by observing high correlations of residuals between items. High correlations were assumed as $\geq \pm 0.3$ in this study (Tennant and Conaghan, 2007). Unidimensionality was tested through the method proposed by Smith (2002) of taking the factor loadings on the first residual component through a PCA. An independent paired t-test comparison verified any difference in the estimates that have been generated. If this value was ≥ 0.05 then the unidimensionality of the scale was deemed acceptable.

After proceeding individual calibration of stimulus items were re-scored and adjusted for a unified, calibrated set for all of the stimuli. The individual data sets that stemmed from calibrated scales were then used as input for the facet approach.

4   RESULTS

A preliminary analysis using the software package RUMM2030® identified significant item-trait interaction, evidencing some misfit to the model. The item fit residual standard deviations for stimulus 1 and the person fit residual standard deviations for stimuli 2, 3, 4 and 5were higher than the arbitrary value of SD = 1.400 established as being maximum acceptable value. Such values along with p<0.05, which pointed to lack of the invariance across the trait, indicated misfit to the model.

4.1   Items' Calibration

The score system was re-coded by applying reversed order for items I1, I11, I13, I14 and I16 for all individual scales. In addition, analyses of individual stimulus scales indicated inconsistent response pattern for some items. The response pattern was identified for each item and every stimulus through disordered thresholds (i.e., transitions between two consecutive categories) (Tennant and Conaghan, 2007). Thus, some items were collapsed to four categories.

The person-item correlation analysis combined with the individual item-fit analysis evidenced items with potential misfit to the model. Furthermore, a common set of items was sought to fit all stimuli. Thus, items I1, I2, I4, I5, I7, I8, I11, I12, I14, I15 and I16 were found to be the most balanced solution (Table 3).

Item bias, called differential item functioning (DIF), was tested for sex and age groups after re-scoring and removing items, indicating non-significance. That is, those person factor groups demonstrated similar ability to endorse an item.

Analysis of individual person-fit statistics indicated patterns of responses with high residuals. Thus, 13 participants for stimulus 1, 12 participants for stimulus 2, nine for stimulus 3, seven participants for stimulus 4 and three participants for stimulus 5 were removed from the analysis.

The fit of data to the model was examined based on the set of items and the sample distributed into three groups of persons ability for each and every stimulus. Rasch analysis identified a non-significant item-trait interaction, which was deemed as an evidence of that the data fit the model (Table 3).

The model's assumption of unidimensionality was met through an independent t-test, which indicated that less than or equal to 5% of observations were expected to fall outside of the confidence interval in the range of ±1.96 for every stimulus.

Table 3 Fit statistics for the calibrated scales

Stimulus	Item-fit residual		Person-fit residual		Chi-square	df	p	N	Items	PSI	<95%CI
	Mean	SD	Mean	SD							
ST1	0.21	0.77	-0.28	1.21	24.6	22	0.32	107	11	0.73	0.02
ST2	0.22	0.86	-0.25	1.15	29.0	22	0.14	108	11	0.71	0.03
ST3	0.01	0.80	-0.36	1.22	22.6	22	0.42	111	11	0.77	0.05
ST4	0.06	0.86	-0.40	1.20	27.7	22	0.18	113	11	0.73	0.03
ST5	0.07	0.96	-0.34	1.15	30.7	22	0.10	117	11	0.66	0.05

4.2 Results from Faceted Rasch Analysis

The faceted Rasch approach allowed comparisons and interpretations of results on a single frame of reference for all stimuli. The effect of different characteristics when comparing containers is demonstrated by the prevailing tendency among individuals of endorsing the affective attribute 'perceptiveness'.

An arbitrary zero was established as the default of the method applied in the analysis. The default for the origin constrained the stimuli facet and the items facet at the centre of the logit scale. That is, either facet had measurement mean of zero. Also, the sum of the category coefficients was constrained to zero. Thus, solely the persons facet floated on the continuum.

The fit statistics for the calibrated scale indicated invariance across the measurement structure, presenting chi-square (df=110) of 121.64, p=0.21.

The facets map (Figure 2) is the representation of the relative locations of all facets on the same logit scale. Person locations are plotted on the scale in the first column. Participant locations that indicate more inclination to endorse the attribute 'perceptiveness' of a light cream product are plotted on the top of the scale and those less inclined to endorse at the bottom. The top of the facets map at the second column indicates items with more difficulty to endorse; i.e., items that were less prone to be endorsed by participants. Finally, the third column in the facets map represents the facet stimuli. Stimuli more prone to be endorsed with regard to

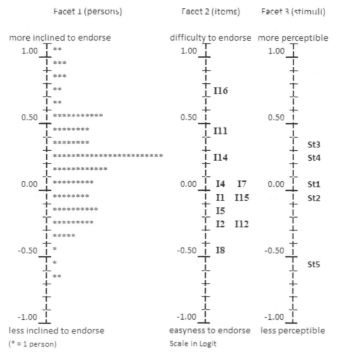

Figure 2 Facets map with a common metric in logit.

'perceptiveness' of a light cream product are on the top of the scale; i.e. stimuli St3 and St4. The scale also indicates lower probability to endorse the affective attribute for stimulus St5. Furthermore, the threshold distribution is widely spread (Figure 3), revealing the respondents are well targeted to the set of calibrated items although the spread of persons on the continuum is comparatively narrow.

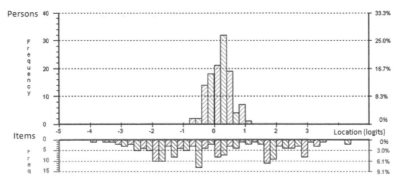

Figure 3 Person-item threshold distribution using faceted Rasch model

5   DISCUSSION

During calibration some of the items were discarded as a consequence of redundancy or inappropriateness in the context. Item I3, 'I might get a bit watery product in this container', and item 12, 'the product in this packaging is likely to flow easily', presented high correlation. In classical theory, for example, items with very similar semantic meanings could inflate the results due their redundancy. Item 13, 'the product in this packaging might seem more medicinal than anything else', presented misfit to the model for four out of five stimuli. In this case, the item could not be contributing meaningfully to the affective attribute being investigated.

The location of stimuli on the continuum demonstrated that the container with lower compliance is located at the bottom of the scale in relation to the stimulus with higher compliance, indicating lower degree of endorsement to 'perceptiveness' for the first. On the other hand, the map indicates that according to participants' perception there is an intermediate range of compliance subject to be more prone to relate a container to a light cream product. Nevertheless, that range does not follow the order of the physical measurement for the stimuli compliance. Furthermore, the degree of endorsement was associated to the group of stimuli without taking into account any relationship regarding whatsoever the product inside the container.

The invariant comparisons property of the Rasch model has been reached within the frame of reference of the study. This can be claimed given that items' difficulty are independent of the distribution of abilities in the relevant group of respondents and person ability estimates are independent of the set of items used for estimation. Because there is a stable relation between items after their calibration, the ratio of their relative endorsement rates will remain statistically equivalent.

6   CONCLUSION

Rasch analysis evidenced the feasibility of a preliminary scale for the "perceptiveness" of some products when squeezing their containers. Because the scale has met the essential properties of additivity, constant unit and invariant comparisons for a measurement structure, it is possible to assume that comparisons between persons, between items and between products through the common metric are valid. Based on this metric, the results suggest there is a range of compliance where consumers can infer   what kind of product they might obtain from the containers.

REFERENCES

Andrich, D. 1978. A rating formulation for ordered response categories. *Psychometrika* 43 (4): 561-573.

Andrich, D. 1988. *Rasch models for measurement*. Sage University Papers series on Quantitative Applications in the Social Sciences, No. 68, London: Sage Publications.

Camargo, F.R. and Henson, B. 2011. Measuring affective responses for human-oriented product design using the Rasch model. *Journal of Design Research* 9 (4): 360–375.

Chen, X., Shao, F., Barnes, C., Childs, T. and Henson, B. 2009. Exploring relationships between touch perception and surface physical properties. *International Journal of Design* 3 (2): 67-77.

Linacre J.M., Wright B.D. 2002. Understanding Rasch Measurement: Construction of Measures from Many-facet Data. *Journal of Applied Measurement* 3 (4): 486-512.

Linacre, J.M. 1989. *Many-facet Rasch measurement*. Chicago: MESA Press.

Masters, G.N. 1982. A Rasch model for partial credit scoring. *Psychometrika* 47: 149–174.

Pallant, J.F. and Tennant, A. 2007. An introduction to the Rasch measurement model: an example using the Hospital Anxiety and Depression Scale (HADS). *British Journal of Clinical Psychology* 46: 1–18.

Rasch, G. 1960, 1980. *Probabilistic models for some intelligence and attainment tests*. (Copenhagen, Danish Institute for Educational Research), expanded edition (1980) with foreword and afterword by B.D. Wright. Chicago: The University of Chicago Press.

Rasch, G. 1961. On general laws and the meaning of measurement in psychology. *Proceedings of the 4th Berkeley Symposium on Mathematical Statistics and Probability*, Vol. 4: 321-334, Berkeley: University of California Press.

Smith, E.V. 2002. Detecting and evaluating the impact of multidimensionality using item fit statistics and principal component analysis of residuals. *Journal of Applied Measurement* 3 (2): 205–231.

Tennant, A. and Conaghan, P.G. 2007. The Rasch measurement model in rheumatology: what is it and why use it? When should it be applied, and what should one look for in a Rasch paper? *Arthritis & Rheumatism* 57 (8): 1358–1362.

Tennant, A., McKenna, S.P. and Hagell, P. 2004. Application of Rasch analysis in the development and application of quality of life instruments. *Value in Health* 7 (1): S22-S26.

Thurstone, L.L. 1959. *The Measurement of Values*. Chicago: The University of Chicago Press.

Wilson, M., 2005. *Constructing measures: an item response modeling approach*. Mahwah, New Jersey: Lawrence Erlbaum.

Section VII

Design and Development Methodology

CHAPTER 55

Invariant Comparisons in Affective Design

Fabio R Camargo, Brian Henson

University of Leeds
Leeds, UK
mnfrc@leeds.ac.uk
B.Henson@leeds.ac.uk

ABSTRACT

The principle of invariant comparisons is a key concept for verifying whether the measurement of affective responses has been achieved. Invariant comparisons show a valid evidence for the relationship between the mathematical properties on numbers and the properties of the affective characteristic that is measured. An empirical investigation demonstrated that the Rasch probabilistic model can implement invariant comparisons across items with free sample distribution, yielding a solid basis for inference. The research aimed to establish a common metric for comparing affective responses to different packaging characteristics through tactile perception when squeezing containers of everyday products. The results point to a range of compliance from which consumers can infer what kind of product they might obtain from the containers.

Keywords: affective design, invariant comparisons, Rasch model, packaging.

1 INTRODUCTION

Integrating established engineering topics with subjective factors has been a challenge. A current problem is that some methods for eliciting affective responses to products have presented shortcomings for violating measurement assumptions. One of the reasons is the difficulty to relate the diverse adjectives that characterize the semantic space to the affective attribute a manufacturer wishes to know about. Another reason is that in current affective approaches frequently the variables are not measured in an interval scale, so that it is not possible to infer reliable meaning to the numbers obtained. The consequence is that various sources of inaccuracy are

1 EMOTIONS AS BASIS FOR PURCHASE DECISIONS

Purchasing decisions do not only depend on rational criteria such as technical and functional compliance. Product quality is accordingly composed of the inseparable dimensions of Protective Quality and Perceived Quality. Whilst Protective Quality is based on the technical functionalities, Perceived Quality describes the realization of product characteristics, which influence the customer's experiences and emotions. The active design of Perceived Quality consequently enables a company to produce a product which technically meets the basic requirements and additionally fosters the development of emotionally inspiring product characteristics (Schmitt and Pfeifer, 2010; Kano, 1984).

The product development process (PDP) is the core process to transfer requirements into products with high Perceived Quality. The PDP defines the product on the basis of an abstract *idea* which adapts desirable product characteristics to customer demands. The subsequent step is the product development conducted on the basis of an approximate *concept* of the product and its stepwise realization. Moreover, *component development* and *process planning* are necessary in order to realize *prototypes* as well as the *pilot production* (Ehrlenspiel, 2009). As the PDP proceeds, the determination of product characteristics restricts the company's possibilities for design alterations and technical adjustments and, subsequently it is crucial to focus early phases since the expenses for the correction of failures are relatively low (Lindemann, 2005; Schmitt and Pfeifer, 2010). Hence, the awareness about customer emotions at early phases increases the possibility to overlap technical innovative proposals generated by the development department with customer emotional perception and avoids costs arising from a lack of emotional design. The early integration of target groups and data about their emotional perception help to choose objectively between alternatives. Evoked stimuli and triggered emotions should be measured and finally used for the development of positively perceived products.

Customers perceive signals and stimuli sent by the product individually with the human senses. These measurable stimuli are transformed into individual patterns which evoke physiological reactions. Subsequently they are processed, pre-selected and transferred to the human brain. Finally, the stimuli are experienced through apperception by the customer, which serves as the premise for the evaluation of a product's Perceived Quality. Apperception and evaluation are triggering an emotion, which – if it is sufficiently positive – enhances the wish to purchase a product (Sigg, 2009; Zeh, 2010; Schmitt and Pfeifer, 2010). This raises the question which decisions are needed in order to design a product, which evokes a positive emotional response during the purchasing decision process.

Many concepts for emotional design are available answering different questions (see for this discussion Jiao et al., 2007). A widespread scientific engineering method is Quality Function Deployment (QFD) that relates customer requirements and technical items (Akao, 1990). A method, which is more relevant in market research, is Conjoint Analysis (Green and Srinivasan, 1978). The combination of emotions and perception with product components is a sophisticated task that requires a differentiated stepwise procedure. Kansei Engineering is a conceptual approach which identifies systematically the most important users' requirements and predicts statistically how emotions are connected to components (Nagamachi, 1989; Schütte, 2005). The procedure starts by correlating a semantic space or ontology (Jiao et al., 2007) with the space of components. Several studies exist to apply Kansei Engineering in several branches and for different product types.

Although those approaches are holistic, they lack the ability of quantifying emotions by measuring physiological signals during the current PDP in order to derive implications for current product development. Commonly, this is considered as too challenging (Schütte, 2005). During early steps of the PDP, the visual impression of product concepts is more important than others (e.g. haptic, acoustic) to evoke emotions. Thus, there is a high potential to concentrate on the evaluation of visual impression, e.g. by using gaze tracking (Duchowski, 2007). Similarly, established approaches lack the systematic and safeguarded integration of emotional information into the PDP by using quality management tools, like Quality Gates (Schmitt and Pfeifer, 2010).

2 A SYSTEMATIC APPROACH FOR OBJECTIFYING USER EMOTIONS IN EARLY PHASES OF PRODUCT DEVELOPMENT

According to the critical review of established approaches for emotional product design, this chapter presents a new methodology containing both the objectified measurement of emotions and the safeguarded integration of objectified information into the PDP. It is a five-stepped approach extending other methods of emotional design by focusing on the early phases of product development (see Figure 1).

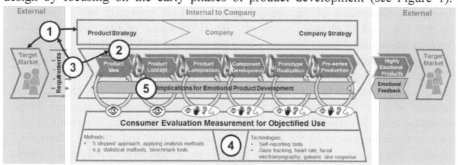

Figure 1: Approach for systematic Consumer Evaluation Measurement for Objectified Use – The five steps for securing Emotional design are located in the Aachen Quality Management Model (From Schmitt and Pfeifer, 2010).

This procedure is developed during the research project CONEMO (Consumer Evaluation Measurement for Objectified Industrial Use). Figure 1 illustrates the five areas of research and locates the approach in an entrepreneurial context. Three levers affecting the survey of customer emotions are elucidated: the internal company orientation, the internal company skills and the external customer perspective (Schmitt and Pfeifer, 2010).

(1) Generating a semantic framework

Already in the *idea* phase of the PDP, it is important that both company and target customer group are "talking about the same things". Therefore, the central goal is to align the company's orientation with the customer perspective. The company strategies for upcoming products are transferred into describing words, called semantic words. Supporting methods are workshops and semi-structured interviews conducted with company representatives. Further, the company has to identify the potential target groups noticing demographic criteria in order to survey external customers' expectations regarding the upcoming product. The semantic customer perspective is defined by using standardized questionnaires and open interviews. The assessment of these words applying semantic scales should be done without knowing the future product. Correlation analysis is used in order to assess the consistency between customers' and company's requirements regarding a future product strategy. Appropriate methods to reduce the amount of words are affinity diagrams, factor analysis and structural equation modeling. The alignment can easily be displayed within strategy requirements catalogues or strategy matrices overlapping both semantic spaces in order to find a common framework and to identify the according product strategy.

(2) Determination of the product structure

The deduction of the product structure from sketches or even computer rendered pictures is the central goal of this step. A technical method for conceptual disassembling a system into its subcomponents is the system and function analysis.

Schmitt and Pfeifer, 2010 provide an approach to subdivide a product beginning from the overall product impression perceived by a customer down to technical parameters: elements perceived as one entity are called perceptive clusters, which again can be divided into quality attributes. These quality attributes are described by descriptors and their correlation with technical parameters leads to physical quantities. Regardless of the applied approach, it is important not to pass an appropriate level of detail. Due to the shown relevance of the early phases of product development (see Chapter 1), both methods should be applied.

(3) Alignment of strategy and structure

The next step consists of aligning the three levers affecting quality. This can be achieved by connecting the derived semantic framework (1) with the deduced product structure (2). Methods that can be applied during this step are brain storming, mind mapping as well as entity-relationship-diagrams. The result of this step is a matrix presenting the linkage between structure and strategy identifying product components that are more related to a semantic word than others. Moreover, sub strategies can be derived for single semantic words as well as for single product components.

To objectify data gained from emotional evaluations, different methods have to be applied for validation. On the one hand self-reporting methods, for instance semi-structured interviews and standardized questionnaires, can be recommended to gather explicit information. On the other hand, methods to analyze visual signals, like eye tracking, should be applied. For a certain time period gaze tracks, the order and the duration of eye fixations are recorded in order to deduce the attention provoked by certain elements. The visual impression is influenceable, as the gaze track is affected by current thoughts. Therefore, while thinking of a semantic word, the attention and the view are focused on the components that are considered as connected to this word (Yarbus, 1967). Thus, it is possible to capture the implicit customer requirements. To validate results, a comparison of both self-reporting-methods and gaze tracking identifies similarities and differences between explicit expressed and implicit recorded customer requirements. The most relevant aspects and data are pointed out with methods like Pareto-Analysis.

(4) Capturing and evaluating emotions
The central aspect of the presented methodology is the objectified measurement of emotions during product development. Therefore, special attention is paid to the application of tools and methods to capture emotions. The application of appropriate methods is part of the skills that a company possesses (see Figure 1). As seen in traditional approaches, self-report tools are possible methodologies to obtain information about the conscious part of customers' emotions. But to generate a holistic knowledge about customers' emotions, this is not sufficient.

Moreover, measuring the changes of physiological signals, simultaneously to the interaction with the products, provides online information about the decision which design option provokes the highest emotional value. In general, emotions can be described with a categorical and a dimensional appraisal. Categorical appraisals interpret emotions as discrete elements. In order to deduce variable technical parameters and specifications it is more suitable to follow the dimensional appraisal which presents emotions in a space built up by two dimensions: the valence and the arousal. The arousal is similar to the emotional intensity and can be measured by e.g. heart rate, blood pressure and galvanic skin response. Whether an emotion is positive or negative is defined by the valence (Cacioppo et al., 1993).

To improve the reliability, to validate and to reinforce the concept of applying methods to capture emotions during the PDP, both modalities are applied. Thus, differences between explicit and implicit requirements as well as emotions can be figured out. In combination with the structure-strategy-matrix from step (3), this provides valid information about the product component provoking a certain emotion and the product component which is linked to a certain strategy. This conclusion also refers to entire products as well as special design rules (Prefi, 2003) for composing product components. Following the recommendations of step (4), hard facts for emotional product design can be deduced.

(5) Implications for the product development process
In this step the most useful information about customer emotions are extracted and integrated into the PDP. The presented concept provides the necessary information regarding perceived product quality to pass the Quality Gate and enter the next phase. The implementation of additional methods and gained information about emotions requires a modification of the current PDP. The extension of Quality Gates to Customer Emotional Quality Gates assures quality and secures emotional value of new products. Step (1) delivers a justified requirement list for determining a strategy. This is the essential information to generate a *product idea* and to start the PDP. After generating this semantic framework, the focus should move on to the

product concept (2). Focusing on the most important structure-strategy-combinations (3) based on explicit meanings and on visual perception, the amount of design alternatives is to be reduced in an early step of the PDP. In addition, analyzing the structure-strategy-matrices is a possibility to identify systematically innovative product alternatives. Step (4) provides valid data to select those design alternatives provoking more positive visual emotions to the customer than others. To exploit the gained emotional information within the overall development process, the current communication structure has to be adapted by methods to analyze internal processes (Prefi, 2003).

3 RESULTS AND INDUSTRIAL BENEFIT

The presented procedure was tested stepwise during a preliminary study. The considered product was an optically refined SUV (Sport Utility Vehicle) and the presented alternative was its base car model. The CONEMO partner, responsible for the body refinement, is a small batch automobile refiner specialized in British car brands. Twenty-five test persons assumed as automobile affine participated.

(1) Generating a semantic framework
A qualitative semantic framework of 14 words describing the product was reduced systematically, by using affinity analysis and the elimination of words that might have caused misunderstandings. For purposes of illustration, only the words *dynamic* and *powerful* are presented in this paper.

(2) Determination of the product structure
As the study concentrated on the visible parts of the car, the vehicle body was divided into clusters of perception and single car components by using the model of subdivision of the overall impression (see Chapter 2). Finally 21 perceivable components in the front of the car and 20 components in the back of the car were found. A technical system analysis could be done to verify the components.

(3) Alignment of strategy and structure
In order to contrast explicit expressions and implicit visual perceptions, firstly the test persons were asked to name the most applicable parts to each semantic word. Secondly, a gaze-tracking-study was conducted. To simulate an early phase of the PDP, 2-dimensional computer-generated photos representing the cars in typical positions like they are used e.g. in marketing brochures were analyzed. To level the visual perception, the performance of the study starts with 15 seconds per picture of tracked free examination. A Pareto-Analysis for the refined car revealed that during 80% of the time nine components were focused; most of them in the front: Grill, Engine Bonnet, Side-Door, Bumper, Number Plate, Signature, Rims, Windscreen and Headlights. A similar result was raised for the base model (see Figure 2).

COMPONENTS	BUMPER	ENGINE BONNET	ENGINE COVER NOSTRILS	FOG LIGHTS	GRILL	HEADLIGHT LEFT	HEADLIGHT RIGHT	MIRROR LEFT	MIRROR RIGHT	NUMBER PLATE
BASE CAR	6,7%	12,2%	0,5%	6,4%	13,7%	7,5%	0,1%	3,3%	0,0%	8,3%
REFINED CAR	10,5%	10,7%	1,4%	5,5%	12,0%	6,0%	0,2%	1,1%	0,1%	9,7%

Figure 2: Relative observation times (free examination)

As gaze trails change depending on a certain task, for the next steps the test persons were instructed to look for elements that apply to one of the semantic words. The comparison of strategy-structure-matrices enables to contrast the results of questioning and eye tracking. The left part of Figure 3 demonstrates the average times each component was looked at. These are normalized and ranked from "no observation at all" (0) to "a very long time of observation" (10). On the right part, the results of the questionnaire asking the subjects to name product components, which they associate with the particular semantic word, are shown. The ranking is similar to the left side: "not named at all" (0) to "named the most" (10).

Relative Observation Time Value

PRODUCT COMPONENTS	DYNAMIC	POWERFUL	SPORTY	YOUNG
SIGNATURE	3	3	3	3
ENGINE COVER NOSTRILS	1	1	2	1
RIMS	3	4	4	4
HEADLIGHTS	8	6	10	9
GRILL	2	2	1	1

Value of Explicit Naming of Product Components

PRODUCT COMPONENTS	DYNAMIC	POWERFUL	SPORTY	YOUNG
SIGNATURE	1	1	3	7
ENGINE COVER NOSTRILS	5	3	2	4
RIMS	10	6	8	10
HEADLIGHTS	7	0	9	7
GRILL	5	7	2	0

Importance Ranking	
0	not at all
1-3	
4-7	
8-9	
10	very much

Figure 3: Strategy-Structure-Matrices

A closer look at the explicit product components regarding the semantic word *dynamic* shows that the rims were indicated quite often (normalized to 10), but the test persons rarely looked at them (normalized to 3). These discrepancies can be recognized all over the tables. Therefore the hypothesis can be stated that there is a significant difference between the subjective answers to an usual questionnaire and perceived components. On the other hand the headlights were implicitly (left) and explicitly (right) rated very similar for the semantic words *dynamic*, *sporty* and *young*. The explicit impression of the headlights is not *powerful* at all, while they were observed for a certain time due to the recording of eye tracking.

(4) Capturing and evaluating emotions

To learn more about the emotional valence for each test person, additional questions, asking for the degree of the product's performance for each semantic word, had to be answered according to a 5-point-scale. The refined SUV was rated very *powerful* and, significantly, it is even more *powerful* than the base model (two-sample-t-test, mean=1.100, p=0.000). Regarding *dynamic*, it is more complicated to draw a conclusion since the graph in Figure 4 is more equaled. In addition, the statistical analysis does not show a significant difference between the *dynamic* impression of both alternatives (two-sample-t-test, mean=-0.240, p=0.298).

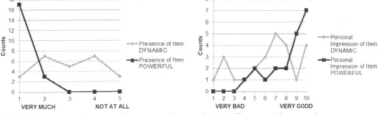

Figure 4: left: Presence of a specific semantic word; right: Personal impression

The results shown in the right part of Figure 4 reflect the personal explicit emotion regarding the degree of the semantic words; rated on a 10-point-scale. For the semantic word *powerful* the graph gives a clear image of the subjects' impressions: the high degree, recognized before, receives a highly positive echo. According to *dynamic* the deduction of a concrete interpretation is more difficult and further objective measurements are required to draw a valid conclusion.

(5) Implications for the product development process

Some findings relevant for the company's PDP are summarized: The gaze tracking results were connected to design rules, which are empirically provoking good valence to users. A centric *dynamic* line was recognized by the test persons, whereas the focus for the *powerful* observation is on the simple and clearly structured geometric forms (e.g. engine bonnet). Special guidelines will be deduced helping the company to be aware of the information needed to pass each Quality Gate. Because of the results from (1) to (4) it can be confirmed that the company's intention to create a *powerful* SUV was achieved, visually recognized (see results of eye-tracking) by customers and evaluated as a good semantic concept. Therefore, *powerful* is a good product strategy for the step of *product idea*. For the step of the *product concept*, the company is equipped with the awareness, that its product alternative evokes a higher emotional response to *powerful* than the base model. Moreover, by using the structure-strategy-matrix, the company can define special component strategies. Nevertheless, physiological measures would validate these findings for *powerful* and would allow more detailed interpretation for *dynamic*.

4 DISCUSSION

In a critical review the design of the study, its results and the five-stepped methodology have to be analyzed. Regarding the study design, some useful lessons learned can be deduced for practical application of the concept. Due to a high number of differences between the refined car and the base model, the interpretation of some results could not be clearly referred to one single component. That is why it can be recommended to apply Design of Experiments for future studies. It has to be investigated in how far the eye tracking calibration process and an unnatural environment during the study influenced the subjects' emotions. Furthermore, the order of the product pictures should vary in each tracking round to prevent boredom and to avoid learning effects.

Showing the product from different angels, or even to track the subjects' views of a product video could analyze the influence of the fact that centrally located product components seem to be focused more than others. Some of the defined components are small, so the tracking system rarely recorded an observation time. One of the advantages of the eye tracking methodology is that the product components can be redefined and the data recalculated. Furthermore the results of the study can be analyzed in different ways like the order, amount and duration of fixations for each component. For some semantic words the collected data of the questionnaires showed lots of overlapping probability distribution. A reason could be a misunderstanding of semantic words. Thus, words for strategy have to be simple and understandable. An online-based questionnaire linked to company's web page and social network is planned in order to gain more semantic words from the target group. Afterwards the results would have to be reduced again by applying multivariate statistical methods like factorial analysis and cluster analysis. These questionnaires could also prevent missing values that were detected and that possibly are caused by a lack of concentration answering to the paper version. In order to analyze equaled distributions (e.g. presence of *dynamic*) and to assess the target group, the application of discriminant analysis is recommended. Another fact to be investigated is that implicit and explicit perception differs for certain components at certain semantic words. Responsible for this could be the subconscious part of human perception. Thus, eye tracking combined with recording of physiological signals give more information about the real perception than usual questionnaires.

The five steps strictly applied, lead to helpful results for the company to achieve a better understanding of the customers' perception. It is not a limitation of the presented approach to start the preliminary study analyzing only visual impressions, since 80% of the overall perception at the point of purchase is covered by the visual perception (Berghaus, 2005). Nevertheless, further research could include the recording of haptic, acoustic and olfactory perception. For reasons of validation and calculation of reliability, empirical studies with a higher amount of test persons will derive more significant data. Furthermore, the different matrices (e.g. semantic-matrix, structure-strategy-matrix) should be connected in order to describe the five steps in a more mathematical manner.

5   CONCLUSIONS AND FUTURE RESEARCH ACTIVITIES

Purchase decisions are highly depending on the emotions evoked by the product. Consequently, companies have to be aware of the customer emotions in a very early phase of product development. To resolve this challenging task, this paper has proposed a systematic and holistic five-stepped innovation methodology to measure objectively the customer emotions provoked by design alternatives. The combined gathering using self-reported and physiological methods delivers new information for the development of emotionally perceived products. In order to secure the appropriate usage of this additional data Quality Gates should be applied, since each step of the PDP contributes to emotional product development.

The presented approach was tested to reveal its practical application. The potential of analyzing visual signals in a very early step of PDP could be highlighted. Further research with empirical studies will be done in order to validate and to generalize the presented concept. Objectifying the achieved results by combining them with the quantification of physiological signals, will be part of future activities for evaluating the reliability of the presented methods. This increases the valid input for quality management methods (e.g. QFD) to safeguard emotional product design.

512

ACKNOWLEDGMENTS

The paper presents results from cooperative research project CORNET CONEMO of the Laboratory for Machine Tools and Production Engineering (WZL), RWTH Aachen University, Germany together with the Biomechanics Institute of Valencia, Spain.

The CORNET promotion plan 47EN of the Research Community for Quality (FQS), August-Schanz-Str. 21A, 60433 Frankfurt/Main has been funded by the AiF within the programme for sponsorship by Industrial Joint Research and Development (IGF) of the German Federal Ministry of Economic Affairs and Technologies based on an enactment of the German Parliament.

The authors would like to express their gratitude to all parties involved.

REFERENCES

Akao, Y. 1990. *History of Quality Function Deployment in Japan.* International Academy for Quality Books Series. Vol. 3. 1990: Hansa Publisher.

Berghaus, N. 2005. *Eye-Tracking im stationären Einzelhandel: Eine empirische Analyse der Wahrnehmung von Kunden am Point of Purchase.* Köln, Germany.

Cacioppo, J.T., Klein, D. J., Berntson, G. G. and Hatfield, E. 1993. *The Psychophysiology of Emotion.* In Lewis, R. and Haviland, J. M. (Eds.). *The Handbook of Emotions* (pp. 119-142). New York, USA.

Duchowski, A. T. 2007. *Eye Tracking Methodology: Theory and Practice,* Vol. 2. London: Springer-Verlag, Great Britain.

Ehrlenspiel, K. 2009. *Integrierte Produktentwicklung: Denkabläufe, Methodeneinsatz, Zusammenarbeit.* Hanser. München/ Germany, Wien/ Austria.

Green, E. P. and Srinivasan, V. 1978. *Conjoint Analysis in consumer research.* Journal of consumer research, Vol. 5, September.

Jiao, R. J., Xu, Q., Du, J., Zhang, Y., Helander, M., Khalid, H. M., Helo, P. and Ni, C. 2007. *Analytical affective design with ambient intelligence for mass customization and personalization.* In: Int. Journ. of Flex. Manufac. Systems, Vol. 19, No. 4, pp. 570-595.

Kano, N., Seraku, N. and Takahashi, F. 1984. *Attractive quality and must be Quality.* Quality, Vol. 14 No.2, p. 39-44.

Lindemann, U. 2005. *Methodische Entwicklung technischer Produkte.* Springer. Berlin, Germany.

Nagamachi, M. 1989. *Kansei Engineering.* Kaibundo, Tokyo.

Prefi, T. 2003. *Qualitätsorientierte Unternehmensführung.* P3 – Ingenieurgesellschaft für Management und Organisation. Aachen, Germany.

Schmitt, R. and Pfeifer, T. 2010. *Qualitätsmanagement. Strategien – Methoden – Techniken.* Hanser. München/ Germany, Wien/ Austria.

Schütte, S. 2005. *Engineering Emotional Values in Product Design: Kansei Engineering in Development.* PhD thesis, Linköping, Sweden.

Sigg, B. 2009. *Emotionen im Marketing. Neuroökonomische Erkenntnisse.* Bern, Switzerland.

Yarbus, A. L. 1967. *Eye Movements and Vision. Plenum.* New York, USA.

Zeh, N. 2010. *Erfolgsfaktor Produktdesign.* Beiträge zum Produkt-Marketing, No. 45, Fördergesellschaft Produktmarketing e.V.. Köln, Germany.

CHAPTER 56

The Effect of Web Page Complexity and Use Occasion on User Preference Evaluation

Shu-Ying Chiang 1, Chien-Hsiung Chen 2

Department of Visual Communication Design
China University of Technology 1
Department of Industrial and Commercial Design
National Taiwan University of Science and Technology 2
Email address: sharron.chiang@gmail.com

ABSTRACT

For years, determining how to design appealing web pages has been taken as an important issue by many researchers and designers. Previous studies have suggested that users' first impression is a key factor affecting their willingness to stay on that particular web page. The feelings resulting from actual usage affect the probability of revisiting that web site (Lee & Koubek, 2010). Several studies on interface design have adopted Berlyne's theory to investigate users' preferences in regard to certain user interface designs (Pandir & Knight, 2006; Tuch et al., 2009). However, in previous research studies, the definitions relating to web page complexity remain deficient. Thus, interface designers have a limited understanding of how web page complexity affects users' preference judgment. Hence, this study aims to investigate the issue of satisfying user preferences. Two main aspects are discussed in this study: use occasion and web page complexity.

The experimental design included menu positions, i.e., upper menu, left menu and right menu. Four column amounts were: one column, two columns, three columns and four columns. The experiment adopted a 3 (menu position) ×4 (amount of column) design. A total of 12 stimuli were applied in the experiment. All of the variables were tested as within-subject variables. The experiment was conducted in two stages. The first stage investigated user preference, based on users' first impression, using a seven-point Likert-type scale (1 denotes most disliked, 7 denotes most liked). The second stage investigated their preference after interacting with the web site user interface. The Semantic Differential Scale was used to help evaluate 9 indexes: like-dislike, complex-simple, beautiful-ugly, familiar-unfamiliar, static-dynamic, light-heavy, rational-irrational, traditional-modern and typical-innovative. After that, all of the participants were required to describe their subjective feelings based on an open-ended questionnaire.

The data were analyzed through descriptive statistics, t-test and factor analysis. The factor analysis yielded three factor names: constitutive property, joyful and contemporary. The descriptive statistics and t-test analyses revealed that both factors of grid complexity and use occasion affected user preference judgment. Before their actual interaction with the web site, users' preference scores rose when the grid amounts increased. After their interactions, the preference scores still rose with the increased grid line. Users' preference scores before their actual interactions were higher than those after the interactions. Moreover, after the actual interactions, participants offered inconsistent opinions in the open-ended questionnaire. Overall, the most popular menu positions were left menu and upper menu, and the most popular amount of columns was three columns. The results disagree with Le Corbusier (1949), who posited that grids cannot influence emotional reactions. The findings of this study proved that grids are able to generate emotions related to light-heavy, static-dynamic, simple-complex and rational-emotional. Moreover, different items can deliver different degrees of emotions by changing the grid amounts.

The preference score of the first impression is significantly higher than that after interacting with the web site user interface. The generated results provide preference understanding of web page complexity and use occasion. The outcomes will form the basis for the application of design guidelines for future web page design.

Keywords: Grid system, Inverted U-curve, User preference, Familiarity

1. INTRODUCTION

The design of appealing web pages that attract users' attention has been an issue concerned by both scholars and designers. In fact, users' first impression is a key factor that affects their intention to stay on a web page. Since the after-use feeling will affect the probability of revisiting the web site, designers should regard the quality of web page under different timing. Recent studies have discussed the preference issue in regard to web pages (Tractinsky & Ikar, 2000; Lindgaard & Dudek, 2003; Park, Choi & Kim, 2004; Lavie & Tractinsky, 2004; Pandir & Knight, 2006; Tractinsky et al., 2006; van Schaik & Ling, 2009; Tuch et al., 2009). Some of researches focused on web page complexity (Pandir & Knight, 2006; Tuch et al., 2009), and some focused on familiarity (Oulasvirta et al., 2005; Santa-Maria & Dyson, 2008). They attempted to find the design guideline on web pages, which can help to enhance both the aesthetics and preference. Despite the fact that familiarity, complexity and preference have long been discussed in the field of psychology (Crandall, 1967; Berlyne, 1970; Shinskey, 2010), they have rarely been mentioned in the field of interface design. Hence, this study aims to investigate the preference improvement issue of users. The two focuses of this study are: complexity of web page and use occasion.

In the field of psychology, Berlyne's (1970) theory suggested that both extreme complexity and extreme simplicity can cause negative feelings, thus, a moderate amount of complexity is able to increase hedonic value. Moreover, both extreme novelty and extreme familiarity can cause negative feelings, hence, a moderate amount of familiarity is able to increase hedonic value. Pandir and Knight (2006) investigated web page complexity according to participants' subjective evaluations. The results revealed that the relationship of complexity and preference cannot totally be explained by the inverted U shape. The finding suggested that while the content factor is influential, participants were led to respond inconsistently. The individual differences of participants may be the main reason that prevents the results from being applied to those of other users. In other words, the content may be a confounding factor affecting the relationship between complexity and preference. Tuch et al. (2009) attempted to control the confounding factor of content on experimental design. The experiment designed the complexity level by the field size of material on web pages. However, the research defined the complexity as remaining inappropriate. Although the field size is related to the image quality, it cannot be adopted as an explanation of web page complexity. The factors of color combination, image shape, text amount and layout are also important factors which can affect the level of complexity. As mentioned above, the experiments adopted subjective evaluation or field size to evaluate the web page complexity, still leaving deficiencies. Thus, the experiment design needed further improvement.

Many studies have realized that use occasion is an important factor in preference judgment; this necessitates considering subjective feeling before and after actual use. Until now, the related studies on the effect of use occasion on preference evaluation of web pages include: usage period and preference judgment before and after use (Saadé & Otrakj, 2007; Lindgaard et al., 2006; Tractinsky et al., 2006; Lee & Koubek, 2010).Users can judge web page aesthetics within an extremely short period (Lindgaard et al., 2006; Tractinsky et al., 2006). Previous studies have adopted exposure period as a dependent variable to investigate the effect of showing time on aesthetic judgment (Tractinsky et al., 2006). In preference research, studies have surveyed preference before actual use, which is based on the inverted U shape to evaluate preference of the whole page (Berlyne, 1970; Pandir and Knight, 2006; Tuch et al., 2009), and investigated preference after actual use, adopting satisfaction as an index to evaluate specific items of function (Hassenzahl, 2004; Tractinsky, et al., 2000). Moreover, most recent studies have separated use occasion into 'before actual use' and 'after actual use' to investigate the relationships among user preferences, perceived usability and perceived aesthetics. They not only found correlations among those three factors, but also conflicting relationships. Before actual use, users' preference and perceived aesthetics had no significant correlation, while users' perceived usability and perceived aesthetics had significant correlation.

On the other hand, after actual use, user preferences was highly correlated with both perceived usability and perceived aesthetics (Lee and Koubek, 2010). Thus, the evidence implied that perceived usability may be affected by use occasion.

2. RESEARCH OBJECTIVE

Although previous studies have attempted to adopt the inverted U shape to explain the relationship between complexity and preference, they did not prove the inverted U shape trend after several trials (Pandir & Knight, 2006; Tuch et al., 2009). Overall, these studies have two crucial faults; the first is the effect of content confounding, such as Pandir and Knight (2006), who adopted various stimuli materials that may interfere with users' preference judgment by personal experience. The second is the effect of the definition of web page complexity, such as Tuch et al. (2009), who adopted the field size to define complexity, which is not a proper approach. Although field size can explain image quality, size cannot represent the web page complexity. Based on the above, this study modifies the previous faults in discussing the two issues. First, this study investigates the relationships among web page complexity, web page familiarity and user preference, and second, it probes into the relationships among use occasion, user preference and perceived usability. The findings can serve as a reference on design guidelines for creating web page aesthetics.

3. RESEARCH METHOD

The experiment was divided into two parts. The first part involved comparing preference performance by use occasion before and after actual use, while the second part was evaluating preference performance after actual use. This study referred to the most important adjectives proposed by previous studies (Hsiao & Chen, 2006). The evaluation, in the second part, adopted nine representative adjectives for evaluating users' feeling. Then, their relationships were analyzed by Factor Analysis.

3.1 PARTICIPANTS

The subjects were 16 undergraduate students (3 males and 13 females), with a mean age of 22.88 years (SD=3.26). All reported 16/20 corrected visual acuity or better. Subjects came to this study with previous computer manipulation experience, using the Web over 14 hours per week. The subjects had never used the web page of Phi Coffman.

3.2 MATERIALS AND EQUIPMENT

The experiment adopted Phi Coffman as the basic frame structure (http://philcoffman.com/work.php). The total width of the browser was 960 pixels. The web page layout consisted of menu column and content column. The menu column had six menu items, and the content column had thirty images. The experiment design included menu positions: upper menu, left menu and right menu. The three column amounts were: one column, two columns and three columns. The experiment used a 3 (menu position) ×4 (column amount) design. A total of 12 stimuli were applied in the experiment. All of the variables were tested as within-subject variables, and shown twice, for 24 evaluations.

3.3 PROCEDURE

A 15-inch LED monitor (HP-L1960) was used for the PC display area. The program window under all of the experimental conditions used the full screen in the IE browser. Participants were allowed to adjust their body positions to a comfortable posture. Before the formal testing, the participants were informed that the objective of the study was to evaluate preferences before and after actual use through subjective intuition via a paper-based questionnaire.

1.   Before actual use evaluation: the number was shown on the center of the screen; the participants were directed to concentrate on the center of the screen, and the researcher read the number aloud. Then, the target stimuli automatically showed for two seconds, controlled by the computer program. After that, participants evaluated them by a seven-point Likert-type scale (1 denotes most disliked, 7 denotes most liked). Only after the participants filled in the score, were they shown the next stimuli. The same procedures were repeated 12 times.

2.   After actual use evaluation: the participants were informed that there was no time limit in executing the task. The task provided 12 links in the IE browser for users to click target icons. The task required participants first to search the target image of love in the Work page, and write the title of the love image in the questionnaire. Then, participants had to search the next target image of a flower in the Photo page, and write the title of the flower image in the questionnaire. After that, they answered nine questions on the Semantic Differential Scale questionnaire, which were: like-dislike, complex-simple, beautiful-not beautiful, familiar-unfamiliar, static-dynamic, light-heavy, rational-emotional, traditional-modern, and imitative-innovative.

4. RESULTS

4.1 PART 1: PREFERENCE BEFORE AND AFTER ACTUAL USE

The descriptive statistics showed the preference score of before and after actual use (Figures 1 and 2). Paired-samples t-test showed no significant different on preference between before and after actual use. Moreover, there was significant difference in regard to the upper menu of three column amounts (t(15)=-2.15, p=0.48). The preference of after actual use (M=6.81, SD=.557) was significantly higher than before use (M=5.5, SD=.548). The results identified a main effect of the menu position on preference (F2,78=9.40, p=0.00). Multiple comparisons showed that the preference for left menu (M=4.36, SD=0.93) was significantly higher than it was for upper menu (M=4.62, SD=0.87) and right menu (M=4.10, SD=0.81). The right menu obtained the lowest score in preference.

4.2 PART 2: SUBJECTIVE EVALUATION AFTER ACTUAL USE

The descriptive statistics revealed the subjective evaluation after actual use (Figures 3-10). The explanations were separated into two parts: column amounts and menu positions. The subjective evaluation of column amounts involved adjectives. Figure 3 shows that for the adjective pair of complex-simple, more column amounts delivered a feeling of higher complexity. Figure 4 shows that for the adjectives beautiful-ugly, one column amount delivered the feeling of consistency, while two, three and four column amounts delivered the feeling of less consistency. Figure 5 shows that there were no significant differences in four column amounts in the adjectives familiar-unfamiliar. Figure 6 illustrates that when the adjectives were static-dynamic, fewer column amounts tended to be more static, while more column amounts tended to be more dynamic. Figure 7 shows that when the adjectives were light-heavy, fewer column amounts tended to be lighter, while more column amounts tended to be heavier. Figure 8 illustrates that when the adjectives were rational-emotional, fewer column amounts tended to be more rational, while more column amounts tended to be more emotional. Figures 9 and 10 show that there were no significant differences in four column amounts for the adjectives traditional-modern and imitative-innovative. The results of the subjective evaluation of column positions in regard to adjectives showed no significant difference in the nine adjectives. The results showed that participants were more familiar with the menu in the upper and left positions, and less familiar with the menu in the right position. Moreover, the adjectives traditional- modern showed that the most traditional layout was one column in the left position, while the most modern layout was four columns in the right position. The adjectives imitative-innovative showed that the most classical layout was one column. The upper menu tended to be more innovative while the left and right menus tended to be more imitative. Overall, participants gave the upper menu of four columns the highest preference score.

REGRESSION ANALYSIS

Regression analysis was performed to determine the correlation relationship in regard to the upper menu between column amount and preference (F1, 2 =25.694, p=0.036). The regression equation showed a 92.8% variance of the column rations in preference (R=0.963, R2=0.928, SE=0.237). The regression equation of column amount describes preference: Y= 4.16+0.537X (X: column amount, Y: preference score).

FACTOR ANALYSIS

This study collected nine indexes to evaluate subjective feelings, which were: like-dislike, complex-simple, beautiful-ugly, familiar-unfamiliar, static-dynamic, light-heavy, rational-emotional, traditional-modern, imitative-innovative. The original data were examined by KMO and Bartlett's test of sphericity. The results showed KMO was .41, Bartlett's was 83.058 and the degree of freedom was 36, reaching the significant level .05, so that these data could be examined by factor analysis. The extraction was executed by Principal Axis Factoring (PAF) with Varimax orthogonal rotation. First, the extracted factor with an eigenvalue greater than 1.0 could be retained. They were 4.357, 2.006 and 1.626. Next, the cumulative percentage of variance explained for all of the factors can be used to choose a suitable number of factors. They were 48.411%, 22.29% and 18.069%, which accounted for 88.77% of the total variance. Following rotation, the loadings of the 9 items on the three factors are displayed in Table 1. For each item, the highest loading is printed in boldface. Factor one, named "Structure", includes light-heavy, static-dynamic, complex-simple and rational-emotional. Factor two named "Joyful", includes like-dislike, familiar-unfamiliar and beautiful-ugly. Factor three, named "Contemporaneity", includes traditional-modern and imitative-innovative.

5. DISCUSSION

Generally, use occasion and web page complexity significantly affect the judgment of subjective preference. The preference score before actual use is significantly higher than it is after actual use. The results showed that the degree of web page complexity is enhanced by column amounts (Figure 3); web page complexity can be defined by column amounts. Therefore, this study described web page complexity by column amounts, with one and two columns belonging to low complexity, and three and four columns belonging to high complexity.

After actual use, participants gave negative judgments for high complexity in the opening questionnaire. Some participants gave positive judgments: "convince, adequate, plenty, easy to search", while other participants declared the negative judgment of "tired, disordered, crowded." One possible reason for the difference is the effect of large information amounts. More grid lines are able to contain more picture amounts in pages, which may contribute to search speed. Meanwhile, more grid lines may cause a crowded feeling which also adds to mental loading. Moreover, Figure 4 shows the trend in aesthetics after actual use. It is found that one column of three menu positions gained the highest aesthetic evaluation in general. However, inconsistent outcomes of two, three and four columns are shown in aesthetic evaluation. This finding implies that higher complexity of web page can generate more information, as well as different feelings among the participants. Thus, the results of subjective evaluations are contradictory as shown above.

In the menu position, the upper menu received the highest preference score, while the right menu received the least preference score. Interestingly, the preference scores increased along with the column amounts, which suggests a linear trend. However, the web page with the least preference did not contain any linear trend or quadratic trend. Before actual use, the more numerous the column amounts, the higher the preference score (Figure 1). Unexpectedly, after actual use, in spite of the fact that more columns also obtained a higher preference score, the score of four columns decreased (Figure 2). This finding suggests that large amounts of information may cause negative feelings during the manipulation period. In use occasion, higher complexity received higher preference scores both before and after actual use (Figures1 and 2). The observer interviewed participants after actual use. Participants perceived a higher degree of usability in searching web page with larger amounts of pictures. Previous studies have indicated that web page complexity and preference score have a correlation relationship in the situation of before actual use (Tuch et al., 2009). Among the data on the upper menu, the results showed that web page complexity and preference not only have a correlation relationship before actual use, but also after actual use (Figure 11). The results of this study differed from those of Lee and Koubek (2010), who suggested that perceived usability and preference have a correlation relationship. This study found that whether participants evaluated before or after actual use, the perceived usability has a correlation relationship with preference. Thus, perceived usability is regarded as a key factor which affects users in regard to preference evaluation.

The factor analysis yielded three factor names: constitutive property, joyful and contemporary. The results showed that grid is a main tool for web page construction which can generate the emotions of light-heavy, static-dynamic, simple-complex and rational-emotional. This finding disagreed with Le Corbusier (1949), who argued that grids cannot affect emotions. This study proved that grids are able to generate the emotions of simple-complex, static-dynamic, light-heavy and rational-emotional (Figures 3, 6, 7 and 8). Moreover, different factors can deliver different degrees of emotion by changing grid amounts. On the other hand, the emotions of like-dislike, familiar-unfamiliar and beautiful-ugly were attributed to the factor 'joyful'. In other words, there are strong relationships among preference, familiarity and aesthetics. This finding is consistent with Berlyne (1970), although the data trend cannot construct the inverted U shape between preference and familiarity. Figure 11 illustrates that upper menu is the most attractive layout before actual use, which can generate a linear trend. During the experiment process, many participants suggested the difficulty in judging their preferences concerning the whole web page. On the other hand, when the observer asked participants to evaluate preference in specific function item, they were able to clearly and quickly respond to their preference degree, i.e., functional item, picture, text, etc. One possible reason for the difference is that participants may have preferred some elements of the web page, but not all. As previous studies may have garnered confounding information, they may have been unable to observe the relationship between complexity and preference (Pandir & Knight, 2006; Tuch et al., 2009). This study found that when a webpage received a high preference score, the regression line could be built; when a webpage received a low preference score, the regression line could not be built, such as the first linear trend shown on the upper menu before actual use (Figure 11). The important issue needing to be reconsidered is that the web page is an information carrier which has abundant data to satisfy each participant's specific needs. Based on the different specific needs of each person, these data may drive users' different individual intrinsic motivation, and users may also exhibit different degrees of preference (Van der Heijden, 2003). This study further suggests that when each participant has inconsistency in perceived preferences, the relationship between preference and complexity can be built in a linear trend. In other words, if researchers would like to build the linear trend for contributing to practical design, the concept of linear trend would conflict with various information quantities. Thus, the application of an inverted U shape on web page would need to be reconsidered.

For participants who were familiar with web pages with a higher degree of complexity, their preference score after actual use was higher than it was before actual use. The results revealed that the process of manipulation enhanced positive emotions. Importantly, for both situations of before and after actual use, the highest preference score was for the menu position in the left, with three columns. Participants offered positive judgment for three column amounts, which are "popular, general, typical, comfortable, no feeling, arrangement, and balance." These descriptions expressed that participants feel familiarity with three columns. The results of this study imply that whether the situation is before or after actual use, the web page with a higher level of familiarity can obtain better positive emotions and perceived usability, which can obtain higher preference scores.

In regard to its theoretical contribution, to further investigate the abovementioned grid viewpoint of Le Corbusier, the factor of perceived usability is the main effect on preference judgment in both situations, before and after actual use. Moreover, in regard to web page familiarity, the high familiarity web page is able to enhance perceived usability and preference degree. This study provides further explanation regarding the inverted U shape, In practical contribution, with the use occasion before and after actual use, the most favorite layout positions are left menu and upper menu. Also, three columns are the most familiar arrangement which can obtain higher preference scores and better search efficiency.

Future studies can survey other aesthetic factors and methods to obtain further knowledge concerning how preference can be generated and lost.

REFERENCES

Lavie, T. & Tractinsky, N. (2004). Assessing dimensions of perceived visual aesthetics of web sites. Int. J. Human-Computer Studies, 60, 269-298.

Lindgaard, G. & Dudek, C. (2003). What is this evasive beast we call user satisfaction? Interacting with Computers, 15, 429-452.

Lindgaard, G., Fernandes, G., Dudek, C. & Brown, J. (2006). Attention web designers: You have 50 milliseconds to make a good first impression! Behavior & Information Technology, 25, 2, 115-126.

Lee, S. & Koubek, R. (2010). Understanding user preferences based on usability and aesthetics before and after actual use. Interacting with Computers, 22, 530-543.

Park, S., Choi, D. & Kim, J. (2004). Critical factors for the aesthetic fidelity of web pages: empirical studies with professional web designers and users. Interacting with Computers, 16, 351-376.

Pandir, M. & Knight, J (2006). Homepage aesthetics: The search for preference factors and the challenges of subjectivity. Interacting with Computers, 18, 1351-1370.

Tractinsky, N., Cokhavi, A., Kirschenbaum, M. & Sharfi, T. (2006). Evaluating the consistency aesthetic perceptions of web pages. Int. J. Human-Computer Studies, 64, 1071-1083.

Tracinsky, N., Katz, A. & Ikar, D. (2000). What is beautiful is usable. Interacting with Computers, 13, 127-146.

Tuch, A. N., Bargas-Avila, J. A., Opwis, K. & Wilhelm, F. H. (2009). Visual complexity of websites: Effects on users' experience, physiology, performance, and memory. Int. J. Human-Computer Studies, 67, 703-715.

van Schaik, P. & Ling, J. (2009). The role of context in perceptions of the aesthetics of web pages over time. Int. J. Human-Computer Studies, 67, 79-89.

Tuch, A. N., Bargas-Avila, J. A., Opwis, K. & Wilhelm, F. H. (2009). Visual complexity of websites: Effects on users' experience, physiology, performance, and memory. Int. J. Human-Computer Studies, 67, 703-715.

Ngo, D.C.L., Teo, L.S. & Byrne, J.G. (2000). Formalising guidelines for the design of screen layouts. Display, 21, 3-15.

Saadé, R. G. & Otrakji, C. A. (2007). First impressions last a lifetime: effect of interface type on disorientation and cognitive load. Computer in Human Behavior, 23, 525-535.

Roth, S. P., Schmutz, P., Pauwels, S. L., Bargas-Avila, J. A., & Opwis, K. (2009). Mental models for web objects: Where do users expect to find the most frequent objects in online shops, news portals, and company web pages? Interacting with Computers, 22, 140-152.

Shinskey, J. L., & Munakata, Y. (2010). Something old, something new: a developmental transition form familiarity to novelty preferences with hidden objects. Behavior Research Methods, 41, 699-704.

Oulasvirta, A., Kärkkäinen, L., Laarni, J. (2005). Expectations and memory in link search. Computers in Human Behavior, 21, 773-789.

Santa-Maria, L., & Dyson, M. C. (2008). The effect of violating visual conventions of a website on user performance and disorientation. How bad can it be? SIGDOC'08, 47-54.

Crandall, J. E. (1967). Familiarity, preference, and expectancy arousal. Journal of Experimental Psychology, 73, 374-381.

Paré, D. E., & Cree, G. S. (2009). Web-based image norming: How do object familiarity and visual complexity ratings compare when collected in-lab versus online? Behavior Research Methods, 41, 699-704.

Hsiao, K. A. & Chen, L. L. (2006). Fundamental dimensions of affective responses to product shapes. International Journal of Industrial Ergonomics, 36, 553-564.

Hassenzahl, M. (2004). The interplay of beauty, goodness, and usability in interactive products. Human-Computer Interaction, 19, 319-349.

Schenkman, B. N., and Jönsson, F. U. (2000). Aesthetics and preferences of web pages. Behaviour & Information Technology, 19, 367-377.

Van der Heijden, H. (2003). Factors influencing the usage of websites: the case of a generic portal in the Netherlands. Information & Management, 40, 541-549.

CHAPTER 57

Effects of Unity of Form on Visual Aesthetics of Website Design

Ahamed Altaboli[1,2], Yingzi Lin[1]

[1] Mechanical and Industrial Engineering Department
Northeastern University, Boston, MA, USA
[2] Industrial and Manufacturing Systems Engineering Department
University of Benghazi, Benghazi, Libya
* Corresponding author: Tel: 1.617.373.8610, Fax: 1.617.373.2921, E-mail:
yilin@coe.neu.edu

ABSTRACT

Unity of form represents the extent to which visual objects on the screen are related in size. High levels of unity of form can be achieved by using objects with similar sizes on the screen and/or by reducing number of objects on the screen. Findings of earlier observational studies suggested that unity of form has significant effects on perceived visual aesthetics of website design in the case of highly symmetrical designs. The purpose of this study is to verify these findings. An experiment was conducted to systematically study effects of number of objects and number of different sizes of objects on perceived visual aesthetics of website design under the condition of high level of symmetry. Results showed that both factors have significant effects on perceived visual aesthetics. Designs with lower levels of both factors were perceived as having better visual aesthetics.

Keywords: perceived visual aesthetics, website interface design, unity of form.

1   INTRODUCTION

With the recent shift in the human factors and ergonomics field towards incorporating aesthetic aspects in the design process (Liu, 2003; Norman, 2004), visual aesthetics of website and interface design became a research discipline of its own. Findings of latest studies regarding aesthetic aspects in interface design showed a clear effect of visual aesthetics on users perception of usability of the system (Kurosu, and Kashimura, 1995; Laviea and Tractinsky, 2004; Lindgaard et al., 2006, Phillips and Chapparro, 2009; Tractinsky et al., 2000; Tractinsky, 1997) and suggested a possible positive effect on performance (Moshagen et al., 2009; Sonderegger, and Sauer, 2010).

One line of research in interface aesthetics is concerned with determining what features in the interface design triggers users' precipitin of aesthetics of the interface. It tries to explore the possibility of expressing changes in such features using numerical values and use these numerical values to assess users' perception of interface aesthetics. One approach argues that physical layout of visual objects on the screen may play a role in users' perception of aesthetics. This approach builds on earlier quantitative measures of aesthetics (e.g Birkhoff's aesthetic measure (Birkhoff, 1933)) and principles of Gestalts laws for visual design (Chand et al., 2002). The procedure involves expressing visual design features (like symmetry, balance, unity ...etc) using mathematical formulas and combine calculated values for all features to build an overall measure that would reflect aesthetic level of the interface design.

The current study follows a similar approach. The purpose of the study is to systematically investigate the effect of unity of form on users' perception of visual aesthetics of website design in the case of high levels of vertical symmetry. The effects of visual symmetry on users' perception of visual aesthetics of website design have been comprehensively studied and reported in the related literature (Bauerly and Liu, 2008; Lai et al., 2010; Ngo and Byrne, 2001; Ngo et al., 2003; Tuch et al., 2010). Most findings indicate that symmetry has a positive effect on perceived visual aesthetics. Unity of form represents the extent to which visual objects on the screen are related in size (Ngo et al., 2003). High levels of unity of form can be achieved by using objects with similar sizes on the screen (Ngo et al., 2003). Ngo et al. (2003) developed mathematical formulas that give numerical estimation of symmetry and unity of computer screens. The formulas were developed as part of a larger formularized model to assess visual aesthetics of computer screens (Ngo and Byrne, 2001; Ngo et al., 2003). This model was tested in many cases and unity of form has been found to be one of the significant terms in the model (Altaboli and Lin, 2010 & 2011a; Ngo and Byrne, 2001; Salimun et al., 2010; Zain et al., 2008). The two input parameters in the unity of form formula are number of objects on the screen and number of different sizes of objects on the screen.

In a previous study (Altaboli and Lin, 2011b), correlation analysis was used to compare objective layout-based measures of visual aesthetics with subjective questionnaire-based measures. Unity, number of objects, and number of different sizes of objects were included in the test. Values for the tested objective measures were calculated for forty-two web pages already used in a previous study (Moshagen and Thielsch, 2010), for which subjective questionnaire scores were already available. Results showed significant correlations between unity, number of objects, and number of different sizes with subjective questionnaire-based measures. Number of different sizes produced the highest correlation coefficients. Findings also suggested that this high correlation between unity of form and perceived aesthetics only occurs at high levels of symmetry. The main purpose of the current study is to confirm these findings. An experimental approach using controlled experiments will be used to verify these findings.

The goal is to test the hypothesis of findings significant effects of unity of form on perceived aesthetic of website design in case of designs with high levels of vertical symmetry. Specifically, the study aims at systematically study effects of number of objects and number of different sizes of objects on perceived aesthetics of website design under the condition of high symmetry.

2   METHOD

2.1   Design of Experiment

An experiment was designed and conducted to test the effects of the two parameters of unity of form, namely: number of objects and number of different sizes of objects on participants' perceived visual aesthetics of website design.

A two- factor within subject design was utilized with the two parameters as the main factors. Each of the two factors was tested at two levels (high and low). Values for the levels of the two factors were chosen based on observations and results of previous studies. Four different designs of a webpage were prepared to represent the four experimental conditions. All four designs have identical styles (colors, fonts ...etc); only visual elements related to the two factors were manipulated. Symmetry in all the four designs was kept at higher levels. Table1 shows factors values and levels associated with the four experimental conditions.

Table 1 The four experimental conditions and the associated factors levels and values.

Condition (design)	Factors values		Factor Levels	
	No of objects	No of different sizes	No of objects	No of different sizes
1	6	3	High	High
2	6	5	High	Low
3	16	3	Low	High
4	16	11	Low	Low

User perception of visual aesthetics was measured using the VisAWI (Visual Aesthetics of Website Inventory) questionnaire (Moshagen and Thielsch, 2010).

The instrument is based on four interrelated facets of perceived visual aesthetics of websites: simplicity, diversity, colorfulness, and craftsmanship. Simplicity comprises visual aesthetics aspects such as balance, unity, and clarity. The Diversity facet comprises visual complexity, dynamics, novelty, and creativity. The colorfulness facet represents aesthetic impressions perceived from the selection, placement, and combination of colors. Craftsmanship comprises the skilful and coherent integration of all relevant design dimensions. Each of the first two facets is presented by five items in the questionnaire, while each of the last two facets has four items. The two factors (number of objects and number of sizes) are the independent variables. Questionnaire scores represent the dependent variable.

Fig 1 shows screen shots of the four designs of the webpage. The webpage represents a homepage of a hypothetical website that talks about the ancient history of a certain region of North Africa. It uses the local language of that region (Arabic). However; the content of the website was irrelevant for the purpose of the study and doesn't relate to any of the items of the used questionnaire. The questionnaire items were deigned to only evaluate visual aesthetics of a website based on its visual appearance.

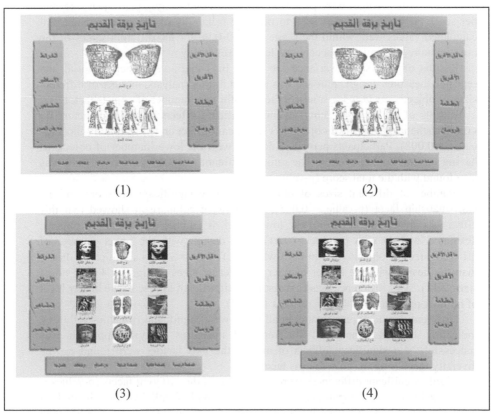

Figure 1 Screen shot images of the four designs of the webpage.

2.2 Participants

Participants were recruited online. Email invitations were sent to a total of 251 individuals with the choice of entering a lottery to win 100 US dollars. A total of 40 responses were received, from which 25 were valid responses. All invitations were sent to audience within the United States. Average age of participants with valid responses is 41.5 years with a standard deviation of 16.4 years. 13 were males and 12 were females.

3   RESULTS

Images of screen shots of the four designs were presented to each participant one at a time with an on screen size of 800X600 pixels. The questionnaire was placed under each image. Each participant had to answer the questionnaire for each design using a seven-point Likert scale. Both images and questionnaire items were presented in random orders for each participant.

Table 2 summarizes average scores per each scale (aesthetic facet) and for the total score. The averages were presented per condition (design), per object size, and per scale. Cronbach's α was used to measure reliability of the questionnaire. All calculated values were within the range of 0.60-0.99 for the different scales of the questionnaire, indicating an acceptable level of reliability.

Results of analysis of variance are given in Table 3. A nested factorial and repeated measures design approach was utilized in this analysis; number of different sizes of objects was tested as a nested factor within the number of objects factor.

Results of the analysis show several significant effects of the two factors on visual aesthetics. Number of objects has significant effect on the simplicity facet (p-value = 0.008). Participants perceived designs with the lower number of objects as having better visual aesthetics. The same significant effect of number of sizes was also found with the total score (p-value = 0.028).

Number of different sizes of objects was only significant on scores of the craftsmanship facet (p-value = 0.025). A pair-wise comparison showed that this effect was significant in case of the lower level of number of objects (16 objects, designs 3 and 4) with p-value = 0.012. The design with the lower number of different sizes of objects (design no 3 with no of sizes = 3) was significantly better than the design with the higher number of different sizes (design no 4 with no of sizes = 11). This difference wasn't statistically significant in case of the higher level of number of objects (6 objects, designs 1 and 2).

Differences among participants in all cases were significant (p-values < 0.001), indicating that the choice of the repeated measures analysis approach was successful. No significant effects of any of the two factors were found on the diversity facet or on the colorfulness facet.

Highly significant differences were found among the different facets (p-values < 0.001). All pair-wise comparisons were significant with simplicity given the highest average score (4.18) followed by colorfulness with an average score of 3.95, than craftsmanship with an average of 3.54, and last diversity with an average score of 3.26.

Table 2 Summary of descriptive statistics for the questionnaire scores.

Scale	Design no	No of objects	No of sizes	Per design	Per no of objects	Per scale	Standard deviation (per screen)
				Average score			
Simplicity	1	6	3	4.31	4.34		1.08
	2		5	4.38		4.18	1.20
	3	16	3	4.08	4.01		1.15
	4		11	3.94			1.17
Diversity	1	6	3	3.37	3.33		1.30
	2		5	3.30		3.26	1.33
	3	16	3	3.28	3.20		1.34
	4		11	3.11			1.13
Colorfulness	1	6	3	4.03	3.99		1.47
	2		5	3.95		3.95	1.60
	3	16	3	3.88	3.91		1.54
	4		11	3.94			1.45
Craftsmanship	1	6	3	3.70	3.59		1.19
	2		5	3.48		3.54	1.28
	3	16	3	3.71	3.50		1.20
	4		11	3.28			1.07
Total	1	6	3	3.85	3.81		1.02
	2		5	3.78		3.73	1.13
	3	16	3	3.74	3.65		1.16
	4		11	3.57			1.05

Table 3. Analysis of variance results

Case	Element	F	P-value
	Objects	7.51	0.008
Simplicity	Sizes within Objects	0.38	0.682
	Participants	9.96	< 0.001
	Objects	1.38	0.24
Diversity	Sizes within Objects	0.63	0.54
	Participants	15.65	< 0.001
	Objects	0.66	0.42
Colorfulness	Sizes within Objects	0.26	0.77
	Participants	25.59	< 0.001
	Objects	0.60	0.44
Craftsmanship	Sizes within Objects	3.87	0.025
	Participants	10.69	< 0.001
	Objects	5.01	0.028
Total	Sizes within Objects	1.66	0.196
	Participants	27.35	< 0.001

4 CONCLUSIONS

The purpose of the study was to verify findings of earlier observational studies. These earlier findings suggested that the visual feature of unity of form has significant effects on perceived visual aesthetics of website design. These findings also suggested that these effects are more evident in case of highly symmetrical webpage designs. An experiment was designed and conducted to systematically study effects of number of objects and number of different sizes of objects (the two parameters of unity of form) on perceived visual aesthetics of website design under the condition of high levels of symmetry.

Analysis of results showed that both factors (number of objects and number of different sizes) have statistically significant effects on perceived visual aesthetics. Designs with lower levels of both factors were perceived as having better visual aesthetics. However, these effects weren't consistent over all the four parts of the questionnaire (representing the facets of visual aesthetics). The effect of number of objects was only significant on the simplicity facet and on the overall perceived aesthetics (total questionnaire score of all facets). Effect of number of different sizes was only significant on the craftsmanship facet, only at the higher level of number of objects.

The next step should be to repeat the experiment in the case of designs with low levels of symmetry. This step would be necessary to completely confirm the results. It would also be helpful in the larger context of developing formularized measures of perceived visual aesthetics of website interface design based on visual features of the design. Better understanding of the combined effects of symmetry and unity of form could help in developing better mathematical formulas and computational models to measure effects of each and the combined effects of both. Such formulas would be very useful in the wider process of developing more comprehensive computational models to predict perceived visual aesthetics of website interface design.

REFERENCES

Altaboli, A. and Lin, Y. 2010. Experimental Investigation of Effects of Balance, Unity, and Sequence on Interface and Screen Design Aesthetics, in: Blashki, K.. (Ed.), *Proceedings of The IADIS International Conference Interface and Human Computer Interaction 2010*, Freiburg, Germany, IADIS Press, pp. 243-250.

Altaboli, A., and Lin, Y., 2011a. Investigating Effects of Screen Layout Elements on Interface and Screen Design Aesthetics. *Advances in Human-Computer Interaction*, vol. 2011, Article ID 659758, 10 pages, 2011. doi:10.1155/2011/659758.

Altaboli, A. and Lin, Y. 2011b. Objective and subjective measures of visual aesthetics of website interface design: the two sides of the coin. In *Proceedings of the 14th international Conference on Human-Computer interaction: Design and Development Approaches - Volume Part I* (Orlando, FL, July 09 - 14, 2011). J. A. Jacko, Ed. Springer-Verlag, Berlin, Heidelberg, 35-44.

Bauerly, M. and Liu, Y. 2006. Computational modeling and experimental investigation of effects of compositional elements on interface and design aesthetics. *Int. J. Human-Computer Studies,* 64 (2006) 670–682.

Bauerly, M. and Liu, Y. 2008. Effects of Symmetry and Number of Compositional Elements on Interface and Design Aesthetics. *International Journal of Human- Computer Interaction*, 24: 3, 275 - 287.

Birkhoff, G. 1933. *Aesthetic Measure*. Harvard University Press, Cambridge, MA.

Chand, D., Dooley, L., and Tuovinen, E. 2002. "Gestalt Theory in Visual Screen Design – A New Look at an Old Subject". Australian Computer Society, Inc. presented at the Seventh World Conference on Computer in Education, Copenhagen, Denmark.

Kurosu, M. and Kashimura, K. 1995. Apparent usability vs. inherent usability: experimental analysis on the determinants of the apparent usability. *CHI '95: Conference companion on Human factors in computing systems,* Denver, Colorado, United States, 292-293.

Lai, C., Chen, P., Shih, S., Liu, Y., Hong, J. 2010. Computational models and experimental investigations of effects of balance and symmetry on the aesthetics of text-overlaid images, *Int. J. Human-Computer Studies.* 68, 41–56.

Laviea, T., and Tractinsky, N. 2004 Assessing dimensions of perceived visual aesthetics of web sites. *Int. J. Human-Computer Studies.* 60, 269–298.

Lindgaard, G., Fernandez, G., Dudek, C. and Brown, J. 2006. Attention web designers: You have 50 milliseconds to make a good impression. *Behavior & Information Technology* 25, 2 (2006), 115-126.

Liu, Y. 2003. The aesthetic and the ethic dimensions of human factors and design. *Ergonomics,* 46, 1293–1305.

Moshagen, M. and Thielsch, M. T. 2010. Facets of visual aesthetics. *Int. J. Hum.-Comput. Stud.* 68, 10 (October 2010), 689-709.

Moshagen, M., Musch, J., Göritz, A.S. 2009. A blessing, not a curse: Experimental evidence for beneficial effects of visual aesthetics on performance. *Ergonomics* 52, 1311-1320.

Ngo, D. and Byrne, J. 2001. Application of an aesthetic evaluation model to data entry screens. *Computers in Human Behavior,* 17, 149-185.

Ngo, D. C. L., Teo, L. S., & Byrne, J. G. 2002. Evaluating Interface Esthetics. *Knowledge and Information Systems* (4), 46-79.

Ngo, D. C. L., Teo, L. S., & Byrne, J. G. 2003. Modelling interface aesthetics. *Information Sciences,* 152(1), 25-46.

Norman, D. 2004. Emotional design: Why we love (or hate) everyday things. Basic Books. New York, NY, USA.

Phillips, C., and Chapparro, C. 2009. Visual Appeal vs. Usability: Which One Influences User Perceptions of a Website More?. *Usability News,* Vol 11(2).

Salimun, C., Purchase, H. C., Simmons, D. R., and Brewster, S. 2010. Preference ranking of screen layout principles. *The 24th BCSHCI'10,* September 6-10, Abertay Dundee.

Sonderegger, A., Sauer, J. 2010. The influence of design aesthetics in usability testing: Effects on user performance and perceived usability. *Applied Ergonomics.* 41, 403-410.

Tractinsky, N., Shoval-Katz, A., Ikar, D. 2000. What is beautiful is usable. *Interacting with Computers,* 13, 127-145.

Tractinsky, N. 1997. Aesthetics and apparent usability: empirically assessing cultural and methodological issues. In S. Pemberton (Ed), *Proceedings of the 1997 Conference on Human Factors in Computing Systems (CHI '97).* New York: ACM Press.

Tuch, A., Bargas-Avila, J., and Opwis, K. 2010. Symmetry and Aesthetics in Website Design: It's a man's Business. *Computers in Human Behavior,* 26(2010), 1831-1837

Zain, J., Tey, M., and Goh, Y. 2008. Probing a Self-Developed Aesthetics Measurement Application (SDA) in Measuring Aesthetics of Mandarin Learning Web Page Interfaces. *IJCSNS International Journal of Computer Science and Network Security,* Vol. 8 No. 1.

CHAPTER 58

Design Principles for Sustainable Social-Oriented Bike Applications

Da Lee[1], Chao-Lung Lee[2], Yun-Maw Cheng[3], Li-Chieh Chen[4],

and Shih-Chao Sheng[5]

Tatung University
Taipei, Taiwan
{d9806003[1], d9806006[2], g9806023[5]}@ms.ttu.edu.tw
{kevin[3], lcchen[4]}@ttu.edu.tw

Frode Eika Sandness

Oslo and Akershus University College of Applied Sciences
Oslo, Norway
Frode-Eika.Sandnes@hioa.no

Chris Johnson

Glasgow University
Glasgow, UK
christopher.johnson@glasgow.ac.uk

ABSTRACT

As the understanding of environmentally green issues grows, HCI researchers have attempted to explore the potential of technology to encourage people to develop healthy and environmentally friendly lifestyles. Recently, promoting biking has become a hot topic. Although there are a number of positive examples known to enhance biking experiences based on augmenting human perception and awareness of their surroundings, there are still questions of where exactly their success lies. This highlights a necessity for designs that encourage biking while addressing its sustainability. This paper proposes a new perspective, inspired in part by the social influence emphasized in previous work. We also looked at the existing bike applications and persuasive research to extract a set of principles and applied them to construct and present a new design of a social-oriented application for biking.

Keywords: Design principles, social influence, bike applications

1   INTRODUCTION

Regular cycling is good for one's health, but many people find it difficult to increase or maintain their motivation in their daily life. Researchers in the HCI community identify many ways in which well-designed bike applications provide promising directions for motivating people to cycle. The *social influence* in particular, is a crucial foundation that structure many well-designed bike applications (Consolvo et al., 2006, Fogg, 2002). These point out a necessity for designs that encourages users to keep up a biking habit. This paper focuses on two specific features of *social influence*: experience-sharing features and fun-oriented features.

Experience-sharing features include *personal experience* sharing (such as physical status and performance) and *environmental experience* sharing (such as route and terrain information) (Eisenman et al., 2010, Reddy et al., 2010). These are embodied in augmenting the users' perception of experience awareness, and encourage bikers to develop social communities via frequent experience sharing. For example, the users share real-time *personal experience* through CenceMe (Miluzzo et al., 2008). BikeNet and Biketastic exploit shared environmental experience by uploading route information on a website after finishing a cycling journey (Eisenman et al., 2010, Reddy et al., 2010). BiSeeCall-AR uses manifold channels to increase the personal experience in support of group activities with different scenarios, making it an augmenter of group social interaction (Lien-Wu et al., 2011). Miluzzo et al. have taken similar approaches about the benefits of experience sharing and strongly noted that experience sharing has positively influenced sustainable social interaction (Miluzzo et al., 2008).

The applications with a fun-oriented feature typically aim to integrate fun into everyday life in form of games. For example, Yasuda et al. explored the urban pleasure of a bike competition in their Bikeware project (Yasuda et al., 2008). UbiGreen proposed by Jon et al. is a mobile application, which links green transportation behavior (such as riding a bike) to feedbacks about how users' behavior affects the global environment after the setting their goals. Although UbiGreen was not inherently social sharing, Jon et al. noted that display contexts acts as conversation starters of daily social interaction (Froehlich et al., 2009). Ubifit Garden utilizes similar approaches by tracking users' activities and encourages the users to focus on achieving their goals through sustainable cycling and walking (Consolvo et al., 2008).

As mentioned, experience-sharing and fun-oriented approaches have shown the potential to encourage and motivate people to cycle. However, there are still opportunities for sustainable interaction design for cycling. Interactions with experience-sharing features may not interest non-cycling people. Viewing others bike experience may not significantly improve the motivation of non-frequent users. Also, applications with fun-oriented features may bore people with non-changing interaction scenarios. This poses a challenge for researchers in this field to explore the potential for technology to encourage people to maintain their cycling habit.

Inspired by the *social influence* of existing applications, this paper looks into the features of the applications literature that focuses on the bike experience with the goal of long-term bike use. The following pages illustrate the commons and its effect on promoting cycling. To conclude the findings, a design of a social-oriented application for cycling, *BikeLine Challengers*, based on the commons are presented.

2   THE DESIGN SPACE OF A SOCIAL-ORIENTED BIKE APPLICATION

Positive experience is a main factor for encouraging bikers to keep bike habit, because bikers' performance and emotions always be affected easily through it (Picard, 1997). As Fogg described (Fogg, 2002), designers should focus on the "planned persuasive effects" rather than on the "effects of technology use". This brings a key concept for designing a new sustainable social-oriented bike application. Since designers start their new designs, they require key functions to inspire user motivation through enhancing user bike experience. However, this design concept also point out a challenge for designers. They must consider critical factors which influence bike experience.

Social influence increases activity experience and motivates user to stay their physical habits. Through interaction with other people or join a group, social influence strengthens users' faith to achieve their goals and better their performance in activities. Recently, the importance of social influence has increased noticeably because the growing of social network and mobile technologies (Fogg, 2002). The existing projects of this kind can be categorized into *fitness game design* (Mueller et al., 2011, Campbell et al., 2008, Fogtmann et al., 2008), and *social experience awareness* (Fogg, 2008, Jacob et al., 2008, Mueller et al., 2011, Fogtmann et al., 2008). *Fitness game design* focuses on promoting people to do exercise more through giving them both of exertion and fun. Users can play together and get achievement through competition and cooperation in a fitness game. *Social experience awareness* values content and benefit of other users' experience in physical activities through technology supporting.

Inspired by Fogg's and Consolvo's theories, there are four elements highlighted for design of social influence: *competition, cooperation, communication,* and *recognition.* All of these implicitly demonstrate how to guide people to form a sustainable habit.

Competition energizes people (it includes single user or group members) to work harder for their physical activity. Honor and achievement are core intrinsic motivators of competition. Honor and achievement promote users facing challenge, overcome physical exertion and challenges, defeat competitors, and achieve their goals. On the other hand, a competition reward is the extrinsic motivator of social influence, and it is a catalyst for raising users' honor and achievement. *Cooperation* is structured by social support and social pressure for group activity. Group members devote in common goal, and receive encouragement from each other. Social pressure drives people to monitor the performance of their activity among the social peers while social support takes place to form group achievement.

Communication not only provides people with the enhanced ability to express the feelings about their activity, but also learn activity knowledge from other people's experience, for example, social networking website (Fogg, 2008), real-time messenger (Lien-Wu et al., 2011), and geowiki (Priedhorsky et al., 2007). *Recognition* relates to interface issue, including how to provide users a clear and natural interface. Designers must setup interfaces that provide their users simple interaction between bike applications during biking.

Previous work on social-oriented bike applications can be analyzed in terms of the elements of social influence. These applications may have *experience-sharing features* or *fun-oriented features*. *Experience-sharing features* highlight *communication*, which promote bikers to share their experience. These specific benefits include increased environmental experiences (Eisenman et al., 2010, Reddy et al., 2010), quantity of social interactions (Lien-Wu et al., 2011, Miluzzo et al., 2008). Also, *fun-oriented features* place value on *competition* and *communication*. Applications with *fun-oriented features* increase social pressure and stimulate users actively participate in their bike activity via competition and sense of achievement. These applications build *communication* through providing users a monitor of their own performance. *Fun-oriented features* are often used in the personal fitness applications and race application (Consolvo et al., 2008, Froehlich et al., 2009, Yasuda et al., 2008). All current bike applications are equipped with *recognition* from different causes. For examples, *experience-sharing features* focus on providing suitable interface for reading route information from other users' experience; *fun-oriented features* place heavy emphasis on performance monitoring for competition (such as self-performance and competitor's performance).

Most of the previous applications simply focus on partial elements of social influence in their design. For example, experience-sharing features lack for the *competition* to stimulate users to be actively engaged in their biking lifestyle. On the other hand, fun-oriented features use the competition to increase bikers' experience in their activity. However, there are few focusing on *cooperation*. Those situations highlight some opportunities for designing more comprehensive and sustainable social-oriented bike applications. Designers need a new perspective on social influence for their designs.

As shown in Fogg's behavior model, designers should build up their designs that provide people positive experience via enhancing their intrinsic motivators (Fogg, 2009). The core of the challenge is how to provide users the positive experience through elements of social influence. *Competitive spirit* is brought up to be the foundation to reason and tackle the challenge. For example, the *competitive spirit* created specific goals to encourage bikers to engage in bike activities such as to defeat their friends, cooperate with his friend to challenge a long-distance bike route, learn more bike-related experience and knowledge, or to push through their own limitations. The bikers will actively seek cycling experience for improving their performance. It is intrinsic to say that whether a social-oriented bike application is designed, largely hinges on *competitive spirit*.

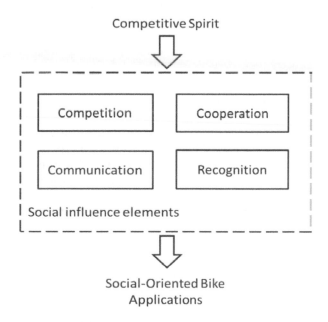

Figure 1. Social-oriented bike applications are based on social influence elements and competitive spirit

In this section, we have proposed a new perception of social influence for developing a social-oriented bike application: *competitive spirit*. *Competitive spirit* works as a foundation for considering how it conditions bikers' biking experience of the world in social interactions through application. This perception offers a view on how to motivate bikers cycling enjoyment to encourage them to maintain their biking habit.

3   DESIGN PRINCIPLES FOR A SOCIAL-ORIENTED BIKE APPLICATION

As stated, the *competitive spirit* works as a foundation for the quality of cycling experience. This idea offers a view on how to motivate and encourage bikers to maintain their biking habit. The following describes a set of design principles relevant to social-oriented bike applications based upon prior work on social-oriented persuasive applications and social influence theories.

3.1   Defining application topics

Owing to the application topic possibly impacting the interactions between the bike application and the user, a designer should define a simple and clear topic for the application. The application topic is the key with which a designer can influence bikers. Designers must consider how to inspire bikers' *competitive spirit* to get fun and to increase the interactions of experience-sharing through the application topic. Therefore, the topics will encourage the bikers to form a sustainable habit for biking.

3.2   Setting up scenarios

The functional design of the bike application should consider the information needed and the interactions under different scenarios. There are three major biking scenarios: non-biking, single biking and group biking. The non-biking scenario requires meaningful information of manifold bike experience for sharing, comparing, and discussing. This scenario is also suitable for complex interactions, such as planning a bike journey, and browsing variety of information from other users' bike experiences. Social experience is brought into the single biking scenario from interactions such as social communication (e.g. monitoring of real-time experience from other users) and *competition* (e.g. set a bike race for users). Group biking focuses on supporting real-time group interaction such as real-time experience sharing which forms the base of competition.

3.3   Developing ambient display interfaces

A biker requires different types of information during a bike activity. The designer should offer different display types of interfaces to the users so that they can access the information under different scenarios. Ambient display interfaces provide users a more natural interaction for *recognition* class. The interface displays necessary information for users through different styles (such as simple colors, icons, earcons, and tactons), and user can receive those information by simple interactions.

3.4   Providing experience awareness

Experience awareness helps users to gauge progress in the bike activity and satisfies users' sense of information control. The bike experiences are often derived from personal experience and environmental experience, both of which fall under the common experience of bike activities. The common personal experiences include personal status, performance, quality of presence, and feedback. The environmental experience comprises of information about surroundings such as route terrain, air pollution, and bike facilities. Bike application provides a level awareness field in which all users' requirements are successfully met in different conditions. In brief, the experience awareness can remind users about their *competitive spirit*, and have them in control of bettering their performance.

3.5   Setting goals

The achievement of goals is an important factor of the *competitive spirit*. To do this, applications must guide the users to set their goals before a biking activity and offer them positive feedback while the users are on the right way to achieve their goals. The applications also provide them with relevant feedback to illustrate their performance and indicate the performance differences between those users. The key behind this approach is to keep the users having fun during the competition process. This factor can be designed for group biking through a real race, or by employing a virtual race for single biking. The personal challenge can provide users set their goals to achieve, or are generated automatically by the applications. A suitable challenge provides users with good experiences and increases their motivation to face the next challenge.

3.6   Identifying emotional feedback

There are three kinds of positive feeling inspired from *competitive spirit*: sense of control, sense of co-presence, and sense of achievement. A sense of control satisfies users' needs to keep track of their bike activities. A sense of co-presence provides bikers with illusion of cycling together even though the difference in physical space and time. The sense of co-presence is constructed by real-time experience awareness. This augments social pressure together with support between bikers. A sense of achievement as a catalyst for bikers' *competitive spirit*, this sense increases their motivation to bike and promotes positive feedback for the activity.

4   THE DESIGN OF A SUSTAINABLE SOCIAL-ORIENTED BIKE APPLICATION

Inspired by the successful encouragement of the long-term bike use from the previous work, a notion of social-oriented bike application, BikeLine Challengers, are proposed based upon the principles. It was designed for enhancing users' motivation to bike, and the application supports user in building social relationships, increases cycling skill, and interaction with other bikers through performance comparison and social communications. These relationships are developed through the users' *competitive spirits* (see figure 2).

The users can create bike groups for group biking, viewing other members' real-time status for *competition*, send messages between group members for *cooperation*, sharing experience on social network for *communication*, and providing users compare their performance for *recognition*. Also, the application can provide the users with terrain information about bike routes. The terrain information is then

540

related to users' physical energy cost. The terrain information is automatically uploaded to a cloud server while users are cycling and shared with all the users.

The terrain information is exploited to establish the users' daily goals through personal challenges (figure 2-c). This equips the users with environmental awareness and enhances their bike activity (figure 2-b). It also provides the users with real-time personal experience information (such as route status and performance, see figure 2-a). The detailed personal experience history is logged for comparisons among group members (figure 2-e, 2-f). Comments regarding to the activities can be share among the members (figure 2-d).

(a) Being aware of Real-time personal performance (b) Real-time environment experience (c) Editing a goal of bike route (d) Group communication

(e) Sharing personal achievement (f) Reviewing personal experience history (g) Performance comparison

Figure 2. The "BikeLine Challengers" applications

To drive sustainable cycling, the application satisfies users' *sense of control* and *sense of achievement*. This application interacts with its user through an *ambient display interface*. This gives the users an effortless sense of control over their cycling activities. In addition, the application can manage *personal challenges* for the users based on their past performance. It can slightly adjust the challenge level to drive user meet the challenges. Thus the users will have positive feedback from the application when they break a bottleneck so the *sense of achievement* is satisfied.

5   CONCLUSIONS

This paper has identified the opportunity for the social-oriented bike application, combining the persuasive technology features for maintaining and sustaining physical activity. By looking to the existing bike application designs, this paper has reported a relatively full picture of a social-oriented bike application design through discussing and comparing many different approaches of bike applications.

The key concept - *competitive spirit*, are elicited from the literature. A set of design principles based on this concept is proposed. The design principles can serve as a tool to guide the designers in the process of social-oriented bike application development. They can use the design principles to get inspiration for their work and differentiate their designs. This paper also proposes BikeLine Challengers, a design for sustainable social-oriented bike application based on the design principles.

ACKNOWLEDGMENTS

This work was funded by Tatung University, Taiwan (B100-I06-035).

REFERENCES

CAMPBELL, T., NGO, B. & FOGARTY, J. 2008. Game design principles in everyday fitness applications. *Proceedings of the 2008 ACM conference on Computer supported cooperative work.* San Diego, CA, USA: ACM.

CONSOLVO, S., EVERITT, K., SMITH, I. & LANDAY, J. A. 2006. Design requirements for technologies that encourage physical activity. *Proceedings of the SIGCHI conference on Human Factors in computing systems.* Montréal, Québec, Canada: ACM.

CONSOLVO, S., MCDONALD, D. W., TOSCOS, T., CHEN, M. Y., FROEHLICH, J., HARRISON, B., KLASNJA, P., LAMARCA, A., LEGRAND, L., LIBBY, R., SMITH, I. & LANDAY, J. A. 2008. Activity sensing in the wild: a field trial of ubifit garden. *Proceeding of the twenty-sixth annual SIGCHI conference on Human factors in computing systems.* Florence, Italy: ACM.

EISENMAN, S. B., MILUZZO, E., LANE, N. D., PETERSON, R. A., AHN, G.-S. & CAMPBELL, A. T. 2010. BikeNet: A mobile sensing system for cyclist experience mapping. *ACM Trans. Sen. Netw.,* 6, 1-39.

FOGG, B. 2008. Mass Interpersonal Persuasion: An Early View of a New Phenomenon Persuasive Technology. *In:* OINAS-KUKKONEN, H., HASLE, P., HARJUMAA, M., SEGERST HL, K. & ØHRSTR M, P. (eds.). Springer Berlin / Heidelberg.

FOGG, B. 2009. A behavior model for persuasive design. *Proceedings of the 4th International Conference on Persuasive Technology.* Claremont, California: ACM.

FOGG, B. J. 2002. Persuasive technology: using computers to change what we think and do. *Ubiquity,* 2002, 2.

FOGTMANN, M. H., FRITSCH, J. & KORTBEK, K. J. 2008. Kinesthetic interaction: revealing the bodily potential in interaction design. *Proceedings of the 20th Australasian Conference on Computer-Human Interaction: Designing for Habitus and Habitat.* Cairns, Australia: ACM.

FROEHLICH, J., DILLAHUNT, T., KLASNJA, P., MANKOFF, J., CONSOLVO, S., HARRISON, B. & LANDAY, J. A. 2009. UbiGreen: investigating a mobile tool for tracking and supporting green transportation habits. *Proceedings of the 27th international conference on Human factors in computing systems.* Boston, MA, USA: ACM.

JACOB, R. J. K., GIROUARD, A., HIRSHFIELD, L. M., HORN, M. S., SHAER, O., SOLOVEY, E. T. & ZIGELBAUM, J. 2008. Reality-based interaction: a framework for post-WIMP interfaces. *Proceedings of the twenty-sixth annual SIGCHI conference on Human factors in computing systems.* Florence, Italy: ACM.

LIEN-WU, C., YU-HAO, P. & YU-CHEE, T. An augmented reality based group communication system for bikers using smart phones. Pervasive Computing and Communications Workshops (PERCOM Workshops), 2011 IEEE International Conference on, 21-25 March 2011 2011. 325-327.

MILUZZO, E., LANE, N. D., FODOR, K., PETERSON, R., LU, H., MUSOLESI, M., EISENMAN, S. B., ZHENG, X. & CAMPBELL, A. T. 2008. Sensing meets mobile social networks: the design, implementation and evaluation of the CenceMe application. *Proceedings of the 6th ACM conference on Embedded network sensor systems.* Raleigh, NC, USA: ACM.

MUELLER, F. F., EDGE, D., VETERE, F., GIBBS, M. R., AGAMANOLIS, S., BONGERS, B. & SHERIDAN, J. G. 2011. Designing sports: a framework for exertion games. *Proceedings of the 2011 annual conference on Human factors in computing systems.* Vancouver, BC, Canada: ACM.

PICARD, R. 1997. *Affective Computing*, The MIT Press.

PRIEDHORSKY, R., JORDAN, B. & TERVEEN, L. 2007. How a personalized geowiki can help bicyclists share information more effectively. *Proceedings of the 2007 international symposium on Wikis.* Montreal, Quebec, Canada: ACM.

REDDY, S., SHILTON, K., DENISOV, G., CENIZAL, C., ESTRIN, D. & SRIVASTAVA, M. 2010. Biketastic: sensing and mapping for better biking. *Proceedings of the 28th international conference on Human factors in computing systems.* Atlanta, Georgia, USA: ACM.

YASUDA, S., OZAKI, F., SAKASAI, H., MORITA, S. & OKUDE, N. 2008. Bikeware: have a match with networked bicycle in urban space. *ACM SIGGRAPH 2008 talks.* Los Angeles, California: ACM.

CHAPTER 59

Applying Microblogs to Be an Online Design Group: A Case Study with Human-Centered Methods

Jui-Ping Ma, Rung-Tai Lin

National Taiwan University of Art
New Taipei City, Taiwan, R.O.C.
artma2010@gmail.com

ABSTRACT

This research attempted to understand the possibility on whether non-deign background users can develop creative design works and activities through microblogs by information from users interact with each other or user-produced. Moreover, the differences on functions and usage of microblogs may be the important factors, which effected design activities. The research combines the perspective and methods of human-centered design, then, a series step of experiment in case study were adopted by recording, investigating, and integrating. All of conversations, posting messages, comments from 15 participants' logging on exchange information were classified by card sorting and analyzing them in four famous microblogs. The results indicate that (1) it's not only the design activities from participants' co-operating will be achieved possibly but also their message, idea can be converted into creativity represented on design works, even the participants with non-design background and no design experience (most of them are students of commercial departments) ; (2) in this case study, the participants are easily familiar with the interfaces of four microblogs even they never use them. Hence, when they exchange absolutely and focus on posting message, comments and conversation, the design works will be generated more easily ; (3) there is an obvious difference on the quantity of design sketches between four microblogs from participants. But there is insufficient evidence on which one element of the user interface in four microblogs, which affected greatly participants' design behavior,

544

quality of user-generated content and creative contents. However, indeed, the microblog can not only be a way of communication same as application of traditional media tools, but also is a main influential factor to induce participants creating user-generated contents for achieving innovation in cyberspace.

Keywords: Microblog, Online Community, Human-Centered Design

1   INTRODUCTION

It is an extreme phenomenon that the growths of online communities (OCs), in particular, blogs have being a novel, easy, and popular form of online communities. Frequently, the blogs are one of wide, acceptable media by users in cyberspace, which contain more virtual spaces named blogosphere. Users of blogosphere are also called 'bloggers', who published and spread information (comment, trackback) anywhere as well as anytime by many ways. Due to microblogs' operation are simple   but multi-function, they increasingly to be a mainstream of online communities, thus, microblogs offer an unprecedented form, opportunity to bloggers by posting many kinds of data and interacting each other easily.

In Taiwan, until 2011 Q2, according to Institute for Information Industry, there are 10.96 million population of internet, most of these internet users are 20-24years students and their main activities top1 is to upload and download files (76.3%), top2 is sending instant short message (64.6%), while top3 is that users own their social network website (63.7%)such as Facebook, Plurk, and so on. Some social network websites marked on the hit parade of top 100 on target ethnic groups is that Facebook (86.39%), Google (81.93%), Plurk (19.96%). Furthermore, another survey of insightxplorer company state that three most important reasons of people who use microblogs are that firstly, "all friends use this microblog service" (44.7%),; secondly, "user interface is friendly"(34.1%), and thirdly, "there are more users in it"(26.5%).

The object of this paper is trying (1) to explore the possibility that whether the design activities in microblogs can generate design works through the participants' behavior each other, even they are non-design background ; (2) to realize what will effect design activities in a microblog participants' behavior by their interacting each other.

For this goal, the study launched to (1) participate and observe the collaborative design activity of bloggers; (2) identify and estimate information within different interface of microblog showing suitable design interaction behaviors by time geography method, and (3) evaluate the effects of these design activities on the precedence and satisfaction degrees achieved by online investigating. The target information will be identified in terms of the human-centered theory and social networking characteristics. The evaluation not only based on the diversion of micro-blogs, but also variables of bloggers' operation. Then, the data resulting from the experiment is analyzed to establish whether any feasibility effect existed from these online design activities when bloggers interacting with each other in microblogs.

The paper is organized as follows. Section 2 reviews the related literatures on online communities, blogs, microblogs, and social network. Section 3 describes case study of microblog adopted for designing. Section 4 presents experimental results, and Section 5 concludes the paper.

2   LITERATURE SURVEY

This section first reviews related literatures of online community. Sub-section then introduces the characteristics of microblog that may be applied to online design group and the perspective of social network is applied to explain the activity of microblogs. Finally, using human-centered design methods within microblogs to be an online design studio context will be discussed.

2.1   Online Communities

Recently, people communicate and share ideas each other by internet media increasingly, that is why global quantity of online communities (OCs) has risen quickly (Smedberg, 2008).

In cyberspace, OCs are virtual, real-time community, whose members can login and enable their existence by taking part in membership trust and norm.

OCs can be a style of an information system, when anyone to be a member, they can post and offer content, just like a Bulletin board, News forum system or a place where only a restricted member can initiate post such as blogs. That is, in a virtual space, OCs can be founded on users, who enabling communication and supporting interpersonal interaction that extends over time (Chu & Chan, 2009; Martínez-Torres, Toral, Barrero, & Cortés, 2010), and forming networks of personal relationships (Jung & Kang, 2010; Wellman et al., 1996). Kim, et al. (2008) stated that OCs are an important resource for people with various interests, goals, and needs.

Since cyberspace become a new frontier in social relationships, people have been used the Internet to obtain colleagues, find lovers, and make friends—as well as enemies (Suler, 2004). Korzeny (1978) pointed out that online communities are formed around interests, and not physical proximity (Korzeny, 1978). Moreover, Wallace (1999) proposed that, based on someone's appearance, meeting in online communities eliminates prejudging and thus people with similar attitudes and ideas are attracted to each other.

2.2   Blog vs. Microblog

Recently, blogs have become a well-known, personal messages published on the World Wide Web. Users input and post journal discretely displayed in reverse chronological order.

Precisely, in technical perspective, blog is a style of website using a dated log format for publishing periodical information. Blogs were designed originally for personal use and are typically used as personal on-line diaries. Blogs are linked together via syndication (Charlie, Elise, 2003). Since microblogs emergeed rapidly, they provide a new communication channel for people to broadcast information that they likely would not share otherwise using existing channels (Dejin Zhao et al, 2009). In general, microblogs to be an informal communication tools that differ from a traditional communication media in their user-generated content and file. Typically, all contents of microblogs are always smaller then other media in both actual and aggregate file size. However, some previous researches have pointed out that informal communication at work may play important roles for collaborative work and organizational innovation (Dejin Zhao et al, 2009). Emotionally, people seem to use microblogging to achieve a level of cyberspace presence, being "out there" and to feel another layer of connection with friends and the world ().

Both blogs and Microblogs allow users to exchange small elements of content such as short sentences, images, or video links.

3   CASE STUDY

The case study named "Explorer Project" was conducted by tracing all messages sent from fifteen participants' login in four microblogs. Time and contents were recorded following each participant's activities, including sending message, posting images, and video chatting. Data were collected by human-centered methods then; they were analyzed by applied statistic.

3.1   Human-Centered Design Methods

In general, the user-centered design (UCD) process can help designers to realize customers' needs and fulfill the goal of a service or product for their users. When design process begins, user requirements are considered and included into a whole product cycle. Designers note and refine these requirements by various investigative methods including: ethnographic study, contextual inquiry, prototype testing, usability testing and other methods.

3.2   Participants

The "Explore Project" was a Creative Design Project launched on microblogs. This activity was promoted during November to December in 2011. The project selected volunteers to attend the activity at random. All participants were from different commercial department of NTCB school, they were ask to demonstrate great interest in creative design, but no one with design background or experience. For smooth data analysis and avoiding they are all the same terms, the project arranged male and female participants to be one group, all of them: (1) age without huge gape; (2) boy and girl numbers were same nearly; (3) have never used four microblog that project adopted. All participants were list by age and gender in Table 1.

Table 1. The Statistics by information of Participants

Department	Participants	Age	Gender	Persona
Accounting Information	2	19	Male	1M
		21	Female	2F
Finance	3	20	Female	3F
		19	Male	4M
		19	Female	5F
Public Finance and Tax Administration	3	20	Male	6M
		23	Female	7F
		21	Male	8M
International Business	2	19	Female	9F
		23	Male	10M
Business Administration	3	24	Male	11M
		20	Female	12F
		23	Female	13F
Information Management	3	24	Female	14F
		24	Male	15M
Total	15	M:21.4(mean)	M:8	15
		F :21.1(mean)	F :8	

3.3 Experimental Platforms

The experimental microblogs in this project are adopted individually in terms of preference of people in common using, so it is Facebook, Google+, Plurk, Twitter. All features of four microblogs were shown in Table 2.

Table 2. The feature of 4 Microblogs

Spec.	Name	Facebook	Google+	Plurk	Twitter
Image	Number	∞	∞	∞	∞
	Resize (%)	9.7%(∗)	15%(∗)	61%(∗◎)	34%(∗)
	Instant Upload	-	V	-	-
Video Link	Post Link	V	V	V	NA
	File Limit	-	-	-	-
Cross-Platform	O.S. Support	Android/iOS/ Win Mango	Android/iOS	Android/iOS	Android/iOS/ Win Mango

Short message	Txt Limit	NA	NA	140	140
	Attach Image	V	V	V	V
	List Style	Wall	Message string	Timeline	Tweets
Special Function	Classified	Social Tag	Circles	Address Book	
	Video Chat	Edit Button	Hangocuts	Timeline	NA
	Message Topic	NA	Sparks	NA	Following/ Follower
	Message umber Marked	Comment Numbers	NA	Karma Value	Tweets
	Track Message	Follow Post	「+1」 Button	NA	Following/ Follower
	Group Chat	Comment	Messenger	NA	NA

(✳) The experimental image is set to 1024x768 pixel, 72dpi and uploaded to 4 microblogs, then, reviewed the Image size percentage.

(◎) The image uploaded to Plurk is resized to 800x600 pixel but the dpi is modified to 28dpi.

3.4   Experimental Procedure

At beginning of experiment, all members of project team were named a code name, the project supervisor selected "Facebook", one of four microblogs randomly, then, all participants login it individually at same time and go about activity.

A creative shape was spread on Facebook(FB). Every participant was encouraged to post messages, replied comments and discussed each other in public area of FB, but at same time, they were asked to developed sketch in private way.

In this paper, several methods will be applied on the case study of project. All procedure of this research was shown in Figure 1.

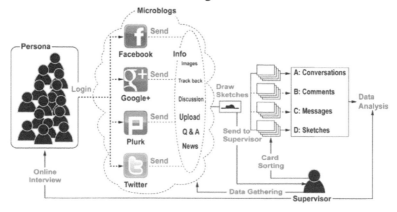

Fig. 1 The experimental flowchart

All messages in experimental procedure will be aggregated by project supervisor. After one hour, the project team logout and transferred to another microblog, "Google+", then, anybody of project team launched any message theme unlimitedly. As anteriorly, the project supervisor asked project members must take idea from all message from blog and draw down the shape. Then, after one hour all members changed to login another different microblog.

3.5 Data Gathering

The information spread from participants in microblogs was gathered by supervisor, and categorized four patterns caused of their attributes: (A) Conversation: Participants' conversation each other by typing txt. (B)Comment: The numbers of message from participants follow and reply anyone's post. (C) Message: The numbers of message from participants' posting message, image or hypertext. (D) Sketch: The design sketch drawing picture. All information calculated was shown in Table 3.

Table 3. The statistics of information from participants

Microblog	Facebook				Google+				Plurk				Twitter			
Data Persona	A	B	C	D	A	B	C	D	A	B	C	D	A	B	C	D
1M	5	3	6	3	7	6	6	4	4	3	2	2	5	4	3	4
2F	2	4	8	2	5	5	6	3	3	2	1	1	3	5	2	2
3F	4	3	7	4	3	3	8	4	6	4	2	3	6	3	2	3
4M	4	2	5	4	5	2	7	5	6	3	4	2	5	4	3	4
5F	5	4	3	4	4	6	5	5	3	2	6	3	4	5	2	2
6M	4	5	4	3	4	4	5	6	3	2	3	3	3	4	1	3
7F	3	6	5	5	6	8	6	4	6	1	3	2	6	2	2	5
8M	5	3	3	2	4	7	7	5	7	2	4	3	4	3	3	4
9F	6	2	5	2	5	6	5	6	4	2	4	2	2	2	4	3
10M	5	4	3	2	1	5	4	3	2	2	2	3	4	4	2	4
11M	6	5	4	5	3	2	4	4	4	5	4	6	8	7	6	4
12F	8	5	6	5	4	5	6	3	2	4	5	6	4	5	4	7
13F	5	6	4	5	4	4	4	4	4	3	5	4	4	4	4	7
14F	4	5	5	6	3	4	6	4	2	4	5	4	4	7	6	8
15M	6	1	4	3	5	7	4	5	2	3	3	1	3	2	3	2
Total	72	58	72	55	63	74	83	65	58	42	44	45	65	61	47	62

(A) Conversation (B)Comment (C) Message (D) Sketch

4. RESULT

This section goes about to show all of data produced form experiment procedure; the data analysis can be divided to three parts to explain. Hence, the paper will discuss as follows 4.1 Correlation coefficient Analysis ; 4.2 Participants Analysis.

4.1   Correlation coefficient Analysis

From the table3, the greater number of messages(C) is Google+, while the fewest than other microblog is Plurk. The sketches (D) can be noticed that Google+ is more than others. However, it seems that participants prefer conversation in Facebook, which talk each other more frequently. Because the researchers wonder whether information posted by the participants can correlate with sketches, therefore, the study chooses the greater number to calculate the correlation coefficient between data A to data C and sketches in every microblog.

The statistic data in correlation coefficient are shown as follows Table 4.

Table 4. The statistics of Correlation analysis on greatest quantity of data

	Facebook conversation	Google+ message	Plurk conversation	Twitter conversation
Pearson Correlation	.135	.537*	-.168	.160
Significant	.633	**.039**	.549	.568
Quantity	15			

Messages from Google+, the correlation coefficient are 0.039<0.1, which is the only one evidence that sketches may be produced by messages from participants.

Table 5. The statistics of information from participants

Anova[b]

ode	sum of squares	df	sum of squares	F	Significant
1 regression	9.231	1	9.231	5.270	.039[a]
residuals	22.769	13	1.751		
Total	32.000	14			

a. Predictor variables :(constant),google+ Message   b. The dependent variable : Sketch3

Hence, the scatter matrix plot of four microblogs by gender within messages and sketches are shown in Figure 2 to Figure 5.

From Fig.2 to Fig. 5, firstly, there are red circle and blue circle marked on figures, which with a only one point of three correlated member (red circle) on message and sketches in Google+ and Plurk, while Facebook and Twitter with not any one. Secondly, four points combined by two members shape a co-related are produced in Google+ and Plurk. Both of Facebook and Twitter own three points composed by two members, in one word, the participants can creative more sketches transferred from messages through Google+ and Plurk.

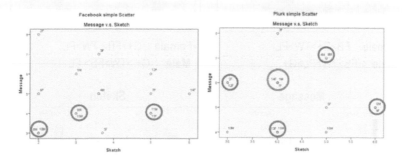

Fig.2 Facebook Simple Scatter Plot   Fig.3 Plurk Simple Scatter Plot

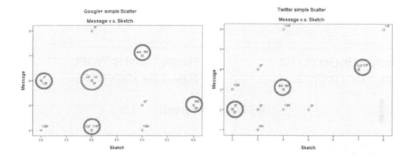

Fig.4 Google+ Simple Scatter Plot   Fig.5 Twitter Simple Scatter Plot

4.2   Participants Analysis

All numbers of individual sending information is including: messages, comments, conversations, and their sketches from persona in Facebook(FB), Google+(G+), Plurk(PL), and Twitter(TW) were recorded by gender. The statistic data shown on Figure 6.

Conversation

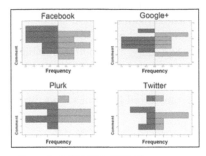

■**Female**：FB>G+>TW>PL
■ **Male** ：FB>TW>PL>G+

Comment

■**Female**：G+>FB> TW>PL
■ **Male** ：G+ >TW>FB> PL

Message

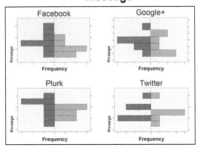

■**Female**：FB>G+> PL>TW
■ **Male** ：G+>FB>TW>PL

Sketch

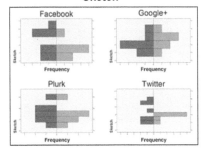

■**Female**：G+>FB>TW>PL
■ **Male** ：G+ >TW>FB> PL

Fig.6 Population pyramid

5   CONCLUSION

The results in the experiment of this case, shown as follows :
(1) There are different kinds functions which is suitable to develop design sketches in four microblogs, and participants select which one, it seems depending on their accessibility and usability of user interface.
(2) All of four microblogs seem to be common tools for users, even most of the users were novices.
(3) Some microblogs were showed their advantages:
   ■ The numbers of conversations in Facebook are more than other microblogs.
   ■ The numbers of comments, messages, and sketches in Google+ are more than microblogs.
(4) The correlation number between messages and sketches on Plurk and Google+ are more greater than Facebook and Twitter.

ACKNOWLEDGMENTS

The authors would like to thank Prof.John G. Kreifeldt for his insightful comments on our ideas. We also thanks our students from National Taipei College of Business for their participating to experiment.

REFERENCES

C. Lindahl, E. Blount, Weblogs: simplifying Web publishing, IEEE Computer 36 (11) (2003) 114–116.

Chu, K.-M. (2009). A study of members' helping behaviors in online community. Internet Research, 19(3), 279–292.

Chu, K.-M., & Chan, H.-C. (2009). Community based innovation: Its antecedents and its impact on innovation success. Internet Research, 19(5), 496–516.

Chun, S. Y., & Hahn, M. (2007). Network externality and future usage of Internet services. Internet Research, 17(2), 156–168.

Chung, K. S. K., & Hossain, L. (2010). Towards a social network model for understanding information and communication technology use for general practitioners in rural Australia. Computers in Human Behavior, 26(4), 562–571.

E. Young Jr., Terrence, Blogs: is the new online culture a fad or the future? Knowledge Quest 31 (5) (2003) 50–51.

Grenny, J., Maxfield, D., & Shimberg, A. (2008). How to have influence. MIT Sloan Management Review, 50(1), 47–52.

Ishii, K., & Ogasahara, M. (2007). Links between real and virtual networks: A comparative study of online communities in Japan and Korea. CyberPsychology & Behavior, 10(2), 252–257.

Kim, H.-S., Park, J. Y., & Jin, B. (2008). Dimensions of online community attributes: Examination of online communities hosted by companies in Korea. International Journal of Retail & Distribution Management, 36(10), 812–830.

Korzeny, F. (1978). A theory of electronic propinquity: Mediated communication in organizations. Communication Research, 5:3–23.

Lin, K.-Y., & Lu, H.-P. (2011). Why people use social networking sites: An empirical study integrating network externalities and motivation theory. Computers in Human Behavior, 27(3), 1152–1161.

Martínez-Torres, M. R., Toral, S. L., Barrero, F., & Cortés, F. (2010). The role of Internet in the development of future software projects. Internet Research, 20(1), 72–85.

McFedries, P. 2007. Technically speaking: All a-twitter. IEEE Spectrum, 44(10), 84.

Rizavi, S. S., Ali, L., & Rizavi, S. H. M. (2011). User perceived quality of social networking websites: A study of Lahore region. Interdisciplinary Journal of Contemporary Research In Business, 2(12), 902–913.

Smedberg, A. (2008). Learning conversations for people with established bad habits:A study of four health communities. International Journal of Healthcare Technology & Management, 9(2), 143–154.

Suler, J. (2004). The final showdown between in-person and cyberspace relationships. Retrieved March 3, 2009 from http://www-sr.rider.edu/~suler/psycyber/showdown.html.

Wallace, P. (1999). The Psychology of the Internet. Cambridge, UK: Cambridge University Press.

http://en.wikipedia.org/wiki/Blog

http://en.wikipedia.org/wiki/Online_communities

http://www.find.org.tw/0105/howmany/howmany_disp.asp?id=302&SearchString=%B3%A1 %B8%A8%AE%E6

CHAPTER 60

Design Guidelines That Keep Users Positive

NAKANE Ai, NAKATANI Momoko, OHNO Takehiko

NTT Cyber Solutions Laboratories
Yokosuka-Shi Kanagawa Japan
nakane.ai@lab.ntt.co.jp

ABSTRACT

ICT devices that connect to the network are becoming common at home, and more people are being required to set them up; a huge psychological burden that hampers ICT adoption. This study introduces design guidelines that encourage users to set up ICT devices.

First, we conducted interviews and analyzed the data to elucidate the stages in negative feeling development. As a result, we could construct a user's psychological process model. We identify 5 factors that cause negative feeling.

Second, we draw the design guidelines from these five factors. The key point of these guidelines is that they are user-centered and are quite unique since they are derived from negative emotions. ICT devices that are designed according to these guidelines will be well received and widely adopted.

Keywords: negative feeling, mental process model, qualitative research, set up task,

1.1 INTRODUCTION

Nowadays, It is becoming commonplace for the home to host ICT (information communication technology) products (e.g. PC, HDD) that connect the Internet. These products have many users who have low ICT skill but are required to set up the devices by connecting them to the network. The set up tasks are a psychological burden on users, who develop strong negative feelings[1]. We find that negative users fail to use the ICT products and service despite the interest in using them [2].

This study focuses on designing set up tasks that don't impose undue psychological burdens on the users and thus encourage the use of ICT products. It is important that user's feeling is made better in set up tasks because it was reported importance of consideration about user's feeling on the design process of products and services (Bitner, Ostrom, and Morgan, 2008; Jordan, 2002; Norman, 2004).

There are two approaches for improving the user's feeling. One is to emphasize positive feelings such as fun. The other is to suppress negative feelings such as anxiety. This study adopts the latter one. Because positive and negative feelings are basically not related (Watson and Clark,1999), emphasizing the former does not suppress the latter. It is essential to identify and tackle the cause of negative feelings about the set up task. Additionally, it has been shown that negative experiences during set up degrade the interest in using the device (Serif and Ghinea, 2005). This indicates that suppressing negative feelings during set up is helpful to not only lower barrier to initial use using but also to encourage the active use ICT products.

What is the set up task for ICT products? For example, the set up task of a Set Top Box (STB) connected to a TV is as follows. First, the user opens the STB box, and takes out manuals, STB, cable, and so on. He/she chooses what appears to be the appropriate manual and attempts to identify the condition of his/her home network from choices in the manual, attaches the cable according to the manual, powers up the STB, and finds the contract document with Internet Service Provider (ISP). He/she watches the TV display and inputs the data necessary by the STB's remote controller. GUI is displayed on the TV display after a few minutes. Finally, he/she configures STB to link other ICT products or set preferences. The users must handle numerous artifacts, which triggers negative feelings. Thus we must design all artifacts handled by the user to suppress negative feelings.

Previous research hasn't identified why users develop negative feelings about the set up task, and failed to show how artifacts should be design to suppress them. In this study we clarify why users feel negative about the set up of ICT products by constructing a user's mental process model, and then propose design guidelines for artifacts.

---

[1] Result of Web-based research from Feb to March, 2010. 792 subjects, 20 to 64 years old, living in the Tokyo metropolitan area and using broadband (optical line, CATV and DSL line). In this result, 65.4% of subjects feel weak point and 84.8% of them feel troublesome in setup.

[2] Result of Web-based research in June, 2008. 116 subjects interacted with video delivery service but didn't take up the contract. They were quried as to why they didn't contract the service.

1.2   CONVENTIONAL STUDIES

Artifact design mainly focuses on usability, which unfortunately has multiple definitions (ISO,1998; Nielsen,1993). The common approach is to focus on the effective achievement of a specialized task. However, ICT users often give up because they feel negative, not only because they can't understand the task. For example, they break off the task when they feel negative such as anxiety, troublesome, and so on. Additionally, they don't begin to set up the ICT products and don't use it, when they have mental hurdle such as think that "I don't want to do the task" or "it is too difficult for me". Usability can't predict such user decisions caused by feelings.

There are some GUI guidelines that focus on users' feelings such as disgusted and frustrated (Cooper, Reimann, and Cronin, 2007), but they are not useful in designing manuals, boxes, products and so on. One method of service analysis that considers the emotion of the customer is named Service Blueprinting (Bitner, Ostrom, and Morgan, 2008). This is a method that visualizes the interaction between a customer and a service and pays attention to their emotional state. This contributes to design services that satisfy customers. However, it isn't showed how to imagine user's emotion in this method, so it leaves the decision on analyzer's skill. Thus none of the studies that focus users' feeling and emotion have addressed the factors that affect user's feeling when considering the set up task.

The goal of this study is that the user can begin to use ICT products without negative feelings about the set up task, and propose design guidelines that can be used when designing the artifacts. Specifically, we conduct interviews and from an analysis of the results establish a user's mental process model that identifies the factors influencing negative feelings.

2   INVESTIGATION

Participants: Fifteen Japanese (average age=37.2, SD=7.27, maximum=56, minimum=28). Five thousand one hundred and eleven people provided answers to a questionnaire on the Web, and we screened their answers. Six of the 15 (positive group) had a positive image of the set up task, and the remaining 9 (negative group) had a negative image of the task. All were rewarded for participating.

Method: We conducted semi-structured interviews. Interview time was 90 minutes per person. The interview targeted their feelings of set up tasks that they had experienced in the past. Four items were found to shared by most participants as imposing negative feelings on users; (1) Physical, temporal, economic and cognitive costs incurred by set up. (2) Reliability of ICT products and network. (3) Own skill and knowledge about set up. (4) Past experience.

Analysis: We used the qualitative research method to analyze the interview protocol (Strauss and Corbin, 2008).

1: We transcribed all interviews. (487 pages in A4 paper)

2: We conducted open coding, with the aim of identifying key themes in the data.

3: We grouped the initial sets of the phenomena described by the open codes into categories.

4: We related the categories and refined them.

3.1   MENTAL PROCESS MODEL

From the results of the interview data analysis, we constructed the process model shown in Figure1. This shows the user's mental process for the set up task and identifies the factors yielding negative feelings. The process model consists of five internal states (squares with rounded corners), 2 external inputs (squares drawn by dotted line) and 5 factors making users negative (squares).

(I) Past experience, knowledge and evaluation of task, product, and artifact, and self-cognition: This internal state has 3 items. One is past experience. This indicates the user' subjective experience such as results and feelings. For example, a result may be "I failed the Wi-Fi set up task" and the feeling was "I became anxious". We show interview data in box.

Next is knowledge and evaluation of the task, product and artifact. This item contains knowledge that they already have about ICT service, product, content of task, artifacts such as cables and routers, and evaluation of status of current set up task. For example, Knowledge is "I have to power up the ICT product and configure it after the connection task" and "I think I can't complete the task so I need someone's help", and evaluation is "I guess connection is failed".

Third one is self-cognition. This contains an understanding of own skill and knowledge needed for the set up task and character and habits. An example is "set up task is my weak point" and "I quickly give up, that is my nature".

The individual differences expected in (I), each user has a different experience and knowledge, influences the subsequent states. For example, a user who has failed set up once will become anxious when presented with many cables. On the other hand, a user who rates his own ability highly won't feel negative in the same situation.

(II) External input: This includes visual and linguistic information. Visual information includes the box containing ICT products, cables in the box, manuals and so on. Linguistic information are words such as "IPTV" and "router".

(III) Estimation of task and result: This means the estimation of the next task or tasks and the results expected when state (I) receives the input of (II). Task estimation consists of a specific task and a brief understanding of the implications of the task. The former identifies the specific action needed for the task and its result. For example, "I'm going to put this cable in that jack", "I'm going to input numbers via the remote controller", "I guess cable will click if it put in completely" and "I think this display will change to black after inputting the number". The implications include cost (e.g. time) taken to complete the task, the number of steps needed, degree of difficulty and so on.

(V) External input triggered by user's action: This is external information triggered by user's action. For example, "display changes to confirmation screen" and "no change" after input of numbers.

558

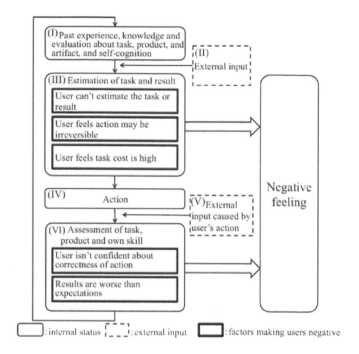

Figure1 user's mental process model on set up task

(VI) Assessment of task, product, and own skill: This state is developed after the user's action made to complete the current task. The information includes correctness of action, degree of difficulty, state of task, recognition of own skill and so on. Examples are "action was easy because task is progressing as expected", "action has failed in some way because ICT product is disabled" and "I have more skill than I thought".

(I) is updated when (VI) is revised. To complete the set up task, the user undertakes various actions based on external input. This mental process is repeated until the task is completed or halted.

3.2   FACTORS CAUSING NEGATIVE FEELINGS

We found that states (III) and (VI) contained several factors that were capable of strengthening negative feelings. We explain these factors and how to suppress the negative feelings.

Factor1 {user can't estimate the task or result}
Our analysis of interview data found that negative feelings are raised when the user can't estimate the task or result in (III). That is, user feels negative if he can't quickly estimate task attributes such as time to complete, degree of difficulty and task complexity. This factor includes the case wherein no specific action is known (which jack should take this cable?) and the case wherein the result of the action is unclear.

> I was at a loss, because I can't imagine at all how long time the set up task will take (about set up task). (ID=13)

> I had a lot of anxiety. I didn't understand at first whether it is OK or NG to put the cable in jack, and how to put the cable. I was at a loss, because I can't image at all how long time the set up task takes (about set up task). (ID=07)

Factor2 {user feels action may be irreversible}

User feels negative if he estimates his actions will lead to irreversible changes in (III). Examples are "product may break if I set it up", "If I change the setting I may not be able to return to the previous menu". However, users who know how recover don't have negative feelings. For example, a participant said she had negative feelings because she was unable to recover the setting of her PC while another expressed her confidence in being able to recover from the same situation.

> I know nothing about set up. So I was worried. It will be serious if I damage the product, such as breaking it or dropping it. (ID=03)

> I suddenly became anxious. There is a case it can't be turned back after wrong operation, isn't it? (ID=05)

Factor 3 {user feels task cost is high}

User feels negative if he thinks that the task has high physical, mental, monetary, and temporal costs in (III).

> (While she looked down the box of STB) I get strongly anxious (when I will look inside the product's box), because it contains various things...There are four (cables). This suggests that it has 4 kinds of tasks for set up. (ID=02)

> Long manual made me think that I may be puzzled when reading it. I'm worried I will take a long time to set up because of mistakes made by this confusion. (ID=11)

Factor 4 {user isn't confident about correctness of action}

There are two issues. First, user becomes negative if he/she is not confident that the action done in (III) was correct, and the second, and related issue, is the inability to assess whether the action was completed successfully. For example, one participant felt negative because she her action yielded no feedback from the product. Another user had quite different response since her knowledge of the task eliminated the need for feedback.

> I was afraid until confirmation of displayed image. Because I didn't have any confidence in my connections. (ID=01)

Factor 5 {results are worse than expectations}

The user feels negative if the action's results, success, cost, and degree of difficulty, are worse than those predicted in (III). An example is "task has more actions than I thought" and "I thought task would be completed already but it is not".

> I thought the task would take about 15 minutes. (But I couldn't finish it.) Then I gave up. (ID=14)

The proposed mental model clearly indicates the key factors that can strengthen negative feelings when attempting to complete the set up task. We now turn to offsetting these factors through the introduction of design guidelines.

4.2 GUIDELINES FOR SUPRESSING USER'S NEGATIVE FEELINGS

Table1 introduces our proposed guidelines for suppressing the negative feelings.

[Design that allows users to easily estimate task]

We should show users the information that makes it possible to estimate the cost of the task, number of task components, and specific recovery processes. For example, it is effective to provide the user with the time needed to complete the set up task and a manual that provides simple descriptions of each action.

In addition, an effective counter to negative feelings is to describe the results of actions. For example, if the cable is connected properly, the instructions should clearly describe the sequence of status lights that should appear.

Prior research emphasized the importance completing individual actions. Results included predicting the action and identifying the means needed to complete the action (Norman,1988), and making it easy to understand the changes expected after the action(Norman,2007). These results are important for simple actions or tasks; i.e. those that don't make the user perplexed, and covered by our factor I in state III. We emphasize here the new finding of the importance of giving the user an over-view of the entire task. This discovery arose due to the focus we placed on keeping the user positive. That is, the user can continue with the task even if no overview is available, but the user's negative feelings will be strengthened in this case.

[Design that makes users confident that recovery is possible]

It is important for suppressing negative feelings to eliminate the worry that no recovery is possible in any situation. To achieve this, the interface should provide a clearly marked button that returns to the prior state and the user should be alerted to existence of this button.

There are many users who think machines such as PCs and routers will be break down easily. We should design the products so that they appear to be indestructible, and the manual should tell the user that the product is robust.

Designing the system to enable restoration from error is present in most usability guidelines. It is "Help users recognize, diagnose, and recover from errors" in Nielsen (Nielsen,1994), "Design for error" in Norman (Norman ,1988) and "Offer simple error handling" and "Permit easy reversal of actions" in Shneiderman (Shneiderman and Park,1987). On the other hand, this guideline emphasizes the user's opinion as to whether he/she can recover error and that he/she shouldn't believe that error will occur; this is a key advance. It is understood that it is important to design a system that can be restored and this basic requirement is satisfied by placing an undo button on the interface. However, this guideline makes a further demand in that the user must be made aware of the possibility of recovery and the means of doing so. A good point of this guideline is to focus attention on issues other than system improvement. For example, products should be designed to look robust in order to keep users positive.

[Design that reduces task cost estimates]

It is important to prevent the user from over estimating the costs of the task, such as time taken, possible troubles, difficult actions and so on. For example, we should make the task easier by reducing the number of parts and writing the manual

	guideline	Underlying idea	concrete example
(III) Estimation	Design that allows users to easily estimate task	Giving information that suggest task's overview (task time, degree of difficulty, steps of task, and so on).	Printing steps of task (e.g. connect and setting) on box containing product.
			Printing average time taken task on manual.
		Expressing clearly specific task performed next.	Printing concrete action which user can understand easily.
			Showing a page number which user should look at when in trouble.
		Making it possible to estimate the change caused user's action or expressing it.	Expressing clearly that it begin to communicate with server when cable is connected.
			Expressing that Information isn't send when code numbers is input, but when "OK" is held down.
	Design that makes users confident that recovery is possible	Making users estimate that product don't break down easily or express it.	Design appearance friendly, not like a precision machine.
			Suggest that products isn't break down easily on manuals.
		Suggest concrete means for restore.	Telling user that he/she can turn back the setting with "undo" on UI.
			Telling user that data can be taken back as long as he/she backup the files.
	Design that reduces task cost estimates	Giving no information which suggest physical, cognitive, monetary and temporal cost is high.	Eliminating the image of difficult task to approach by simple manual.
			Making users think that he/she can't complete the task by enhancement of support.
	Design that gives users confidence about action's success	Giving user confidence on the action by intuitional interface.	Coloring a set of cables and jacks in same color.
		Making users think task is moving ahead by user's action.	Express as "NEXT", not "UNDO" in UI when error occured.
	Design that encourages agreement between expectations and actual results	Making it possible to estimate task's overview accurate.	Showing task's overview accurately.
		Making it possible to understand user's status.	Showing progress bar reflected actually status of task accurately.
(VI) Assessment	Design that gives users confidence about action's success	Giving feedback that suggest right or wrong of task.	Giving sound click or sense when cable is put in jack completely.
			Making UI Change when system receives input.
	Design that encourages agreement between expectations and actual results	Making task's cost lower.	Making the manual be comprehensible easily.
			Lessoning amount of action of user by supplement the user's action by system.

Table 1 Guidelines for suppressing user's negative feelings

in the most simple manner possible. One idea is to move the full parts list to the back of the manual so that the front shows only those parts needed in the most common configuration.

[Design that gives users confidence about action's success]

There are 2 ways of encouraging the user's confidence in the correctness of an action. One involves state (III) and the other way state (VI). The former involves ensuring that the user estimates the action clearly. In fact, the interface must unambiguously indicate what the next action is. For example, color coding the plug and associated jack ensures that the user feels certain about where to insert the plug.

The latter demands that the user be provided with appropriate feedback when the action is performed. Feedback should be given even if user's action is false. This is because strong negative feelings are aroused if an action that the user thought was performed successfully later turns out be have been a failure. This is indicated by factor V in state VI. Appropriate feedback is shown in Norman (2007).

[Design that encourages agreement between expectations and actual results]

Here also there are 2 issues, one involved with state (III) and the other with state (VI). The former attempts to block overly-optimistic estimates. The information given to the user must not wrongly claim that complex tasks are simple, for example. The latter reduces the difficulty or complexity of the actual task. In addition, making the user create accurate estimates (e.g. "task is approaching its end" and "task is making progress") is effective in suppressing negative feelings. This can be realized by showing a progress bar that reflects the actual status of the task and suggesting user's mistake if user is wrong.

The basic idea underlying this guideline, the importance of minimizing the gap between expectations and reality, is quite novel and has been noted in the literature (Nielsen,1994; Norman,1988; Shneiderman et al.,1987; Kamper,2002). It was created because we focused user's negative feelings, not just usability.

This last guideline appears to conflict with the guideline [Design that reduces task cost estimates]. The reduction in negative feelings created by an "optimistic" estimate of task cost (state III) will overcome when the user perceives that the expectations did not match the actual experience. This problem is avoided because of our instance that the guideline [Design that reduces task cost estimates} must produce accurate estimations.

4.2   CONCLUSION

The goal of this study is to suppress the negative feelings users commonly experience when setting up the ICT services so as to encourage the usage of said services. We constructed the user's mental process model (Figure1) for the set up task, and elucidated the stages in the development of negative feelings. Based this model, we developed seven design guidelines (Table1).

The literature is full of proposals that attempt to make artifacts easy and effective to use without confusion or error, Examples include Nielsen's "Ten Usability Heuristics" (Nielsen,1994), Norman's checklist and four principles (Norman,1988), Shneiderman's eight golden rules of interface design (Shneiderman et al.,1987), and Kamper's 18 heuristics grouped under 3 general principles (Kamper,2002). Our guidelines complement these techniques by focusing on eliminating the factors that cause negative feelings. Rather than addressing "how to construct systems", they express "how the user feels and think about undertaking tasks involving artifacts". This difference lead to the creation of these new and novel guidelines. For example, we proposed that the user should be supported in making accurate estimates of not only specific actions, but also the entire task (time, degree of difficulty and so on). Of course, the artifacts should be designed to give the user confidence in setting them up. In addition, we introduced [Design that minimizes the gap between estimation and assessment task attributes] as a guideline.

These guidelines are comprehensive in that they direct the design process of artifacts even before they are used. Examples include simple packaging and the appearance of the manual. As such they differ markedly from guidelines on usability, which apply only when the artifacts are used.

Our research shows that to suppress user's negative feelings demands more than improving the system. For example, most users develop negative feelings (anxiety) when inputting authorization codes. Our work indicates that providing the user with reassurance of easy recovery will be very effective in suppressing anxiety, and well supplements the regular techniques proposed for simplifying user input. These innovations appear rather minor but their impact is significant and well addresses the problem of user's negative feelings.

Our goal is to encourage the user to actively consume ICT services, and a key requirement is that the user be able to set up the equipment. For this, all artifacts handled by the user meets before and during set up must be designed appropriately. That is, these guidelines should be applied to not only manuals, devices, cables, and software, but also ICT service commercials and packaging.

To maintain the user's motivation, it is very important that all artifacts must be designed around the user in a comprehensive manner, regardless of the number of groups involved in development. This is becoming an urgent need given the penetration of ICT services.

While the guidelines introduced here are specific to the set up process, the concept of targeting the user's feelings is very powerful and is expected to lead to other studies on different subjects. Given the restricted age group who were interviewed, further work is needed to confirm that these guidelines cover other groups, for example the elderly.

We introduced design guidelines using the task of set up as our example, but there is the possibility of applying them to more complex tasks that involve several sub tasks. An example is "recording a certain TV program on HDD". We will continue our research to elucidate the applicable scope of the guidelines in terms of users and tasks.

REFERENCES

M.J.Bitner, A.L.Ostrom, and F.N. Morgan. Service blueprinting: A practical technique for service innovation. California Management Review, Vol. 50, No. 3, p.66, 2008.

A.Cooper, R.Reimann, and D.Cronin. About Face 3: The Essentials of Interaction Design. Wiley, 2007

ISO 9241-11 ergonomic requirements for office work with visual display terminals (vdts) – part 11, guidance on usability, 1998

P.W.Jordan. Designing pleasurable products. CRC, 2002.

R.J. Kamper. Extending the usability of heuristics for design and evaluation: Lead, follow, and get out of the way. International Journal of Human-Computer Interaction, Vol.14, No.3, pp. 447–462, 2002.

J.Nielsen. Usability Engineering. Morgan Kaufmann, 1993.

J.Nielsen. Enhancing the explanatory power of usability heuristics. In Proceed- ings of the SIGCHI conference on Human factors in computing systems: celebrating interdependence, pp. 152–158. ACM, 1994.

D.A. Norman. The Psychology Of Everyday Things. Basic Books, 1988.

D.A.Norman. Emotional Design: Why We Love (or Hate) Everyday Things. Basic Books, 2004.

D.A. Norman. The Design of Future Things. Basic Books, 2007.

T.Serif and G.Ghinea. Hmd versus pda: a comparative study of the user out-of-boxexperience. Personal and Ubiquitous Computing, Vol.9, No.4, pp. 238–249, 2005.

B.Shneiderman and College Park. Instructional TelevisionSystem Maryland. Designing the user interface. Addison-Wesley Reading, MA, 1987.

R.Spencer. The streamlined cognitive walkthrough method, working around social constraints encountered in a software development company. In Proceedings of the SIGCHI conference on Human factors in computing systems, pp. 353–359. ACM,2000.

A.L.Strauss, J.M.Corbin. Basics of Qualitative Research: Techniques and Procedures for Developing Grounded Theory Sage Publications, Inc. 2008

D.Watson and L.A. Clark. The panas-x: Manual for the positive and negative affect schedule-expanded form. 1999.

CHAPTER 61

Affective Evaluation and Design of Customized Layout System

*Chun-Cheng Hsu, **Ming-Chuen Chuang

* Department of Communication and Technology, National Chiao Tung University, Taiwan
**Institute of Applied Arts, National Chiao Tung University, Taiwan
* chuncheng@mail.nctu.edu.tw, **cming@faculty.nctu.edu.tw

ABSTRACT

Since the introduction of Web2.0, blog has become the virtual space where people express their uniqueness and define themselves in relationship to social groups. Nowadays blog platforms offer a choice of ready-designed interface layout templates, thus lowering the technological and cost threshold for setting up personal blogs. However, little research has been undertaken analyzing users' experience when using blog interfaces. This article uses the Semantic Differential (SD) method to evaluate affective responses and then used Multidimensional Scaling (MDS) algorithms PREFMAP to explore main dimensions influencing users' affective responses to personalized blog interfaces. In conclusion, this study: (1) investigates the relationship between blog interfaces and design features; (2) provides guidance for blog design and applications of customized templates system.

Keywords: Blog; Affective evaluation; Multidimensional Scaling (MDS); Layout templates system

1   INTRODUCTION

The American IT blog columnist Gilmer (2004) points out that 1.0 was old or traditional media, 2.0 is what people normally call new or cross-media, and 3.0 is We Media, or personal media, of which blogs are a trend. The blog is an author-centered communication medium that emphasizes absorbing information and sharing it, and representing the blogger's value system and convictions. Research has found that the two most important considerations when choosing a blog platform are (1) which platform their friends use, and (2) good interface design (InsightXplorer, 2011). Blogs were originally mainly textual, but later increasingly emphasized visual modes of expression; bloggers too attempted to find a way to express themselves within the constraints of the blog form (Fullwood, Sheehan, & Nicholls, 2009; Zheng, 2005). Moreover, Herring, Scheidt, Bonus, & Wright (2004) found that bloggers want their blogs to be individualistic and reflect the self-identity, and for these reasons it is important, therefore, to examine blog interfaces from an emotional perspective.

Nowadays blog platforms offer a choice of ready-designed interface layout templates, thus lowering the technological and cost threshold for setting up personal blogs. However, little research has been undertaken analyzing users' experience when using blog interfaces; furthermore, blog platforms offer layout template choices but choosing many templates can be very time-consuming for bloggers. Moreover many blogs still use the same blog templates, and lack distinctiveness. Therefore how to make a blog interface easy to use, individualistic, and that stands out, are issues strongly emphasized by both bloggers and developers. The purposes of this study are: (1) to investigate the relationship between blog interfaces and design features; (2) to provide guidance for blog design and discuss possible applications.

2 LITERATURE REVIEW

2.1 LAYOUT TEMPLATES OF BLOG

Blogs differ in type according to the differing goals of bloggers and the scale of the blog. Hsu and Lin (2008) divided blogs into two main categories: personal interest blogs, including diaries and opinions; and specialist forums, such as economic polls and surveys or reports about using new technologies. The object of this current research is personal blogs, as with personal blogs there is more emphasis on the arrangement and editing of visual layout, and layout changeability, whereas group or forum-type blogs tend to emphasize clarity of information layout, and are normally visually less creative.

Nowadays blog platforms offer a choice of ready-designed interface layout templates (Figure 1), thus lowering the technological and cost threshold for setting up personal blogs. However, little research has been undertaken analyzing users' experience when using blog interfaces; furthermore, blog platforms offer layout template choices but choosing many templates can be very time-consuming for bloggers. Moreover many blogs still use the same blog templates, and lack distinctiveness. Therefore how to make a blog interface easy to use, individualistic, and that stands out, are issues strongly emphasized by both bloggers and developers.

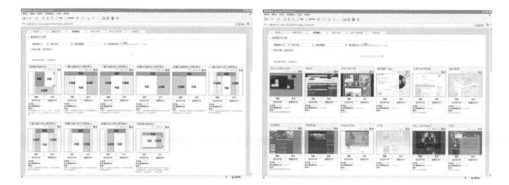

Figure 1   Ready-designed interface layout templates systems of Wretch (Source: http://www.wretch.cc/admin/style/?func=template_new&source=B)

Bloggers regard blogs as their own publically accessible virtual space, and so besides textual content, also utilize arrangements of visual features such as images and color to make that space unique. Lin (2006) explored personal blog design from the blogger's perspective, indicating that, apart from currently preferring simple and clear interface designs, bloggers also require their blogs to have a personalized look. This emphasis on personalization means that bloggers wish their blogs to have their own unique personality and style, which they set about expressing chiefly through visual layout design. This means that blog personalization is very closely related to visual expression; for example, in order to attract users, bloggers put a great deal of thought into decorating their blogs, and a good visual design can increase users' sense of enjoyment when browsing a blog. The goal of this research is to understand exactly what kind of feelings the different visual features of blog interfaces give users. It also seeks to understand which main design factors can be manipulated to arouse users' affective responses.

2.2   Affective evaluation and design of visual interface

As blogs are themselves a form of webpage, past research relating to webpages can also be used to discuss and compare blog interface designs. Researchers suggested two approaches to identifying high-level concepts of website design which may be used as focuses for improving user experience (Noam Tractinsky, Cokhavi, Kirschenbaum, & Sharfi, 2006; van Schaik & Ling, 2008). First, an affective approach focuses on users' processing of the attributes of webpages (Hassenzahl, 2004; Lavie & Tractinsky, 2004; Pandir & Knight, 2006; Schenkman & Jönsson, 2000). Second, a design-based approach focuses on how to find the quantitative relationship between design factors and affective responses to webpages (Kim, Lee, & Choi, 2003; Park, Choi, & Kim, 2004).

Tending towards the affective approach, Schenkman and & Jönsson (2000) carried out research into aesthetic appeal and user preferences in webpages. Using multidimensional scaling (MDS) and factor analysis, this study summarized seven property vectors for webpage preferences: complexity, legibility, order, beauty, meaningfulness, comprehension, and overall impression. Hassenzahl (2004) discussed how overall quality or 'goodness' of an interactive interface is formed. Lavie & Tractinsky (2004) used factor analysis to investigate the perceived visual aesthetics of websites, and found that factors influencing user perception include classical aesthetics and expressive aesthetics. Pandir and Knight (2006) found that among aesthetic preferences for webpages there was a correlation and consistency between complexity, pleasure, and interest, and evaluations of the aesthetics of webpages. Of these, pleasure and interest were relatively subjective aesthetic evaluations, while complexity was an objective, non-aesthetic evaluation.

The above research shows that the feeling created by a website is of considerable importance. Social networking blogs emphasize communication between people more than ordinary websites; however there is little research into the relationship between visual expression and user emotion in blogs. This study will use a design-based approach to explore the theme of blog interface design.

3 Research Method

3.1 Experiment Design and Relevant techniques

This research first used cluster analysis to select a representative sample of blog interfaces. An Semantic Differential (SD) survey into affective response of subjects was undertaken, and the data analyzed using MDS, in order to understand the relationship between visual features of blog interfaces and affective responses. A brief introduction to the characteristics of the MDS algorithms PREFMAP follows.

MDS analysis is a set of statistical techniques used to explore potential relationships in human perceptions of a stimulus and generate a perceptual space. In this study, subjective affective response data is transposed onto a geometrical space to allow easier interpretation. MDS represents objects as points in a Euclidean space so that the perceived distances between points can reflect dissimilarity between objects. For practical reasons, the number of dimensions of the projected space is usually kept as low as possible.

3.2 Sample

Blog Sample Selection

The stimuli of the experiment in this research are blogs. Initially, a wide-ranging sample group of 315 blogs were collected from Taiwan's top 5 most popular blog platforms (in terms of number of users), i.e. Wretch, Yahoo!Kimo, PIXNET, Xuite, and Yam blogs (InsightXplorer, 2010). In order to reduce the burden on the subjects in the experiment, one representative blog from each group was chosen, producing a final sample of 18 blogs.

Selection of Affective Adjective Pairs

Researchers first searched for affective adjectives suitable for describing images of blog interfaces in books and research relating to webpage or blog interface, initially obtaining 200 such words. The 8 experts who had selected the blog samples then discussed these words, arranging them in to pairs, eliminating those that were too similar, until finally 60 pairs remained. These words were written on small cards, and then the same 30 subjects from the blog selection stage were asked to divide them into groups according to similarity. Then hierarchical cluster analysis (Ward's method) was used to produce a cluster dendrogram of the 60 affective adjective pairs. The 8 experts discussed the results of the cluster analysis and decided by majority decision to choose a division of 11 groups. They then selected a single adjective pair from each group that best represented that group. The final selection is 11 adjective pairs. The aim of this screening process was to reduce the number of criteria of the SD method, so that subjects did not spend an excessive amount of time carrying out the evaluations.

4   RESULTS AND DISCUSSION

The 127 subjects evaluated each of the 18 blogs on a 9-point semantic differential scale for each of the 11 adjective pairs. PREFMAP analysis plots the structure of affective response data collected from the subjects onto the similarity space previously obtained from Hsu & Chou (2010)'s research. This allows affective response and similarity structures to be simultaneously displayed in the same space. In the PC-MDS software, the PREFMAP can be divided into four stages (Phases I-IV). Phase IV's Vector Mode shows the adjective pairs as vectors, and the projection of each stimulus point along each vector represents the degree to which that stimulus point possesses the qualities of that adjective pair.

The 3-dimensional space (R2=0.79) obtained using PREFMAP's Vector Mode is now split into two separate maps, showing Dimension 1 against Dimension 2 (Figure 2a) and Dimension 1 against Dimension 3 (Figure 2b) respectively.

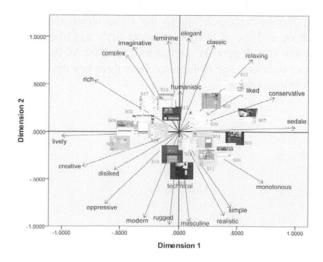

Figure 2a   Dimensions 1 and 2 of the affective perceptual space produced by the PREFMAP analysis

570

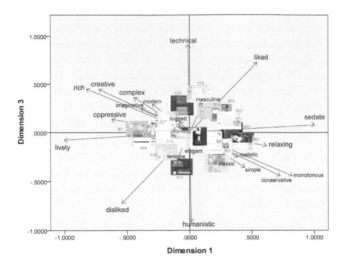

Figure 2a   Dimensions 1 and 3 of the affective perceptual space produced by the PREFMAP
analysis

This research used the SD method and MDS analysis to construct an affective
perceptual space, labeling the 3 axes 'Liveliness & Creativity,' 'Refinement,' and
'Technicality.' This space enabled visualization of users' affective responses to
each blog interface. MDS's PREFMAP algorithm could plot affective response data
into a perceptual space of blog interface similarity (ALSCAL). Examining how the
visual features of the sample blogs are distributed along the dimension axes of the
space allowed more objective identification of the important design factors that may
be used to manipulate those visual features. We could also predict which affective
responses will be aroused by manipulating a given design factor. The results show
that the design factors corresponding to the three dimensions of the affective
perceptual space were 'Type of Images,' 'Layout Style & Color,' and 'Text-Image
Ratio,' respectively (see Figure 3). Designers can refer to this perceptual space
when designing an interface, and manipulate these three key design factors to create
a variety of different moods.

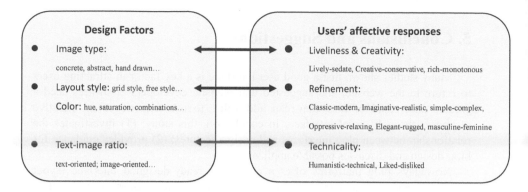

Design Factors	Users' affective responses

- Image type:

  concrete, abstract, hand drawn…

- Layout style: grid style, free style…

  Color: hue, saturation, combinations…

- Text-image ratio:

  text-oriented; image-oriented…

- Liveliness & Creativity:

  Lively-sedate, Creative-conservative, rich-monotonous

- Refinement:

  Classic-modern, Imaginative-realistic, simple-complex,

  Oppressive-relaxing, Elegant-rugged, masculine-feminine

- Technicality:

  Humanistic-technical, Liked-disliked

Figure 3 The corresponding design factors to each affective response (Source: this study)

We have developed a prototype, dynamic visualization template choice interface based on MDS where bloggers can choose templates directly and interactively (see Figure 4). Therefore, bloggers who do not come from a design background would no longer have to rely solely on intuitive choice of blog interface. By just keying in an adjective or directly clicking in the perceptual space, the computer will help select the interfaces that best match the required affective responses, and allow bloggers to further manipulate the key factors to design templates. It also might be generalized or modified into similar template selection systems for other websites or social networking media in the future.

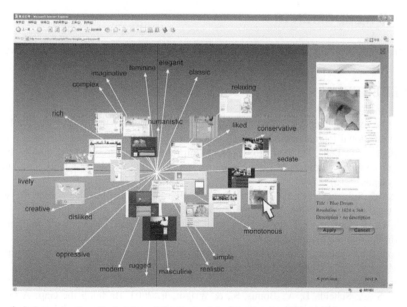

Figure 4 A dynamic visualization template choice interface (source: this study)

5. Conclusions and Suggestions

Many studies shown that a good user interface is a key factor in attracting users to return to the website, and suggested that website developers or owners need a greater understand of how they can add value to user experience or affective responses through visual interface. In conclusion, this study: (1) investigates the relationship between blog interfaces and design features; (2) provides guidance for blog design and discusses possible applications.

Nowadays blog platforms offer a choice of ready-designed interface layout templates, and bloggers could both express their own uniqueness and define themselves in relationship to social groups from the visual interface. However, many blogs use the same blog templates and lack distinctiveness. If developers or bloggers could know the key factors influencing users' affective responses and the corresponding design factors in advance, they may manipulate the design factors effectively. That is to say, time and cost can be reduced and quality can be enhanced, especially in the initial stages of the platform's development, and platform could provide many possible alternatives for bloggers to choose their likings. Also, the result of this research data could be used to set up a database for blogs, or a computerized custom blog template selection system, whereby a blogger could input an adjective describing the style of blog they desire, and a choice of blog templates of this style could be retrieved from a database and offered to the bloggers.

ACKNOWLEDGMENTS

This study was supported in a grant from the National Science Council, Taiwan.

REFERENCES

Du, H. S., & Wagner, C. 2006. Weblog success: Exploring the role of technology. International Journal of Human-Computer Studies, 64: 789-798.

Everard, A., & Galletta, D. 2006. How presentation flaws affect perceived site quality, trust, and intention to purchase from an online store. Journal of ManagementI nformation Systems, 22(3): 55–95.

Fang, X., & Salvendy, G. 2003. Customer-centered rules for design of e-commerce Web sites. J Commun. ACM, 46(12): 332-336.

Fornell, C., & Larcker, D. F. 1981. Evaluating structural equation models with unobserved variables and measurement error. Journal of Marketing Research 18: 39-50.

Fullwood, C., Sheehan, N., & Nicholls, W. 2009. Function revisited: A content analysis of Myspace blogs. Cyberpsychology & Behavior, 12(6).

Gillmor, D. 2004. We the Media, Vol. 2010.

Herring, S. C., Scheidt, L. A., Bonus, S., & Wright, E. 2004. Bridging the Gap: A Genre Analysis of Weblogs. System Sciences,Proceedings of the 37th Hawaii International Conference on January 5th-8th: 101-111.

Hsu, C. C. & Chou, E. T. (2010). Multidimensional Scaling Analysis of Aesthetic Preferences for Blog Interfaces. International Joint Conference - Asia Pacific Computer Human Interaction (APCHI) ERGOFUTURE 2010. Bali, Indonesia.

Hsu, C. L., & Lin, J. C. C. 2008. Acceptance of blog usage: The roles of technology acceptance, social influence and knowledge sharing motivation. Information & Management, 45(1): 65-74.

InsightXplorer. 2010. InsightXplorer Report 2010. Taipei: InsightXplorer Limited.

InsightXplorer. 2011. InsightXplorer Report 2011. Taipei: InsightXplorer Limited.

Lavie, T., & Tractinsky, N. 2004. Assessing dimensions of perceived visual aesthetics of web sites. International Journal of Human-Computer Studies, 60(3): 269-298.

Lin, Y. C. 2006. The Research of Crucial Factors in the Users' Selecting,Using,and Evaluating of Blog Service Platform. National Chung Cheng University, Chiayi.

Pandir, M., & Knight, J. 2006. Homepage aesthetics: The search for preference factors and the challenges of subjectivity. Interacting with Computers, 18(6): 1351-1370.

Rousseau, D. M., Sitkin, S. M., Burt, R. S., & Camerer, C. 1998. Not so different after all: across-discipline view of trust. Academy of Management Review, 23(3): 393–404.

Schenkman, B. N., & Jönsson, F. U. 2000. Aesthetics and preferences of web pages. Behaviour & Information Technology, 19(5): 367-377.

Tractinsky, N., Shoval-Katz, A., & Ikar, D. 2000. What is beautiful is usable. Interacting with Computers, 13(2): 127-145.

Zheng, G. W. 2005. Blog, Research, and Beyond., Vol. 2010: E-Soc Journal.

CHAPTER 62

Tactical Scenarios for User-based Performance Evaluations

Linda R. Elliott (1), Elizabeth S. Redden (1), Elmar T. Schmeisser (2)

Angus Rupert (3)

(1) US Army Research Laboratory
Human Engineering and Research Directorate
Fort Benning, GA, USA
(2) US Army Research Office,
LS/Neurosciences
Research Triangle Park, NC, USA
(3) US Army Aeromedical Research Laboratory
Fort Rucker, AL, USA

ABSTRACT

While laboratory-based experiments are critical to research, development of new systems also requires context-driven evaluation of technology to assure generalizability to realistic operator task demands. The capture of Soldier performance demands is particularly challenging in this regard, as many factors converge to impact performance "in the wild", such as interactions with other soldiers also performing operational tasks, tactical conditions and additional workload that is multifaceted, dynamic, and often uncertain and stressful. Thus, equipment or principles that work well in laboratory conditions need follow-up evaluations within controlled scenario-based situations that have critical characteristics of core work demands. Scenarios having well-defined boundary conditions (e.g., realistic and well articulated task demands, user experience, etc.) thus yield assessments of performance and operational relevance that are arguably valid. This report describes issues and outcomes related to development of scenarios and measures for the assessment of devices for dismount soldier teams. Procedures and results are discussed within the context of cognitive task analytic techniques

Keywords: Cognitive task analysis; Multisensory systems; Scenario validation

1   INTRODUCTION

U.S. military operations involve ever-increasing interactions with advanced technology—from vehicle consoles and wearable displays to robot control interfaces. Soldiers are now expected to master an array of sensors, controls, and displays. These capabilities can enable Soldiers to see in the dark, maintain map-based situation awareness, control semi-autonomous air and ground vehicles, and network up and down communication channels. However, it should be noted that the Soldier/user can also easily be overcome with technology complexity, high workload, information overload, and distraction from critical mission events. It is critical that these systems be evaluated in situ for effectiveness and ease of use. Lab-based performance criteria are not sufficient. Engineering developers focused on a particular capability may indeed succeed to performance criteria (e.g., ability to detect and identify a human threat at 200 meters using night vision) and yet still produce a device that is unsuited for operational use.

User-based evaluations should be conducted from the time of concept development of a new device or system. Concepts for operational use can be generated from formal organization-based analyses of military requirements or they may arise from less structured processes, such as responses to calls for proposals or concepts previously developed for other users. Whether formal or informal, steps should be taken to more specifically identify the user group of interest and the operational context.

2   SCENARIO-BASED EVALUATIONS

In some circumstances, requirements for advanced technology are identified and specified in engineering terms. For example, the performance requirements for a night vision device may be very explicit, detailing criteria with regard to range, resolution, and operational context (e.g., ability to "see through" fog or smoke for 50 meters). While these specifications can be demonstrated in the lab, these engineering-based evaluations are useful only up to a point. At some time, users must take the system into the field in order to identify unexpected issues of use or misuse. Soldiers should be provided with performance task demands that reflect to some degree, operational scenarios and task demands. These task demands are identified through discussions with user subject matter experts who translate the requirements to an operational context. Various cognitive task analysis techniques such as structured interviews, focal group discussions, and decision analysis, or combinations thereof, can be used to answer, in detail, basic questions such as:

-    Who needs this technology / who is the user

-    Why do they need this technology

-    When / Where will this technology be used

Different methods have arisen to better answer these questions, and each technique has its advantages, with suitability for different purposes (Adams, Rogers, & Fisk, 2012). The more fully these questions can be answered, the better. These techniques should result in detailed descriptions of the users and of the context of use. For example, for night vision systems, discussions with operators would identify task categories that would typically include (a) stationary long range target detection (e.g., with or without fog/smoke), (b) target detection while navigating through wooded terrain, and (c) target detection while performing urban patrol (Redden, Bonnett, & Carstens, 2007). These operations provide quantitative measures of performance with regard to requirements and specifications. They also provide a context of use that enables the Soldier/user to identify and evaluate additional issues. Systems may be found to be difficult to train, difficult to use, or difficult to wear while walking cross country over rough terrain. Such systems may have small parts that are easily lost or broken, may be unbalanced if helmet-mounted systems, or may have complex interfaces that are unsuitable for use at night or use with gloves. Such systems may have excelled during lab-based performance specifications, only to be fully or partially rejected when evaluated in context.

3   SCENARIO-BASED APPROACHES AS AUGMENTING COGNITIVE TASK INTERVIEWS

Many techniques have arisen to enable technology developers to better identify issues related to effective human-technology interactions. These cognitive task analysis (CTA) techniques use methods such as structured interviews to answer certain fundamental questions with regard to the user, the technology, and the context of use (Adams et al., 2012; Crandall, Klein, & Hoffman, 2006). In fact, Crandall and her colleagues listed over thirty CTA techniques that used interview as the basic approach, and many more that used other approaches. As another example, the Army Research Laboratory Human Research and Engineering Directorate uses cognitive task interviews as a first step in building task analytic models of human performance in context (Mitchell, 2007). While specific methods differ, all attempt to answer core questions, starting from the generic and probing to the desired level of detail.

While cognitive task analysis techniques can identify the core tasks, with very detailed descriptions of task elements, information gained from interviews can only go so far. Once core tasks are identified, scenarios must be developed that reflect these core tasks. Devices can then be assessed by potential users for the purpose for which they would use it, in a representative context. For office equipment, that is not such an issue, and assessing the usability of a new website can be easily arranged. However, for performance in more challenging context (e.g., terrain, physical tasks, time pressure, etc.), the development and implementation of a scenario can be effortful. It provides a structured yet realistic means by which the potential user can use the device in a way that yields performance measures as well as user-based feedback.

User scenarios are particularly important with emerging technology that is not driven by specifications. For example, advanced wearable display concepts result in prototypes that are not yet fully developed. Requirements are not specified; indeed, the user groups are not yet identified. In such situations, more generic questions may be explored, such as:

- How might this technology change the cognitive demands of work (mission ops)

- Might there be misapplications or unanticipated consequences?

What functions would you want in such a device?

4 SCENARIO-BASED ENVISIONING

Scenario-based techniques can be used very early in the development of a device; even when the device is still a concept, with no prototypes in hand. To answer these questions, use case scenarios can be constructed, where the user is "walked" through a scenario and encouraged to think aloud and respond to guided interview questions. This was the approach taken to evaluate flexible display concepts that were not yet engineered. A scenario was identified and developed, to fully detail a mission that is common to most Soldiers and expected to be a realistic context where flexible displays would be useful, that of cordon and search (Hoffman & Elliott, 2010). The scenario becomes a context for structured interview of soldiers, to guide them through scenario events and articulate when and where particular capabilities would be most useful. In this example, the scenario centered around a "story", generated by preliminary interviews with a user group, to identify a scenario context that is typical, detailed, and expected to benefit from the device capability. In this example the story begins with the description that a division intelligence cell received information about the location of the terrorist. The scenario had two phases, a planning phase that described how company-level leaders produced a mission plan, and a mission phase that described how the platoons set up the cordon and conducted the search. Once the scenario was detailed, cognitive task interviews were scheduled with a separate group of Soldiers having real-life experience with this kind of mission. Each cognitive task interview was conducted one-on-one, for 90-120 minutes. Eight experienced soldiers, having at least one tour of combat duty, and qualified to command a Bradley vehicle, were interviewed. The interviewer explained project goals and began by collecting very detailed background information from each Soldier. This phase is not simply to collect information, it also serves as a means to establish trust and rapport. The interviewer then presented a general overview of the scenario and the Soldier is asked to "think aloud" while the interviewer presented each scenario event. As each scenario event is described, the Soldier is encouraged to comment on the event itself (e.g., realism, whether they would do it the same way, etc.) and also what they would be thinking or doing at the time, what capabilities would assist, or what information would be most useful. This scenario-based envisioning technique generated much information and insight with regard to when and why a flexible wearable display might be used, the functions that would be most useful, and possible problems or misapplications of the technology (Hoffman & Elliott, 2010).

5   IMPORTANCE OF SYSTEMATIC SCENARIO GENERATION

Figure 1 provides a simple conceptualization representing three core dimensions of information: (a) User, (b) Technology, (c) Operational Demands/Context.

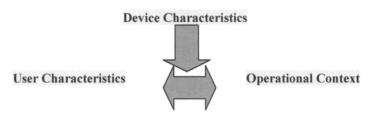

Figure 1 Interacting factors relevant to scenario generation

For the purpose of scenario generation, it is critical to consider the interaction of all three dimensions. The goal is the identification of a scenario that is not simply realistic, but one where the capability would most likely benefit performance. Given the three dimensions, knowledge of any two will help define the third.

Ideally, one would begin with the operational context and users in order to determine device characteristics that would be most helpful. However ideal, the more usual circumstance is that one begins with a device and must identify the users and then specify the context. For example, if the capability were camera-based, one would identify a mission context where the camera capability would be needed (e.g., having requirements for target/threat detection) through discussion with Soldiers representing the user group. It may be necessary to first ascertain the relevant Soldier user group (e.g., dismounted infantry, vehicle commanders, company commanders, etc.,) through preliminary interviews. Subsequent interviews would then focus on scenario generation, to identify the type and number of events that sample the performance of interest (e.g., types of targets, target distance, target placement, etc.) and enable quantitative measures (e.g., detection, identification, as measured by accuracy, speed).

A systematic approach, using iterative interviews with more than one or two representative users, that identify the users AND the context, is needed, yet often skipped. Given a user population and a device, it would seem relatively straightforward to generate a scenario, and often a scenario is generated on the opinion of the researcher or a single subject matter expert. However, this can often lead to error. Scenarios should be crafted for the particular device and user group – "one scenario does not fit all". There are mistakes of omission, where not all capabilities of the device are used in the scenario. It is not sufficient for the scenario to be "realistic"—it must also contain the task demands suited for each capability.

As an example, Soldiers may be asked to use and evaluate two different robot platforms with regard to target detection and identification. One platform may have a more flexible camera control system that allows for easier horizontal and vertical scanning. In such a case, care must be taken to place targets that will more likely distinguish the two robot/controllers (e.g., that require the operator to look up, down, and around, rather than having all targets on the ground—even if ground targets are more typical). That is, evaluators must identify a realistic scenario where in fact the enhanced scanning capability would be relevant.

Similarly, there may also be mistakes of commission, where developers want to use capabilities that are not relevant to the chosen scenario—perhaps for convenience (e.g. while you are at it, get information on this capability too; use this scenario, it is already developed, etc.). For example, a video-based targeting system may have a recording capability that logs all video and allows replay at any time. It may be tempting to allow free use of such a function during a reconnaissance mission scenario; however, such free use may in fact distract and diminish the mission performance, particularly if the mission scenario has threat targets that occur very close together in time. While tempting to allow free play of a device within a generic but realistic scenario, such free play should be a first step, to lead ultimately to more controlled scenario-based events linked to defensible expectations with regard to device use and enhanced performance. Given Soldier performance, events must be in line with established tactics and procedures.

6   DEVELOPMENT OF A SCENARIO: EXAMPLE

Here, we describe the efforts taken to identify relevant task demands for a multisensory navigation and communication display.

Description of Device. The Active Tactile Array Cueing Navigation and Communication (ATAC-NavCom) system integrates android-based smartphone and tablet technology with GPS-driven tactile cues presented through an elasticized torso belt. The smartphone and tablet displays provide map-based information with regard to location of soldiers, assets, and threat. Location information can be transmitted from one person (e.g., using the tablet) to another (e.g., having the smartphone and tactile belt). Iconic information is transmitted to the smartphone map, while immediate direction information is provided by activation of one of 8 tactors located around the belt. Thus, the user would not have to see the visual display in order to immediate move in the recommended direction (toward the activated tactor). This eliminates the effort required to translate map-based information (you are here) to direction information (turn left now). In addition, patterns of tactors can be activated to communicate information. For example, all tactors can be activated in off-on sequence, to signify a command, such as "stop". This capability allows communication of direction and alerts without having to look at the visual display (hands-free and eyes-free) and without audio communications (covert communications).

Description of User. Several different types of Soldier groups are likely to find this capability for handsfree, eyesfree, and silent communications useful. For example, introduction of this concept to a group of Snipers was well received, as they appreciated the covert nature of critical communications. Other groups appreciated the capability to navigate easily at night without having to look at a display. Examples of use cases included (a) quickly converging on a rally point after being dropped to terrain (Airborne), (b) dismounted from a vehicle (Infantry), or (c) scouting terrain on a reconnaissance mission. Thus, identification of the user group in turn will help to narrow the appropriate scenario context.

Scenario development. For initial development and evaluation, we chose a user group that is perhaps most common—that of dismounted Soldiers (e.g., dismounted from a vehicle). The ATAC-NavCom was then presented to a small group of Soldiers with extensive combat experience in dismount missions (e.g., reconnaissance, patrol, cordon and search, movement to enemy contact, etc.). Each soldier was trained to use the equipment. They were encouraged to comment at all times, with regard to device usefulness, ease of use, when they would use it, and when they would not. This interaction is critical at this early stage of development, so that engineers and developers understand critical concerns for operational use (e.g., how much does it weigh, how rugged is it, how long does the battery last, etc.). Soldiers were then asked to describe, assuming the device was in fact combat-ready, the kinds of situations that they would expect the device to be useful, and why. In this case, several situations were described. One was chosen, route reconnaissance, because it is a common type of mission, and not urban. At this time the GPS feature needs further enhancement for effective urban use.

Given this scenario context, further interviews with Soldiers identified some core communications that would be most useful for the tactile communication – that is, communications that are critical, in situations where other communication systems may not be effective (e.g., too noisy for radios, not enough visibility for hand signals, etc.). This resulted in a set of core commands, in addition to 8 direction cues, to include "Stop", "Shift fire left"; "Shift fire right", "Take cover", and "Look at me". Further interviews are planned, in order to flesh out scenario details and gain more feedback, as device characteristics evolve. These interviews will serve to justify the experiment-based evaluation. The experiment will include the task demands generated from scenario development. For example, the Soldiers will wear and use the device while standing and also while walking, as on patrol. They will receive signals indicating changes in direction (rerouting). They will also receive critical communications via tactile patterns, as suggested by Soldiers. For each of these tasks, performance will be assessed in quantitative terms when possible (i.e., speed, accuracy). Finally, experiment Soldiers will be asked to provide feedback through rating scales and discussion. This feedback serves to inform device developers, and also to refine scenario development for subsequent experiment-based evaluations.

7   DISCUSSION

Scenario-based evaluation is a critical aspect to user-based development of equipment. While seeming straightforward, a systematic approach is necessary and yet often skipped. Errors, deficiencies, and assumptions are common. To use a common saying -- "the map is not the terrain". In the same way, the scenario is not "reality". One cannot justify just any scenario with the simple explanation that it is "realistic". Instead, the scenario must be carefully crafted, to utilize the device characteristics in situations identified by the user, with events that can be quantitatively measured

8   REFERENCES

Adams, A., Rogers, W., Fisk, A. (2012).   Choosing the right task analysis tool. Ergonomics in Design: The Quarterly Journal of Human Factors Applications. January 2012, 20, 1, 4-10.

Crandall, G., Klein, G., Hoffman, R. (2006). Working Minds. Cambridge, MA: MIT Press.

Hoffman, R., & Elliott, L. (2010).   Wearable flexible displays: a scenario-based usability evaluation suggests Army applications for an emerging technology.   Ergonomics in Design: The Quarterly Journal of Human Factors Applications. October 2010, 18: 4-8.

Mitchell, D., Henry, S., Animashaun, A. & Abounader, B. (2009). A procedure for collecting mental workload data during an experiment that is comparable to IMPRINT workload data. ARL-TR-5020. Aberdeen Proving Ground, MD: Army Research Laboratory.

Mitchell, D. K. (2007). Please Don't Abuse the Models, Proceedings of the 51st Annual Meeting of the Human Factors Engineering Society, Baltimore , MD October 1-5, 2007.

Redden, E. S., Bonnett, & C. C., Carstens, C. B. (2007).   Enhanced Night Vision Goggles, Digital: Limited User Evaluation II (Technical Report ARL-TR-4233).   Aberdeen Proving Ground, MD: Army Research Laboratory.

Section VIII

Diverse Approaches: Biosignals, Textiles, and Clothing

CHAPTER 63

Psychological Factor in Color Characteristics of Casual Wear

*Chiyomi Mizutani[1], Kanae Kurahayashi[1], Momoe Ukaji[1], Tetsuya Sato[2],
Saori Kitaguchi[2], Gilsoo Cho[3], Soonjee Park[4], Kanji Kajiwara[5]*

[1]Faculty of Domestic Science, Otsuma Women's University, Japan
[2]Center for Fiber and Textile Science, Kyoto Institute of Technology, Japan
[3]College of Human Ecology, Yonsei University, Korea
[4]School of Textiles, Yeungnam University Fiber, Korea
[5]Innovation Incubator, Shinshu University, Japan
Email address: mizutani@otsuma.ac.jp

ABSTRACT

A human character plays a role in choosing a color of everyday dressing. The human character test (YG test) confirmed a majority of the youth as the well-rounded type or the aggressive type with a moderate psychological stress. The eccentric type suffers from psychological stress in large extent, but the administrative type is relatively stable for psychological stress. The eccentric type has a specific color preference for purple but obtains fashion information casually. Since RP (red-purple) and P (purple) indicate the nobility, self-expression and offensiveness, the subjects of the administrative type and the eccentric type lead the fashion scene but within the published information. Considering the Japan cool consists of more individualistic styles such as Lolita and Goth-Loli, the Japan cool could have been emerged from the private clothes of the aggressive type and the placid type, and not from the fashion leaders. The black-list type tends to be an "otaku" type and we notice this type reacts with fashion in a characteristic way by expressing more individualistic taste. The majority of the Korean subjects are classified as the aggressive type. The Korean youth count their fashion information on the US/European fashion magazine and internet, so that the street fashion in Korea is rather ubiquitous and less distinct regardless of personal characters.

Keywords: character test, color trend, street fashion

1   INTRODUCTION

People wish to differentiate themselves from others by clothing, but at the same time are obliged to indicate where they belong as a member of a particular group of the society and subculture. In consequence people are always on the tense balance between imitation and discrimination for daily clothing. For Japanese, the European style is not a traditional way of clothing, but the Japanese youth now adapt the European style as granted. As the cultural background is different from that in Europe, the Japanese youth never consider the social significance of the European style clothes, and coordinate their casual clothes without any traditional restriction (Jiratanatiteenun, Jung, and Kitaguchi, et al., 2010, Jiratanatiteenun, Kitaguchi, and Sato, et al., 2010). In these 150 years, the Japanese have been unconsciously balancing their mentality between the European rational logics and the Japanese traditional ambiguity. In other words, we are always under psychological pressure by dressing ourselves, and in consequence are obliged to create the European-style clothes with the Japanese sense of value, leading to the birth of Japan cool. The Japan cool is in a sense a resistance against the rational civilization based on the European logics. This fact explains why the phenomena of so-called Japanism appear periodically at the time of the industry revolution.

The street fashion plays a significant role for social networking of young females in Japan. The Japanese street fashion has taken its own way from around 2000, when the fast fashion industry has established its global single-cycled supply chain to distribute high-sense fashion items with a reasonable price. New ubiquitous fashion products appear every two weeks everywhere on the globe. Against the fast fashion, the comic-maniac girls started making their own costumes according to their imagination. A geek-chic fashion (such as Lolita and Goth-Loli) emerged from the comic maniacs. Those individualistic fashions gradually influenced Japanese street fashion developing in a characteristic way as observed as particular fashion groups in particular districts in Tokyo.

The clothing situation is similar in Korea, where the European style was introduced at the same time as in Japan. However, the Koreans seem to be rather indifferent about the street fashion of their own. Here the street fashion has no function for social networking in Korea.

2   EXPERIMENTAL PREPARATION

2.1   Fashion Conscious

The survey was conducted by questioning the undergraduate students aged from 18 to 22 at a traditional women's university in central Tokyo. A similar survey was conducted at a women's university in Seoul, and private universities in Seoul and Daegu with a slightly modified questionnaire. The first survey was conducted in 2006, and followed biannually till 2010 and in 2011. Approximately 350 subjects in each year consisted of female undergraduates aged from 18 to 22, and were asked to answer two sets of questionnaires with respect to their fashion interest and lifestyle. The first set of questionnaire was intended to clarify the present fashion styles among young girls and their fashion information sources. The second questionnaire consists of the questions concerning the subjects' lifestyle and fashion interest. The colors of the subjects' clothes were measured by the spectra colorimeter as described below. Additionally the subjects were asked to fill the questionnaire to evaluate their basic character.

2.2   YG (Yatabe-Gilford) Character Test

The YG (Yatabe-Gilford) character test (Yagi, 1987) classifies the human characters into 15 subtypes in 5 primary types; Type A (the average type), Type B (the aggressive type), Type C (the even-minded type), Type D (the administrative type) and Type E (the eccentric type). The character profile of the subjects was analyzed by scaling the intensity of 12 elementary factors (D (depression), C (cyclic tendency), I (inferiority feelings), N (nervousness), O (lack of objectivity), Co (lack of cooperativeness), Ag (lack of agreeableness), G (general activity), R (rhathymia), T (thinking extraversion), A (ascendance), and S (social extraversion)) evaluated by a set of questionnaire.

2.3   Color Measurements

The color characteristics of the clothes of the subjects were evaluated by the Konica-Minolta CM2600D spectrum colorimeter. The results were analyzed in terms of hue, value and chroma.

3   RESULTS AND DISCUSSION

3.1   Color Trend

Traditionally the Korean preferred brighter color than the Japanese. In fact, the Korean youth has worn the clothes of a wider variety of hue and higher chroma in comparison with the Japanese youth in 2005 as shown by the analysis of color measurements (Fig. 1). The Korean youth wear more colorful clothes than the Japanese in winter, 2005. Green and purple seem to be a favorite color in Korea. However, the color difference is not much remarked in winter, 2008.

The color trend in 2010 and 2011 is similar as in 2008. The Korean youth prefer more achromatic color and chic (or matured) styles than the Japanese youth in 2011. The Korean youth are vulnerable for European fashion as they rely mostly on "Vogue" for the fashion information, while there are so many fashion magazines available in Japan for respective street fashions.

588

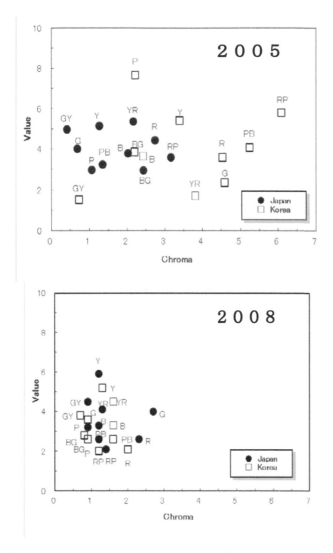

Figure 1 Color trend in winter clothes of Japanese and Korean youth in 2005 and 2008.

3.2   YG Test and Color Characteristics

The YG (Yatabe-Gilford) character test evaluates a human character in terms of five types as summarized in Table 1. Although many Japanese are classified as Type A (the well-rounded type) and Type C (the placid type), 22 %, 45 %, 4 %, 15 % and 14 % of the Japanese subjects (the majors in clothing and apparels) in the present survey are Type A, Type B, Type C, Type D and Type E, respectively. The Korean subjects (also the majors in clothing and apparels) are mostly Type B.

Table 1 5 Basic human characters

Type	Character	Emotionally	Social attitude	Remarks
A	well-rounded	neutral	neutral	average
B	aggressive	unstable	social	May cause problems with others
C	placid	stable	unsocial	Stable, but passive and dependent
D	administrative	stable	social	Aggressive, and leading others
E	eccentric	unstable to neutral	unsocial to neutral	Passive and cocooned, but self-content with own cultural activity

Type B (the aggressive type) and Type E (the eccentric type) are vulnerable to psychological stress, which suppresses the subjects to become more distinct and isolated from others. They may choose a lower chroma and a lower value under the psychological stress, but normally prefer a more distinct color (such as "purple", see Table 2) to give an active or unique impression. The color preference of each character type is summarized in Table 2, together with the favorite fashion magazines. Although Type D and Type E prefer distinct colors such as red-purple and purple, their favorite magazines are casual or girly casual (amalgamated casual + gal with a tint of Lolita), indicating those types are always fashionable within the published information.

Table 2 Favorite fashion magazines for each human character

Character Type	Preferred Color	Favorite Fashion Magazine
A	Y, YR	"non-no", JILLE", "ViVi", "SEDA", "Soup", "mina", "Ray", "Zipper"
B	YR	"non-no", "ViVi", "Ray", "Zipper", "JELLY", "mina", "JJ", "S-Kawaii", "KERA", "Soen", "Vogue"
C	Y	"non-no", "SEDA", "Sweet", "Gothic & Lolita", "Elle"
D	RP	"ViVi", "Zipper", "Sweet", "Ray", "Nadeshiko"
E	P	"non-no", "Soup", "CanCam", "mina", "PS", "Ray"

Type B (the aggressive type) and Type C (the placid type) read the fashion magazines for individualistic fashion (KERA, Gothic & Lolita) or more elaborated fashion (Soen, Vogue, Elle). Type B (the aggressive type) is outward-minded and confident of her fashion sense. Considering the Japan cool consists of more individualistic styles including Lolita and Goth-Loli, the Japan cool could have been emerged from the private clothes of a minority Type C (the placid type) but not from the fashion leaders. Type C tends to be an "otaku" type and reacts with fashion in a characteristic way by expressing more individualistic taste.

The character of Korean subjects was classified mostly into Type B according to the YG character test. Although whether the YG character test could be adapted to the Koreans is not clear, the Korean subjects are thus more aggressive and outward-minded than the Japanese subjects. In fact a favorite fashion magazine is "Vogue" for the Korean subjects, and they look more the US and Europe for their fashion information. In consequence the street fashion in Korea is rather ubiquitous and less distinct regardless of personal characters.

3.3   Fashion Conscious and Lifestyle

The following figures are the summaries of the questionnaire concerning the fashion conscious and lifestyle. The aim of this section is to discriminate the fashion life in Japan and Korea from the analysis of the questionnaire.

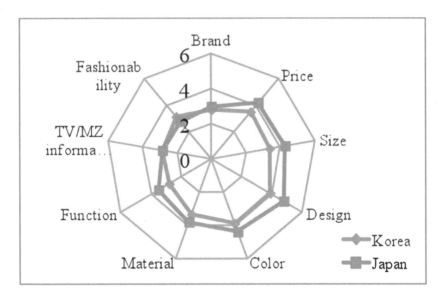

Figure 2 How much emphasis do you place on the purchase of fashion products?

The shopping behavior of young girls is similar in Japan and Korea. Here they consider all the factors for shopping. Fashionability seems less important, indicating that they consider the long-term use of clothes. However, there are remarkable differences in the confidence about themselves as shown in Figs. 3 and 4.

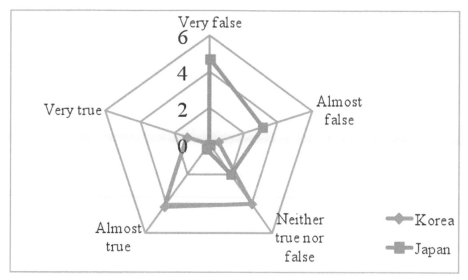

Figure 3 I am confident about my appearance.

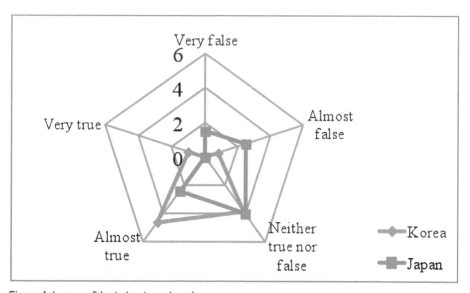

Figure 4 I am confident about my dressing.

The Korean girls are quite confident about their appearance and dressing, while the Japanese girls have no self-confidence. Thus the Japanese girls rely on others' opinion on their daily life including dressing. Their information source is compared in Fig. 5. The Japanese girls ask the opinion of the shop attendants, but do not rely on internet. The Korean girls do not listen to the shop attendants' opinion and seek information through internet. The fashion magazine is a main information source in both countries, but "Vogue" seems to be an almost sole source of fashion in Korea.

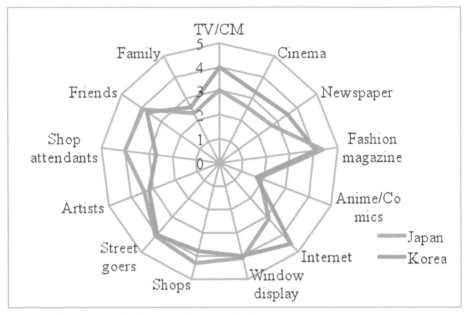

Figure 5 Where do you obtain information about fashion?

Comparatively the Korean girls spend more money for their clothes, and frequently shop more luxury clothes at department stores and brand boutiques. They also use internet for shopping. The Japanese girls seldom use internet shopping, and prefer face-to-face selling. The shopping behavior reflects the psychological state as the Japanese girls feel more comfortable when they wear a similar type of the clothes as their friends, while the Korean girls are more independent. Although they are fond of wearing accessories, the Japanese girls wear more accessories and seem to distinguish themselves by small fashion items from others (Fig. 6).

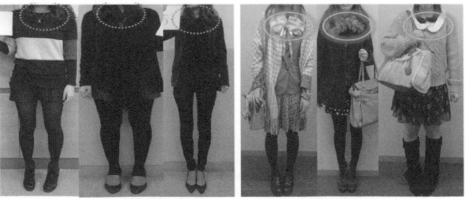

Figure 6 Typical styles of the Korean girls (left) and the Japanese girls (right) in 2011. The Korean girls wear simple and chic clothes, but the Japanese girls put more fashion items. See the collars and small patterns of the skirts.

4 CONCLUSIVE REMARKS

Japan and Korea are experiencing a similar process of post-colonialization (or restructuring of traditional subculture) with respect to dressing behavior. The Japanese are more placid and feel assured when they belong to a group. Here the clothes function as a kind of a uniform to identify the group where they belong. The social isolation in Japan is the worst in the world, and the sense of estrangement promotes the grouping by external appearance. The Korean girls are more self-confident and the group is formed not by external appearance but on the basis of an existing community. A rural-type community is still functioning in Korea. In Japan the community is neither a rural-type nor an urban-type, so that the Japanese girls long to belong to a group even superficially.

ACKNOWLEDGMENTS

The authors would like to express their gratitude to Aliyaapon Jiratanatiteenun, Jung Hyojin, Nami Suzuki, Kanae Sudo for their help in collecting and analyzing survey data.

REFERENCES

Jiratanatiteenun, A., H. Jung, and S. Kitaguchi, et al. 2010. Characteristic Feature of Japanese Street Fashion: Lolita, Gothic-Lolita and Costume Play. *Proceedings of 22nd IFATCC INTERNATIONAL CONGRESS*, Stresa, Italy.
Jiratanatiteenun, A., S. Kitaguchi, and T. Sato, et al. 2010. Dynamics of Street Fashion in Japan Represented by Cosplay and Lolita. *Proceedings of The Textile Institute Centenary Conference*, Manchester, UK.
Yagi, T, 1987. *YG Character Test*, Chiba, Japan: Nihon Shinrigijutu Kenkyusho.

CHAPTER 64

Sound Characteristics and Auditory Sensation of Combat Uniform Fabrics

Eunjung Jin, Jeehyun Lee, Kyulin Lee, and Gilsoo Cho

Dep. of Clothing and Textiles, Yonsei University, Seoul, Korea
{ e.j.jin, gscho }@yonsei.ac.kr

ABSTRACT

The objectives of this study were to investigate the sound characteristics of combat uniform fabrics and to provide the condition for arousing positive sensibility by measuring psychological sensation for the frictional sound of the fabrics. The frictional sounds of six combat uniform fabrics were generated by the Simulator for Frictional Sound of Fabrics(SFSF). The frictional speeds of fabrics were determined with the speeds of walking, jogging, and running. The frictional sounds were recorded by the Data Recorder of the Pulse System(Type 7700, B&K), then the SPL and Zwicker's parameters were calculated by the Sound Quality Program(ver. 3.2, B&K). Variables for psychological sensibility are loudness, highness, sharpness, roughness, hardness, obscurity, changeableness, and pleasantness and these were measured with the 7-point Semantic Differential Scales(SDS). The values of SPL, loudness(Z), roughness(Z), and fluctuation strength(Z) were significantly increased by frictional speed. As increasing speed level, the frictional sounds were evaluated as louder, higher, harder, more changeable and unpleasant. As the results of regression analysis, combat uniform fabrics would have subjective pleasantness by reducing SPL, loudness(Z), and roughness(Z).

Key words: Sound characteristics, Psychological sensibility, Frictional sound of fabric, SPL, Zwicker's parameter, SDS

1   INTRODUCTION

The noise of combat uniform is a key factor that affects auditory camouflage performance. That is, the importance of reducing noise of military clothing cannot be understated since soldiers can be discovered by the sense of hearing.

Frictional sound of fabric affects clothing sensorial comfort. Products considering auditory sensibility have recently been expected to broaden the textile and clothing market because many consumers are gradually interested in the sensorial quality of clothing (Yi, E. et al., 2002). The coated fabrics with vapor permeable water repellent function usually generate the noise over 70dB, which causes unpleasantness to wearers. To solve this problem, several studies have been made on the relationship between fabric sound and auditory sensory response. However, the fabric sounds of combat uniform have never been examined so far except in the case of Cho, J. (2006), which used old military fabrics.

To develop combat uniform fabrics with auditory camouflage performance, it is necessary to find decisive sound variables affecting audible sensibility. The objectives of this study are to analyze the sound characteristics at various frictional speeds by simulating wearer's movement, to investigate relationship between sound characteristics and psychological sensibility, and to predict the condition of positive sensibility, that is, pleasantness. This study will be helpful to design and produce more appealing fabrics with auditory comfort.

2   EXPERIMENTAL

2.1   SPECIMENS

The specimens are six combat uniform fabrics, which divided into two groups – combat uniforms with woodland camouflage patterns and new combat uniforms with digital pattern. The characteristics of the specimens are presented in Table 1.

Table 1 Characteristics of specimens

Specimen	Fiber composition	Yarn type	Fabric construction	Fabric count (w x f/inch)	Thickness (mm)	Weight (mg/cm²)	Fabric name
W-S	65/35 P*/R*	Staple	Ripstop	92 X 92	0.49	20.11	Ripstop
W-W	65/35 P/C*	Staple	Twill	83 X 83	0.66	25.18	Gabardine
W-WR	80/20 P/R	Staple	Plain	125 X 125	0.49	17.88	Taffeta
D-F	68/32 P/C	Staple	Ripstop	100 X 100	0.58	22.55	Ripstop
D-UDT	75/25 P/R	Staple	Twill	75 X 75	0.76	25.57	Gabardine
D-FR	50/50 AD*/FR*	Staple	Ripstop	67 X 67	0.55	22.84	Ripstop

* P – polyester, C- Cotton, R- Rayon , AD - Aramid, FR- Flame Retardent

2.2   RECORDING FABRIC SOUND

Fabric sound was generated on a Simulator for Frictional Sound of Fabrics (SFSF) (Patent, No. 10-2008-0105524, 2008) (Figure 1) (Yang, Y. et all., 2009) in a soundproof room (anechoic room, background noise 20dB). The fabric sound was recorded through high performance microphone (Type 4190, B&K) which was connected to the Data Recorder of the Pulse System (Type 7700, B&K). The sound of each fabric was recorded three times. Fictional speeds and time by motion analysis are presented in Table 2, which are the usual frictional speeds and time between wearer's arm and trunk.

Table 2 Frictional speeds and time by motion analysis

	Direction	Walking	Jogging	Running
Frictional speeds	Front → back	0.64	0.99	1.71
(m/s)	Back → front	0.62	1.21	2.25
Frictional time	Front → back	0.19	0.12	0.07
(sec)	Back → front	0.19	0.10	0.05
Non-frictional time	Front → back	0.19	0.12	0.13
(Sec)	Back → front	0.22	0.13	0.14

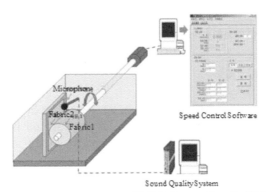

Figure 1 Simulator for Frictional Sound of Fabrics (SFSF) (Patent, No. 10-2008-0105524, 2008)

2.3   ANALYZING SOUND CHARACTERISTICS

Recorded sounds were converted into sound spectra and the sound characteristics such as sound pressure level (SPL) and Zwicker's psychoacoustic parameters were calculated using the Sound Quality Program(ver. 3.2, B&K). The recorded fabric sounds were analyzed by FFT over a frequency range from 0 to 17,000 Hz. The sound characteristic values such as (SPL), loudness(Z), sharpness(Z), roughness(Z), and fluctuation strength(Z) were calculated by equations mentioned in the previous study (Cho, J., 2006).

2.4   EVALUATION OF PSYCHOLOGICAL SENSIBILITY

2.4.1 PARTICIPANTS

There were 20 right-handed female participants aged from 20 to 30 years. They were screened for normal hearing according to Houghson-Westlake, "5dB up and 10 dB down" procedure (Morill, J., 1984).

2.4.2 EXPERIMENTAL PROCEDURE

Psychological sensibility for fabric frictional sound was elvaluated with the AQEEST(Affective Quality Evaluation and Estimation System for Textile)(Park, J. et all, 2010). Prerecorded sound of specimens was presented to participants wearing a powerful headphone (Philips, SBC HP 110) through a computer with the AQEEST. The frictional sound of each fabric was presented randomly. The participants were asked to answer the questionnaires on the system of AQEEST. The questionnaires were consisted of eight different bipolar descriptors with Sematic Differential Scales (SDS) as presented in Table 3. The experimental protocol was shown in Figure 2. Participant who entered the lab took a rest for 2 minutes and then answered the questionnaire after listening to each fabric sound for 30 seconds. If the participant wanted to listen to the frictionl sound repeatedly, it was replayed as much as she did.

Table 3 Adjective descriptors for audible sensibility evaluation in AQEEST

Descriptors		
-3	•••	+3
Quiet	•••	Loud
Low	•••	High
Dull	•••	Sharp
Smooth	•••	Rough
Soft	•••	Hard
Clear	•••	Obscure
Monotonous	•••	Changing
Pleasant	•••	Unpleasant

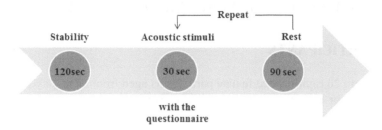

Figure 2 Experimental Protocol

2.5   STATISTICAL ANALYSIS

Data were analyzed using the SPSS package (Ver. 18.0). To investigate the difference of sound characteristics among specimens, one-way ANOVA was conducted and then post-hoc test was carried out with Duncan's multiple range test. To predict the psychological sensibility with the sound characteristics, regression analysis was carried out.

3   RESULTS AND DISCUSSION

3.1   FFT SPECTRA OF THE FABRIC SOUNDS

The sounds of six specimens were converted into frequency-to-amplitude waveform by FFT (Fast Fourier Transform) method. The shapes of sound spectra changed by fabric types and frictional speeds are shown in Figure 3. The fabric sound's amplitudes were ranged between 7.81~74.63 dB at 0~17,000Hz. The level of amplitude was gradually decreased from low to high frequency band. It was similar with the previous studies (Kim, C., Yang, Y., and Cho, G., 2008) in that the sound spectra's peak of amplitude was appeared frequently in low-frequency band and amplitude was gradually decreased from low to high frequency band. The amplitude of W-WR had the highest value, while that of D-FR showed the lowest value over most ranges of frequencies at walking, jogging, and running speed. Therefore, it can be expected that the sound of W-WR at running speed has the highest sound level among specimens. The wave of spectrum was placed the lowest at walking speed and it was increased at jogging and running speed. It can be predicted that the SPL showed a tendency to increase according to the frictional speed.

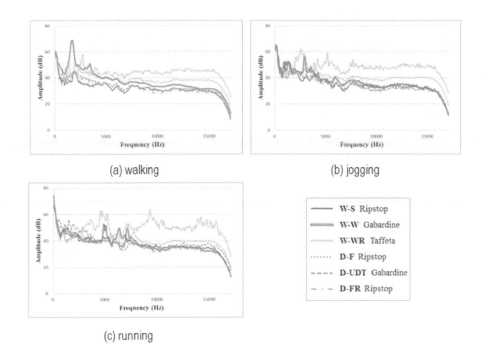

(a) walking   (b) jogging

——— W-S Ripstop
▬▬▬ W-W Gabardine
——— W-WR Taffeta
········ D-F Ripstop
---- D-UDT Gabardine
— · — D-FR Ripstop

(c) running

Figure 3 Sound spectra of specimens under three different frictional speeds

3.2   SOUND CHARACTERISTICS OF COMBAT UNIFORM FABRICS

To investigate the difference of sound characteristics to movement speeds, one-way ANOVA analysis was conducted and then post-hoc test was carried out with a Duncan's multiple range test. The SPL, loudness (Z), roughness (Z), and fluctuation strength (Z) were significantly affected by frictional speeds (p<.05). Therefore, the sound level, the roughness, and the vibration were signicantly increased by increasing speeds. The sounds of jogging and running speed were louder, rougher, changeable than walking speed (p<.05) (Figure 4). The SPL values had significant differences among frictional speeds. They were ranged from 64.73 to 77.98dB (Figure 5). Most of the combat uniforms showed SPL values more than 65dB, which were about the same noise level as an automobile interior. It indicates that the noise of combat uniforms was very loud. Also, combat uniforms with woodland camouflage pattern had higher noise than those with digital pattern. The range of loudness (Z) was 8.07-17.4 sone, values at jogging and running speed were higher than that at walking speed (Figure 6(a)). The result provided that it was perceived as the noisiest psychoacoustically. Unlike loudness (Z), sharpness (Z) tended to decrease according to frictional speed, which indicated that frictional sound at running was recognized as the least sharp (Figure 6(b)). Roughness (Z) ranged from

2.83 to 5.99asper and had significantly low values by decreasing speed (Figure 6(c)). In case of Fluctuation strength (Z), values of all specimens were quite similar to one another (Figure 6(d)). Values at walking had significantly lower than those at jogging and running. It can be interpreted that it could be felt more changeable when arms get faster by increasing speed. By fabric types, W-WR had the highest values of SPL, loudness(Z) and sharpness(Z), which meant that W-WR was recognized as the loudest and sharpest psychoacoustically.

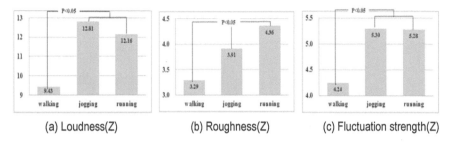

(a) Loudness(Z)   (b) Roughness(Z)   (c) Fluctuation strength(Z)

Figure 4 Result of Post-Hoc test for Sound Characteristics

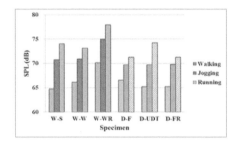

Figure 5 SPL according to Specimens and Frictional Speed

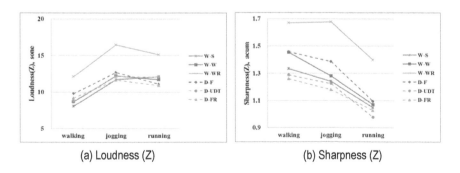

(a) Loudness (Z)   (b) Sharpness (Z)

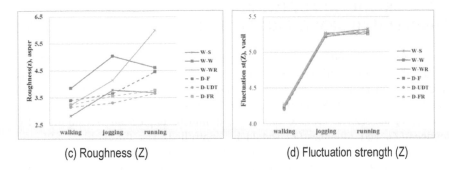

(c) Roughness (Z)   (d) Fluctuation strength (Z)

Figure 6 Zwicker's Psychoacoustic Parameters according to Frictional Speed

3.3   PSYCHOLOGICAL SENSIBILITY FOR THE FABRIC SOUND

The auditory sensibilities with the exception of 'dull – sharp', 'clear – obscure' tended to increase as changing frictional speed. As increasing speed level, the frictional sounds of fabric were evaluated as louder, higher, harder, more changeable and more unpleasant. The psychological response evaluation for 'quiet – loud' is presented in Figure 7(a). The old combat uniform fabrics with high values of SPL, were evaluated as 'loud' sounds than new combat uniforms with digital pattern. Especily, W-WR with the high values of SPL, loudness(Z) was estimated as the loudest sound at walking, jogging and running. It means that sensibility loudness is influenced by the sound characteristics such as SPL and loudness(Z). According to Figure 7(b), all of specimens were considered to be 'rough' sound at jogging and running speed, but there were differences among specimens at walking speed. D-UDT and F-FR were close to neutral sensibility, but W-W and D-F were assessed to 'rough' sound. It appears that the specimens with high roughness(Z) are felt to be 'rougher' sound. According to specimens, W-WR was evaluated as the loudest, highest and most unpleasant at walking, jogging and running. It is thought that W-WR is the type of a vapor permeable water repellent fabric which makes an exceptional noise level arousing negative sensibility. Thus, it is possible to design fabrics with positive sensibility by controlling fabric tyes, sound characteristics and frictioanl speed.

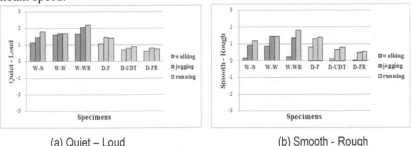

(a) Quiet – Loud   (b) Smooth - Rough

Figure 7 Psychological Sensibility for Specimens

3.4 PREDICTION OF PSYCHOLOGICAL SENSIBILITY OF COMBAT UNIFORM FABRICS WITH SOUND CHARACTERISTICS

To predict overall psychological satisfaction, regression analysis between 'pleasant – unpleasant' sensibility and sound parameters was conducted. The 'pleasant' sensibility was predicted by SPL (Y= - 0.12 SPL + 7.60, R^2=0.50), Loudness(Z) (Y= - 0.18 Loudness(Z) + 1.27, R^2= 0.38), and Roughness(Z) (Y= - 0.60 Roughness(Z) + 1.45, R^2=0.51). To diminish 'unpleasant' sensibility, SPL, loudness(Z), and roughness(Z) should be controlled. The threshold of SPL was 63.33dB (Figure 8(a)), loudness(Z) was 7.05sone (Figure 8(b)), and roughness(Z) was 2.41asper (Figure 8(c)) for 'pleasant' sensibility. Accordingly, combat uniform fabrics would have 'pleasant' sensibility by reducing the SPL below 63.33dB, loudness(Z) below 7.05sone, and roughness(Z) below 2.41asper.

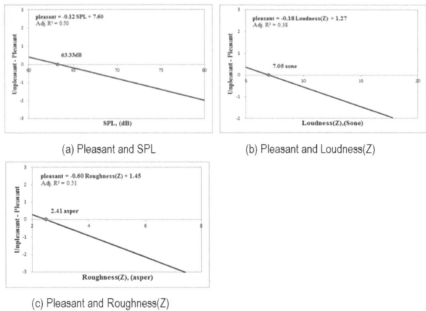

(a) Pleasant and SPL   (b) Pleasant and Loudness(Z)

(c) Pleasant and Roughness(Z)

Figure 8 Prediction of Pleasant Sensibility with Sound Characteristics

4 CONCLUSION

In this study, we have quantified the sound characteristics and evaluated the psychological sensibility to provide prediction model for satisfaction of auditory sensibility. SPL and the psychoacoustic parameters such as loudness (Z), roughness (Z), and fluctuation strength (Z) were significantly increased by frictional speed. Unlike other psychoacoustic parameters, sharpness(Z) decreased according to frictional speed. The auditory sensibilities with except 'dull – sharp', 'clear –

obscure' tended to increase as changing frictional speed. At running speed, fabric sounds were estimated as the loudest, highest, loudest, hardest, most changeable and unpleasant. As a result of regression analysis for subjective pleasantness, sound characteristics such as SPL, loudness(Z), and roughness(Z) were important predictors. Therefore, combat uniform fabrics would have subjective pleasantness by lowering the SPL, loudness(Z), and roughness(Z).

This study can be used widely in manufacturing process to textile and apparel fields for auditory-sensible fabrics as well as the development of combat uniform with auditory camouflage. Further study is needed to evaluate physiological responses for the fabric sound. Physiological responses can be used as an objective index to human sensation because physiological resposnses mediated by autonomic nervous system is difficult to change on purpose. Besides making auditory-sensible fabrics, it is needed to deduce the condition related with mechanical properties for minimizing the noise of combat uniforms.

ACKNOWLEDGMENTS

This research was supported by Basic Science Research Program through the National Research Foundation of Korea (NRF) funded by the Ministry of Education, Science and Technology (No. 2011-0015658).

REFERENCES

Cho, J. 2006. Effect of Sound Characteristics of Combat Uniform Fabrics at Different Frictional Speeds and Mechanical Properties on the Distance of Audibility. PH. D. Dissertation, Yonsei University, Seoul.

Kim, C., Yang, Y., and Cho, G. 2008. Characteristics of Sounds Generated from Vapor Water Repellent Fabrics by Low-speed Frictions. Fiber. Polym, 9(5): 639-645.

Morill, J. 1984. Audiometric Instrumentation: Equipment Choices, Techniques, Occupational Health and Safety, 53(10): 78-84.

Park, J., Kim, S., and Yang, Y., el al. 2010. H., Development of an Affective Quality Evaluation and Estimation System for fabric frictional sound. The Ergonomics Society of Korea, 29(2): 217-224.

Kim, C., Yang, Y., and Park, J., et al. 2008. Determination of Frictional Speeds by Arm Movement and Simulation of Frictional Sounds of Fabrics. Proceedings of the 2nd International Conference on Applied Human Factors and Ergonomics, 14-17.

Yang, Y., Kim, C., and Park, J., et al. 2009. Application of the Real Fabric Frictional Speeds to the Fabric Sound Analysis using Water Repellent Fabrics. Fiber. Polym, 10(4): 557-561.

Yi, E., Cho, G., and Na, Y., el al. 2002. A fabric Sound Evaluation System for Totally Auditory-Sensible Textiles, Textile Res. J. 72(7): 438-944.

CHAPTER 65

Effect of Color on Visual Texture of Fabrics

Eunjou Yi, An Rye Lee

Jeju National University
Jeju, Korea
ejyi@jejunu.ac.kr

ABSTRACT

This study was aimed to investigate if visual texture could be affected by color variables such as CIE values and hue/tone categories as well as mechanical properties of fabrics and to establish prediction models for fabric visual texture using both mechanical properties and color variables. Six different silk and cotton fabrics were colored into 3×3 hue (Red, Yellow, and Green)/tone (pale, vivid, and grayish) for chromatic shades and Gray for neutral using digital textile printing system and were measured by Kawabata Evaluation System for their mechanical properties. Subjective sensation of visual texture was evaluated with human participants using modified magnitude estimation. As results, sensory descriptors of fabric visual texture were primarily influenced by mechanical properties. Moreover there were also significant differences in each of them among tones as well as between chromatic shade and Gray while hue categories didn't give any effect. Finally prediction models for visual texture were provided by employing both mechanical properties and color variables.

Keywords: visual texture, color, mechanical properties, subjective sensation, fabric, prediction model

1   INTRODUCTION

Clothing and textiles provide diverse senses simultaneously to humans, which leads consumers to evaluate fashion products and to determine if they purchase them. Visual attributes including texture, color, and pattern play more important roles than any other sensory aspects (Kwon, Kim, and Na, 2000) especially at the first sight of fashion goods. Among them, texture has been regarded as an isolated object property and its relationship with only mechanical or geometrical properties has been provided by some earlier studies (Roh and Ryu, 2004; Burn et al., 1995; Lee and Shin, 2003). As for color, its sensory effects have been focused on in textile field by lots of fruitful works (Gao and Xin, 2006; Ou et al, 2004; Yang and Yi, 2010; Yi and Rhee, 2009) of which most also rarely considered other visual cues such as texture. However, in real life, textile objects likely to others are carrying a variety of visual stimulus such as texture, color, pattern, and shape and furthermore such sensory stimulus may modify or enforce another sensory attributes, which is explained as multisensory integration (Goldstein, 2002). For that reason, some of efforts have been made to explore the influence of texture on color sensation for textiles (Xin, Shen, and Lam, 2005; Kandi and Tehran, 2009; Lee and Yi, 2010) and they successively found that color sensation may differ according to texture of fabrics. Moreover many attempts have been carried out to investigate even the relationship between tactile hand attributes and color perception (Yenket, Chambers IV, and Gatewood, 2007; Nishimatsu and Sakai, 1987$_a$; Nishimatsu and Sakai, 1987$_b$; Nishimatsu and Sakai, 1988; Lee and Yi, 2011). However, the role of color in visual texture has relatively received only little attention. A logical next step in visual texture evaluation for textiles is the extension from a uniquely solid or gray-colored fabric samples toward variously colored textile specimens in order to explore color effect on visual texture. Therefore this study was initiated in an attempt to investigate the effect of color on visual texture for some selected fabrics. To achieve this, it was firstly examined if neutral Gray and chromatic shades as objective color variables are differentiated to each other in subjective visual texture following the primary investigation of mechanical properties for their dominance over visual texture of fabrics. Secondly, hue/tone categories were observed for their influence on visual texture evaluation. Finally it was explored to provide prediction models of visual texture using both mechanical properties and color variables of fabrics.

2   EXPERIMENTAL

2.1   Stimuli

Six different apparel fabrics including cotton and silk were selected as stimuli for this study. Cotton included canvas, oxford, and gaberdine while silk consisted of chiffon, georgette, and serge. Their characteristics were summarized in Table 1. Prior to subjective assessment of visual texture, each of them was prepared into sixty differently colored specimens by being colored using digital textile printing system in 3 hues (Red, Yellow, and Green) x 3 tones (pale, vivid, and grayish) combinations for chromatic shades and neutral Gray. Eventually a total of sixty different stimuli with combination of six fabrics and ten colors were used for visual assessment. Each stimulus was framed with neutral gray mat board in size of 20cm×20cm respectively for presentation.

Table 1 Fiber composition and structural characteristics of fabric specimens

Abbre.	Fiber	Weave	Yarn Number (warp·weft)	Density (warp·weft/cm²)	Thickness (mm)	Weight (mg/cm²)	Fabric Name
CC	Cotton 100%	plain	10.4·11.0	149.6·85.0	0.91	25.73	Canvas
CO		plain	22.2·11.2	197.0·75.4	0.68	18.98	Oxford
CT		twill	43.1·41.0	315.0·156.0	0.64	13.47	Gaberdine
SC	Silk 100%	plain	33.4·16.2	51.68·27.12	0.10	2.22	Chiffon
SG		plain	33.6·34.5	40.28·32.72	0.16	2.63	Georgette
ST		twill	21.2·98.0/2	113.92·61.92	0.42	12.15	Serge

2.2 Objective Measurement

KES-FB system (Kato Tech, LTD. Co) was used to measure a total of 17 mechanical properties including tensile, bending, shear, compression, surface properties, thickness and weight. The measurement was based on standard condition (Kawabata, 1980). As for objective color variables, the CIE color properties of each stimulus were measured by a spectrophotometer (CM 2500D, Minolta, Japan) with CIE D65/10°illuminant-observer reference conditions and they were converted into using Munsell conversion software (version 9.0.1) in order to determine PCCS(Practical Color Coordination System) tone notation.

2.3 Subjective Assessment of Visual Texture

Thirty college students with normal vision aged 18 to 26 who have taken a least one of courses in clothing and textiles were invited as participants in the visual assessment. The stimuli were presented one by one in a viewing cabinet (Gretagmacbeth, the Judge II) illuminated by a D65 simulator. Each participant was asked to see the stimulus while not allowed to touch and to fill out a questionnaire about subjective sensation for visual texture by using modified magnitude estimation (Mackay, Anand, and Bishop, 1999) with the range of scores from -10 to +10 without any reference stimulus. The questionnaire dealt with six aspects of fabric visual texture including smooth, buoyant, heavy, warm, thick, and stiff.

2.4 Data Analysis

In order to investigate the relationship between visual texture and CIE color properties as well as mechanical properties of fabric stimulus, Pearson's correlation coefficients were obtained. For the differences between Gray and chromatic color in visual texture, paired t-test was used. One-way ANOVA was employed to figure out the differences of visual texture according to hue or tone categories. Finally stepwise linear regression was used to establish the prediction models for visual texture by both mechanical properties and objective color variables.

3   RESULTS AND DISCUSSION

3.1   Effect of Mechanical Properties on Visual Texture

First of all, relationship between mechanical properties and subjective visual texture of fabric stimulus was investigated by calculating Pearson's correlation coefficients as shown in Table 2. All of visual texture descriptors were found to be significantly related with mechanical properties. Among them, both 'smooth' and 'buoyant' were positively affected by tensile recovery and compression recovery while negatively related with many mechanical properties including tensile energy, friction coefficient, and geometrical roughness. The other descriptors for visual texture such as 'heavy', 'warm', 'thick', and 'stiff' were positively correlated with bending rigidity, shear stiffness, thickness, and weight. From these results, Subjective visual texture of fabrics seemed to be determined primarily by fabric mechanical properties.

Table 2 Correlation coefficients between subjective visual texture and mechanical properties

Mechanical properties		Visual texture					
		smooth	buoyant	heavy	warm	thick	stiff
Tensile	EM	-0.251**	-0.442**	0.414**	0.365**	0.447**	0.368**
	WT	-0.303**	-0.505**	0.471**	0.411**	0.508**	0.424**
	RT	0.512**	0.665**	-0.619**	-0.515**	-0.657**	-0.577**
Bending	B	-0.201**	-0.451**	0.414**	0.364**	0.464**	0.377**
Shear	G	-0.419**	-0.533**	0.508**	0.425**	0.534**	0.454**
Surface	MIU	-0.207**	-0.106**	0.092**	0.060**	0.085**	0.088**
	SMD	-0.402**	-0.519**	0.497**	0.412**	0.526**	0.445**
Compression	WC	-0.475**	-0.69**	0.636**	0.536**	0.685**	0.598**
	RC	0.393**	0.580**	-0.522**	-0.444**	-0.569**	-0.494**
Thickness	T	-0.572**	-0.755**	0.699**	0.576**	0.747**	0.659**
Weight	W	-0.584**	-0.76**	0.704**	0.572**	0.749**	0.670**

** p<.01

3.2   Effect of Color Variables on Visual Texture

It was investigated if there is any difference between neutral Gray and chromatic color in each visual texture descriptor using paired t-test. The results were presented in Table 3 in which among visual texture descriptors, 'warm', 'thick', and 'stiff' were found to be differently felt between Gray and chromatic on fabrics. As for 'warm', fabrics in Gray seemed to be rated to be lower than those in chromatic did. On the contrary, chromatic color was thought to give an effect on getting higher rates for 'warm' in terms of visual texture so that the colored fabrics were perceived as warmer. While all of fabric stimulus were rated negatively in both 'thick' and 'stiff', Gray-colored fabrics tended to be given much less scores for both 'thick' and 'stiff' than their counterparts. Most of previous studies dealt with visual texture using fabrics shaded grayish or white as mentioned above (Roh and Ryu, 2004; Burn et al., 1995; Lee and Shin, 2003). From the finding that neutral color seemed to cause differences in human perception of visual texture from chromatic shades, it could be thought that color needs to be considered as a significant factor to determine fabric visual texture.

Table 3 Differences of subjective visual texture according to neutral Gray/chromatic

Visual texture descriptors	Gray/chromatic	Mean	t-value
smooth	Gray	-0.514	-0.498
	chromatic	-0.311	-0.533
buoyant	Gray	0.015	0.745
	chromatic	-0.314	0.780
heavy	Gray	-1.569	-1.853
	chromatic	-0.749	-1.945
warm ***	Gray	-0.150	-4.603
	chromatic	0.347	-4.554
thick *	Gray	-1.171	-2.090
	chromatic	-0.237	-2.127
stiff **	Gray	-1.804	-3.004
	chromatic	-0.510	-3.340

*p<.05 ** p<.01

According to chromatic categories of hue and tone, the difference of each aspect for visual texture was examined using one-way ANOVA as shown in Table 4. As for hue, any significant difference did not appear in all of visual texture descriptors, which means that hue is little influential on subjective fabric visual texture. On the contrary, all of visual texture descriptors were found to be significantly differentiated by chromatic tones. Precisely, both 'smooth' and 'buoyant' were more strongly felt in pale so that fabrics colored in pale had positive mean scores for both two descriptors whereas those in grayish gave negative feeling for them. This caused statistically higher rates of pale than those of grayish. On the other hands, 'heavy', 'thick', and 'stiff' seemed to be given higher scores in grayish than in pale. Especially grayish fabrics showed positive values for 'thick' and 'stiff' whereas pale or vivid ones were scored negatively. Vividly toned fabrics were found to be perceived as significantly more warm in their visual texture than palely fabrics. Moreover they tended to be positively evaluated in the descriptor unlikely to palely ones. From these results, it could be concluded that chromatic tones play a meaningful role in human perception of fabric visual texture while hue has little effect on visual texture.

Table 4 Differences of subjective visual texture according to hue and tone

hue/tone	type	smooth	buoyant	heavy	warm	thick	stiff
hue	Red	-0.229 A	0.052 A	-0.945 A	0.300 A	-0.317 A	-0.614 A
	Yellow	-0.528 A	-0.382 A	-1.008 A	0.719 A	-0.257 A	-0.450 A
	Green	-0.178 A	-0.611 A	-0.293 A	0.022 A	-0.137 A	-0.467 A
	F-value	0.585	1.589	2.189	2.125	0.115	0.118
tone	pale	0.348 B	0.190 B	-1.486 A	-0.217 A	-0.781 A	-1.158 A
	vivid	-0.491 AB	-0.254 AB	-0.637 AB	0.814 B	-0.093 AB	-0.493 AB
	grayish	-0.791 A	-0.878 A	-0.123 B	0.438 AB	0.163 B	0.120 B
	F-value	5.730**	4.038*	6.642**	4.753**	3.291*	5.959**

*p<.05 ** p<.01
A and B mean scheffe's multiple comparison result.

The relationship between CIE color properties and subjective visual texture of fabric was explored using Pearson's correlation coefficients as given in Table 5. All of visual texture descriptors showed significant correlations with some of CIE color properties. Color lightness, CIE L^* was likely to give positive effects on both 'smooth' and 'buoyant' whereas negative on 'heavy', 'warm', 'thick', and 'stiff'. This means that fabrics with higher L^* seemed to be rated as smoother and more buoyant while as less heavy, less warm, less thicker, and less stiffer in terms of visual texture. On the contrary, CIE C^* indicating color saturation was revealed to have negative correlations with 'smooth' and 'buoyant' whereas positive with the other visual texture descriptors. This result implies that more saturation of fabric color tended to evoke less sensation for 'smooth' and 'buoyant' while more feeling for 'heavy', 'warm', 'thick', and 'stiff'. It could be worth to note that CIE b^*, yellowness gave effects on subjective visual texture of fabrics. Precisely, it was positively correlated with 'smooth' and 'buoyant' while negatively with 'heavy', 'warm', 'thick', and 'stiff' similarly to CIE C^*. In summary, some of CIE color variables like L^*, b^*, and C^* could contribute significantly to determining visual texture of fabrics.

Table 5 Correlation coefficients between subjective visual texture and CIE color properties

CIE color properties	smooth	buoyant	heavy	warm	thick	stiff
L^*	0.216**	0.269**	-0.285**	-0.192**	-0.259**	-0.258**
a^*	-0.010	0.014	-0.007	0.048	0.022	0.018
b^*	-0.066**	-0.109**	0.096**	0.183**	0.15**	0.123**
C^*	-0.085**	-0.128**	0.134**	0.209**	0.175**	0.151**
h	-0.052	-0.097**	0.074**	0.002	0.055**	0.010

** $p < .01$

3.3 Predicting Visual Texture by Mechanical Properties and Color Variables

We attempted to establish prediction models of visual texture using both mechanical properties and color variables. Table 6 shows the stepwise linear regression equation for each visual texture of fabrics including those in Gray and in chromatic shades. All of visual texture descriptors provided prediction models in which both mechanical properties and tone categories entered. As for 'smooth', fabrics colored in pale were likely to be felt as smoother as well as some mechanical properties seemed to arouse smoother in visual texture. Tone of grayish was the significant predictor of visual texture 'buoyant' which was affected negatively. It was also a positive determinant for 'heavy' whereas pale and Gray were negative ones. Visual texture 'warm' and 'thick' were found to be less felt when the fabric colored in Gray or pale. As for 'stiff', Gray and pale were negative predictors while grayish positive one. These results were supported by the above mentions that there were significant differences in fabric visual texture between Gray and chromatic color as well as among chromatic tones. Furthermore tone categories could be applied to explain human subjective sensation of fabric visual texture in practice.

Table 6 Prediction models for visual texture by mechanical properties and tone

Visual texture	Prediction model	R^2
smooth	$Y=62.279 \cdot B+60.981 \cdot MIU+0.970 \cdot pale-63.279 \cdot 2HB+0.154 \cdot RC-22.422 \cdot MMD-18.710$	0.393
buoyant	$Y=-107.042 \cdot MMD-1.383 \cdot 2HG5-0.853 \cdot grayish-11.740 \cdot LC+14.838$	0.606
heavy	$Y=0.145 \cdot W+55.028 \cdot MMD+0.665 \cdot 2HG5+0.515 \cdot grayish+3.823 \cdot 2HB-0.848 \cdot pale-1.040 \cdot Gray-7.378$	0.518
warm	$Y=7.123 \cdot T-2.239 \cdot Gray+35.103 \cdot MMD-0.845 \cdot pale+0.346 \cdot SMD-5.721$	0.362
thick	$Y=80.322 \cdot MMD+0.944 \cdot 2HG5+7.699 \cdot 2HB-1.195 \cdot Gray-0.816 \cdot pale-7.293$	0.598
stiff	$Y=0.195 \cdot W+49.625 \cdot MMD-1.466 \cdot Gray-0.665 \cdot pale+0.941 \cdot 2HG+0.613 \cdot grayish-6.721$	0.478

As another ultimate goal of this study, prediction models of visual texture by inviting both CIE color properties and mechanical properties as explanatory variables were established. As presented in Table 7, all of visual texture descriptors provided significant equations which were described by both mechanical properties and CIE color properties. Lightness, L^* entered equations for 'smooth' and 'buoyant' respectively as a positive predictor while for 'heavy', 'thick', and 'stiff' it did as a negative one. This implies that human seemed to feel fabric visual texture as smoother and more buoyant whereas as less heavy, less thick, and less stiff as fabric color was lighter. This result is also supported by fabrics in pale showing stronger feeling of 'smooth' and by those in grayish for more buoyant. Color saturation, C^* was another significant contributor to prediction models of fabric visual texture in that it was included for explaining 'buoyant' and 'warm' positively. It could be interpreted that as fabric color was more saturated 'buoyant' and 'warm' as visual texture tended to be evoked more strongly. On the other hands, redness, a* and yellowness, b* were also found as helping to get more sensation of 'heavy' in fabric visual texture. These results provide a meaningful potential to application of CIE color properties for explaining human perception of fabric visual texture.

612

Table 7 Prediction models for visual texture by mechanical properties and CIE color properties

Visual texture	Prediction model	R^2
smooth	Y=43.931·B+0.026·L*-43.566·2HB+0.051·RC-25.461·MMD-4.670·T-1.984	0.395
buoyant	Y=-105.451·MMD-1.368·2HG5+0.025·L*-12.515·LC+0.013·a*+0.017·C*+12.940	0.608
heavy	Y=-0.037·L*+62.103·MMD+1.623·2HG+4.178·B-0.012·a*+0.011·b*-4.565	0.520
warm	Y=6.807·T+0.025·C*+34.063·MMD+0.343·SMD-6.698	0.357
thick	Y=76.788·MMD+0.905·2HG+7.770·2HB-0.022·L*+0.019·b*-6.190	0.598
stiff	Y=0.281·W+34.992·MMD+0.594·2HG-0.036·L*-0.009·h-0.022·a*+0.018·b*-3.658	0.481

4   CONCLUSIONS

This study was performed to investigate if objective color variables could be affective on human subjective visual texture of fabrics and to predict visual texture by the color variables as well as by traditionally used mechanical properties of fabrics. As results, fabric visual texture descriptors were primarily influenced by mechanical properties and they also showed significant relationships with color variables such as tone categories and CIE color properties. Finally each of fabric visual texture descriptor was significantly predicted by employing both mechanical properties and color variables.

These results lead us to the conclusion that objective measurements of fabric color are helpful to predict human subjective visual texture. This approach could be a useful starting point for integrating visual information of texture and color for designing visually sensible textiles. In a future study, more variety fabrics need to be investigated in order to provide powerful and reliable predictions for visual sensation of fabric texture. In addition, other visual cues such as pattern and shape could be employed for explaining visual texture of fabrics.

ACKNOWLEDGMENTS

The authors would like to thank the Korea Research Foundation which provided financial support to this work (KRF-2008-331-H00007).

REFERENCES

Burns, L. D., D. M. Brown, B. Cameron, and J. Chandler. 1995. Sensory interaction and descriptions of fabric hand. *Perceptual and Motor Skills*, 81: 120.

Gao, X. and J. H. Xin. 2006. Investigation of human's emotional responses on colors. *Color Research and Application*, 31: 411-417.

Goldstein, E. B. 2002. *Sensation and perception -6th edt.-*. Belmont: Wadsworth Publishing.

Kandi, S. G. and M. A. Tehran. 2009. Investigating the effect of texture on the performance of color difference formulae. *Color Research and Application*, 35: 94-100.

Kawabata, S. 1980. *The Standardization and Analysis of Hand Evaluation*, Textile Machinery Society of Japan, Osaka, Japan, 2nd edt.

Kwon, O. K., H. E. Kim, and Y. J. Na. 2000. *Fashion and science of emotion & sensibility.* Seoul: Kyomunsa.

Lee, A. R. and E. Yi. 2010. Prediction models for fabric color emotion factors by visual texture characteristics and physical color properties. *Journal of the Korean Society of Clothing and Textiles*, 34: 1567-1580.

Lee, A. R. and E. Yi. 2011. Prediction models for tactile sensation/sensibility image of silk fabrics by mechanical properties and color characteristics. *Korean Journal of the Science of Emotion and Sensibility*, 14: 127-136.

Lee, J. S. and H. W. Shin. 2003. The sensibilities of cotton fabrics. *Journal of the Korean Society of Clothing and Textiles*, 27: 800-808.

Mackay, S., S. C. Anand, and D. P. Bishop. 1999. Effects of laundering on the sensory and mechanical properties of 1 × 1 rib knitwear fabrics -part II: changes in sensory and mechanical properties-. *Textile Research Journal*, 69: 252-260.

Nishimatsu, T. and T. Sakai. 1987$_a$. Quantitative investigation of the surface color's effect on the hand evaluation of pile fabrics. *Sen'i Gakkaishi*, 43: 553-557.

Nishimatsu, T. and T. Sakai. 1987$_b$. Significance of the influence of the sense of sight on the hand evaluation of pile fabrics. *Sen'i Gakkaishi*, 43: 211-217.

Nishimatsu, T. and T. Sakai. 1988. Application of the information theory to hand evaluation and proposal of the design equation of the visual and tactual sense values of pile fabrics. *Sen'i Gakkaishi*, 44: 88-95.

Ou, L., M. R. Luo, A. Woodcock, and A. Wright. 2004. A study of colour emotion and colour preference -part 1: colour emotions for single colours-. *Color Research and Application*, 29: 232-240.

Roh, E. and H. S. Ryu. 2004. Effect of Constituent characteristics of cotton fabrics on the visual perception and image scale. Journal of the Korean Society of Clothing and Textiles, 28: 1142-1152.

Xin, J. H., H. L. Shen, and C. C. Lam. 2005. Investigation of texture effect on visual colour difference evaluation. *Color Research and Application*, 30: 341-347.

Yang, Y. and E. Yi. 2010. Color sensibility image of naturally dyed silk fabric. *Korean Journal of the Science of Emotion and Sensibility*, 13: 403-412.

Yenket, R., E. Chambers IV, and B. M. Gayewood. 2007. Color has little effect on perception of fabric handfeel tactile properties in cotton fabrics. *Journal of Sensory Studies*, 22: 336-352.

Yi, E. and Y. Rhee. 2009. A psychophysical approach to color sensory evaluation of yellowish natural dye fabrics. *Fibers and polymers*, 10: 200-208.

CHAPTER 66

The Individual Adaption Module (iAM): A Framework for Individualization and Calibration of Companion Technologies

Steffen Walter, Kerstin Limbrecht, Stephen Crawcour, Vladimir Hrabal,
Harald Traue

University of Ulm
Ulm, Germany
Steffen.Walter@uni-ulm.de

Keywords: companion system, individual specificity, feature selection, individual Adaption Module (iAM)

ABSTRACT

The authors of the present study argue for the notion that assistant systems of the future will become "companion systems", i.e., systems that adapt to individual-specificity of the user (i.e., to one's emotion patterns, personality, preferences, etc.). Due to the individual-specificity of emotion patterns, individualization and calibration processes will become necessary. In the present paper, an introduction to companion systems will be given, while placing special focus on the significance of automatic emotion recognition. Resulting problems for a transsituationally robust feature selection of signals and consequences on classification rates will be presented. In the end, a procedure for a potential structural individualization or calibration process for companion systems will be presented to find robust individual features. The vision is that robust individual-specific features and multiple personality trait variables will be stored in an individual Adaptation Module (iAM). Such module will be outlined in the present article.

1 FROM ASSISTANT SYSTEMS TO COMPANION TECHNOLOGIES

In the future, conventional assistance systems will continue to be developed to become Companion Systems (CS). In the specialized research area Transregio-62 (http://www.sfb-trr-62.de/): "A Companion Technology for Cognitive Systems" the basics for future CS will be explored. Basically, CS can be regarded as user-adaptive individualized assistance systems. Particularly, this notion has been articulated in the contribution collection "Close involvement with Artificial Companions" by Wilks (2010) that was issued as a result of an extensive seminar at the Oxford Institute. Specifically, the authors suggest the following features to be characteristic for CS:

(1) CS sensitivity to the user's emotions. Because of the importance of emotions in setting priorities, decision-making and action control, situational aspects and emotional processes are to be made available for the concept of CS. Two main aspects should be mention regarding the necessary of emotion recognition and CS.

- Emotion induction in the interaction with the CS (e.g.: anger vs. happiness of the CS)
- emotion induction regarding the exogen and endogen process independent of the direct interaction with the CS (e.g.: anger about the home technology, anger about a person, fear of a thunderstorm)

In this respect, audio, video and psychobiological parameters (i.e., Skin Conductance [SCL], Blood Volume Pressure [BVP], Respiration [RSP], Electromyography [EMG], Elektroencephalography [EEG]) are relevant. Emotion recognition proceeds with so-called classification algorithms. Because the emotional patterns for each user are highly individual-specific (Fahrenberg and Foerster, 1982; Lacey, Bateman and Vanlehn, 1953; Stemmler and Wacker, 2010), a systematic individual-specific calibration - in form of a feature selection - should proceed.

(2) CS sensitivity to the environment and situations. In other words, CS can reliably capture situational and environmental parameters, such as, e.g., time (year, month, day of week, time of day), space (location) and temperature.

(3) CS verbal interaction in a user-adaptive manner. In this respect, robust speech recognition and speech synthesis is required. The user's emotional state and contextual information should influence the selection of the dialogue strategy and increase the usability of a CS.

(4) Assuming that the points 1, 2 and 3 are given, CS may ultimately support the user in decision making and, when appropriate, elicit changes in policy. Simply put, a companion system should be able to work-out solutions for complex tasks - either completely independently or in cooperation with the user, while proceeding in a goal-oriented and comprehensible manner, as well as to provide decision support

and recommendations for action in dialogue with the user (Buindo, and Wendemuth, 2010).

However, the way in which CS should be implemented remains mostly unclear. The options that would be conceivable are as follows:

- CS would be implemented in stationary technologies and applied in a specific domain. Its implementation could occur, e.g., into a personal computer, a ticket vending machine, a beverage vending machine, kitchen appliances, etc.

- CS are mobiley applied in highly complex situations and remain omnipotent. That is, CS are applied in multimodal surroundings (e.g., multi-modal stress management). The implementation could occur, e.g., into a *Smartphone*. Such systems could network with people (e.g., family members), institutions (hospital) and technical equipment (including robots) (see Figure 1).

In order for a CS to optimally ensure the points 1-4, a so-called personalization or calibration will be necessary. In this respect, the transsituationally robust calibration is an absolute priority. In this respect, a transsituational feature selection shall be presented in the next chapter. Regarding the audio and video parameters, etc., similar studies will be required.

2   PRELIMARY RESULTS

For emotion induction a human-computer-scenario was selected. With such scenario a simulation of a naturalistic context with the following events was intended: (1) delay of the command, (2) non-execution of the command, (3) incorrect speech recognition, (4) offer of technical assistance, (5) lack of technical assistance, (6) request for termination and (7) positive feedbacks.

These technical conditions allowed the inducement of different emotions in the user in a theory-based manner. The procedure of emotion induction included differentiated experimental sequences during which the user passed through specific valence/ arousal/ dominance octants of the PAD model in a controlled fashion. It was a key priority to induce the emotion of "negative valence/ high arousal/ low dominance" in the participant starting from the state of "positive valence/ low arousal/ high dominance".

In the present study, we performed a combined analysis by using 4-channel peripheral physiological measurements including blood volume pulse (BVP), skin conductance level (SCL), and 2-channel electromyography (EMG) for classifying two emotional states, i.e., positive valence/ low arousal/ high dominance and negative valence/ high arousal/ low dominance. Both emotional states are corresponding to the reciprocal octants with "anchor points" (i.e., extreme points) for valence (positive versus negative), arousal (low versus high), and dominance (control versus loss of control; see Fig. 1).

2.1   Participants

A total of 20 participants participated in the experiment, who received an expense allowance. The study was conducted according to the ethical guidelines of Helsinki and approved by the ethics committee of the University of Ulm/Germany (C4 245/08-UBB/se).

2.2 Experimental Design

The simulation of the natural verbal human-computer interaction was implemented as a Wizard-of-Oz (WOZ; Kelley, 1984) experiment. The WOZ experiment allows the simulation of computer or system properties in a manner such that participants have the impression of experiencing a completely natural verbal interaction with a computer-based mental trainer. The design of the mental trainer followed the principle of the popular game "Concentration". The variation of the system behavior in response to the subjects was implemented via natural spoken language, with parts of the subject's reactions taken automatically into account. This method allows simulating human-computer interactions that technically are not yet possible (or only rudimentarily so), such as a reliable recognition of natural spoken language. In the following we describe details of our experiment: Instruction from the experimenter at the beginning: "You will communicate with a computer and complete a memory test. The experiment will be carried out in this room. You will receive instructions from the system. The system can react to you on the basis of language, gestures, and facial expressions. For that purpose, a microphone and a camera will be employed to record the experimental session. The system includes a variety of functions. At the beginning of the test you will be required to answer some questions. When you exceed certain thresholds with regard to the psychobiological parameters, the system will ask you whether you wish to terminate the task. Poor language recognition may lead to technical delays. Try to treat the system like you would a human being."

Structure of the experimental sequences (ES):

The experiment involves a two part mental training (Fig. 1). Both parts consist of identical ES; Part 2 also includes a debriefing sequence at the end of the scenario. Between round one and round two, 24 images with a mix of valence, arousal and dominance (International Affective Picture System, IAPS [Lang, 1995]) are shown to the subjects, so that they could distance themselves emotionally from the first round. The presentation time for each image is 6 s, with a break of 4 s to 10 s.

618

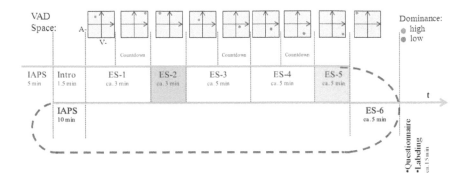

Figure 1 Experimental design, including the expected position in the VAD (Valence, Arousal, Dominance) space. After a five minute initialization with IAPS images, the experiment starts with a short introduction, followed by the first round of the mental training (ES-1 - ES-5). The second round of the mental training contains six games (ES-1 - ES-6) and starts after a ten minute-presentation of IAPS images. The experiment is finalized with the completion of standardized questionnaires. Our classification categories were ES-2 (orange) vs. ES-5 (turquoise).

2.3   Feature Extraction

We used 4-channel biosignals (2 x EMG, BVP, and SCL) for the emotion recognition. The signal length is about 3 min for ES-2 and 5 min for EP-5, representing antithetic emotions, positive valence/ low arousal/ high dominance vs. negative valence/ high arousal/ low dominance respectively. Using a 10-second window length with an overlap of 5 seconds, we segmented the signals into about 20 samples of ES-2 and 20-30 samples of ES-5, depending on different experiment durations of each participant. For the classification of the two emotional states, we extracted a total of 13 statistic features based on mean, minimum, and maximum values from each signal.

2.4   Classification

For classification, a feed-forward Artificial Neural Network (ANN) in a 13-40-20-1 architecture with two hidden layers was employed. Using the Matlab NN-Toolbox, the network was trained with the gradient descend algorithm with momentum (traingdx) for 1000 epochs. The training vectors were normalized (prestd) in order to present means of zero and standard deviations of 1. Since testing the performance of Support Vector Machines (SVM) didn't find any significant differences, we decided to use the ANN based on our positive experience with the emotion classification problem.

To test its performance for individual classification, the ANN-classifier was trained with preprocessed data of ES-2 (target = 0) and ES-5 (target = 1). Each of the vectors was left out once, then used for testing and the output was recorded. We rated an output x > = 0.5 as 1 and x < 0.5 as 0.

- An individual-specific feature selection was implemented for round one (ES-2 vs. ES-5). We trained round one (ES-2 vs. ES-5) and recognized for round two (ES-2 vs. ES-5) with a classification rate of 53.0 % (SD = 17.4 %).
- We applied an individual-**transsituationally**-specific feature selection on round one and two. Next, we trained round one and recognized for round two with a classification rate of 70.1 % (SD = 17.8 %).

To verify which descriptive emotional states were actually perceived by the subjects during the experiment (Valence ES-2: M = 7.7; Arousal ES-2: M = 4.3; Dominance ES-2: M = 7.1/ Valence ES-5: M = 3.4; Arousal ES-5: M = 5.9; Dominance ES-5: 4.4). All participants stated that they experienced the same emotion during the first and second round. This was relevant for the first and the second round of the experiment, as the participants' emotional response could be identified in the same VAD octants. This was a crucial prerequisite for the implementation of a transsituational classification.

2.5   Conclusion

The study shows what problems a transsituational feature selection has posed thus far.   This ultimately implies that a transsituational classification of emotions has to date not been possible, thus not being able to guarantee emotion recognition for companion technologies. The transsituational feature selection was applied only on biosignals in the study described. However, to the best of our knowledge, there are no studies solving the transsituational problem for audio and video data. The weakness in prior data sets lies in the lack of sufficient or even existent transsituational datasets. Hence, the solution in the present problem may be found solely in the collection of larger transsituational data sets. Such a potential scenario will be displayed in the concluding chapter.

3   THE INDIVIDUAL ADAPTION MODULE (iAM)

In the following, a concrete application of a user-adapted individualization of companion technologies shall be presented. The individual Adaptation Module (iAM) is the core element of a companion technology. It is possible that the user's individualization occurs via an avatar in a private setting, or rather that the individualization processes are carried out by a service company with help of a "companionologue", which would be considerably time- and cost-expensive. The complete data set, which is structurally assessed by the individualization or calibration, is located in the iAM. The iAM consists of several submodules, which are as follows:

- (1) the individual General Diagnostic Module (**iGDM**), (2) the individual Need State Module (**iNSM**), (3) the individual Cognitive Module (**iCM**) , (4) the individual Dialog Preference Module (**iDPM**), and
- **the calibration of individual affective features (audio, video, psychobiological data) via the individual Emotion Module (iEM).**

While it is highly probable that several modules will be needed, the options presented here should only represent priorities for an individualization and calibration. In prospect, it can already be shown that the quality of a companion system will be dependent on the complexity of the modules and their networks. For example, the application of companion feature for kitchen utensils will require a lower number of modules relative to its implementation (i.e., support) in the medical and psychotherapeutic realm. However, a companion system should tend to include a maximum of user information. Relative to module networking, it appears relevant to discern the weight/priority the modules will carry (see Fig. 2). In this respect, the focus on individuality is highly relevant.

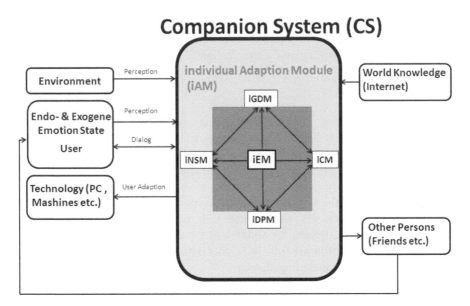

Figure 2: Suggested companion architecture with an intergrated individual Adaptation Module (iAM); individual General Diagnostic (iGD), individual Need State Module (iNSM), individual Cognitive Module (iCM), individual Dialog Preference Module (iDPM), individual Emotion Module (iEM)

iGDM: Here the following variables will be assessed: age, gender, and personality (e.g. NEO-Five Factor Inventory [NEO-FFI], Costa and McCrae, 1992; Behavioral inhibition, behavioral activation [BIS/BAS], Carver, C.S. and T.L. White, 1994), attention (d2, Brickenkamp, 1962), memory (wordlists, Mittenberg et al., 2009), attitudes towards technical utensils and user types, emotion regulation (Emotion Regulation Questionare [ERQ], Gross and John, 2003), attachment (Adult Attachment Interview [AAI], Main and Goldwyn, 1996; Adult Attachment Projective [AAP], George and West, 1999; Relationship Questionnaire [RSQ], Bartholomew and Horowitz, 1991) could be implemented.

iNSM: Need states shall be individually assessed via a questionnaire or interview: technical affinity, taste for music, color, means of transportation, literature, and more.

iCM: Cognitive preferences shall be assessed via a questionnaire or interview. In this respect, a dynamic adaptation could proceed via structural questions related to priming, e.g., "Have you ever experienced something like that before? Does this seem familiar to you?"; attribution, e.g., "What would you think if a technology would react in the following manner? To what extent do you see the cause of the reaction?"; projection, e.g., "Do you think a companion technology will be able to solve a personal social problem? Please try to give a reason for your answer."

iDPM: Data collection can be performed via questionnaires. Here the voice (i.e., male vs. female, young vs. old, pertaining to the user, user's friend, mother, etc.) of the CS will be assessed. Furthermore, phrases and feedback that are experienced as helpful by the user in certain situations will be queried.

iEM: In this respect, the aim of calibration would be to gain a maximum of information about the individual-specific features from all signals (audio, video, psychobiological data) collected. This information would be used to guarantee accurate and robust recognition rates of emotions in real time. In consequence, processes shall be described on how an individual-specific calibration should proceed. A distinction is made between a laboratory calibration and field calibration. It is conceivable that the decision would not be mutually exclusive (i.e., an "either-or" decision), but would rather allow the combination of both approaches. The dimensional perspective of valence/arousal/dominance and basic emotions shall be used as a model of emotion.

1. Laboratory Calibration: The persons are requested to appear to at least 5 sessions in the laboratory. Psychobiological parameters (EEG, BVP, SCL, RSP, EMG, pCO_2, PEP), audio and video data will be recorded. The advantage of using laboratory calibration is a controlled recording of all parameters. The disadvantage is that laboratory situations are always artificial, and the induction of emotions succeeds only partially. A laboratory procedure could proceed as follows:
• At the beginning of the calibration, a baseline recording in the form of a relaxation exercise would be performed.
• The International Affective Picture System images (IAPS; Lang, 2005) would be presented in blocks (i.e., same octant, 10 pictures positive valence/low arousal/high dominance, 10 negative valence/low arousal/high dominance and so on)
• Three short movie clips (of about 4 minutes in duration) for the basic emotions (fear, anger, disgust, sadness, and happiness) would be shown.
• An entire feature film would be presented to the user. The occurring emotions of the viewer would be labeled by the user himself during the movie.
• So-called case vignettes would be generated, presenting the valence, arousal, dominance room with eight octants as a basis. The content of the vignettes would include individual-specific or standardized scenarios. The scenario for the octant negative valence / high arousal / low dominance could thus be described as follows:
"You are unexpectedly pressed for time and need to boot up your PC quickly and

require some information from the Internet. The connection to the Internet is not possible and you are extremely nervous and trembling with rage, and ultimately you want to kick the damned computer out the window. "With the data, an individual-specific and transsituational feature selection across context would be performed (Transsituational Feature Selection Algorithm). Hence, a search for the so-called "best feature" would take place. A decision would then be made about the classification algorithm (SVM, KNN, NN, etc.) or hybrid architectures of data fusion that work best individually-specific (Fig. 3).

2. Regarding field calibration under realistic conditions in everyday life, the same parameters that are measured in a laboratory situation should be tested. The person would then have to constantly label situations during the course of the day or create such labels post hoc. The individually-specific analysis would be similar to laboratory data collection.

Featureselektion

t1	t2	t3	t4	t5
1	1	1	1	1
2	2	2	2	2
3	3	3	3	3
4	4	4	4	4
5	5	5	5	5
6	6	6	6	6
7	7	7	7	7
8	8	8	8	8
9	9	9	9	9
10	10	10	10	10

Figure 3: Transsituational feature selection. The lines represent single features and the columns single time points. Bold letters imply that the corresponding feature is transsituationally robust.

4   OUTLOOK

Companion technologies will only be able to react in a user-adaptive manner once the technological basis of the functioning of an individual Adaptation Module is provided. To this end, however, the first problem to solve would be the transsituationally robust selection of bio-, video, and audio signals. In this respect, it will be necessary to test such transsituational feature robustness in an "artificial context" (laboratory) and further specify the robustness in a naturalistic (real) context. This may be time- and cost- expensive, but still vital for the companion system's ability to function. It may be possible that whole service industries, the so-called companiologues, will be created in the future in order to perform such calibration process.

ACKNOWLEDGEMENTS

This research was part of the SFB/Transregio 62 - Companion Technology for Cognitive Systems project - funded by the German Research Foundation.

REFERENCES

Bartholomew, K. and L.M. Horowitz. 1991. Attachment styles among young adults: A test of a four-category model. *Journal of Personality and Social Psychology* 61: 226-224.

Biundo S. and A. Wendemuth. 2010 Von kognitiven technischen Systemen zu Companion-Systemen. *Künstliche Intelligenz* 24: 335-339.

Brickenkamp, R. 1962, 1975. Test d2, Aufmerksamkeits-Belastungs-Test [attentional load test]. Göttingen: Hogrefe.

Carver, C.S. and T.L. White. 1994. Behavioral inhibition, behavioral activation, and affective responses to impending reward and punishment: The BIS/BAS scales. *Journal of Personality and Social Psychology* 67: 319–333.

Costa, P.T. and R.R. McCrae. 1992. NEO PI-R. Professional manual. Odessa, FL: Psychological Assessment Resources, Inc.

Fahrenberg J. and F. Foerster. 1982. Covariation and consistency of activation parameters. *Biology Psychology* 15: 151–169.

George, C. and M. West. 1999. Developmental vs. social personality models of adult attachment and mental ill health. *British Journal of Medical Psychology* 72. 285-303.

Gross, J.J. and O.P. John. 2003. Individual differences in two emotion regulation processes: Implications foraffect, relationships, and well-being. *Journal of Personality and Social Psychology* 85: 348-362.

Kelley, J.F. 1984. An iterativ design methodolgy for user-friendly natural language office information applications. *ACM Transaction on Office Information Systems* 2: 26–41.

Lacey, J.I. and D. Bateman. R. Vanlehn. 1953. Autonomic response specificity; an experimental study. *Psychosomatic Medicine* 15: 8-21.

Lang, P. 1995. The emotion probe: Studies of motivation and attention. American Psychologist 50: 372–385.

Main, M. and R. Goldwyn. 1996. Adult Attachment Classification System. University of California, Berkeley.

Mittenberg, W. and R. Azrin, C. Millsaps, R. Heilbronner. 1993. *Psychological Assessment:* 5, 34-40.

Stemmler, G. and J. Wacker. 2010. Personality, emotion, and individual differences in physiological reponses. *Biology Psychology* 84: 541–551.

Wilks, Y. 2010. Introducing artificial Companions. In. *Close Engagements with Artificial Companions,* ed. Y.Wilks. Amsterdam: J. Benjamins Publishing Company (p. 11-22)

CHAPTER 67

Self-Adaptive Biometric Signatures Based Emotion Recognition System

Yuan Gu, Su-Lim Tan, Kai-Juan Wong

School of Computer Engineering
Nanyang Technological University
Singapore 639798
guyu0006@e.ntu.edu.sg, ASSLTan@ntu.edu.sg, ASKJWong@ntu.edu.sg

ABSTRACT

Robust and efficient emotion recognition is very important and challenging for many physiological and augmented applications. These applications however, usually suffer from the non-emotion-related varieties, the "individual varieties", especially when cope with multiple subjects. In this paper, we proposed and presented an improved grouping structure called the Self-Adaptive Biometric Signatures Based (SABSB) emotion recognition system. The system is unique in that it transforms a traditional subject-independent problem into several subject-dependent cases to remove the influence of individual varieties. Also, the self-adaptive procedure allows a flexible model generating process which is more practical for real applications. The proposed framework is implemented and tested using mixture multivariate t-distributions (mmTD). Results are compared with conventional emotion recognition system as well as the original BSB system.

Keywords: emotion recognition, human-computer interaction

1   INTRODUCTION

As the pioneer in the area of affective computing, Picard et al. (Picard, 1995) introduced a complete emotion recognition procedure, which can be generally divided into three sections: (1) the experimental design and gathering of good affective data, (2) signal preprocessing and feature extraction, and (3) feature selection/transformation and classification. Since then, several research groups have followed the lead and published their results based on various experiments (Picard, Vyzas and Healey, 2001, Nasoz et al. 2003, Haag et al. 2004, Wagner et al. 2005, Gu et al. 2008(1), Gu et al. 2008(2)). Among all these studies, one particular issue has gradually raised the attentions, which was described as "the intricate variety of non-emotional individual contexts among the subjects, rather than an individual ANS (autonomous nervous system) specificity in emotion" (Kim and André, 2008). Basically, the intentention to discover the inner trend of signal variation during human's emotional changes yields a fundamental knowledge that the physical properties of all the signals should at least follow the same patterns. Though the exact patterns of emotional changes could be buried by the other distinct patterns brought by non-emotional varieties. Upon such cases, subject-independent (meaning all signals mixed together for different subjects) approach tends to perform worse than subject-dependent (single subject) approach due to the effects of non-emotion-related varieties.

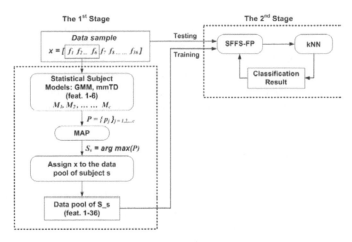

Figure 1 Diagram of the BSB emotion recognition system

In two of our previous publications (Gu et al. 2010(1), Gu et al. 2010(2)), we proposed a 2-stage Biometric Signature Based (BSB) emotion recognition system aiming to solve the problem of "individual varieties" when multiple subjects are involved. The basic idea is to transform a traditional subject-independent problem into several subject-dependent cases (separate the mixed data samples into subject groups based on their biometric characteristics) by conducting an additional subject classification process prior to the usual emotion recognition procedure. During this process, data from each subject are trained into separated statistical subject models. A new incoming sample will be classified into one subject model that best suits its inner structure. In other words, this data sample shows the most similar characteristics to the assigned statistical model. In other words, this data sample shows the most similar characteristics to the assigned statistical model. Then in the succeeding stage, a traditional feature selection plus classification procedure is applied to fulfill the task of emotion recognition. Compare with the usual structure, the BSB system eliminates the influence of the "individual varieties" therefore enhances the recognition performances of the whole system. Prior result shows great improvement compared to conventional subject-independent procedure.

One limitation of the BSB system lies in the fact that it requires each and every subject to train one subject model. On one hand, this structure could increase the complexity of the system when more subjects are involved. On the other hand, some of the subject models can be "useless" during the subject classification process. In other words, it may not necessary for the system to train all the subject models. To further validate the idea, in [8], two types of experiments are designed with close-set (all of the testing data are chosen from the data pool of subjects that build the subject models) and open-set (not all of the testing data are chosen from the data pool of subjects that build the subject models) settings to examine whether it is necessary for the system to learn all the subjects beforehand. The results from the close-set experiment showed that although quite a few data samples are misclassified into other subject models rather than themselves in the 1st stage, the system still performed well (roughly 93% accuracy) in the 2nd stage of emotion recognition. The open-set experiment also received comparable results (roughly 90% accuracy) with using the close-set settings. The results showed that it is not required for the system to prepare statistical models for each individual subject, as long as there are enough representative models known (trained) by the system.

In this paper, we improved the BSB system using an adaptive method and named it as the Self-Adaptive BSB (SABSB) system. The following section presents all the details.

2   METHODOLOGY

The SABSB system is different from the original BSB system in that, instead of directly creating several models based on individual subject, the self adaptive procedure first assumes the whole data pool as one single statistical model, and successively splits the model into two new ones based on their statistical structures, until the model number reaches a pre-defined value. By doing so, the models in the first stage are built in a more flexible way rather than limited to individual subject. The system is implemented using the mixed multivariate T-distribution (mmTD).

Before going into details of the SABSB system, let's first introduce the term, BIC. In statistics, the BIC is a criterion for model selection among a class of parametric models with different numbers of parameters. It has been applied for determining the number of components in a model and for deciding which among two or more partitions most closely matches the data for a given model. The BIC is depicted as:

$$2\log p(x\,|\,M) + const. \approx 2l_M(x,\hat{\theta}) - m_M \log(n) \equiv BIC$$

where $p(x\,|\,M)$ is the (integrated) likelihood of the data for the model M, $l_M(x,\hat{\theta})$ is the maximized mixture log likelihood for the model, and m_M is the number of independent parameters to be estimated in the model. Accordingly, the larger the value of BIC, the stronger the evidence for the model is. The Self-adaptive procedure is described as follows:

Assume $\phi_s = \{\mu_s, \sigma_s, v_s\}$ are the parameters of the sth component.

- Initialization:
  (1) Build an mmTD1 model based on the whole data (one single multivariate T-distribution in the mixture model). Calculate the parameters ϕ_1: μ_1 = sample mean; σ_1 = sample covariance matrix; v_1 = const. (set to 25 in out experiment). Let $\omega_1 = p(\phi_1) = 1$.
  (2) Perform EM learning on the data for modeling as mmTD2 using ϕ_1 as the initial parameters. Return: $\tilde{\phi}_1, \tilde{\phi}_2, p(\tilde{\phi}_1) = \tilde{\omega}_1, p(\tilde{\phi}_2) = \tilde{\omega}_2$.
- Loop:
  Take mmTDk for example, meaning there are k clusters now ($clstN = k$).
  (1) Decide which cluster ($clst$) to split ($splitC$):
  - Calculate Δ_{BIC} for each cluster:
    $$\Delta_{BIC_i} = BIC(mmTD2, clst_i) - BIC(mmTD1, clst_i),$$
    where $i \in [1,k]$.
  - $splitC = \arg\max_i \Delta_{BIC_i}$.

  (2) Perform EM learning on $splitC$ for modeling as mmTD2 using ϕ_{splitC}.
  (3) Update parameters for $splitC$ as
    $$\tilde{\phi}_1, \tilde{\phi}_2,\ \tilde{\omega}_1 = p(\tilde{\phi}_1) * \omega_{splitC}, \tilde{\omega}_2 = p(\tilde{\phi}_2) * \omega_{splitC}.$$
  (4) If $clstN < \max Loop, clstN = clstN + 1$; else end loop.

To better explain the self-adaptive procedure, Figure 2 displays a simulate process by means of a group of randomly generated data using Matlab.

The original data set consists of four clusters of data, shown in Figure 2(a) . Assume the whole data set can be modeled as one single mmTD (mmTD1) using the mean and covariance matrix from the original data set (Figure 2(b)). Split mmTD1 into mmTD2 ($clstN = 2$) using EM learning (Figure 2(c)). Decide the $splitC$ (the cluster to split) by calculating the Δ_{BIC}, and perform EM learning on it (now $clstN = 3$ as shown in Figure 2(d)). Continue the process until reaching a satisfactory result (eventually $clstN = 4$, Figure 2(e)).

628

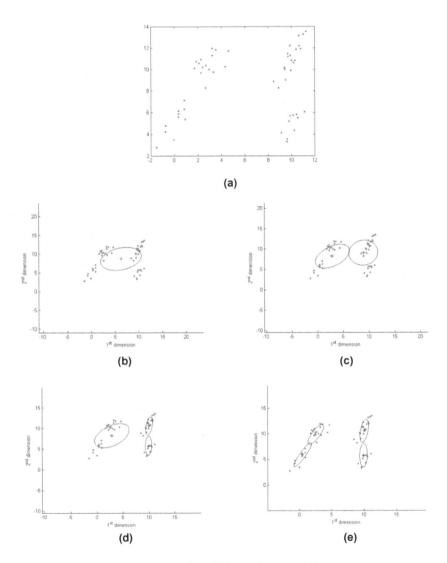

Figure 2 Diagram of the BSB emotion recognition system

3   CLASSIFICATION RESULTS

In this section, the results of using four methods are presented including the conventional architecture, the BSB system implemented by Gaussian Mixture Models (GMM), the BSB system implemented by mmTD, and the SABSB also implemented by mmTD, for a multi-subject emotion recognition task. The experimental data are gathered from two rounds of our experiments. In order to evaluate the consistency of the results, we run the data separately from each experiment.

Collected raw signal data are first trimmed from the beginning of displaying the picture stimulus and ended just after showing the black screen. A series of high and low pass filters are applied to remove noises, also down-sampled by a factor of 8 for further feature evaluation. Instead of directly using the ECG signals, the HR (heart rate) information are deduced from the intervals between successive QRS complexes (the most striking waveform within an ECG signals). Details of the QRS detection can be found in our previous report (Gu et al. 2008(1)). Each 6 physiological signal (HR, BVP, SC, EMGz, EMGc and Rsp) provides 6 features including 1) the mean of the input signal, 2) the standard deviation of the input signal, 3) the mean of the abs (absolute value) of the 1st differences of the input signal, 4) the mean of the abs of the 1st differences of the normalized signal, 5) the means of the abs of the 2nd differences of the input signal and 6) the mean of the abs of the 2nd differences of the normalized signal. In all, there are 36 features prepared for each data entry.

The classification goal is to simultaneously differentiate four areas of the arousal-valence plane: positive &high arousal (EQ1), negative & high arousal (EQ2), negative & low arousal (EQ3), and positive & low arousal (EQ4). The sizes of accordingly divided data set of Experiment I and II are shown in Table 1. Considering the relatively large size of data samples and computational requirement, a 10-fold cross-validation is determined to evaluate the prediction performance of the classification methods. The whole data set is divided into 10 subsets (no overlap upon the validation subsets). For each round, one out of the ten subsets acts as the test set while the other nine subsets are combined together as the training set. The final prediction accuracy is calculated by averaging the results from 10 rounds of tests.

		Arousal		
		Rating 1-5	Rating 6-9	$\sum$
	Rating 1-5	187	294	481
Valence	Rating 6-9	93	210	303
	$\sum$	280	504	784

		Arousal		
		Rating 1-5	Rating 6-9	$\sum$
	Rating 1-5	283	261	544
Valence	Rating 6-9	134	190	324
	$\sum$	714	154	868

(a) Experiment I   (b) Experiment II

Table 1 Data distributions

Figure 3 shows the comparison of results using the general emotion recognition system implemented by SFFS hybrid with kNN (k = 5)(GenSys), the BSB system with GMM (BSB-GM), the BSB system with mmTD (BSB-TD), and also the SABSB implemented by mmTD (SABSB). Figure 3 (a) and (b) respectively depict the results from Experiment I and II. The left graph of each figure draws the curves of recognition accuracies based on each and every fold of cross validation (10-fold therefore 10 units for the X-axis). And the right graph shows the final accuracies corresponding to each recognition method by averaging the results of the 10 cross validations. As expected, both the BSB and the SABSB systems triumph the SFFS-kNN structure over an average 30% accuracies for Experiment I, a bit less but still consistent for Experiment II. The results strongly support the basic idea of the BSB system by separating the mixed multi-subject data samples into individual groups based on their biometric characteristics (6 features from the ECG signal specifically). Also, using the mmTD model performs better than the GMM models. The reason can be lie in the fact that Student's t-distribution (STUDENT, 1908) has been proven to be more robust to real data sets since it has loner tail than traditional Gaussian distributions (Liu and Rubin, 1995).

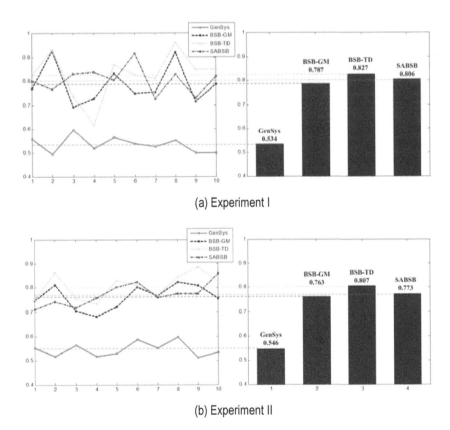

(a) Experiment I

(b) Experiment II

Figure 3 Diagram of the BSB emotion recognition system

Compare the results of BSB-TD and the SABSB, both use the mmTD model with the parameter $\nu = 25$ for all 10 cross validations. Although the SABSB yields a bit less performance to the BSB-TD, we can see from Table 2 that apparently each cross needs the number of subject clusters no more than 16 (Experiment I) and 13 (Experiment II). As we have explained, the BSB system requires exclusively that each subject has to learn its own statistic model for the stage of subject learning, therefore 28 (subjects) for Experiment I and 31 (subjects) for Experiment II. However, the SABSB is able to achieve a relatively comparable results with less constraint in the number of subject models to be built. In other words, the SABSB method can potentially reduce the complexity of the whole system without losing much of the recognition performance.

Validation	1	2	3	4	5	6	7	8	9	10
Experiment I	7	6	6	16	10	16	5	11	6	12
Experiment II	7	5	10	10	8	10	10	13	11	12

Table 2 Number of clusters for each round of validation using a fix $v = 25$

Figure 4 illustrates the performance of the SABSB system in a different way by fixing the number of subject clusters generated. Noted that the parameter v are different for every cross so to ensure the performance of each round of validation. The trend of accuracies shows a positive proportion to the number of clusters. It is quite understandable that suppose in an extreme case, each data entry could form its own model, the learning process would be extremely exclusive to the individuals. However, the more realistic way is to set up a reasonable boundary, which is not too low so to assure the recognition accuracy, but also not too high so to limit the computational expenses.

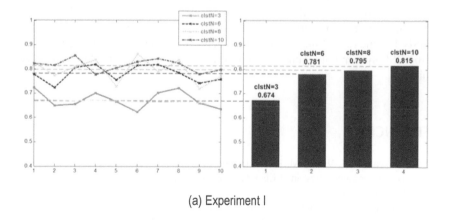

(a) Experiment I

632

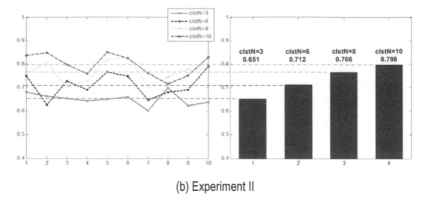

(b) Experiment II

Figure 4   Comparison of results with different number of clusters generated by the SABSB

4   CONCLUSION

This paper introduced the SABSB system as an improved version of the BSB system. Instead of directly creating several models based on individual subject, the self adaptive procedure first assumes the whole data pool as one single statistical model, and then successively splits the model into two new ones, until the model number reaches a pre-defined value. Hence, the total number of statistical models in the first stage is reduced, also the models are generated in a more adaptive and flexible manner. By comparing SABSB with conventional SFFS-kNN, BSB-GM and BSB-TD, the results show that SABSB achieves a relatively comparable results with the BSB system but requires less constraint in the number of subject models to be built.

REFERENCES

Gu, Y., S. L. Tan, K. J. Wong, M. H. R. Ho, and L. Qu, 2008. Emotion aware technologies for consumer electronics, in IEEE International Symposium on Consumer Electronics, Portugal, pp. 1–4.

Gu, Y., S. L. Tan, K. J. Wong, M. H. R. Ho, and L. Qu, 2008. Using GA-based feature selection for emotion recognition from physiological signals, in International Symposium on Intelligent Signal Processing and Communication Systems, Thailand, 2008, pp. 1–4.

Gu, Y., S. L. Tan, K. J. Wong, M. H. R. Ho, and L. Qu, 2010. A GMM based 2-stage architecture for multi-subject emotion recognition using physiological responses, in Proceedings of the 1st Augmented Human International Conference, no. 3, France, pp. 1–6.

Gu, Y., S. L. Tan, K. J. Wong, M. H. R. Ho, and L. Qu, 2010. A Biometric Signature based System for Improved Emotion Recognition using Physiological Responses from Multiple Subjects, in 8th IEEE International Conference on Industrial Informatics.

Haag, A., S. Goronzy, P. Schaich, and J. Williams, 2004. Emotion recognition using biosensors: first step towards an automatic system, in Affective Dialogue Systems, Tutorial and Research Workshop, Kloster Irsee, Germany, pp. 36–48.

Kim, J. and E. André, 2008. Emotion recognition based on physiological changes in music listening, IEEE Transactions on Pattern Analysis and Machine Intelligence, vol. 30, no. 12, pp. 2067–2083.

Liu, C. and D. B. Rubin, 1995. Ml estimate of the t distribution using em and its extensions, ecm and ecme, Statistica Sinica, vol. 5, pp. 19–39.

Nasoz, F., K. Alvarez, C. L. Lisetti, and N. Finkelstein, 2003. Emotion recognition from physiological signals for presence technologies, International Journal of Cognition, Technology and Work, Special Issue on Presence, vol. 6, no. 1.

Picard, R.W. Affective Computing, 1995. MIT Media Laboratory Perceptual Computing Section Technical Report No. 321.

Picard, R.W., E. Vyzas, and J. Healey, 2001. Toward machine emotional intelligence: Analysis of affective physiological state, IEEE Transactions on Pattern Analysis and Machine Intelligence, vol. 23, no. 10, pp. 1175–1191.

STUDENT, 1908. The probable error of a mean, Biometrika, vol. 6, no. 1, pp. 1–25.

Wagner, J., J. Kim, and E. André, 2005. From physiological signals to emotions: implementing and comparing selected methods for feature extraction and classification, in Proceedings Of IEEE ICME International Conference on Multimedia and Expo, pp. 940–943.

CHAPTER 68

Inferring Prosody from Facial Cues for EMG-based Synthesis of Silent Speech

Christian Johner, Matthias Janke, Michael Wand, Tanja Schultz

Cognitive Systems Laboratory
Karlsruhe Institute of Technology, Germany
christian.johner@student.kit.edu,
{matthias.janke, michael.wand, tanja.schultz}@kit.edu

ABSTRACT

In this paper we introduce a system which is able to detect prosodic elements in a spoken utterance based on signals from the facial muscles. The proposed system can augment our surface electromyography (EMG) based *Silent Speech Interface* in order to make synthesized speech more natural. Having shown in (Nakamura, Janke, Wand, & Schultz, 2011) that it is possible to produce understandable synthesized speech from EMG signals, our current interest is to improve the quality and expressivity of the synthesis.

We show that a standard phonetically balanced German speech corpus with only a few additional utterances is sufficient to train a system that can discriminate yes/no questions from normal speech and also distinguish between normal and emphasized words in an utterance.

For the detection of prosodic information in facial muscle movement we extend our EMG based speech synthesis system with two additional EMG channels, recording the movements of the facial muscles *musculus corrungator* and *musculus frontalis*. Our classification method uses a frame-based SVM classification, followed by a majority vote to classify a whole word.

Our system achieves F-scores of up to 0.68 for the recognition of emphasized words and 1.0 for the classification between questions and normal utterances although the results show large variations depending on the feature combination used for training.

Keywords: EMG, synthesis, prosody, speech recognition

1   INTRODUCTION

Speech is the most convenient and natural way for humans to communicate. Beyond face-to-face talk, mobile phone technology and speech-based electronic devices have made speech a wide-range, ubiquitous means of communication. Unfortunately, voice-driven communication systems suffer from several challenges which arise from the fact that the speech needs to be clearly audible and cannot be masked: first, understanding the speech becomes very difficult in the presence of noise, for both humans and computers. Second, confidential communication in public places is difficult if not impossible, and even if privacy is not an issue, the audible speech frequently disturbs bystanders. Third, speech-disabled people may be unable to talk to other persons or to use voice-controlled systems.

These challenges may be alleviated by *Silent Speech Interfaces*, which are systems enabling speech communication to take place without the necessity of emitting an audible acoustic signal, or when an acoustic signal is unavailable (Denby, Schultz, Honda, Hueber, & Gilbert, 2010). Over the past few years, we have developed a Silent Speech Interface (Schultz & Wand, 2010) based on surface electromyography (EMG). This technique captures articulatory activity related to speech production from the speaker's face using surface electrodes. Rather than recording acoustics, EMG captures muscle activity and therefore does not require any kind of acoustic signal: The speaker can speak *silently*, which means that the words are mouthed without any sound.

In this paper we report on using this technique to directly synthesize speech based on the electromyographic signals. This approach is particularly suited to human-human communication. However, speech carries more information than the pure meaning of words. Prosodic information codes whether an utterance should be understood as an urgent request, a question or a mundane statement. Current text-to-speech systems are capable of modulating the produced voice to match different sentence types based on prosodic annotations (Silverman, et al., 1992). The goal of our work is to empower our EMG-based speech synthesis system with the same capabilities – i.e. speech synthesized from EMG signals should exhibit varying intonation, emphasis, and prosody, reflecting the speaker's intentions.

In (Toth, Wand, & Schultz, 2009) we showed that understandable speech may be synthesized from electromyographic signals, using a voice conversion technique. However, these initial systems suffered from an unnatural quality of the synthesized voice. This paper draws on (Nakamura, Janke, Wand, & Schultz, 2011), where we showed for the first time that F0 contours may by recognized from EMG signals, and that this information may be incorporated into an EMG-to-speech system. Beyond the F0 contour generation, in this paper we focus on two aspects of prosody which we perceive to be of utmost importance for conveying meaning by speech: Firstly, we recognize emphasized words from EMG signals of spoken utterances, and second, we distinguish questions from normal statements.

In this work we only report results on recordings of audibly spoken speech. For our experiments we extend the setup used by (Toth, Wand, & Schultz, 2009) with two additional electrodes placed on the *musculus corrungator* and the *musculus frontalis*, using a total of seven EMG channels.

This paper is organized as follows: In Section 2 we describe our experimental setup, the data corpus, and EMG feature extraction and SVM training setup. Section 3 details our experiments for question and emphasis classification. In Section 4, we draw conclusions and outline future work.

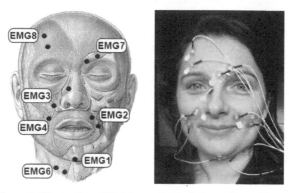

Figure 1: Illustration of facial muscles and electrode positions

2 EXPERIMENTAL SETUP

2.1 Data Acquisition and Corpus

Our data corpus is based on the Berliner Sätze (Sotschek, 1984), which is a phonetically balanced German speech corpus containing 102 utterances. We extended this corpus with 13 additional utterances to be realized as normal statements, 35 utterances with one emphasized word and 20 questions, both of which were recorded twice, and 10 exclamations, which were not used for this study, so that each subject, in each recording session, recorded 235 sentences. The sentences beside the Berliner Sätze corpus are not phonetically balanced.

The EMG signals were recorded with 14 Ag/Ag-Cl surface electrodes attached to the skin, as depicted in figure 1. The 14 electrodes were positioned in order to pick up the signals of major articulatory muscles: the *levator angulis oris* (EMG2,3), the *zygomaticus major* (EMG2,3), the *platysma* (EMG4), the *anterior belly of the gastric* (EMG1), the *tongue* (EMG1,6) and relevant facial muscles: *corrungator* (EMG7) and *frontalis* (EMG8). Channels 2, 6, 7, and 8 used bipolar derivation, whereas channels 1, 3, and 4 were unipolarly derived, with the reference attached to either the nose (EMG1) or to both ears (EMG3,4). Note that the electrode positioning follows (Maier-Hein, Metze, Schultz, & Waibel, 2005) for the articulatory (lower facial) muscles and (Pruzinec, 2010) for the upper facial muscles. EMG channel 5 remains unused.

EMG data was recorded with a multi-channel EMG recording system (Varioport, Becker Meditec, Germany). EMG responses were differentially amplified, filtered by a 300 Hz low-pass and a 1Hz high-pass filter and sampled at 600 Hz.

We recorded seven sessions with 4 different speakers (three male, one female) all of them native Germans. In order to avoid inconsistencies due to slightly different electrode positioning and speaker properties, we trained *session-dependent systems*. Since all male speaker recorded two sessions, we nonetheless have a means of asserting that classifier settings remain consistent across recording sessions.

The recording protocol was as follows:

In a quiet room, the speaker read German sentences in normal audible speech, which were recorded with a parallel setup of an EMG recorder and a USB soundcard with a standard close-talking microphone attached to it. An analog marker signal was used for synchronizing the EMG and the speech signals.

We gave each speaker a short introduction about the purpose of the recording and encouraged them to be as natural during the recording as possible. It should be noted that none of the subjects was a professional actor or speaker.

2.2   EMG Feature Extraction

The feature extraction is based on time-domain features (Jou, Schultz, Walliczek, Kraft, & Waibel, 2006). Here, for any given feature **f**, M_f is its frame-based time-domain mean, P_f is its frame-based power, and z_f is its frame-based zero crossing rate. $S(\mathbf{f}, n)$ is the stacking of adjacent frames of feature **f** in the size of $2n+1$ (-n to n) frames.

For an EMG signal with normalized mean x[n], the *nine-point double-averaged* signal w[k] is defined as:

$$w[n] = \frac{1}{9}\sum_{k=-4}^{4} v[n+k], \text{ where } \quad v[n] = \frac{1}{9}\sum_{k=-4}^{4} x[n+k].$$

The high-frequency signal is p[n] = x[n] - w[n], and we define the rectified high-frequency signal r[n] = |p[n]|. The final base feature is:

$$TDn = S(f_2, n), \text{ where } \quad f_2 = \left[\overline{w}, P_w, P_r, z_p, \overline{r}\right]$$

We evaluated the final feature TDn with a stacking width of n = 1, 3, 7, 10 and 15. Frame size and frame shift were set to 27ms respectively 10ms. The feature extraction was applied for each EMG channel separately. In contrast to (Schultz & Wand, 2010) we did not apply any dimension reduction.

2.3   SVM Training

The experiments described in this paper require a word segmentation of the data to function properly. For this study, we obtained these word alignments by forced-aligning the simultaneously recorded audio data with a German speech recognizer trained with the Janus Recognition Toolkit (JRTK). We note that such time-alignments may be obtained just as well based on purely EMG data (Schultz & Wand, 2010).

638

For the training we split the set of recorded utterances into a training set and a test set. These sets were formed separately for the two recognition tasks.

Yes/no-questions have a significant different prosody compared to normal statements. Other types of questions do not necessarily use the same prosody. From

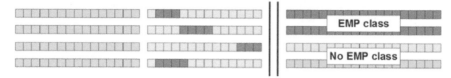

Figure 2: Illustration of training set generation. Each bar represents an utterance and each cube a frame. The red frames align with an emphasized word. The red frames were used as training data for the SVM for the emphasis-class (EMP class) and the blue and green ones for the no-emphasis-class (No EMP class)

the 40 questions in our corpus only 20 are yes/no-questions. We split this set into 14 questions for training and six questions for testing. From the remaining utterances we selected 16 sentences for the training. For testing we used the six yes/no questions and eight sentences from the normal utterances. Using only the last word of each utterance, this results in 30 words for training (14 questions/16 normal) and 14 words for testing (6 questions/8 normal).

For the training of the emphasis detector, we took 46 emphasized words and 36 normal words from the emphasized sentences, and 15 sentences containing of 81 words from the remainder corpus. For testing we used the remaining 24 sentences of the emphasized and 15 from the remaining corpus with a total of 226 words (24 emphasized/202 normal).

As described in 2.2, all our data is divided into frames with a frame shift of 10 ms. Every frame represents a data point for one of the two classes. We now train an SVM (Support Vector Machine) to perform a *frame-based* classification of the two classes in question, i.e. to discern questions from normal utterances or emphasized words from normally pronounced words. Figure 2 depicts, as an example, how the training data for the emphasis classifier is constructed, and Table 1 shows the number of training frames for each session. The results of the SVM classification are used as an input for a majority vote on a word base as described in Section 3.

Table 1: Number of frames for each of the seven sessions and classification tasks. Each frame has a duration of 10ms

	spk1-ses1	spk2-ses1	spk3-ses1	spk4-ses1	spk1-ses2	spk2-ses2	spk3-ses2
Emphasis	5500	5800	8366	5999	4748	5581	8396
Question	1972	2118	2375	2052	1902	1998	1858

3   EXPERIMENTS

We examined a large number of different parameter settings to find the best values for the classification of each task. For EMG-based speech recognition (Schultz and Wand, 2010) TD15 features have been shown to give good results. (Pruzinec and Schultz 2011) examined TD0 to TD5 features for EMG-based facial expression classification. We initially studied TD1, TD3, TD7, TD10 and TD15 features, but discarded TD1 and TD3 due to the poor results on the first classifications. We used all seven EMG channels for classification, but also investigated the impact of reducing the set of channels.

Our classifier is a combination of a SVM which was trained with frames for each class and a suffixed majority vote. We use the MATLAB 2011b SVM implementation with a radial basis kernel. As parameters we examine every combination of *sigma* in $\{1, 20, 50\}$ and *boxconstraint* in $\{1, 10\}$ and allow a violation of the Karush-Kuhn-Tucker condition of 5%. For convenience, parameter combinations of sigma and boxconstraint will be written in the remainder of this paper e.g. as s10-b1.

For the emphasis classifier every frame of every non-silent word in the test set the SVM returns a probability to which class the frame belongs. We sum the class probabilities and divide the result with the number of frames for a word. This results in a value between $[0,1)$ which can be understood as the probability that a word is emphasized. We examine different thresholds above which a word is classified as emphasized. For each of these values or thresholds we count the number of *true positives* (TP) and *true negatives* (TN) and use the F-score (F1-score to be accurate) as an evaluation criteria.

For the question classifier our approach works analogously, but we only test the frames of the last word. We consider this a reasonable simplification because after a question the speaker normally stops talking, waiting for an answer or giving a listener a short pause to think about the said.

We did all the experiments using the first session of every speaker. The three additional sessions of the male speakers are used as an evaluation set.

3.1   Question Classification

Our test set contains 14 words, 6 from questions and 8 from normal utterances. The question words will be regarded as true positives and the normal words as true negatives for the purpose of F-Score calculation.

In our first series of experiments we investigate the performance of different SVM training parameters and TDn features using a stacking of all seven EMG channels for the final feature vector. As shown in Table 2, the classification achieves a very good overall average F-score of 0.87. First, these experiments indicate that SVM parameter combination s50-b1 and s50-b10 lead to slightly better results than the other parameters. Not shown in the table are the results for the combination s1-b1 and s1-b10, which did not lead to a classification better than chance. This behavior is consistent throughout the experiments. Second, the results indicate that prosodic information can be extracted by using the TD7 and TD10 features which achieved 7 respectively 5 percentage points more than TD15.

Table 2: Question classification - Average F-Scores of the test sessions separated for different TDn features and SVM parameters using all seven EMG channels

	s10-b1	s10-b10	s50-b1	s50-b10	AVG
TD7	0.88	0.90	0.92	0.92	0.90
TD10	0.84	0.85	0.92	0.89	0.88
TD15	0.76	0.76	0.92	0.87	0.83
AVG	0.83	0.84	0.92	0.90	0.87

We now examined the impact of different thresholds. As explained in 3 each word will be assigned a value between zero and one. Depending on the threshold the word will finally be labeled as part of one of the two classes. As it can be seen in Table 3 best results are achieved with a threshold around 0.5. This is encouraging because it allows the conclusion that the SVM training results in good classifications for the frames.

Table 3: Question classification – Average F-scores over all test sessions with different thresholds, SVM parameters s50-b1 and TD10 features using all seven EMG channels

0.6	0.55	0.5	0.45	0.4	AVG
0.79	0.87	0.89	0.89	0.86	0.86

As demonstrated by (Wand & Janke, 2011) the EMG signals of the articulatory muscles differ strongly when changing the speaking mode from audible to silence. Achieving good classification results with the EMG channels from the non-articulatory muscles (EMG channels 7 and 8) would be desirable. With SVM parameters s50-b1 and TD10 features, the best average F-score when using only EMG channels 7 and 8 is achieved for a threshold of 0.4. The F-score for this parameter combination is 0.69, which is a difference of 0.17 compared to the best results with using EMG channels 1 to 8. Results improve when using SVM parameters s10-b10 to an average F-score of 0.77.

This indicates that using EMG channels 7 and 8 is not sufficient to get high recognition rates. Speaker 3 performs extremely poor using this EMG channel combination with a best F-score of 0.67. Speaker 4, on the other hand, achieves a perfect F-score of 1.0 for the parameter combination s10-b10, threshold 0.5 and 0.55 and TD7 feature.

Table 4: Question classification - F-scores for test and evaluation set of TD10 feature with threshold 0.45 and EMG channels 2, 6, and 8 (upper table) and all seven EMG channels (lower table)

	spk1-ses1	spk2-ses1	spk3-ses1	spk4-ses1	AVG	spk1-ses2	spk2-ses2	spk3-ses2	AVG
s50-b10	0.85	0.92	0.8	0.92	0.87	0.73	0.92	0.86	0.84
s50-b1	0.86	0.91	0.86	0.92	0.89	0.73	0.73	0.73	0.73

	spk1-ses1	spk2-ses1	spk3-ses1	spk4-ses1	AVG	spk1-ses2	spk2-ses2	spk3-ses2	AVG
s50-b10	0.77	0.92	0.91	0.83	0.86	0.77	0.86	0.8	0.81
s50-b1	0.77	0.83	0.83	0.83	0.82	0.73	0.83	1.0	0.85

To find a good parameter combination for all speakers, we examined in a last experiment all EMG channel combinations with up to four different channels. Using as a constraint that one of the EMG channels should be either EMG7 or EMG8, the combination with EMG channels 2, 6, and 8 showed the most promising results. The best average F-Score for the four sessions was 0.89 with SVM parameters s50-b1 and 0.87 for s50-b10 both times with TD10 features and a threshold of 0.45, which were indicated by the previous experiment with seven channels to provide good results. We evaluate these parameters with our three unused sessions. As shown in Table 4 the average F-score for SVM parameter s50-b10 of the three sessions is nearly as good as the one with the four test sessions.

3.2 Word Emphasis Classification

The classification of emphasized words in an utterance showed to be a challenging task. We have 24 emphasized words as possible true positives and 202 words as true negatives. It should be noted that this discrepancy will normally lead to low F-Scores.

Table 5: Emphasis classification - Average F-scores for stacking all seven EMG channels

	s10-b1	s10-b10	s50-b1	s50-b10	AVG
TD07	0.45	0.45	0.46	0.46	0.45
TD10	0.38	0.41	0.45	0.47	0.43
TD15	0.37	0.35	0.47	0.5	0.42

We approached the task in the same way we did for classifying questions. First we investigated the results for EMG stacking of all seven EMG channels. TD1 and TD3 showed bad results from the start and were discarded. The behavior of our four test sessions was not consistent. Speaker 3 performed very poorly on TD10 with a maximum F-score of 0.3. Using only the other three sessions, best average F-scores were achieved with s50-b10 as can be seen in Table 5 and TD15 features.

Table 6: Emphasis classification - F-scores for EMG channels 7 and 8 with threshold 0.5, TD15 features

	spk1-ses1	spk2-ses1	spk3-ses1	spk4-ses1	AVG
s50-b1	0.4	0.38	0.25	0.33	0.34
s50-b10	0.42	0.32	0.23	0.38	0.34

EMG channels 7 and 8 showed lower performance than the full channel set, as for the question classification. The average F-score decreases to 0.34 for the best parameter combination, as shown in Table 6. Again Speaker 3 performs poorly.

We did not obtain a channel/parameter combination with good classification rates over all sessions. As can be seen in Table 7, the parameters for the best F-score results with TD15 features vary for each session. Compared to the best results with seven channels the sessions gained between 0.01 for session 1 of Speaker 2 and 0.15 for session 1 of Speaker 4. A positive conclusion, however, is that for all best results either EMG7 or EMG8 are involved, which indicates that we are on the right track.

Table 7: Emphasis classification - Best F-scores for each session with TD15 features and up to four different EMG channels

	spk1-ses1	spk2-ses1	spk3-ses1	spk4-ses1	spk1-ses2	spk2-ses2	spk3-ses2
Channels	2-6-8	6-8	1-3-8	1-5-7	1-2-3-7	1-6-7-8	1-2-7-8
Parameters	s50-b10	s10-b10	s10-b1	s50-b1	s50-b1	s10-b1	s50-b10
Threshold	0.5	0.5	0.6	0.55	0.55	0.55	0.5
F-Score	0.67	0.44	0.42	0.64	0.68	0.55	0.4

4   CONCLUSION AND FUTURE WORK

We showed that it is possible to detect prosodic information in EMG signals. Our approach achieves high classification rates for yes/no questions, and we showed that for this task, the optimal parameter combination remained stable across different speakers and sessions. On the evaluation set, the average F-score is 0.86.

The detection of emphasized words in a complete sentence showed to be a somewhat more challenging task. The best F-score was achieved for Speaker 1 with 0.68, but large variations over different speakers could be noticed.

Having a large discrepancy between true positives (#24) and true negatives (#202), a weak classifier improving the ratio could yield significantly higher recognition rates. A quick examination of our results with low thresholds showed that while preserving over 20 of the true positives, more than 50 percent of the true negatives could be discarded. The best result achieved was for session 2 of Speaker 2. For a specific parameter combination 24 true positives and 124 true negatives could be investigated. This is a reduction of the true negatives of over 60 percent. Defining a classifier for the remaining data could be something worth researching.

Both classifications showed that using only EMG channels 7 and 8 for classification is not enough to get good results. Our future work includes reevaluating the results of this paper on EMG signals of *silent* speech. EMG channels 7 and 8 may become more important for silent speech EMG.

It should be stated that both classifiers still need some kind of labeling process providing the word boundaries even if used with EMG-based voice conversion. To get rid of the necessity of word boundaries is an issue for further research.

REFERENCES

Black, A., & Lenzo, K. (2000). Building voices in the Festival speech synthesis system. http://festvox.org/bsv/.

Chan, A., Englehart, K., Hudgins, B., & Lovely, D. (2001). Myoelectric Signals to Augment Speech Recognition. *Medical and Biological Engineering and Computing , 39*, 500-506.

Denby, B., Schultz, T., Honda, K., Hueber, T., & Gilbert, J. (2010). Silent Speech Interfaces. *Speech Communication , 52* (4), 270-287.

Dupont, S., & Luettin, J. (2000). Audio-Visual Speech Modeling for Continuous Speech Recognition. *IEEE Transactions on Multimedia , 2*, 141-151.

Eide, E., Aaron, A., Bakis, R., Hamza, W., Picheny, M., & Pitrelli, J. (2004). A corpus-based approach to <AHEM/> expressive speech synthesis. *5th ISCA ITRW on Speech Synthesis*, (pp. 79-84).

Ekman, P., & Frisen, W. (1978). The facial action coding system (FACS): A technique for the measurement of facial action. *Consulton Psychologists Press* .

Grice, M., & Baumann, S. Deutsche Intonation und GToBi.

Jou, S.-C., Schultz, T., Walliczek, M., Kraft, F., & Waibel, A. (2006). Towards Continuous Speech Recognition Using Surface Electromyography. *Proc. Interspeech*, (pp. 341-353).

Maier-Hein, L., Metze, F., Schultz, T., & Waibel, A. (2005). Session-independent non-audible speech recognition using surface electromyography. *Proc. ASRU*, (pp. 331-336).

Nakamura, K., Janke, M., Wand, M., & Schultz, T. (2011). Estimation of fundamental frequency from surface electromyographic data: EMG-to-F0. *Proc. ICASSP*, (pp. 573-576).

Pruzinec, M. (2010). Facial Expression Recognition using Surface Electromyography. *Diploma Thesis, Karlsruhe Institute of Technology* .

Schultz, T., & Wand, M. (2010). Modeling coarticulation in large-vocabulary EMG-based speech recognition. *Speech Communication , 52* (4), 341-353.

Silverman, K., Beckman, M., Pierrehumbert, J., Ostendorf, M., Wightman, C., Price, P., et al. (1992). ToBI: A Standard Scheme for Labeling Prosody. International Conference of Spoken Language. *Proc. of International Conference of Spoken Language*, (pp. 867-869).

Sotschek, J. (1984). Sätze für Sprachgütemessung und ihre phonologische Anpassung an die Deutsche Sprache. *Tagungsband DAGA: Fortschritte der Akustik*, (pp. 873-876).

Toth, A., Wand, M., & Schultz, T. (2009). Synthesizing Speech from Electroyography using Voice Transformation Techniques. *Proc. Interspeech.*

Walliczek, M., Kraft, F., Jou, S.-C., Schultz, T., & Waibel, A. (2006). Sub-Word Unit based Non-Audible Speech Recognition using Surface Electromyography. *Proc. Interspeech.*

Wand, M., & Schultz, T. (2011). Session-Independent EMG-based Speech Recognition. *Proc. Biosignals.*

CHAPTER 69

Multi-Modal Classifier-Fusion for the Classification of Emotional States in WOZ Scenarios

Martin Schels[1], Michael Glodek[1], Sascha Meudt[1], Miriam Schmidt[1], David Hrabal[1], Ronald Böck[2], Steffen Walter[1], Friedhelm Schwenker[1]

[1]University of Ulm
Ulm, Germany
{firstname.lastname}@uni-ulm.de

[2]University of Magdeburg
Magdeburg, Germany
ronald.boeck@ovgu.de

ABSTRACT

Learning from multiple sources is an important field of research in many applications. Amongst of the benefits of such an approach is that different sources can correct each other or that a failure of a channel can be easier compensated. The emotion of a subject can give helpful cues for a computer in a human machine dialog. The problem of emotion recognition is inherently multimodal. The most intuitive way of inferring a user state is to use facial expression and spoken utterances. However, bio-physiological readings can be helpful in this context. In this study, a novel information fusion architecture for the classification of human emotions in a computer interaction is proposed. We use information from the three modalities above mentioned. It turned out that the combination of different sources can be helpful for the classification. Also, a reject option for the classifiers is evaluated and yields promising results.

Keywords: multi-modal emotion recognition, multiple classifier systems

INTRODUCTION AND MOTIVATION

Multi classifier systems (MCS) aim at improving the robustness and overall classifier performance by fusing the results of different classifiers. According to Kuncheva et al. such a classifier system should be capable to outperform the single classifier performs in case they are accurate and diverse (Kuncheva, 2004). While the requirement to the classifiers to be as accurate as possible is obvious, diversity roughly means that classifiers should not agree on the set of misclassified data. Various modalities and feature views can be utilized on the data to achieve such a set of diverse and accurate classifiers.

Research in affective computing has made many achievements in the recent time. Emotions begin to play an important role in the field of human-computer interaction, allowing the user to interact with the system more efficiently (Picard R. , Affective computing: challenges, 2003) and in a more natural way (Sebe, Lew, Sun, Cohen, Gevers, & Huang, 2007). Such a system must be able to recognize the users' emotional state which can be done by analyzing the facial expression (Ekman & Friesen, 1978), taking the body posture and the gestures into account (Scherer S. , Glodek, Schwenker, Campbell, & Palm, 2012) and by investigating the paralinguistic information hidden in the speech (Sato & Morishima, 1996; Schuller, Rigoll, & Lang, Hidden Markov model-based speech emotion recognition, 2003). Furthermore, bio-physiological channels can provide valuable information to conclude to the affective state (Cannon, 1927; Schachter, 1964).

However, the emotions investigated so far are in general acted and the larger part of research has been focused on a single distinct modalities, albeit the problem of emotion recognition is inherently multimodal. Obviously, the entire emotional state of an individual is expressed and can be observed in various different modalities, e.g. through facial expressions, speech, prosody, body movement, hand gestures as well as more internal signals such as heart rate, skin conductance, respiration, electroencephalography (EEG) or electromyogram (EMG). Recent developments aim at transfering the insights already made to more natural settings using multiple modalities (Chen & Huang, 2000; Caridakis, et al., 2007). The uncontrolled recording settings of non-acted data and the manifold of modalities make emotion recognition a challenging task: subjects are less restricted in their behaviour and the emotional ground truth is more difficult to determine.

In our work the emotion recognition problem is considered as a multimodal pattern recognition problem, including bio-physiological features, prosodic features, and high-level facial expression features. A MCS has been developed, in which for every channel independent classifiers are trained. However, the proposed approach extends the classical techniques by allowing the rejection of classifier results in case confidences are to low and additionally fuses the remaining results temporally to increase the overall robustness. The fused outputs of all channels are then combined using late-fusion. The proposed architecture is studied on recordings obtained from a Wizard-of-Oz set-up and makes uses of audio, video and physiological channels.

The remainder of this work is structured as follows. In Section 1 we introduce the proposed multiple classifier architecture. The experiment used to validate the model and the features extracted are described in Section 2. The proposed classifier system is evaluated in Section 3 and discussed in Section 4.

METHODOLOGY

In the following, the basic concepts of the proposed method for the classification from multiple sources are described. Further, the proposed method is introduced.

Base Classifiers

In our architecture we propose the utilization of two classifiers, namely a linear perceptron and a Support Vector Machine (SVM). In machine learning one of the most basic classifiers is the perceptron, which is implemented by a linear combination of the feature sets to the labels. Since the function to be learned is extremely simple the method is applicable for very noisy data, since an over-fitting is not possible. We obtained the mapping by directly computing the Moore-Penrose pseudo-inverse function. The Support Vector Machine (SVM) is a supervised learning method following the maximum margin paradigm. The kernel trick increases the dimensionality of the feature space and therefore allows non-linear hyper-planes. Within our study we used the Gaussian Radial Basis Function (RBF) kernel, which transforms the input data into the Hilbert space of infinite dimensions and is calibrated by the parameter γ. However, due to noise or wrong annotations it is convenient to have a non-rigid hyper-plane, being less sensitive to outliers in training. Therefore, an extension to the SVM introduces a so-called slack term that tolerates the amount of misclassified data using the control parameter C. A probabilistic classification output can be obtained using the method proposed by Platt et al. (Platt, 1999). Detailed information of the algorithm can be found for instance in (Bishop, 2006).

Information Fusion using Multiple Sources

In order to solve a classification problem, it is possible to create a single classifier that has all available information at hand. However, it turned out, that creating multiple individual classifiers and to combine them appropriately, is a very successful strategy. This approach is called multiple classifier system (MCS) (Kuncheva, 2004). The performance of a MCS not only depends on individual accuracy of the individual classifiers, but also on the diversity of the classifiers, which roughly means that classifiers should not agree on the set of misclassified data. MCS are highly efficient multimodal pattern recognizers that have been studied by various numerical experiments and mathematical treatment, and led to numerous practical applications such as action recognition (Glodek, Bigalke, Schels, & Schwenker, 2011), EEG analysis (Schels, Scherer, Glodek, Kestler, Palm, & Schwenker, 2011) and classification of bioacoustic signals (Dietrich, Palm, & Schwenker, 2003) to mention just a few of them. There are different techniques in the literature to attain diverse classifier teams: The individual classifiers can be trained on different subsets of the training data(Breiman, 1996). Another way is to conduct multiple training runs on the data using different base models or different configurations of a model (model averaging). Furthermore, different subsets of the available feature space (so called feature views) are often used to construct individual classifiers. This somehow emulates the situation where truly independent data streams, e.g. from different physical sensors are at hand.

Such a situation has to deal with several issues: Different modalities often imply different sample rates of the respective sensors. Also, in the real world it is not certain that all the sensors are available at every time.

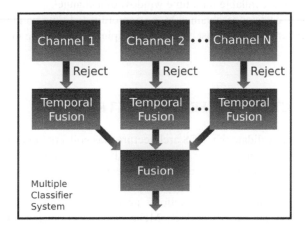

Figure 1 Proposed Multiple Classifier Architecture for emotion recognition. The classification result of each channel has to pass a rejection step, in which decisions with low confidences are filtered out. The outcome is temporally fused on combined to form the final class decision.

Classifcation with Reject Options

Since it can be assumed that the automatic recognition of emotional states is difficult task, the MCS has to deal with weak classifiers. We used confidences of the classifier outputs estimated on some predefined time windows to reject uncertain decisions. Therefore, the rejection is about deciding (yes or no) whether a certain confidence level has been achieved or not. Various attempts have been made to introduce confidence-based rejection criteria. Commonly threshold based heuristics are used on probabilistic classifier outputs utilizing a distinct uncertainty calculus, for instance doubt and conflict values computed through Dempster's Rule of Combination in the very well known Dempster-Shafer theory of evidence (Thiel, Schwenker, & Palm, 2005).

Proposed Algorithm

These concepts already introduced are used to propose a new architecture, which is capable to classify from multiple channels having potentially noisy signals and to take the failure of one or more channels into account. The architecture, illustrated in Figure 1, treats every available input channel separately and classifies on a single feature vector. The output of each classifier is monitored and rejected in case the confidence fall below a certain threshold. Since the remaining results are often still relatively weak and in order to make a decision for the time segments of the rejected samples, a temporal integration over a set of individual frames is conducted. Due to the rejection and failure of channels it is possible that no classification result is available for a time window of a channel.

The final class decision is obtained by a superordinated time window (Dietrich, Palm, & Schwenker, 2003)the least common multiple number of the individual time windows.

DATA COLLECTION AND FEATURE EXTRACTION

The data used to validate the MCS architecture was collected in a Wizard-of-Oz experiment where human-computer interaction (HCI) is simulated (Kelley, 1984). Within the study, the computer interacts as a mental trainer of the popular game "Concentration" and the subjects are able to control the system using short speech commands. The set-up induces emotions according to the Valence-Arousal-Dominance (VAD) model (Russell & Mehrabian, 1977) using the following affective factors:

- Delaying the response of a command
- Non-execution of the command
- Simulate incorrect speech recognition
- Offer of technical assistance
- Lack of technical assistance
- Propose to quit the game ahead of time
- Positive feedback

The procedure of emotion induction is structured in differentiated experimental sequences (ES) in which the user is passed through VAD octants by the investigator. Within this study we focus on the recognition of the emotional octants in ES-4 and ES-6 ("*positive valence, low arousal, high dominance*" versus "*negative valence, high arousal, low dominance*"). The database used comprised of 8 subjects with an average age of 63.5 years.

Audio, video and physiological data (namely EMG, blood volume pulse, respiration and skin conductance level) was recorded. In the following the features extracted from the respective channels are explained.

Audio channel classification

In order to extract the speech non-speech segments from the recorded audio, an energy-based threshold was defined. The energy was determined using a window of 40ms. From this signal, Mel-Frequency Cepstral Coefficients (MFCC) and Modulation Spectrum (ModSpec) features were calculated(Scherer S. , Glodek, Schwenker, Campbell, & Palm, 2012). The MFCC were calculated for two configurations using 40 and 200 milliseconds. The ModSpec features were solely extracted using 200 milliseconds windows. For all features the windows are shifted with an offset of half the respective window size.

Video channel classification

The video channel was recorded using a frame rate of 25Hz and a resolution of 320x240. The camera was directly positioned in front of the subject. We used the Computer Expression Recognition Toolbox (CERT) to extract the following values as features for the further processing: "Basic Emotions 4.4.3", "FACS 4.4", "Unilaterals" and "Smile Detector" (Littlewort, et al., 2011).

Physiological channels classification

The physiological signals were acquired using a NEXUS-32 polygraph, a flexible 32 channel monitoring system. Five physiological channels were recorded: the electromyogram (EMG) of the corrugator supercilii and zygomaticus major, and the skin conductance (SC). In general, a slow low- or band-pass filter is applied together with a linear piece-wise detrend of the time series at a 10s basis. From the subject's respiration the following features (Boiten, Frijda, & Wientjes, 1994) are computed (low pass filtered at 0.15Hz): mean and standard deviation of the first derivatives (10s time window), breathing volume, mean and standard deviation of breath intervals, Poincaré plot (30s time window each). The EMG signals was used to compute the following features (bandpass filtered at 20–120 Hz, piecewise linear detrend): mean of first and second derivatives (5s time window) (Picard R. , Affective computing: challenges, 2003), power spectrum density estimation (15s time window) (Welch, 1967). The following features are extracted from the skin conductance (SCL) (low pass filtered at 0.2Hz): mean and standard deviation of first and second derivative (5s time window).

STATISTICAL EVALUATION

In this section, the statistical evaluation of the architecture is descried for the described data. First the individual channels are evaluated and in a second part the over all combined results are presented. All reported results originate from leave one subject out experiments.

Statistical Evaluation: Unimodal Evaluation

Classification of Spoken Utterances

The MFCC, which have been calculated using 40ms windows, have been averaged to 200ms blocks such that all features have the same alignment. The three available features were separately classified using a SVM with a RBF kernel function and a probabilistic output function. For the individual feature accuracies from 52.5% to 57.9% were accomplished. Furthermore, these results were combined using the average of the confidence values and a temporal fusion of 10s was conducted. This resulted in an accuracy of 55.4% (compare Table 1)

Classification of Facial Expressions

We classified the facial expressions using a multivariate Gaussian. In order to render a stable classification, a bagging procedure was conducted and a reject option was implemented: hereby 99% of the test frames were rejected with respect to the confidence of the classifier. An accuracy of 54.5% was achieved and only 52.3% without reject option (compare Table 1).

Table 1 Accuracies for each unimodal classifier. Results in percent with standard deviation.

FEATURE	ACCURACY	ACCURACY (reject)
MFCC 40ms	57.1% (2.4%)	n/a
MFCC 200ms	57.9% (2.1%)	n/a
ModSpec	52.5% (2.2%)	n/a
Fusion of Audio channel	55.4% (4.6%)	n/a
EMG (5s)	56.5% (10.7%)	69.5% (14.0%)
EMG (20s)	50.2% (11.0%)	52.4% (24.6%)
SCL (5s)	52.8% (4.5%)	44.1% (3.4%)
Respiration (20s)	52.6% (8.4%)	44.6% (17.0%)
Fusion of physiology channel	55.6% (14.5%)	59.7% (13.0%)

Classification of Biophysiological Signals

The described features were partitioned into different sets defined by the feature type and window size. This results in four different feature set. Each of the sets was classified by a perceptron using a bagging. Furthermore, a reject option of 80% is used. The setting results in accuracies ranging from 69.5% to 50.2%. Without reject option the accuracies were lower.

These intermediate results were combined based on a new superordinate time window of 60s with an offset of 30s: first the confidences of the individual channels are averaged followed by the combination of the channels. Resulting in accuracies of 59.7% (55.6% without reject; compare Table 1).

Statistical Evaluation: Multimodal Combination

The intermediate results of the unimodal classifiers are combined to obtain the final class membership. The final time granularity was set to 60s to evaluate all combinations of modalities. If no decision can be made (e.g. due to rejection of samples) the over all window is rejected. The weights for the classifiers have been chosen equally for all classifier combinations. An additional study putting a weighting with focus on audio and physiology according to the respective performance was conducted. The results are shown in Table 2 and vary around an accuracy of 60%. The highest accuracy is achieved using solely audio and video. Generally the standard deviations are high. When all available sources are combined, the mean accuracy slightly drops, but on the other hand the standard deviation decreases.

Table 2 Accuracies of every multimodal combination. Results in percent with standard deviation.

COMBINATION	ACCURACY
Audio (1) + Video (1)	62.0% (15%)
Video (1) +Physiology (1)	59.7% (13.4%)
Audio (1) +Physiology (1)	60.2% (9.3%)
Video (1) + Audio (1) + Physiology (1)	60.8% (9.1%)
Video (1) + Audio (2) + Physiology (3)	61.5% (12.6%)

CONCLUSION AND FUTURE WORK

Classifying the emotion is generally a difficult task when leaping from overacted data to realistic human computer interaction. In this study the problem was investigated with respect to different modalities. The result of the evaluation is that the usage of different modalities can reduce the testing error. On the other hand the variances of the classification are relatively high.

Rejecting samples when classifying such kind of data turns out to be a sound approach. Especially when the distribution of the classes in the data is heavily overlapping. On the other hand it is crucial to define a reliable confidence measure so that the classifier outputs are meaningful. For future work, it could be promising to implement an iterative classifier training procedure, were the training data can be rejected.

Acknowledgment

This research was supported in part by grants from the Transregional Collaborative Research Centre SFB/TRR 62 "Companion-Technology for Cognitive Technical Systems" funded by the German Research Foundation (DFG). Miriam Schmidt is supported by a scholarship of the graduate school Mathematical Analysis of Evolution, Information and Complexity of the University of Ulm.

BIBLIOGRAPHY

Bishop, C. (2006). *Pattern recognition and machine learning* (Bd. 4). springer New York.

Boiten, F. A., Frijda, N. H., & Wientjes, C. J. (1994). Emotions and respiratory patterns: review and critical analysis. *International Journal of Psychophysiology , 17* (2), 103-128.

Breiman, L. (1996). Bagging predictors. *Machine learning , 24* (2), 123-140.

Cannon, W. B. (1927). {The James-Lange Theory of Emotions: A Critical Examination and an Alternative Theory}. *The American Journal of Psychology , 39* (1/4), 106-124.

Caridakis, G., Castellano, G., Kessous, L., Raouzaiou, A., Malatesta, L., Asteriadis, S., et al. (2007). {Multimodal emotion recognition from expressive faces, body gestures and speech}. In *Artificial Intelligence and Innovations 2007: from Theory to Applications* (Bd. 247, S. 375-388). Springer Boston.

Chen, L., & Huang, T. (2000). Emotional expressions in audiovisual human computer interaction., *1*, S. 423 -426 vol.1.

Cowie, R., Douglas-Cowie, E., Tsapatsoulis, N., Votsis, G., Kollias, S., Fellenz, W., et al. (2001). {Emotion recognition in human-computer interaction}. *IEEE Signal processing magazine , 18* (1), 32-80.

Dietrich, C., Palm, G., & Schwenker, F. (2003). Decision templates for the classification of bioacoustic time series. *Information Fusion , 4* (2), 101-109.

Ekman, P., & Friesen, W. (1978). *Facial action coding system: investigator's guide.* Consulting Psychologists Press.

Gilroy, S. W., Porteous, J., Charles, F., & Cavazza, M. (2012). Exploring Passive User Interaction for Adaptive Narratives. (S. 119-128). ACM.

Glodek, M., Bigalke, L., Schels, M., & Schwenker, F. (2011). Incorporating uncertainty in a layered HMM architecture for human activity recognition. *Proceedings of the 2011 joint ACM workshop on Human gesture and behavior understanding* (S. 33-34). ACM.

Glodek, M., Tschechne, S., Layher, G., Schels, M., Brosch, T., Scherer, S., et al. (2011). Multiple classifier systems for the classification of audio-visual emotional states. *6975*, S. 359-368. Springer.

Gunes, H., & Pantic, M. (2010). Automatic, dimensional and continuous emotion recognition. *International Journal of Synthetic Emotions , 1* (1), 68-99.

Kelley, J. (1984). An iterative design methodology for user-friendly natural language office information applications. *ACM Transactions on Information Systems (TOIS) , 2* (1), 26-41.

Kuncheva, L. I. (2004). *Combining Pattern Classifiers: Methods and Algorithms.* Wiley.

Littlewort, G., Whitehill, J., Wu, T., Fasel, I., Frank, M., Movellan, J., et al. (2011). The computer expression recognition toolbox (CERT). *IEEE*, (S. 298-305).

Picard, R. (2003). Affective computing: challenges. *International Journal of Human-Computer Studies , 59* (1), 55-64.

Platt, J. (1999). Probabilistic outputs for support vector machines and comparisons to regularized likelihood methods. *Advances in large margin classifiers , 10* (3), 61-74.

Russell, J. A., & Mehrabian, A. (1977). Evidence for a three-factor theory of emotions. *Journal of Research in Personality , 11* (3), 273-294.

Sato, J., & Morishima, S. (1996). Emotion modeling in speech production using emotion space., (S. 472-477).

Schachter, S. (1964). The interaction of cognitive and physiological determinants of emotional state. (L. Berkowitz, Hrsg.) *Advances in Experimental Social Psychology , 1* (Bd. 1), 49-80.

Schels, M., Scherer, S., Glodek, M., Kestler, H. A., Palm, G., & Schwenker, F. (2011). On the Discovery of Events in EEG Data utilizing Information Fusion. *{Computational Statistics: Special Issue: Proceedings of Reisensburg 2010 .*

Scherer, S., Glodek, M., Schels, M., Schmidt, M., Layher, G., Schwenker, F., et al. ((accepted)). A Generic Framework for the Inference of User States in Human Computer Interaction: How patterns of low level communicational cues support complex affective states. *Journal on Multimodal User Interfaces, special issue on: Conceptual frameworks for Multimodal Social Signal Processing .*

Scherer, S., Glodek, M., Schwenker, F., Campbell, N., & Palm, G. (2012). Spotting laughter in naturalistic multiparty conversations: A comparison of automatic online and offline approaches using audiovisual data. *ACM Transactions on Interactive Intelligent Systems: Special Issue on Affective Interaction in Natural Environments ,* (accepted).

Schuller, B., Rigoll, G., & Lang, M. (2003). Hidden Markov model-based speech emotion recognition. *Ieee, 2,* S. II--1.

Sebe, N., Lew, M. S., Sun, Y., Cohen, I., Gevers, T., & Huang, T. S. (2007). Authentic facial expression analysis. *Image Vision Comput. , 25,* 1856-1863.

Simon, H. (1999). *Neural networks: a comprehensive foundation.* Prentice Hall.

Thiel, C., Schwenker, F., & Palm, G. (2005). Using Dempster-Shafer Theory in MCF Systems to Reject Samples. *Multiple Classifier Systems* (S. 959-961). Springer.

Welch, P. (1967). The use of fast {F}ourier transform for the estimation of power spectra: A method based on time averaging over short, modified periodograms. *IEEE Transactions on Audio and Electroacoustics , 15* (2), 70-73.

Young, S., Kershaw, D., Odell, J., Ollason, D., Valtchev, V., & Woodland, P. (1999). The HTK book, version 2.2. *Entropic Ltd .*

Zeng, Z., Pantic, M., Roisman, G., & Huang, T. (2009). A survey of affect recognition methods: Audio, visual, and spontaneous expressions. *Pattern Analysis and Machine Intelligence, IEEE Transactions on , 31* (1), 39-58.

CHAPTER 70

ATLAS – An Annotation Tool for HCI Data Utilizing Machine Learning Methods

Sascha Meudt[1], Lutz Bigalke[1], Friedhelm Schwenker[1]

[1]University of Ulm
Ulm, Germany
{firstname.lastname}@uni-ulm.de

ABSTRACT

ATLAS is a graphical tool for the annotation of multi-modal data streams. In our application, the data is collected in human computer interaction (HCI) scenarios. Basically, any type of input data can be processed by the ATLAS annotation tool, including speech (multi-channel), video (multi-channel), EEG, EMG, ECG data. In addition to the raw data, intermediate data processing results such as extracted features, and even (probabilistic or crisp) outputs of pre-trained classifier modules can be displayed. Tools for the annotation and transcription of HCI scenes are integrated as well. In this paper ATLAS's basic architectures and features are described, furthermore it is explained how different types of data and label information can be presented. Besides these basic annotation features, an active learning module (active data selection) has been integrated . Currently, probabilistic Support Vector Machines (SVM) are available in the ATLAS system.

Keywords: Multimodal Pattern Recognition, Partially Supervised Learning.

INTRODUCTION

In today's Wizzard of Oz (WOz) experiments not only audio and video data is collected, but also a variety of bio-physiological measurements are available (Walter, et al. 2011). For a human expert, the analysis of this kind of multimodal sensory data is difficult, time consuming, and exhausting, and therefore interactive, intelligent data analysis tools might be useful. What is an intelligent tool? An intelligent tool is based on machine learning techniques, particularly unsupervised or partially supervised learning methods are of interest in our settings. The idea of interactive data analysis is that pre-trained machine learning models are applied to select smaller subsets of informative data out of the huge pool of available data. This small set – possibly just a single data point - is then propagated forward to an human expert for data annotation or data labeling. Subsequently to this labeling process the model is re-trained by using this newly expert-labeled data set.

Unfortunately, in most annotation tools such types of intelligent data analysis components are not implemented yet. Even more, standard annotation tools are rather limited in regard to the number and type of possible input modalities, typically only single video and audio streams can be processed (Harper 2001).

While developing the ATLAS tool we focus on these two aspects: On the one hand we want to develop a graphical tool with very high usability, even usable for non computer experts. On the other hand an assistive system should be included using active learning techniques in order to select data and to suggest label information for unlabeled (unseen) parts of the data collection, so that the annotator is able to evaluate (accept/change/reject) the suggested label information more efficiently than to analyze and label all data from scratch.

As another important feature we took into account was that in multimodal WOz data usually different kinds of labels should be used on the same dataset, depending on the modality and application. For instance, some labels might cover more global aspects, e.g. on sentence level in speech data or some complex user activities, whereas others focus on events or short term behavior, e.g. on word or phonetic level, or some multimodal events.

BASIC FRAMEWORK

Figure 1: Screen shot of the ATLAS annotation and analysis tool. The most important windows are shown (Line-Track window, Video windows, Main Controls, Label details window, Tabular label view and the Legend window).

This section describes the basic technical structure of ATLAS framework and

656

gives an insight to its data representation. ATLAS implements two different kinds of data stream structures called *tracks*. A *track* is a representation of a single modality or label type. *Media-Tracks* containing audio or video data. These tracks are implemented using standard video and audio tools and are of course the most important structures. All the other – non-audio/non-video – data is represented through so-called *Line-Tracks*, which are divided into three sub-categories, namely *Scalar Tracks, Vector Tracks*, and *Label Tracks*:

1) *Scalar-Tracks* consisted of a sequence of one dimensional values, represented as a common (t, y_t) axis plot.

2) *Vector-Tracks* are used for multivariate data, and are represented using the well-known *MATLAB's imshow* style, a color image where each row stands for a certain feature dimension, each column is representing a time step, and the measurement is encoded through a color map.

MATLAB, CSV or XML files can be imported into these two types of tracks. In addition to the standard imports some special import filters, for instance import of biophysical data from the NeXus32 have been implemented (Mind Media B.V. 2011).

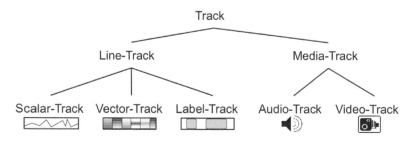

Figure 2: Hierarchical track structure of ATLAS

3) *Label-Tracks* are be used for all kinds of label information. Labels are displayed as colored boxes. Labels can be contain additional information such as comments, values, text and information about the labeling status, e.g. whether the label has been manualy entered or generated automatically during by a machine learning procedure. Labels are stored in XML format, one file per Label-Line. Import filters to import labels from other annotation tools are available as well, e.g. for the FOLKER transcription tool (Schmidt und Schuette 2010).

The overall ATLAS framework is implemented in Java, Linux, Windows and Mac versions are available on request. ATLAS is highly efficient, even for multi-video streams, e.g. we successfully tested datasets up to 100 GB per hour using a modern standard PC. Problems to display the data have been observed in case of hard disk I/O limitations, and so we suggest to use SSD hard disk drives at least in multi-video applications.

USER INTERFACE

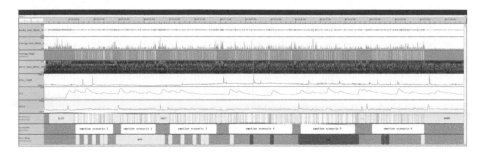

Figure 3: Screen shot of the Label-Line window. From top to bottom the lines contain: Audio raw samples, energy value extracted from audio, marked parts where energy is over a given threshold, MFCC features extracted from audio, EMG of corrugator muscle, skin conductance, respiration, labels which mark the activity in the experiment, labels which mark the six parts of the experiment and finally labels that mark the users feeling (positive or negative).

The graphical frontend consists of a variety of windows to display all kinds of information. The most important window type is the Line-Track window, which represents all three kinds of Line-Track data (see previous section). For audio and video playback the Java library of VLC (VideoLAN 2011) has been applied, thus numerous audio and video codecs are supported .

In the Line-Track window labels could be created, changed and delete by drag and drop. Exact positioning of labels is possible either by zooming the Line-Tracks or by editing details in a separate label property window. The label property window can be used to enter additional specific information or comments of the annotator.

PARTIALLY SUPERVISED LEARNING

In this section we describe the basic ideas of partially supervised learning. Some of these methods have been realized in ATLAS. In supervised learning a large amount of labeled training data must be available in order to construct models of acceptable prediction accuracy. But in many applications such as remote sensing image classification (Shahshahani und Landgrebe. 1994), automated classification of text documents (Nigam, et al. 2000), or affect recognition (Piccard 2003) labeling data is difficult, tedious, expensive, or time consuming, as it requires the efforts of human experts and/or special devices, and therefore labeled data is rare, but sometimes huge pools of unlabeled data are available.

In order to integrate unlabeled data in a supervised machine learning procedure two different partially supervised learning approaches have been applied, namely, *semi-supervised learning*, and *active learning. Semi-supervised learning* refers to group of methods that attempt to take advantage of unlabeled data for supervised learning (*semi-supervised classification*) or to incorporate prior information such as class labels, pairwise constraints or cluster membership (*semi-supervised clustering*). *Active learning* or *selective sampling* (Settles 2009) refers to methods where the learning algorithm has control on the data selection, e.g. it can select the most important/informative examples from a pool of unlabeled examples, then a human expert is asked for the correct data label. Here is the aim is to reduce annotation costs. In our application - the recognition of human emotions in human computer interaction - we focus more on active learning (Schwenker and Trentin 2011, Abdel Hady and Schwenker 2011)

The ATLAS learning procedure is based on LIBSVM library (Chang und Lin 2011) An example of this iterative annotation process is shown at the bottom line of Figure 3. The goal in this example experiment was to annotate the user's emotional state based on multi-modal data. The details of the experiment are described in (Walter, et al. 2011). In the example the automatic annotation is based on MFCC audio features (Imai 1983) and are shown in Line-Track 4. A SVM with RBF-kernel has been applied. Suggestions for new labels have been generated by voting, weighted with the confidence of a single feature classification of all decisions within the time boundaries of the suggested label.

CONCLUSION AND FUTURE WORK

We presented a tool for visualization and annotation of multi-modal data. Because of the generic structure of ATLAS it is applicable for data sets generated in various experimental settings. The ATLAS user is supported by active learning algorithms in order to reduce the annotation costs.

Future work on the ATLAS system includes the integration of classifiers for sequence recognition, e.g. Hidden Markov Model (HMM), and the integration of trainable confidence-based fusion methods in order to improve the flexibility of the ATLAS tool.

ACKNOWLEDGMENT

The presented work has been developed within the Transregional Collaborative Research Centre SFB/TRR 62 "Companion-Technology for Cognitive Technical Systems" funded by the German Research Foundation (DFG).

REFERENCES

Abdel Hady, Mohamed Farouk, and Friedhelm Schwenker. Partially supervised learning. In *Monica Biancini and Marco Maggini and Lakim Jain, Handbook on Neural Information Processing* 2012 (to appear)

Chang, Chih-Chung, and Chih-Jen Lin. "LIBSVM: A library for support vector machines." *ACM Transactions on Intelligent Systems and Technology*, 2, 27:1-27, 2011.

Glodek, M., Tschechne, S., Layher, G., Schels, M., Brosch, T., Scherer, S., et al. Multiple classifier systems for the classification of audio-visual emotional states. *6975*, S. 359-368. Springer, 2011

Harper, A. L. "Emerging requirements for multi-modal annotation and analysis tools." 2001.

Imai, S. "Cepstral analysis synthesis on the mel frequency scale." *Acoustics, Speech, and Signal Processing, IEEE International Conference on ICASSP '83*, 93-96, 1983

Nexus32, http://www.mindmedia.nl/german/nexus32.php, 2011

Nigam, K., A. K. McCallum, S. Thrun, and T. Mitchell. "Text Classification from Labeled and Unlabeled Documents using EM." *Machine Learning* 39, no. 2-3, 103-134, 2000

Picard, R. Affective computing: challenges. *International Journal of Human-Computer Studies , 59* (1), 55-64, 2003

Schels, M., Scherer, S., Glodek, M., Kestler, H. A., Palm, G., & Schwenker, F. On the Discovery of Events in EEG Data utilizing Information Fusion. *{Computational Statistics: Special Issue: Proceedings of Reisensburg 2011*

Scherer, S., Glodek, M., Schels, M., Schmidt, M., Layher, G., Schwenker, F., et al. A Generic Framework for the Inference of User States in Human Computer Interaction: How patterns of low level communicational cues support complex affective states. *Journal on Multimodal User Interfaces, special issue on: Conceptual frameworks for Multimodal Social Signal Processing*, 2012.

Scherer, S., Glodek, M., Schwenker, F., Campbell, N., & Palm, G. Spotting laughter in naturalistic multiparty conversations: A comparison of automatic online and offline approaches using audiovisual data. *ACM Transactions on Interactive Intelligent Systems: Special Issue on Affective Interaction in Natural Environments* , 2012 (accepted).

Schmidt, Thomas, and Wilfried Schuette. "FOLKER: An Annotation Tool for Efficient Transcription of Natural, Multi-party Interaction." *Proceedings of the Seventh conference on International Language Resources and Evaluation (LREC'10)*. 2010.

Schwenker, Friedhelm, and Edmondo Trentin. *Parially Supervised Learning (PSL 2011)*. Springer LNAI 7081, 2012.

Settles, B. "Active Learning Literature Survey." Tech. rep., Department of Computer Sciences, University of Wisconsin-Madison, Madison, WI, 2009.

Shahshahani, B., and D. Landgrebe. "The effect of unlabeled samples in reducing the small sample size problem and mitigating the hughes phenomenon." *IEEE Transactions on Geoscience and Remote Sensing, Vol* 32, no. 5, 1087-1095, 1994

VideoLAN. http://www.videolan.org/vlc/ 2011

Walter, S. Scherer, S., Schels, M., Glodek, M., Hrabal, D., Schmidt, M., Böck, R., Limbrecht, K., Traue, H.C. & Schwenker, F. "Multimodal Emotion Classification in Naturalistic User Behavior." Edited by Julie A. Jacko. Springer, 603-611, 2011

Section IX

Novel Devices, Information Visualization, and Augmented Reality

CHAPTER 71

Pleasurable Design of Haptic Icons

Wonil Hwang

Soongsil University
Seoul, Korea
wonil@ssu.ac.kr

Jihong Hwang

Seoul National University of Science & Technology
Seoul, Korea
hwangjh@snut.ac.kr

Taezoon Park

Nanyang Technological University
Singapore
TZPark@ntu.edu.sg

ABSTRACT

Haptic icons are the icons based on haptic stimuli, from which users can intuitively perceive information, such as abstract concepts and emotional contents. In this study, we focused on haptic icons as a medium of conveying emotions, and tried to find whether haptic icons can deliver to users a specific emotion, such as pleasure or unpleasure, distinctively by investigating user's perception to different types of vibrotactile stimuli in term of pleasure. For experiments, we generated twenty different types of vibrotactile stimuli. During experiments, each of twenty participants perceived each of those twenty types of stimuli and responded to four measures for pleasure index. As a result of experiments, we found the most pleasurable vibrotactile stimuli and the least pleasurable vibrotactile stimulus among twenty stimuli, and thus, we concluded that those stimuli were effective haptic icons that were distinguishable from each other in terms of pleasure.

Keywords: haptic icons, pleasure index, emotional design

1   INTRODUCTION

Currently, haptic interfaces design and interactions based on haptic interfaces have been actively studied (Hwang and Hwang, 2011; Pasquero, 2006). In order to improve haptic interfaces and interactions based on haptic interfaces, many researchers have tried to develop haptic icons (Brewster and Brown, 2004; MacLean and Enriquez, 2003; Rovers and Van Essen, 2004). Haptic icons mean icons, from which users can intuitively perceive information, based on haptic stimuli. Haptic icons are often compared to visual icons that are used in GUI systems and auditory icons that are named earcons.

Haptic icons are designed to deliver information, such abstract concepts and/or emotional contents. Especially, haptic icons as a medium of conveying emotions attract attention for affective interface designs, due to their applicability to various domains, including interfaces of games, SNS, and so on. However, current research has not reached satisfactory levels of studies for affective design of haptic icons. For instance, Shin et al. (2007) developed haptic icons, named TCONs, whose role was to supplement emotional aspects of text-based messenger. Thus, it is needed to develop haptic icons that can convey emotional contents independently of other information media, such as texts, pictures and sounds.

In this study, we try to develop candidates of haptic icons by generating a variety of vibrotactile stimuli, and to find whether haptic icons can deliver to users a specific emotion, such as pleasure or unpleasure, distinctively by investigating user's perception to different types of vibrotactile stimuli in term of pleasure.

2   METHODS

2.1   Apparatus

The vibration excitation system was designed to generate sinusoidal vibration with a range of frequency and amplitude. The key component of this vibration excitation system is the mini-shaker (Brüel and Kjær Type 4810). The mini-shaker generates sinusoidal vibration whose frequency and amplitude are determined by those of the electrical, sinusoidal voltage input signal. This input signal is provided by a programmable function generator (Tabor Electronics WW5062) that was controlled by a waveform creation software (ArbConnection 4.x) in PC. The operating frequency range of the mini-shaker is DC - 18,000 Hz, and the maximum bare table acceleration amounts to 550 m/s2 (55.1 g) in the frequency range of 65 Hz - 4,000 Hz. The mini-shaker was mounted onto the main body made up of aluminum frames, aluminum plates, joints and fixtures (see Figure. 1).

Figure 1 Picture of experiments (Left: experimenter, Right: participant wearing earmuff)

2.2   Experiments

Twenty different types of vibrotactile stimuli were generated by the vibration excitation system that was described in the above apparatus section. These vibrotactile stimuli were made by changing duration and amplitude in a given time (i.e., 1 second) based on sine waves with 160 Hz of frequency and 20dB of sensation level. Four rhythm patterns were generated by changing duration of stimuli, such as 1 second, 3/4 second, 1/2 second and 1/4 second, in a given 1 second. Five macro wave types were generated by changing amplitude of stimuli, such as fixed amplitude (rectangular shape), increasing amplitude (triangle shape), decreasing amplitude (triangle shape), increasing & decreasing amplitude (diamond shape), and decreasing & increasing amplitude (butterfly shape), for a duration of stimuli. Table 1 summarizes four rhythm patterns and five macro wave types, whose combination resulted in twenty different types of vibrotactile stimuli.

During experiments, twenty different types of vibrotactile stimuli were randomly presented to each of twenty participants. After a participant perceived each of vibrotactile stimuli, he/she responded to four pleasure-related adjectives that were adopted from Park and Hwang (2005) according to his/her agreement based on Likert type 7-points scale. Participants conducted experiments wearing an earmuff to avoid noisy environments, and perceived vibrotactile stimuli by their index fingers (see Figure 1).

Table 1 Rhythm patterns and macro wave types for experiments

Rhythm Patterns	Macro Wave Types
1. Continuing for 1 s	A. Fixed (like rectangular)
2. Continuing for 3/4 s and pausing for 1/4 s	B. Increasing (like triangle)
3. Continuing for 1/2 s and pausing for 1/2 s	C. Decreasing (like triangle)
4. Continuing for 1/4 s and pausing for 3/4 s	D. Increasing and decreasing (like diamond shape)
	E. Decreasing and increasing (like butterfly shape)

3   RESULTS

3.1   Pleasure index

Based on the measurement of four pleasure-related adjectives, which are 'satisfied', 'enjoyable', 'likable' and 'pleased', the pleasure index was made by arithmetic average of scores from four pleasure-related adjectives. To verify the internal consistency of those measures, Cronbach's alpha value (0.94) was calculated (see Table 2). As a result, those four measures showed satisfactory level of internal consistency for the pleasure index, because the internal consistency is considered as satisfied when Cronbach's alpha value is more than 0.7.

Table 2 Measures for pleasure index

Measures	Cronbach's alpha
Satisfied	
Enjoyable	0.94
Likable	
Pleased	

3.2   Most pleasurable haptic icons

ANOVA (analysis of variance) was conducted to find whether twenty different types of vibrotactile stimuli were perceived as significantly different levels of pleasure. As shown in Table 3, there is significant difference among twenty stimuli based on their pleasure index scores ($F(19, 379)=1.94$, $p=0.0108$).

In order to find where significant difference exists among twenty stimuli, Duncan test was conducted as a multiple comparison test. As shown in Table 4, the result of Duncan test indicates that 'decreasing stimulus for 1/4 second and pausing it for 3/4 second (C4)' and 'decreasing & increasing stimulus for 1/2 second and pausing it for 1/2 second (E3)' are the most pleasurable stimuli, or haptic icons, and 'continuing stimulus with fixed amplitude for 1 second (A1)' is the least pleasurable stimulus, or haptic icon.

Table 3 ANOVA results for pleasure index

Source	DF	F value	Pr > F
Vibrotactile stimuli	19	1.94	0.0108
Error	379		
Total	398		

Table 4 Multiple comparison using Duncan grouping

Duncan grouping				Mean	Haptic icons
		A		3.8250	C4
		A		3.7500	E3
B		A		3.6711	E4
B		A		3.6500	E2
B		A	C	3.4375	C3
B		A	C	3.2750	D3
B		A	C	3.2625	D2
B	D	A	C	3.1625	D4
:	:	:	:	:	:
:	:	:	:	:	:
B	D	A	C	2.9375	D1
B	D		C	2.6625	A3
B	D		C	2.6250	A2
	D		C	2.4000	B2
	D			2.1125	A1

4   CONCLUSIONS

Twenty different types of vibrotactile stimuli represent candidates of haptic icons, which are designed to convey various emotional contents. The most pleasurable haptic icons are 'decreasing stimulus for 1/4 second and pausing it for 3/4 second' and 'decreasing & increasing stimulus for 1/2 second and pausing it for

1/2 second'. It means that decreasing or decreasing & increasing amplitude of vibrotactile stimulus for short duration (i.e., 1/4 or 1/2 second) and pausing it for short duration (i.e., 3/4 or 1/2 second) give us relatively more pleasure than others. By the way, the least pleasurable haptic icon is 'continuing stimuli with fixed amplitude for 1 second', which means that continuous vibrotactile stimuli with fixed amplitude without any pause give us the least pleasurable experience.

In conclusion, the experiments of this study show (1) we can make a variety of candidates of haptic icons using the combination of rhythm patterns and macro wave types; and (2) we can find effective haptic icons that are distinguishable from each other in terms of pleasure. For further study, more design parameters for haptic icons besides rhythm pattern and macro wave type need to be investigated and other emotional contents than pleasure need to be considered.

ACKNOWLEDGMENTS

This research was supported by Basic Science Research Program through the National Research Foundation of Korea(NRF) funded by the Ministry of Education, Science and Technology(2011-0013713).

REFERENCES

Brewster, S. A. and L. M. Brown. 2004. Tactons: Structured tactile messages for non-visual information display. *Proceedings of 5th Australasian User Interface Conference* (AUIC2004), 15–23.

Hwang, J. and W. Hwang. 2011. Vibration perception and excitatory direction for haptic devices. *Journal of Intelligent Manufacturing* 22: 17-27.

MacLean, K. and M. Enriquez. 2003. Perceptual design of haptic icons. *Proceedings of Eurohaptics.*

Park, T. and W. Hwang. 2005. Derivation and validation of optimal vibration characteristics for pleasure and attention: Implication for cell phone design. *Proceedings of the 11th International Conference on Human-Computer Interaction.* Las Vegas, NV.

Pasquero, J. 2006. *Survey on communication through touch* (Tech. Rep. TR-CIM 06.04). Montreal, Canada: McGill University, Department of Electrical and Computer Engineering, Center for Intelligent Machines.

Rovers, A. F. and H. A. Van Essen. 2004. Design and evaluation of hapticons for enriched instant messaging. *Proceedings of Eurohaptics.*

Shin, H., J. Lee, J. Park, Y. Kim, H. Oh, and T. Lee. 2007. A tactile emotional interface for instant messenger chat. *Lecture Notes in Computer Science* 4558: 166-175.

CHAPTER 72

Conscious and Unconscious Music from the Brain: Design and Development of a Tool Translating Brainwaves into Music Using a BCI Device

Raffaella Folgieri, Matteo Zichella^*

* Dip. Di Scienze Economiche, Aziendali e Statistiche,
^ CdL Comunicazione Digitale, Università Statale di Milano
Milan, Italy
Raffaella.Folgieri@unimi.it

ABSTRACT

Music plays a fundamental role in games, Virtual Reality and digital entertainment design, due to the impact of music in humans' experiences, emotions and cognitive processes. An interesting question is if it could be possible to design and develop a tool reproducing conscious and unconscious music by subjects' brain activity. This work presents the first step of a challenging research about the effect of different combinations of perceptual and cognitive aspects in individuals' ability to create music consciously by brain. In fact, thanks to information collected in literature and in our previous experiments, we designed and developed a prototype of a software tool allowing users to play music by unconscious or conscious brain activity. This first prototype is currently working, allowing us to perform more experiments, in order to refine the tool and to deeper understand more complex emotional and cognitive phenomena in music active listening and performing.

Keywords: brain computer interaction, BCI, music

1   BACKGROUND

The reliability of commercial non-invasive BCI (Brain Computer Interface) devices and the low cost of these EEG-based systems, compared to other brain image techniques, such as fMRI, determined the increasing interest in their application in different Research fields (Friedman et al., 2007; Nijholt et al., 2008; Pfurtscheller and Neuper, 2001), also thanks to the portability of the equipment.

This last feature makes BCI devices particularly suited for experiments involving virtual (Friedman et al., 2007) and real (Nakamura et al., 1999) situations and a larger number of subjects, especially when evaluating emotional or cognitive response of individuals. In fact, during EEG measures, anxiety induced by invasive devices could influence the emotive response of individuals. Commercial BCI (Allison et al., 2007) devices consists in a simplification of the medical EEG equipment, communicating EEG response to stimuli by wi-fi connection, allowing to people to feel relaxed, to reduce anxiety and to move freely in the experiment environment, acting as in absence of the BCI devices.

In researches related to music or to relationship between music and brain, BCI devices applications concern mainly the psychological implications area (Skaric et al. 2007; Wickelgren, 2003) or the neurofeedback-based therapeutic applications (Minsky, 1981; Pascual-Leone, 2001). Few studies treat the conversion of EEG signals into music notes (Dan et al., 2009). Moreover, among the considered works, none tries to individuate a specific characteristic of human brain allowing to make subjects consciously able to reproduce the same single sound. Specifically, we are interesting in discovering if individuals, subject to the same situation, i.e. in the same environment and with the same tool at disposition, are able to reproduce requested single sounds, reducing propaedeutic musical or brain training time.

An interesting question is, in fact, if humans may consciously translate their brainwaves into music, selecting specific sounds to reproduce and changing, in this way, environments, game or training sessions, multimedia and Human-Computer Interaction or communicating current emotions, for example during a psychological or neurological therapy.

In our work we show the results of some preliminary experiments performed with the aim to investigate the possibility to make users able to play a specific sound consciously. The final goal of our experiments, in fact, has been to make individuals able to play, through the control of their own brainwaves, a specific single music note. To do this, we used a BCI device, reading subjects' brainwaves, after a "as-short-as-possible" training session, necessary both to enhance the ability of the developed software to recognize the desired sound by the collected EEG signals, and, especially, for subjects to understand their own brain mechanism in selecting the appropriate imaginative or memory recalling process to play a specific selected sound.

In the second section we describe the materials and methods adopted in our experiments. Section 3 concerns the experiments performed and the corresponding results. Finally, in section 4 we discuss the obtained results and present future developments.

2   MATERIALS AND METHODS

EEG data have been collected using a Neurosky *Mindwave*™ BCI device. The *Mindwave*™ is widely used in several commercial and research applications (Chuchnowska and Skala, 2011; Yasui, 2009). It consists of a headset mounting an arm equipped with a single dry sensor acquiring brain signals from the forehead of the user at a sample rate of 512 Hz, transmitted via bluetooth to a host computer. Before choosing this device, we compared it to other BCIs, such as, for example, Emotiv *Epoc*™. Comparison analysis showed that the *Mindwave*™ BCI results more comfortable for users, both for the easiness of positioning the device on the scalp, and because it uses a dry sensor instead of wet ones. Moreover brain functions interesting our work, are specifically related to the premotor frontal cortex area, that is the area on which the *Mindwave*™ sensor is positioned. Many studies, in fact, confirm that the signals from the frontal lobes (Blood et al., 1999) are linked to higher states of consciousness and emotions stimulated by music. Another advantage, convincing us about using *Mindwave*™, consists in the wireless communication between the BCI device and the computer during the collection of data, making comfortable wearing the BCI during the experiments.

BCIs collect several cerebral rhythms grouped by frequency. For our purpose, we decided to exclude only theta rhythms, for their low presence in wake state, while we concentrate on alpha, beta and gamma bands. In fact, activity in the alpha band (7 Hz - 14 Hz) is usually related to relaxed awareness, meditation, contemplation, etc., beta band (14 Hz - 30 Hz) is associated to active thinking, active attention, focus on the outside world or solving concrete problems. Finally, activity in gamma band (30 Hz - 80 Hz) is considered to be related to cognitive processes involving different populations of neurons, and to the processing of multi-sensorial signals. Delta band (3 Hz – 7 Hz) has been used, as shown in next paragraphs, to modulate, in amplitude, the output signal of our music application.

Brainwaves registered through the BCI device have been sent to Processing[1], an open source programming environment containing a java library allowing the development of a Java program realizing the connection between Processing and the *Mindwave*™ BCI and, after, among Processing and Max 6[2], a popular environment for visual programming, specifically developed by Cycing '74 for applications in music and multimedia.

To perform the experiments, we chose 30 subjects, 15 women and 15 men, aged between 14 and 49. Difference in age has been considered potentially relevant for the variability of the results. Each subject took the test separately, in a comfortable environment, to reduce variation influenced by external diseases. Each EEG registration session had the duration of two minutes, during which subjects were rest, recommended to not close their eyes, do not speak or move.

---

[1] http://www.processsing.org
[2] http://cycling74.com/products/max/

3   EXPERIMENTS DESCRIPTION

We performed two experiments, finalized to investigate the different response of individuals to music and their ability to reproduce music with a "as-short-as-possible" training: (1) Unconscious production of music. In this first experiment we registered EEG signals from brainwaves spontaneously (without any stimulus) produced by the different subjects. The registration, collected by the BCI devices, has been after processed by an application developed specifically for the experiment, with the aim to transform the registered brain waves into sounds. The experiment has had a specific objective, consisting in verifying if we could individuate characteristics or value ranges allowing the following experiment 2. (2) Conscious production of music. The experiment has consisted in verifying if subjects could be trained to reproduce a single specific note, through a BCI device and the developed interpretation software. Subjects have been invited to reproduce more single notes, supported by an audio stimulus presented alone or in association with other reinforcement stimuli.

3.1   Experiment 1: unconscious production of music

In this experiment we have created an application able to interpret EEG signals of subjects, transforming their brainwaves into sounds. Individuals have been immersed in the same relaxing environment and in absence of specific stimuli. Also if people differs in EEG response to the same stimuli (Mauri et al., 2010), we were looking for a mix of characteristics or common patterns helping us to create an environment allowing users to play, consciously, a specific music note.

To obtain a music trace corresponding to brainwaves, we have used the cerebral waves of the 30 subjects as the input for the application created with Max. With the configuration described in the paragraph 2, we have collected the subjects' brainwaves sending in real time these input to Max, to be processed and translated into sound overlapped to a loop.

We have written few lines of code reading the data, drawing the EEG graphic and sending signals to Max that, through a patch specifically created, reads and plays the music. A groove object has been introduced, to regulate the reproduction of the loop and allowing to set its start and end point. The second step has consisted in reading the EEG signals sent by the Processing sketch and after pass them to the object *mxj jk.link*. The information is so acquired jointly to the parameter set in the sketch. We also introduced a keyboard, necessary to send the right value to the object *makenote* that sends the value to the object *noteout*, followed by the name of the midi synthesizer creating the note according to MIDI notation. All the components are shown in the figure 1.

Fundamentally, a MIDI note, through the object *kSlider*, is generated by the alpha waves, while the eye blinks, if present, regulate the delay of the note. The MIDI signal, sent by the *kSlider*, is manipulated by the object *ddg.mono*. The MIDI note is after converted into its corresponding frequency, through the object *mtof*, whose output represents the input of the object *phasor ~* and the frequency is used to generate a saw-toothed signal.

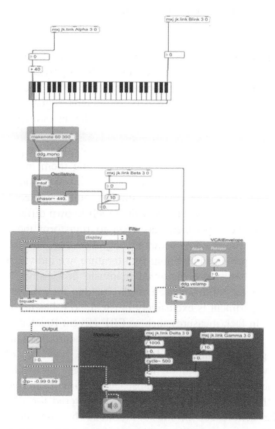

Figure 1 The part of the patch transforming brainwaves into an acoustic signal

The beta band is used to modify the phase of the signal, using the second input of the object *phasor~*. The signal is after sent to a filter in notch configuration (used to eliminate the central band of the signal and allowing the listening of the low band and the high frequencies), implemented by the object *biquad~*. After filtered, the signal is sent to the input of a variable gain amplifier, realized by the object *ddg.velamp*, allowing the control of the attack and release of the signal, that is the delay controlled by a curve, after what the signal reaches the maximum amplitude and the delay (after what it results attenuated to the minimum amplitude) and by the object *~, a simple multiplier that applies the amplitude gain of the signal itself. The following phase consists in sending the signal to an object *gain~*, to regulate the general reproduction volume, linked to the object *clip~*, needed to limit the amplitude of the signal to avoid distortions on the output audio signal.

At the end, the output signal from the object *clip~* is modulated in amplitude, also in this case through the operator * ~, by a signal resulting by the modulation of two signals regulated by delta and gamma waves. The last object (*ezdac~*) is a digital-analog converter transforming the digital signal produced by Max in an analogic signal, reproducible by the amplifiers.

The last part of the application has been designed to manage alpha, beta, gamma and delta brainwaves, passing them through a filter, a variable gain amplifier and a modulator.

Some audio track produced by EEG rhythms could be listened on the website collecting experiments' sources and results[3]. The significant diversity among subjects, both in the EEG graphic representation and in the produced sounds, corresponds also to a significant diversity in sounds registered from the same subject in other repetitions of the experiment in different moments, also under the same conditions.

The experiment, *per sé*, has the only objective to explore the possibility to create a tool able to read cerebral brainwaves transforming them into sound, to verify effective differences among different individuals' brainwaves and the corresponding music sound. In fact, if we had obtained similar sounds, we could argue that the software was not enough sensitive to individual well-known differences.

The objective of the experiment has been to transform the changes in EEG registered in persons subjected to a motor, cognitive and sensorial relax state, in music unconsciously produced by brain. By literature, we individuated in alpha and beta band the main brainwaves involved in music processing, but we aimed to evaluate the effectiveness of this chosen bands, looking for evident activity in them. The result of the experiment is exclusively musical, so it has not been subjected to specific measures. In fact, measures such as ERP (Event Related Potentials), Event Related Phase Resetting (ERPR) and Desynchronization/Synchronization (ERD/ERS) are generally introduced to verify the response of individuals to sensorial, cognitive or motor stimuli inducing changes in EEG. In our case the experiment has been based just on the absence of stimuli, and performed only to empirical verify the difference in EEG patterns registered for all the subjects and, consequently, the differences in music produced by the interpretation of the collected signals.

Main result consists in observing that great part of variations in EEG signals occurs in beta (associated to active thinking, active attention, focus on the outside world or solving concrete problems) and alpha (related to relaxed awareness, meditation, contemplation) bands. This fact, and results of other experiment about the correspondence between electroencephalographic signals and music found in literature (Bhattacharya et al., 2001; Dan et al., 2009), induced us to perform the following experiment 2, that is the conscious reproduction of single music note, using, as main factor to determine the single sound, beta and alpha waves for their strong relation to evocative power, active thinking and cognitive processes.

We could also observe that there are some similarities among subjects' brainwaves, so we decided to proceed searching for EEG waves values ranges corresponding to specific sounds (i.e. single music notes) listened by the subjects.

---

[3] http://www.bside.unimi.it/brainmusic/bm.html

3.2 Experiment 2: conscious reproduction of music

The second experiment has been performed to realize two objectives: (1) test the developed application, to refine the performance of the developed software tools; (2) train the subjects to make a task consisting in mentally evoke a single music note, on the basis of the received stimuli, and transmit the brain signal, through the BCI device, to the developed application, reproducing the listened sound.

The second point concerns also the Research field and the investigation of the functioning mechanisms of human brain and, especially, of the subjects' ability in training, to make them able to control the production of specific EEG signals.

In the experiment we considered the beta rhythms for reproducing the music note. This band, in fact, is the most related to music-related problem solving brain function (Bhattacharya et al., 2001; Zumsteg et al., 2004). The experiment 2 involved 20 subjects, among which 9 female. The subjects have been immersed in a familiar environment, wearing headsets, completely isolated by the external world remaining concentrate on the sound listened in the headsets (the single note). In the first phase of the experiment we created a patch in Max, consisting in a simple metronome sending to the MIDI keyboard the value 69, corresponding to the note A, for 1000 milliseconds. After we followed the same procedure for the value 37, corresponding to the note C, as for the other 7 music notes.

Through the same sketch Processing used in the previous experiment, we proceeded to collect beta rhythms data, reading the oscillations induced by the note A and C on the beta rhythm. To do this, we collected data twice: the first time in absence of any acoustic stimuli and in absolute silence, to have a neutral baseline for each subject; the second time with the only presence of the notes A or C or others played for one second.

For all the music notes, we compared the curves corresponding, respectively, to the "listening of the silence" and the "listening of a music note". We after performed power average measures and row data values range close to the acoustic stimulus. Measures showed that in presence of the note, beta waves often give a specific range of raw data values. For example, for the note A, raw data values often were in the range 32000 and 35000, while for the note C the signal tended to give values in the interval (23000; 27000). Moreover, the first attempt to make subjects able to reproduce the listened notes revealed that the needed training time was too much. The difficult, in fact, was not by the computer side in interpreting the subjects' brain signal, but on the subjects side for the difficulties to focus only on the sound, without distracting.

Other attempts to reach the success in making subjects able to reproduce with their mind the target sound (the single note) demonstrated that the subjects needed to try more than 3-4 times, before having success. Then we though to apply a reinforce stimulus to help subjects. Reducing time, in fact, is fundamental to obtain an easy-to-use tool (Wolpav, 2006). We chose, so, to associate a visual and a motor stimulus to the listening of the note, also considering some results in literature (Peretz and Zatorre, 2005; Zatorre et al., 2007). The solution consisted in requiring to subject to make a gesture while observing an image and listening the note for one second, choosing for each note a specific gesture and a specific image. Adopting this solution, the training time has been reduced, in all the cases, of at least 40-50%. The following step, to refine the developed tool, has been to setup a sketch in Processing to use the discovered characteristics.

Once prepared the software with the individuated refinements, the subjects have been invited to think to the observed colours and to the note listened previously, associating the corresponding motor stimuli, following the instructions.

Table 1 Examples of notes and associated visual and motor stimuli.

Note	Visual stimulus	Motor stimulus
A	orange	Knock the first finger of the right hand on the thumb of the left hand
C	blu	Knock all the fingers of the left hand on the palm of the right hand

Max has been used to experiment the difference among the listening of the notes, while Processing for its capability in writing on a file the data collected by the BCI device. Such data have after been processed and transferred in *Matlab*™ for measures and comparisons. The aim of this phase has been to understand what values range correspond to the listening of the notes, in presence of visual and motor stimuli. After, we created a sketch in Processing that, when beta waves reached the value in the defined range, send to the Max patch, developed to receive the signal from Processing, through a MaxLink, the number corresponding to the MIDI values (Dan et al., 2009) for the selected notes, for example the value 69 for A or 37 for C. The patch, when the value has been received, sends the value to the controller MIDI, playing the corresponding note.

Figure 2 A user playing notes using the software application.

The association of other stimuli to the listening of the note makes subjects easy to concentrate. Also in presence of the reinforce stimulus the results have shown that for each note the same specific row data values range correspond to the listening of the note: for the note A the raw data values were in the range (32000; 35000), while for the note C the range was (23000;27000), as previously discovered. We obtained similar (ranges) results for all the subjects. After a few minutes training phase, subjects are able to reproduce the notes just thinking to them, having immediate success in the 40-50% of the cases. There is sometimes a latency due to the BCI for the temporal difference from the will to reproduce the note and the effective reproduction.

From experiment 2 we obtained that, after a relatively short training and with the help of a visual and a motor stimulus, the subjects have been able to reproduce the requested notes. This technique is often used in a similar way to train the users in executing virtual actions on a computer, such as, for example, the rotation of a cube or, linking a BCI to an electronic device, to control it. A part from obtaining the first prototype for creating sound by brainwaves, the experiment demonstrates a correlation between the execution of an action and the will to execute it, also when the action is mainly non-motor. Moreover, shortening the training, by a combination of visual and motor reinforce associated to the listening of the note, induces to think that the same approach could be successful in many cases of neurological disease, for which the cognitive rehabilitation could accelerate the time needed for the recovering of functions lost for traumas or pathologies (Varela ct al., 2001).

4. CONCLUSIONS AND FUTURE DEVELOPMENTS

The main aim of our experiment consisted in developing a tool for unconscious and conscious production of music by brain, in the second case reducing, through appropriate stimuli, the training time needed by subjects and allowing to a generic user to reproduce any single note. The results of the experiments gave us the possibility to verify that with the alone EEG signal a subject could need a long training. In fact, only after 4-5 listening of the note, every 3-4 attempts, subjects were able to correctly reproduce the note. Better results have been obtained if to the listening of the note we associated a motor and a visual stimulus. In such a way, in fact, we potentiated the subjects' ability to concentrate on the task, with a consequent increasing and differentiation in beta waves. Consequently, a subject is able to reproduce the specific target note. The developed software included this results, so the needed training time has been strongly reduced.

The application implements the following characteristics: (a) each note (to be reproduced after) is listened just one time by the users; (b) before the reproduction of the target note, the program gives to the user the instruction asking to associate a simple gesture to the listening (the software suggests a different gesture for each note); (c) at each listening of a note the software shows the associated image (the name of the note on a different note-specific colour background).

Thanks to this system a generic user can be trained listening all the seven notes in the same session, and, after, correctly reproduce them just evoking the associated sounds, images and gestures. For the most part, users are able to reproduce from three to all notes at the first use of the software. The work also opens new scenarios for further developments, for example to obtain complex melodies, but also for future Research scenarios, for possible investigation in subjects' brain training to perform a specific answer to a stimulus.

REFERENCES

Allison B.Z., Wolpaw E.W., Wolpaw J. R. 2007. Brain-computer interface systems: progress and prospects. *Expert Rev Med Devices*, 4(4):463-74.

Bhattacharya J., Petsche H., Pereda E. 2001. Interdependencies in the spontaneous EEG while listening to music. *Internationl Journal of Psychophysiology*. 42:3, 287-301.

Blood, A.J. and Zatorre, R.J. and Bermudez, P. and Evans, A.C. et al. 1999. Emotional responses to pleasant and unpleasant music correlate with activity in paralimbic brain regions. Nature neuroscience, NATURE AMERICA. 2, 382-387

Chuchnowska, I. and Skala, A. 2011, An innovative system for interactive rehabilitation of children at the age of three. Archives of Materials Science, 50.

Dan W., Chao-Yi L., De-Zhong Y. 2009. Scale-Free Music of the Brain. *PloS ONE*.

Friedman D. et al. 2007. Navigating Virtual Reality by Thought: What Is It like? *Presence*. 16:1, 100-110.

Mauri M., Magagnin V., Cipresso P., Mainardi L., Brown E.N., Cerutti S., Villamira M., and Barbieri R. 2010. Psychophysiological Signals Associated with Affective States. *Conf Proc IEEE Eng Med Biol Soc.* 3563–3566.

Minsky M. 1981. Music, Mind and Meaning. *A.I. Memo*, M.I.T. A.I. Laboratories. 616

Nakamura S., Sadato N., Oohashi T., Nishina E., Fuwamoto Y., Yonekura Y. 1999. Analysis of music – brain interaction with simultaneous measurement of regional cerebral blood flow and electroencephalogram beta rhythm in human subjects. *Neurosci.* Lett. 275, 222–226,

Nijholt A, Tan D., Pfurtscheller G., Brunner C., et Al. 2008. Brain–computer interfacing for intelligent systems. *IEEE Intell. Syst*. 23, 72–9,

Pascual-Leone, A. 2001. The brain that plays music and is changed by it, *Annals of the New York Academy of Sciences*, Wiley Online Library. 930:1, 315-329.

Peretz I., Zatorre R.J. 2005. Brain Organization For Music Processing. *Annu. Rev. Psychol*, Annual Reviews. 56, 89–114.

Pfurtscheller G. and Neuper C. 2001. Motor imagery and direct brain-computer communication. *Proceedings of the IEEE*. 89:70, 1123-1134.

Skaric L., Tomasevic M., Rakovic D., Jovanov E., Radivojevic V., Sukovic P., Car M., Radenovic D. 2007. Electrocortical (EEG) correlates of music and states consciousness. *Neuropsychological Trends*.

Varela F., Lachaux J.P., Rodriguez E. and Martinerie J. 2001. The brainweb: phase synchronization and large-scale integration. *Nature reviews*, Macmillan Magazines Ltd. 2, 229.

Zatorre R.J., Chen J.L., Penhune V.B. 2007. When the brain plays music: auditory-motor interactions in music perception and production. *Nat.Rev neurosci*, 8, 547-558.

Zumsteg D. Hungerbuhler H., Wieser H. 2004. Atlas of Adult Electroencephalography, hard cover, 178

Yasui, Y. 2009. A brainwave signal measurement and data processing technique for daily life applications. Journal of physiological anthropology,J-STAGE. 28:3, 145-150.

Wickelgren I. 2003. Tapping the Mind, *Science 24 January*, 299:5606, 496-499.

Wolpav J.R., Birbaumer N., McFarland DJ, Pfurtscheller G., Vaughan T.M. 2006. The Berlin Brain-Computer Interface: EEG- based communication without subject training.

CHAPTER 73

Digital Museum Planner System for Both Museum Administrators and Visitors

Stefan Ganchev, Kegeng Liu, Lei Zhang

Iowa State University
Ames, USA
sganchev@iastate.edu

ABSTRACT

Large museums are often overwhelming for first time visitors. For families and individuals with limited time, museum touring becomes a daunting task. Without professional assistance or an in-depth planning, it becomes impossible for them to narrow down the most significant pieces to see. Also, managing and organizing enormous collections needs systematic approaches.

The purpose of this paper is to present *Digital Museum Planner*, a system designed to improve the experience of both museum administrators and visitors. The visitor interface is designed to assist the needs of museum guests with limited time to tour the museum. For museum staff, the administrator interface will provide a web accessible application for sorting artwork according to age, interest, and educational background. For museum visitors, the system will use touch-screen kiosks for easy and quick interaction. Visitors will be asked to input their age, interests, background, and/or time constraints. *Digital Museum Planner* will then generate a recommendation list and a locations map of artworks that can be printed or delivered to the user's phone.

The methods used to design this system include interviews with museum staff and art historians, prototype development, and user testing. Through user studies and task analysis, we discovered the proper design practices and strategies needed for the development of the *Digital Museum Planner*. Eight users were tested with both interfaces. The results were used to analyze issues with the system and make appropriate changes.

Keywords: Museum Planner, Usability, Visitors, Staff, Touring Planner

1. INTRODUCTION

Technology advances in computing and wireless systems have great potential to improve a person's experience when visiting a museum. For more than 50 years museum touring has been accompanied by handheld electronic technologies, from the shortwave "ambulatory lecture" of the Stedelijk Museum, introduced in 1952, to contemporary "unofficial" podcasts, developed by students and distributed online to the public. During this time, technology, museums, and visitor expectations have changed, along with our understanding of those three (Atkins, L.J., 2009, P1150).

Digital guides have great potential for museum touring. Many prominent museums (MoMA, American Museum of Natural History, San Francisco Museum of Modern Art, National Gallery in London) are working on practical digital tour platforms (Tedeschi, B. 2010, P6). They are focusing their efforts on iPhone, iPod and Android applications with enhanced multimedia experiences that include pictures, text, audio, video and mapping systems. The Museum of Natural History Explorer, for example, features a navigation system that helps users find exhibits and museum facilities more easily than with a printed map (Tedeschi, B. 2010, P6).

In the book "Digital Technologies and the Museum Experience: Handheld Guides and Other Media", the authors detailed studies of a variety of handheld technologies in different settings, including audio tours, cell-phone technology, personal data assistants. The devices address several key problems: aiding the visitor in customizing the visit, providing contextualizing information, linking the museum with visitors' daily lives, enhancing group interactions, engaging young children, and allowing visitors the opportunity to add their experiences and interpretations to the museum's curatorial voice. (Tallon, L. 2008, P238)

Reflections on the past 50 years of innovation around technologies in these diverse museums present common themes not of design and innovation, but how the museum, visitor, and objects can and should interact to construct meaningful experiences and deeper learning opportunities; experiences that structure but do not prescribe visitor engagement in the museum (Atkins, L.J., 2009, P1150).

The purpose of this research was to design and test a system that addresses these issues and delivers a positive experience to the users. Through interviews with Museum staff and Art History faculty members, several key issues related to museum visits were found:

1) The staff needs to plan exhibition programs and help visitors plan their tour through analyzing them by their ages and educational backgrounds. However, they are not able to cover all specific needs of every visitor. This becomes a daunting task for large museums where visitors are from diverse backgrounds and have limited visiting time.

2) Museum staff is not extremely tech-savvy. They are mostly familiar with Microsoft Office products.

3) The educational background of visitors is an important concern when a touring plan is devised.
4) Museum visitors often have limited time and may want to see a museum without any pre-planning.
5) Large museums can be overwhelming for visitors and they might not feel comfortable seeking help.
6) Large museums often become very busy and lack the staff to support museum tours without an extensive waiting period.
7) Even if visitors get a tour guide, the tour can take significant time and might not be customized to their interests, age group or background.
8) Museums provide marketing materials, such as brochures, used by visitors for a guide when they tour the complex. However, these materials are designed for the general public, not the specific individuals. In large museums, visitors are very diverse and come from different backgrounds.

2. DESIGN DEVELOPMENT PROCESS

To meet the challenges stated earlier, a new system for museum touring, called *Digital Museum Planner*, was designed. It is divided into two sections, a visitor and a museum administrator side. The visitor side is an interface that museum guests can use to filter exhibits based on age groups, interests, and educational backgrounds. The interface also allows for the users to select a time frame which is used by the system to calculate the appropriate amount and choices of artworks. When the system completes this process, it provides the users with a list of their tour, as well as a map showing where the selected artworks are located. The second section of the system, the museum administrator side, is designed to facilitate tour data input. This interface is managed by the museum staff and allows users to manage the categories (age, interests, education background) related to an artwork. The main goals pursued in the development of the *Digital Museum Planner* system were to alleviate the work of museum education and marketing departments and to improve the visitors' experience when touring museums.

The process of design of the new *Digital Museum Planner* system included interviews with museum staff and art historians, research of existing museum software, prototype development and user testing.

2.1 Interviews

Several interviews were conducted. Researchers approached the Iowa State Brunnier Art Museum staff and Iowa State Art History faculty members. Several key points were recognized:
1) Staff members of large museums and galleries have background in art history and interdisciplinary subjects.
2) Staff members need to plan exhibition programs and help visitors plan their tour by considering their ages and educational backgrounds. However, they are not able to cover all specific needs for every visitor.

3) Museum staff is not extremely tech-savvy. They are mostly familiar with Microsoft Office products.

4) They have a specific computer system to manage all of their collections: The Museum.

5) In large museums and galleries, there are several departments, such as Education and Marketing, managing the museum collections.

6) Museum staff that is responsible for the tours would rate works based on the canon, culture, personal preferences, aesthetics and other perspectives (ex. Politics).

2.2 Existing Software Research

The research team analyzed several software systems developed for the two user groups: museum administrators and museum visitors. Tools for administrative purposes focus on collections, data management, data query and report. Touring systems for museum guests are mainly in the form of a portable device called Audio Guide/Mate which plays pre-recorded audio introductions of collections during the users' browsing period.

The Canadian Heritage Information Network (CHIN) conducted several editions of in-depth evaluations on major commercial museum collections management system/software. In one of their 2003 evaluation edition, they compared 16 museum management systems that had been in use worldwide at that time. Of the top three museum management systems, The Museum System (TMS) is one of the most popular museum collections management systems. TMS is also in use by the University Museum at Iowa State University. The system allows museum staff to capture, manage and access their collection information and facilitates their daily activities such as cataloging, media tracking, and coordinating exhibitions. TMS is an open architecture and its collection data can be easily and seamlessly integrated with other management systems. Although TMS is a powerful collections management and data organization tool, it does not provide any functions that help museum staff to plan tours for visitors.

2.3 Prototype Development

Based on research analysis and ideation sessions, two functional prototypes (visitor and administrator interfaces) of the *Digital Museum Planner* system were developed to conduct user testing. For the administrator interface, the research team developed a labeling system for easy category management of artworks. The interface allows users to quickly search artwork collections and use drag-and-drop functionality to attach category labels (age, educational background, location) to each work. Once they complete the operation, the data is saved to the *Digital Museum Planner* system database.

For the visitor interface, the team developed a system that allows museum guests to input their ages, interests and time available. After completing this, the interface provides them with a list of recommended works and a map showing the location of each work. The users can choose whether to print this information or send it to their smart phone.

Both of the prototypes were developed for standard computer displays. This was done to facilitate easy screen capture during usability studies.

2.4 User Testing

For usability testing, the researchers recruited volunteers through email and poster announcements. Eight participants from three target groups (graduate students, faculty, professionals) were tested in Fall 2011 to examine the usability of the *Digital Museum Planner* prototypes. This test included three non-native speakers. All participants signed voluntary informed consent release agreements before testing was conducted. Participants were asked to complete a demographic questionnaire before the usability test. A total of thirteen tasks were given to all participants: ten tasks were related to the administrator interface and three tasks were related to the visitor interface.

The tasks were designed to have specific end goals and to be completed in a reasonable length of time. The entire usability test, including the exit interview and survey, was designed to be finished in forty minutes. The tasks were as follows:

1) Log into the system with user name and password
2) Search for all works under medium "Ceramics" and "Sculpture"
3) Add label "First Floor" to the resulting images
4) Under labels, delete the one that says "Modern"
5) Under labels, create a new label for "avant-garde"
6) Do another search for only movie collections
7) Apply the label "avant-garde" to the resulting images
8) Change label "fourth floor" to "third floor-tier"
9) Filter search results for all images with label "6-12"
10) Apply "Classic" label to all images resulting from the previous search
11) Find the works that you think are appropriate for children in elementary school and select 1 hour as your time limit
12) Review the list of works that are displayed
13) Print the list from the device

Tasks one through ten were related to the administrator interface, tasks eleven through thirteen were related to the museum visitor interface. When participants spent more than five minutes on a task and repeated the same error while conducting the task, they were asked to move to the next task. Except to terminate a task, the researchers only observed the participants' performance without intervening during the usability test. Also, participants were asked to speak out their thought process while they were performing each task.

Most of the tests were completed in a quiet classroom, two were completed in quiet offices in Iowa State University. All of the tests were supervised by one or two team members. The participant's voice and mouse movements were recorded by a screen capture technology. After usability testing, an exit interview and survey were conducted.

3. USABILITY RESULTS AND DISCUSSION

Usability data was analyzed and evaluated based on participants' success rate and time spent per task. This analysis allowed the research team to establish the rate of effectiveness of the interfaces and also recognize problems with the system that need to be addressed. Participants' navigation paths were also observed and analyzed. The team also gathered and analyzed data from exit interviews and voice recordings from the videos to get more direct feedback about the system.

3.1 Participants Information

Participants in the study were college graduates, working on a graduate degree, or had a graduate degree. There were three non-native speakers. All users were comfortable using computers.

Table.1 Participant Information

Age					Gender		Language		Education	
18-23	24-29	30-35	36-41	Over 42	Male	Female	Native	Non-native	College Grad	Adv. Degree
12.50%	37.50%	25.00%	0.00%	25.00%	25.00%	75.00%	37.50%	62.50%	62.50%	37.50%

3.2 Performance Data

The performance data was analyzed to examine the time spent to complete each task and the success rate for correctly accomplishing a task. The overall success rate shows that most of the tasks, except three of them, were finished successfully. There was one participant who failed to follow the proper procedures on most tasks. This is the reason why the overall score for the majority of them is at 87.5%. Tasks 2, 6, 7 and 9 proved most difficult with overall success rates at or below 50%. Task 6 also took significantly more time. The results are shown in Table 2.

Table. 2 Participants Performance Data

	Task 1	Task 2	Task 3	Task 4	Task 5	Task 6	Task 7
Time Spent	25.63	62.63	60.5	42.88	53.63	86.38	46.5
Success	100.00%	50.00%	87.70%	87.70%	87.70%	50.00%	50.00%
	Task 8	Task 9	Task 10	Task 11	Task 12	Task 13	Task 7
Time Spent	66.5	48.38	27.13	59.75	13	18.5	46.5
Success	87.70%	37.50%	87.70%	100.00%	100.00%	100.00%	50.00%

3.3 Navigation Path and Video Analysis

To assess how test participants navigated the interfaces, a navigation path analysis was conducted. After analyzing navigation paths, researchers recognized the tendency for some participants (3 out of 8) to use "View All works" as a preferred option when looking for artworks. This option will still take them to the results screen but it will show all artworks instead of the specific ones they are interested in. In addition, task 2 took on average significantly more time to complete than other tasks in the study.

After analyzing video data, participants were observed struggling to complete some of the label actions. They needed to scroll up and down the screen in order to select the label and then click on the desired action button. The analysis of user navigation paths for task 3 showed that participants confused "apply" label functionality with adding and editing of labels. Only one participant completed this task as expected. Hierarchy issues were also observed. Participants were confused with the location of some action buttons. From analyzing user navigation paths for task 3, 7 and 10, a tendency was observed for participants to drag the labels to all of the artworks (task 3: only 2 participants checked all; task 7: 1 participant used check all; task 10: none) instead of using the check all button and apply, which is the faster way to complete the task. After analyzing the navigation path for task 9 (filtering by label), a serious design problem was observed. None of the participants completed the task directly as expected. In total, only 3 participants were able to complete the task after several trial and error clicks. The analysis of user navigation paths for task 6 (using the advanced search drop down) showed a pattern of confusion among participants. Only 3 of them saw the search expand button at the very beginning. Others conducted several trial and error clicks until they got to the search drop down. There were 2 participants who never used it. From video analysis and the error rate in task 3, the research team recognized there were too many steps in applying a label without dragging it. Through navigation path analysis of task 11 (participants were asked to select an age group and time), the team observed that a lot of participants did not use the select button to confirm their choice (only 3 participants used select for age group). After video data analysis, the research team recognized that many participants could not see the "Next" button unless prompted to use the scroll bar of the browser.

3.4 Exit Survey

Five exit interview questions were given to participants after the user test. On the question of what is their anticipation for future development of the system, participants responded that ease of use can be improved. Also, several participants saw the need of such system for museums. When asked what suggestions they have to improve the system, participants responded that some of the directions and button actions were unclear. One interesting response commented on the concept of *Digital Museum Planner*. The participant saw the value in customizing tours based on demographic data but also believed that some people might find it offensive. On the question of what they liked most about the product, participants responded that they enjoyed the visual design, simplicity of navigation, and the concept of a personalized tour. When asked what was most frustrating about the system, participants responded that they had hard time finding and using some of the buttons and labels. On the question of will they use the system in their everyday life, most participants saw a benefit of this product in a museum environment. Some participants did not see how this system will be applicable to their everyday life.

4. DESIGN RECOMMENDATIONS

After analysis of the gathered user test data, the research team created the following recommendations for changes of the system:

1) Change the titles of each section of the homepage screen and increase font sizes for better legibility of the headings. Add an icon for each title (ex. magnifying glass next to the heading above the search form).

2) To decrease the response time of the participants, the research team had a general recommendation for revising the hierarchy of the labels section. Under labels section in the sidebar, categories will be more prominent and more separated (button or box style). When the user has a category selected there will be a change of state in the graphic.

3) To avoid losing visibility of the action buttons, the team recommended moving the buttons under each category above all labels.

4) To avoid confusion about button actions, the team recommended improvement of the icon designs for each button and changing some of the button names (add $\rightarrow$ create new).

5) To minimize confusion about selecting all works, a recommendation was made to convert "Check All" from a button to a check box.

6) Filtering by label was one of the main challenges for the user test participants, the team recommended simplifying this functionality.

7) Several participants were not able to locate the search drop down and how to activate it. The team concluded that "advanced search" button is not prominent enough. The recommendation was to enlarge it, and make it more prominent for the user.

8) The research team had a general recommendation of increasing the size of the "go" button on the visitor side interface.

9) Confusion with the use of the "select" button on the visitor interface prompted a recommendation to remove it.

10) Several test participants could not locate the "Next" button on the visitor interface. The team recommendation was to move the "Next" and "Previous" buttons at the top of the interface.

CONCLUSION

Digital technologies have greatly affected our lives in recent years. Digital guides have a great potential to improve the museum experience for visitors. This paper introduced *Digital Museum Planner*, a system that addresses this challenge and provides museum staff with a platform that can potentially alleviate their work.

This research discussed the methodology used to develop *Digital Museum Planner*, including usability studies. The usability testing allowed the research team to observe how participants interact with the system and also to receive comments related to their experience. The data gathered was analyzed and some key issues were detected. This process helped the team make design recommendations for the next version of the system. Iteration between interface design and usability testing will provide a better user experience and service to *Digital Museum Planner* users.

ACKNOWLEDGEMENT

We would like to thank Sunghyun R. Kang for her guidance during this research.

REFERENCES

Atkins, L. J. (2009). Digital technologies and the museum experience: Handheld guides and other media. Science Education, 93(6), 1149-1151.

Grobart, S. (2011, March 17). Multimedia Tour Guides on Your Smartphone. New York Times. p. 1.

Schneider, K. (2010, March 18). The Best Tour Guide May Be in Your Purse. New York Times. p. 7.

Tallon, L (2008). DigitalTechnologies and theMuseum Experience: HandheldGuides and Other Media. Altamira Press. xxv + 238 pp.

Tedeschi, B. (2010, September 9). Apps as Tour Guides Through New York Museums, Step by Step. New York Times. p. 6.

http://www.chin.gc.ca/cmsr/CMSR/index.cfm?lang=en

Affective Interactions: Developing a Framework to Enable Meaningful Haptic Interactions over Geographic Distance

Martina Balestra, Caroline G. L. Cao

Tufts University
Medford, MA
{Balestra.Martina, Cao.Caroline}@gmail.com

ABSTRACT

Affective haptics describes the physical experience of connecting emotionally with another person. Research on affective haptics and its application is sparse, so a framework comprised of the affective components of touch as well as their sensory scales is needed as the foundation for a reference tool to inform interaction designers. Such a framework requires first understanding which gestures and emotions are relevant followed by the structure of the relationship between gestures and emotions. Two pilot surveys were conducted to identify relevant gestures and emotions, and informed a third, formal survey to explore the perception of affective haptic components in consolatory interaction. This is needed due to the notable lack of vocabulary in the literature to describe affective touch. This work will ideally inform interaction design models for affective haptics and provide a basis for additional studies examining the relationships between affective and discriminative haptic components. Future work could quantify how different tactile variables of a gesture alter the perception of the emotional content.

Keywords: Haptics, Affective Haptics, lexicon, mapping

1   INTRODUCTION

Despite evidence indicating that haptic interaction is a primal means for conveying emotion, research on affective haptics is sparse, preventing designers and engineers from being able to simulate social haptic feedback in digitally mediated and socially sensitive or impoverished interactions (MacLean, 1999). A framework detailing how gestures and emotions are related as well as the relationships between their descriptive and affective components is needed as the foundation for a reference tool to inform the design of such interactions. In this exploratory study we try to first identify a vocabulary to describe the sensory scales of affective components of haptic gestures to inform future studies that can detail the relationships with discriminative components.

1.1 Haptics

The *haptic sense* – broadly defined as the sense of touch and the tactile bodily experience - evolved to permit us to complete tasks, probe objects for their state or qualities, communicate messages, elicit reactions, enjoy aesthetic pleasures or comforts, or to connect physically or emotionally with another person or living thing (MacLean, 1999) (Montagu, 1971) (Turner, 2000) (Loomis & Lederman). In general haptic sensation can be divided between its kinesthetic and tactile components. The kinesthetic attributes are triggered by the movement and position of the limbs and originate in the muscles, tendons and joints. Conversely, the tactile attributes originate on the dermis and epidermis of the skin and account for sensations of texture, temperature, hardness, moisture and (to a certain extent) movement against the skins surface.

Human touch has the unique ability to be either passive or active, that is – we experience haptic sensation both *in* and *out* of our control (Loomis & Lederman). The autonomic physiological changes we experience (heart beat rate, body temperature, etc.) are latent and uncontrollable, whereas physical stimulation (e.g. tickling, hugging) can be actively controlled and experienced (Loomis & Lederman). Moreover, physiological changes and physical stimulation are not mutually exclusive: autonomic physiological changes can occur as a result of, or simultaneous to, stimulation and vice versa. The ability to *consciously engage in haptic interaction* provides the most intentional and bi-directional sensory channel, thereby making it the most intimate and emotionally loaded (Tsetseukou & Neviarouskaya) (MacLean, 1999) (Montagu, 1971) (Turner, 2000). In addition to quantifying the objective somatic components of haptic gestures, we believe it may be possible to anticipate the emotional intent of a social haptic gesture by studying a sensory framework for its affective components.

1.2   Discriminative & Affective Haptics

The haptic sensations experienced in both latent and active experiences may be comprised of affective and discriminative haptic experiences. "Affective" touch pertains specifically to the emotional response or social content of the haptic gesture whereas "discriminative" touch refers to the quantitative somatic physical attributes of the object or interaction (e.g. shape of device, material, texture) (Essick, McGlone, & Dancer, 2010).

The discriminative components of touch define and parameterize the specific tactile experience for context. Karon Maclean (MacLean, 1999) describes some of these characteristics and their respective perceptual scales:

Discriminative Component	Sensory Scale
Information Direction	Manipulate ←→ Perceive
Continuity	Continuous ←→ Discrete
Rate	Static ←→ Rapid
Freedom	Constrained ←→ Free
Effort	Strong ←→ Light
Resolution	Precise ←→ Fuzzy
Passivity	Active ←→ Passive
Location	Focused ←→ Whole Body
Multimodality	Multisensory ←→ Haptic
Mediation	Mediated ←→ Direct
Grounding	Portable ←→ Fixed

Table 1 Tactual components and corresponding sensory scales

There exist established numerical ranges identifying the extreme thresholds for each sensation and characteristic, as well as the *Just Noticeable Difference (JND)* for incremental perception across the scales (Srinivasan & Basdogan, 1997) (Sherrick, 1985). These allow researchers and designers to apply salient and informed tactile profiles to their product designs. Conversely, there is a lack of research specifying affective attributes relevant to specific social situations and their sensory scales. To begin to understand if such sensory scales can be isolated and refined for the *affective components* of haptic interactions we attempted to identify emotions and gestures relevant in a specific social situation, then began to establish a lexicon describing the possible sensory ranges to inform future studies.

2   METHODOLOGY

To determine how people perceive social content in touch we focused on exploring positive interaction or benevolence experienced in consolation with survey participants. The most satisfying and meaningful interactions tend to be those providing positive emotional intent (Gemperle, DiSalvo, Forlizzi, &

Yonkers). Benevolent interactions in which the emotions conveyed through touch are intended to improve the disposition of, or elicit a positive response from, the receiver consists of several of the most universal and culturally powerful haptic gestures, like a hug or a kiss. We expected consolation to therefore act a strong case study to isolate specific affective components and descriptions.

Two pilot questionnaires were used to identify the most salient emotions relayed between consoler and consolee in consolatory interactions and were inspired by Marco Rozendall et al.'s work on eliciting and using specific lists of adjectives to study pleasantness in tactility (Rozendaal & Schifferstein, 2010). Results from the pilot study indicated that the most prevalent affective components in consolatory interactions were: strength, support, reassurance, understanding, shared experience, and sympathy. The most commonly used gestures to communicate these sentiments included: hugging, arm around the back, hand on the shoulder, hand rubbing, hand holding, back rubbing, kissing, and hair stroking. There were eighteen anonymous responses from men and women without disability representing a wide age and socio-economic range. These results formed the basis for a formal survey distributed to 32 participants. The questionnaire tested the pilot results and attempted to solicit a collection of adjectives describing the perceptual extremes of each predetermined affective component. In particular, the questionnaire prompted the free-text description of how a participant felt when they were experiencing both the high and low sensory extremes for the collection of relevant affective attributes. The formal survey also served to provide insight into the most frequently used gestures to communicate each affective component.

3   RESULTS

An average of 45 terms were used to describe each extreme on the sensory scales. Redundant terms and synonyms were collected to identify the most common and salient sensory extremes. Table 2 depicts the dominant responses for sensory scale extremes and gestures for each affective component.

Affective Component	Sensory Scale	Gesture
Strength	Weakness ←→ Invincibility	Hand on Shoulder
Support	Loneliness ←→ Solidarity	Hugging
Reassurance	Uncertainty ←→ Calm	Hugging
Understanding	Isolation ←→ Connection	Hand on Shoulder
Shared Experience	Loneliness ←→ Connection	Hugging
Sympathy	Loneliness ←→ Loved	Hugging

Table 2 Affective components and corresponding sensory scales

Strength

Participants recalled very specific sensations when asked about communicating emotional strength. In addition to "weakness" as the low extreme of the scale, they also frequently noted feelings of "worthlessness", "depression", and "helplessness". Similarly, participants also identified "empowerment" and "confidence" as relevant sensations at the high extreme. Placing the hand on the shoulder was noted as the most commonly used gesture to communicate emotional strength to another person, however hugging and placing an arm around the back were also frequently mentioned.

Support

"Loneliness" and "solidarity" were by far the most commonly referenced emotions to describe the sensory scale extremes for communicating emotional support. The majority of participants referenced "hugging" as the most appropriate way to communicate support, however placing an arm around the back was also cited.

Reassurance

In addition to "uncertainty", participants equally cited "hopelessness", "discouragement" and "insecurity" to describe the low sensory extreme for communicating reassurance. Similarly, "confidence", "security" and "solidarity" were also cited almost as frequently as "calm" to describe experiencing the high sensory extreme of being reassured. The majority of participants cited hugging as the most appropriate method to convey reassurance, however back rubbing was also represented.

Understanding

"Frustration", "misunderstood", and "confusion" were cited in addition to "isolation" to describe how participants feel when they perceive they are not understood. While "connection" (or empathy) was widely cited as the most salient emotion associated with high understanding, participants also frequently mentioned "relief". Placing a hand on the consolee's shoulder was the most frequently identified gesture for relaying understanding, however placing an arm around the back was also referenced.

Shared Experience

The majority of participants cited "loneliness" to describe the way they felt when they perceived others could not relate to their experiences. "Connection" was cited most frequently to describe perceiving a high level of perceived shared experience. Participants cited "hugging" as the most common gestural response, however "hand holding" was also represented by a narrow margin.

Sympathy

In addition to "loneliness", participants frequently cited "others' apathy" and "contempt" to describe how they felt when they perceived low sympathy from others. While the majority of participants also cited feeling "loved" as the overwhelming sentiment when they perceived sympathy to be high, they also frequently mentioned "empathy" and "thankfulness". The majority of participants cited hugging as the most frequent mechanism for conveying sympathy, however back rubbing was also mentioned.

4 CONCLUSION

Affective haptics describes the physical experience of connecting emotionally with another person and is believed to be among the most visceral and high fidelity from birth throughout life. By initiating physical touch with another human being, one commits an *intentional* and *invasive* act with social implications intended to convey a gamut of emotions ranging from the abusive to the marvelous. There is relatively little research on affective haptics, so a framework describing common gestures and affiliated emotions is needed as the foundation for a reference tool to inform interaction designers. This study attempts to lay the foundation for an interaction model by isolating specific affective components of social interaction and beginning to investigate corresponding sensory scales analogously to the perceptual framework for discriminative haptic components detailed by (Maclean, 1999).

The vocabulary elicited from survey participants describing the extremes of each sensory scale begins to describe how participants perceive the different facets of consolation. Additional studies should be conducted to refine distinctions between the terminology identified, and to begin to explore incremental differences on the sensory scales between the emotional extremes. We also attempted to provide a preliminary perspective on the specific gestures participants associate with relaying the emotions described in the sensory scales. We hope that the information provided by this investigation can inform future studies quantifying how different tactile variables of a gesture alter the perception of the affective sensory scale.

REFERENCES

Essick, G., McGlone, F., & Dancer, C. (2010) Quantitative Assessment of Pleasant Touch. *Neuroscience and   Biobehavioral Reviews,* 192-203.

Gemperle, F., DiSalvo, C., Forlizzi, J., and Montgomery, E. (2003). The   Hug: An Exploration of Robotic Form for Intimate   Communication. *Proceedings   of   RO-MAN03.*

Loomis, J., Lederman, S. (1986). *Tactual Perception.* Handbook of   perception and human performance, Vol. 2: Cognitive   processes and performance. Oxford, England: John

694

Wiley & SonsBoff, Kenneth R. (Ed); Kaufman, Lloyd (Ed); Thomas, James P. (Ed),

MacLean, K. (n.d.). Application-Centered Haptic Interface Design. *Human and Machine Haptics* .

Montagu, A. (1971). *Touching: the Human Significance of the Skin.* New York: Harper & Row Publishers.

Rozendaal, M., & Schifferstein, H. (2010). Pleasantness in Bodily Experience: A Phenomenological Inquiry. *International Journal of Design,* 55-63.

Sherrick, C. (1985). A Scale for Rate of Tactual Vibration. *Journal of the Acoustical Society of America* , 218-219.

Srinivasan, M., & Basdogan, C. (1997). Haptics in Virtual Environments: Taxonomy, Research Status and Challenges. *Haptic Displays in Virtual Environments.*

Tsetserukou, D., Neviarouskaya, A., Prendinger, H., Kawakami, N., and Tachi, S. (2009). *Affective haptics in emotional communication.* Proceedings from the InternationalConference on Affective Computing and Intelligent Interaction. Amsterdam, Netherlands.

Turner, J. (2000). *On the Origins of Human Emotion: A Sociological Inquiry into the Evolution of Human Affect.* Stanford: Stanford University Press.

CHAPTER 75

For the Emotional Quality of Urban Territories - Glazed Tiles Claddings Design

Carla Lobo 1, Fernando Moreira da Silva 2

Faculty of Architecture, Technical University of Lisbon; CIAUD – Research Centre in Architecture, Urban Planning and Design
Lisbon, Portugal
carla.a.lobo@gmail.com

ABSTRACT

Ceramic claddings, as visual and physical characteristics of architecture, interact with the environment, changing with the light source, intensity, and direction, inviting Human contact, and actively participating in the construction of a vital and stimulating urban space. Their intrinsic qualities (clay's plasticity, glaze's characteristics), extrinsic qualities (colour, gloss, texture), and emotional characteristics (easily recognized, familiarity, organoleptic qualities), can provide aesthetic pleasure, and emotional comfort.

This paper focuses on a significant and sustained renewal of ceramic claddings by substituting the conventional project development model, which is product centred, for a user-centric model considering the level of function and perception (emotional and sensorial): a new model of thinking related to the project, project development and management, where the perceptual aspects are considered as key elements in the product analysis, development, production and use.

Keywords: Glazed tiles, perception, emotional features, design model, urban user.

1   CONTEXT

The essence of cities is centred on the diversity of human activities, establishing itself as the space par excellence for meeting, and interaction, a place to trade experiences, sensations, and information. According to Abbasy-Asbagh (2011) the urban fabric functions as an agent of cultural preservation, reinforcing the importance that design, architecture, and urban planning have in the construction of spaces that strengthen the social memory, through the preservation of culture, social awareness and individual structures, connecting generations in a process of the transmission of cultural traditions and knowledge (W. Lim *apud* Ahmadi & Ardakani 2011).

The relationship between people and places are at the root of the "Sense of Place" (Weston 2008), so spaces that promote ties between different generations, contribute to the dissemination and consolidation of traditions, and the perpetuation of cultural identity, creating a sense of belonging in the inhabitant, and conferring qualities to the spaces that appoint it a Place. The need to see and think of the city as a space of social and emotional life, compels the (re)creation of the atmospheres of life that provide a quality of life, through the creation of meaningful and integrated forms that enthral the citizen/user.

Enhancing the quality of urban landscapes, to a large extent, depends on the diversity of materials used, and the colours and textures of the natural and artificial elements, and the balanced coexistence of the permanent and the transitory. As Birren (1978:38) stated "people require varying, cycling stimuli to remain sensitive and alert to their environments", underlining that neither overstimulation or monotony, are desired, as both promote the inability to respond to the stimulus. The allocation of sensory, emotional, and aesthetic qualities to the elements that make up the space, which take advantage of the proximity and coexistence of floor coverings, claddings, and textures and colours contributes significantly to the creation of memories and emotional ties (Lobo 2006). As Kuller (1980:87) noticed, architecture and the built environment do provoke emotions, "not the strong emotions that are easy to notice and identify, but rather the delicate result of a persistent everyday influence, which is probably of far greater impact than we might at first have thought or be willing to acknowledge". In this scenario it is important to recognise the pivotal role of the designer and of design, in the recreation of friendly atmospheres that reflect the needs and desires of the local population.

Since the dawn of humanity, the walls of the spaces occupied by man have served as a canvas for individual and collective expression. Glazed tiles claddings, as many other architectural surfaces, have been privileged spaces for communication; their secular durability and chromatic stability imbue them with a presence which is unique in urban spaces: they embody the tangible (the longevity of the materials and their visual language) and the intangible (meaning, symbolism and culture), they act as a humanising element in the spaces that withstands the test of time, bringing past memories to the present.

Figure 1 Portuguese glazed tiles as religious communication.

Cultural differences leave their stamp of individuality on local production, which is characterised by size variations, chromatic palette, the localisation of the areas where they are applied, diversity in the laying patterns and joints, as well as the multiplicity of patterns and motifs, or the tactile quality of the surface. These attributes, which are associated to production processes, which are specific to different geographical regions, imprint ceramic glazed tiles with characterising elements which express the local cultural identity. Knowledge and traditions are carried forward in the material, leaving marks and memories, the marks of the craftsman, unlike modern technological standardised materials, which are used on a massive scale, and contribute to the dehumanisation of human spaces (Shum 2009).

1.2   Glazed Tiles Claddings Production

Ceramic claddings, as visual and physical characteristics of architecture, actively participate in the construction of vital and stimulating urban space, as defend by Shum (2009), Birren (1978) and Kuller (1980).

698

Figure 2 Visual diversity of ceramic glazed tiles on urban environments as an emotional value.

Traditionally, the development and production of glazed tiles, was based on the use of locally available raw materials, which were more suited to the local landscape and the requirements arising from specific geographical and climatic factors, contributing simultaneously to the preservation of the identity of the locale, and the reduction of the environmental impact associated with the production. The adoption of local production technology and labour preserve traditional knowledge, stimulate economic growth, and maintains the self esteem of the population; a natural incorporation of creative solutions, that characterise the cultural uniqueness of the place, the process of production and use of glazed tiles allow updating of communicative languages, and preserve the sense of belonging to the place both culturally and emotionally.

Figure 3 Azulejos (Portuguese glazed tiles) as emotional and iconic references.

The current development process of ceramic tile claddings is usually an integrated process where the technical constraints (material, production and application), are linked to the objectives established by the manufacturer, depending on their market position, demand and market trends, and also the supply and availability from ceramic colours manufacturers, particularly in terms of graphics, paints, and glazes. Pieces are designed as independent units, not as continuous skin, characterized by their type of application, technical characteristics and quality of surface treatment.

1.2   Glazed Tiles Claddings (Re)Design

The importance of perceptual variations, both synchronic and diachronic, in the mental processes of interpretation of patterns and meanings, and spatial organization (Golledge 1999, Allen 1999, Humphrey 1980), as user/environment interaction facilitator's (Lockton), associated to the awareness of the importance of the human factor in the design process of public spaces, particularly those factors related to perception, consolidates the importance of visual and emotional quality of the materials used in the construction of these spaces.

After the analysis of the importance of visual and emotional quality of the living space, and evaluation of current models of tile development, it became apparent that the need to focus on the user, and on their relationship with the spaces and objects, as opposed to the model based on the product and production, was essential. The development of a strategy where issues related to perception are taken as fundamental for addressing the problems inherent in the analysis and generation of solutions, may contribute to the visual, emotional and functional quality of ceramic tiles, and consequently to an enhancement of the living space.

Figure 2 The influence of light quality, viewing angle and distance of observation on the perception of colour, gloss, surface quality and grid application.

This article focuses on a significant and sustained renewal of ceramic claddings by substituting the conventional development model, which is product centred, for a user-centric model considering the level of function and perception. This is a new model of thinking related to the project, where the perceptual aspects are considered as key elements in the analysis, development, production and use of the product.

A holistic approach to the subject object was conducted, including literature review, direct observation and photographic record (analytical-descriptive study of ceramic claddings, its visual and symbolic features, including relations with the user and the surrounding area), case studies analysis, and trials with two groups (Sample - students and Focus – professional designers), as well as ceramic glazed claddings psychometrics measurements.

3   A NEW STRATEGY FOR GLAZED TILES DESIGN PROCESS

The proposed model include in the project methodologies practiced by the various participants involved in the development of ceramic tiles, factors related to perception in the analysis, development, production and application of ceramic claddings, assuming its value not only as an enhancer of understanding of the product, but also as a major factor in product design decisions themselves, and living space.

Aspects considered:
- Countering the secondary role of the user given the importance of the product, which we believe is possible to reverse by the application of perceptive principles – designed to be enjoyed;
- Three areas of perceptive variables were considered:
a. Colour, texture and glossiness;
b. Distance, viewing angle and incidence of light;
c. Relation with the architectural/user space to which the identity is associated (of the locale, and of the product).
The perceptive variables traverse the whole process, and its relative impact is represented by the proportionality of the chromatic area that represents them.

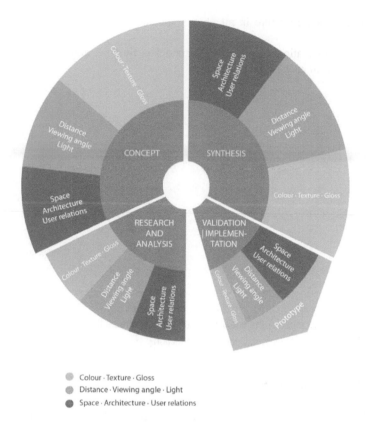

Figure 1 First model proposal – Perceptive variables are represented regarding is relative impact in development process phases.

Considering all the factors and process phases that we pretended to emphasise, the initially proposed model didn't demonstrate the synergies and the fundamental and differentiating interrelations, reason why it was decided to draw a more detailed model.

- A linear, evolutionary representation, was opted for, with cycles of reanalysis in the critical phases of the process, which seemed the most appropriate way to illustrate the existing and necessary dynamics in the process;

- In each phase the rhythm of divergence (analysis) and convergence (synthesis) of action is represented;

- The model begins with the identification of the problem and not the objective, and the conclusion of the application of the product, and not the final product;

Likewise the beginning and the end of the central creative process is marked by the definition of the objective and by the solution = prototype;

- The intermediate stages, represented by circles, are moments of reflection/evaluation, which provide constructive criticism of the interim solutions, defining the rethinking of the questions or the linear evolution of the process;

- The model allows a reformulation of the product up until the validation phase,

702

defending a dynamic posture in all phases and actions to achieve more effective results.

- The implementation/adoption of the model is flexible, able to be moulded to the requirements of each project / designer.

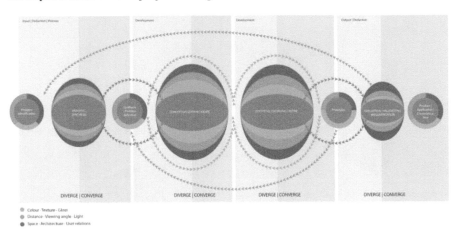

Figure 3 Final design model – the linear representation illustrates the dynamics in the process: |Diverge | Converge, Input | Output, Deduction | Induction.

4   CONCLUSIONS

Validation undertaken with focus groups and sample groups allow the following conclusions to be made:

Importance of the perceptual characteristics of ceramic materials as a significant factor in design project decisions;

The design process can be used to respond to technical and functional requirements and, at the same time consider the user as the motivation of the project.

Recognition of the importance of light variation, viewing angle and distance of viewing in the perception of the ceramic cladding, including colour, texture, motifs; understanding of the functional, perceptive, haptic, and symbolic value of the ceramic material in the design decisions;

Identification of texture, colour, glossiness or lack thereof, shape, and consequently, joint implementation, as tools for the characterization of the product and space.

Integration of perceptual factors, as well as the experiences of users in ceramic claddings design processes contribute to establishing a valid and effective solution, not only at the functional level, such as a protective architectural skin, but also as an element that gives visual and emotional quality to the living space.

Intrinsic characteristics such as - strength and plasticity of the ceramic material, quality of the glaze; extrinsic characteristics - colour, texture and gloss; and emotional characteristics - atavistic familiarity, easily recognized, are all inherent in ceramic tiles, as well as the flexibility of the production processes (greater or lesser depending on the firm's flexibility), allow their customization in accordance with the objectives of their application, so it is possible to create specific solutions for each location.

It is possible to promote the development of multifunctional ceramic cladding materials with added value that suit both architecture and contemporary urban spaces (whether at a cultural, emotional, ergonomic, ecological, or social and economic level).

ACKNOWLEDGMENTS

The present study as been developed with the support from CIAUD - Research Center in Architecture, Urban Planning and Design, Faculty of Architecture, TU Lisbon.

REFERENCES

Abbasy-Asbagh, G 2011, Where "all that is solid melts into the air", there is an ancient land: Urban fabric as agent of cultural preservation. Dolkart, A Al-Gohari, O Rab, S (Ed.), *Conservation of Architecture, Urban Areas, Nature & Landscape Vol.II*, Heritage 2011 – The Second International Conference on Conservation of Architecture, Urban Areas & Landscape Proceedings, pp.143-156, CSAAR Pres. [Jordan].

Ahmadi, SS & Ardakani, M 2011, The role of collective memory in linking the old parts of a city: a case of Ardakan, Dolkart, A Al-Gohari, O Rab, S (Ed.), *Conservation of Architecture, Urban Areas, Nature & Landscape Vol.II*, Heritage 2011 – The Second International Conference on Conservation of Architecture, Urban Areas & Landscape Proceedings, pp.173-189, CSAAR Pres. [Jordan].

Allen, G. 1999, Spatial Abilities, Cognitive Maps, and Wayfinding: bases for individual differences in spatial cognition and behavior. In Golledge, R. (ed.). *Wayfinding behavior: cognitive mapping and other spatial processes*, pp. 46-80. Baltimore, The Johns Hopkins University Press.

Birren, F 1987, *Principles of Color*, Schiffer Publishing, Atglen.

Golledge, R. (ed.) 1999, *Wayfinding behavior: cognitive mapping and other spatial processes*. Baltimore, The John Hopkins University Press.

Humphrey, N 1980, Natural Aesthetics. In Mikellides, Byron (ed.). *Architecture for the People*. London, Studio Vista.

Lobo, C 2006, *Matéria brilho e cor: características do azulejo e sua importância na percepção espacial. Para uma reabilitação do azulejo como elemento qualificador do espaço publico urbano*, Unpublished M.Sc Dissertation, Colour in Architecture, Faculty of Architecture, Technical University of Lisbon.

Lockton, D, Harrison, D. et al. *Designing with intent: 101 patterns for influencing behaviour trough design.* (Internet) http://www.danlockton.com/dwi/Download_the_cards. Retrieved 24.03.2011

704

Kuller, R 1980, Architecture & emotions, In, Mikellides, Byron (ed.), *Architecture for People*. Studio Vista, London.

Shum, W 2009, Desde la tierra a la cerâmica, una construcción viva, In Guisado, J. *Ensayos sobre Arquitectura y Cerámica – Vol. 02,* Mairea Libros. Pp 19-36, Madrid.

Weston, R 2008, *Materials, form and architecture,* Laurence King Publishing Ltd., London.

CHAPTER 76

A Study on the Perception of Haptics in In-cockpit Environment

Kwangil Lee[a], Sang Min Ko[b], Daeho Kim[a], Yong Gu Ji[b]

[a]Republic of Korea Air Force Safety Management Wing
Pyeongtaek-si, Korea

[b]Information and Industrial Engineering
Yonsei University, Seoul, Korea

ABSTRACT

Haptic technology is being used in various areas recently. Especially, the delivery of various information to users with Haptic technology accomplishes many performances. However, Haptic technology in aviation area is being used very restrictedly. Nevertheless it can provide various information for pilots in cockpit to prevent Human Error, it cannot get out from a form of information transmission depended on vision and audition. This way for information transmission is limited to deliver various information to pilots intuitionally and correctly in urgent aviation environment. In this situation, importance of Haptic technology which is a new form of information transmission has increased. This research proposes a method of design guideline development for Haptic technology application considering aviation environment. This study aims to provide practical guideline for designing the haptic interface in cockpit by investigating i) Thresholds of Intensity, ii) Satisfaction Intensity levels, iii) Satisfaction for position, and iv) Satisfaction for Rhythm. Twenty-eight participants took part in the experiments that were conducted at aviation simulation environments, which consisted on real flying, commercial vibration actuators (i.e., the eccentric motors) attached on Anti G-suit and 3D monitor that shows scenes of flying. This study recommended methodology for haptic interface guideline, and showed the characteristics of haptic interface in cockpit.

Keywords: Haptic, User Interface, Guideline, Human Error, Aviation Safety

1   INTRODUCTION

The word "haptic" is originated from a Greek term "touch" and it designates stimulation recognized by hands, feet, and skin (Iwata, 2008). Recently, technical invention has been attempted with 'Haptic'. This technology is highly rated in many areas such as medical science, IT, game, etc. Specially, it gains much recognition since this new type technology conveys many kinds of information. This new information-conveying method is being researched in different spheres and it is related to our routine life such as using mobile phone or home appliances. Mobile phones and game players which use touch screen are a good example (Koskinen, 2008; Liu, Dodds, McCartney, & Hinds, 2004; Mayer et al., 2007; Montfort, 2002; Rupert, 2000; Sklar & Sarter, 1999).

However, haptic technology is not used widely in aviation area. Through development of aviation technology, various safety technologies have been improved and it attributed to safety a lot. Nevertheless, accidents owing to human factors are still occurring and it possess 60~80 % of total accidents cause (Mouloua, Gilson, Deaton, & Brill, 2003). Providing visual & audial information is not enough to prevent the human errors leading to an accident. Visual & audial information cannot give pilots information precisely and intuitively in an emergency situation. Under this circumstance, need for haptic technology has increased. However, there is no specific guideline or criteria for Haptic technology in aviation area. Therefore, this paper attempts to give methodology of aviation haptic technology.

2   RELATED WORKS

2.1   Cockpit Information System

Pilot information interaction in cockpit design could be divided into visual and audial information. Visual information often deals with status of an airplane such as posture, speed, direction, and defect data while audial information gives warning under irregular situation (J. Van Erp & Self, 2008). Although both of methods are used in the same time occasionally, pilots are often exposed to serious workload while performing many missions simultaneously. As fatigue and stress can distract a pilot from visual and audial information, it has limitation (J. B. F. Van Erp & Van Veen, 2004). Accordingly, Haptic Interface for preventing human error is being discussed sprightly (J. Van Erp & Self, 2008).

2.2   Haptic Interface

Haptic interface is a form that delivers information to a pilot by using haptic technology. While visual and audial information delivering method is only focused on eye and ear which would occur workload, Haptic interface is focused on all parts of the body (Iwata, 2008). This characteristic makes personalizing information possible. For example, it can deliver important information necessarily or monopolizing specific information so as to make others not be recognized about information. It also delivers information to different parts of the body so it prevents the workload. Above all, this method gives information directly to a pilot. Haptic interface can be improved considerably by using audial method together (Robles-De-La-Torre, 2008).

Interests in Haptic Interface have been grown in the aviation area and many experiments are under process. For example, U.S army developed Tactile Torso Display (TTD) which allows helicopter pilots to recognize missile attack or flying direction. U.S air force invented SORD(Spatial Orientation Retention Device) to prevent pilots' SD. Providing information from special jacket through haptic technology is concept of this experiment (Ercoline & McKinley, 2008; J. Van Erp & Self, 2008). Haptic technology applied in the aviation area seems promising.

2.3   Haptic Interface properties

B.P Self et, al. (2008) defined characteristics of haptic interface as following 9 properties. 1) Size: size means a range of vibration occurring. 2) Form: each person feels differently about the same stimulation. 3) Orientation: A pilot can distinguish the direction by recognizing different stimulated areas. 4) Position: Human skins have different threshold and responses are very different about the same amount of stimulus. 5) Moving Pattern: when stimulated at few seconds interval, a pilot can recognize the sense of mobility. 6) Frequency: Change of frequency gives different feelings to the pilot under the same degree and form of stimulus. 7) Amplitude: amplitude of vibration is often related to the power of stimulation. As amplitude gets bigger, a pilot feels the stimulation stronger. 8) Rhythm: Rhythm means a fixed signal form designed from a mixture of vibration stimulus time and a gap between the stimulation. People can distinguish various rhythms. 9) Waveform: Waveform means designing a fixed form of vibration signal by the change of frequency or amplitude.

Table 1 Properties of Haptic Interface (B.P. Self et, al., 2008)

Haptic Properties	Characteristics
SIZE	- Limited number of distinctive levels - Large difference between sizes preferable - A clear boundary is needed
FORM	- Fair number of distinctive levels - Similar tactile shapes should be avoided - A clear boundary is needed
ORIENTATION	- Limited number of distinctive levels - The shape should not be rotational symmetric - A clear boundary is needed

POSITION	- Many distinctive levels possible
	- Large distance between displays preferable
MOVING PATTERN	- Any distinctive levels possible
	- The moving patterns should be quickly recognizable after their start
FREQUENCY	- Limited number of distinctive levels
	- Low feasibility for simultaneously displayed frequencies
AMPLITUDE	- Limited number of distinctive levels
	- Low feasibility for simultaneously displayed amplitudes
RHYTHM	- Many distinctive levels possible
	- The rhythms should be quickly recognizable after their start
WAVEFORM	- Includes square, triangular, saw tooth, and sine waves
	- Requires sophisticated hardware

3   EXPERIMENTS

3.1   Apparatus

This research developed Haptic device for experiment of Haptic Interface factors. The Haptic system of this research is developed to deliver aviation information by attaching small motor on Anti G-suit for fighter pilots. The size of small motor attached on this system is small so that it does not affect usability of pilots. T-50 simulator which is being applied for pilot training is used for experiment in flight environment. The experiment is proceeded after pilots wore Anti G-suit with Haptic system and get on the simulator. The major composition of the device is as Figure 1 (a). Figure 1 (b) shows the image of the participant wearing Haptic Anti G-suit in simulator environment.

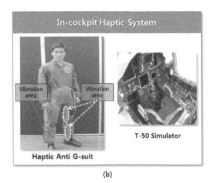

(a)   (b)

Figure 1 Experiment apparatus (a) vibration tactor and (b) Gsuit & Simulator

3.2   Method

This research extracted Haptic Interface factors with considering main factors in flight environment among Size, Form, Orientation, Position, Moving pattern, Frequency, Amplitude, Rhythm, Waveform which are attributes of general Haptic Interface provided by B.P. Self et al. (2008) for Haptic guideline. The expert group that consists of 3 pilots, 2 engineering doctors and 2 psychology doctors examined to extract proper Haptic Interface factors for flight environment. So, extracted Haptic Interface factors are Intensity, Position and Rhythm. 1) Intensity, appropriate strength users feel, is combination of Frequency and Amplitude. 2) Position is the most appropriate part for information transmission. 3) Rhythm is to decide a form of vibration with time between stimuli and deliver information efficiently. Thus, experiment of these 3 Haptic Interface factors should be performed for Haptic guideline development.

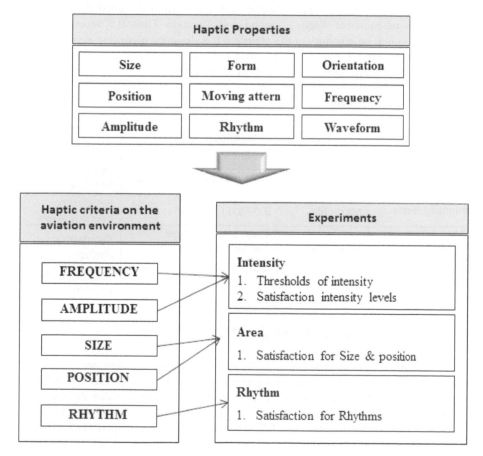

Figure 2 Experiment method

3.3   Experimental structure

It conducted experiment on 28 fighter pilots in the Air Force and divided two groups, beginners and experts. The standard of group division was flight hours. Novice group has 280 flight hours on average, expert group has 900 flight hours on average respectively. They consist of 26 males and 2 females. And average weight is 69.5Kg, average height is 172cm. 1) Thresholds of Intensity get average with each 3 estimation of experiment subjects' stimuli recognition by increase and decrease of voltage. Also, this experiment divided Maximum status and Minimum status considering air expansion and contraction in Anti G-suit. 2) Satisfaction Intensity levels measured satisfaction by each voltage classified a unit of 1V for 2~5V. This experiment also divided Anti G-suit status into Maximum status and Minimum status. 3) Satisfaction for position measured satisfaction of 5 areas such as thigh inside/thigh outside, calf inside/calf outside and back. Similarly, this experiment divided Maximum status and Minimum status considering air expansion and contraction in Anti G-suit. 4) Satisfaction for Rhythm designed Rhythm by combination of Stimulus Time 1 second and Duration Time 0.5 second. This research measured satisfaction for this Rhythm. It is designed according to results of Kirman's study (1974) that the combination of ST and DT makes recognition reduced when stimulus interval is larger than stimulus time.

Table 2 Experiment designs

Experiment properties	Condition of Anti G-suit	Independent Variable	Dependent Variable
Thresholds of Intensity	Max' / Min'	Increase×3 / Decrease×3	Detection of stimulus
Intensity levels	Max' / Min'	2.0 / 3.0 / 4.0 / 5V	Satisfaction Intensity levels
Position	Max' / Min'	Thigh inside/outside, Calf inside/outside, Back	Satisfaction for position
Rhythm	Min'	ST 1' × DT 0.5'	Satisfaction for Rhythm

4   RESULTS AND FURTHER STUDY

In this study, three factors (Intensity, Position, Rhythm) were developed through the characteristic of Haptic Interface to apply to aviation environment. To do the experiment about three factors, we developed Haptic system that can deliver the information in the aviation environment through Anti G-suit and small motor. And then, we progressed the experiment following as the factors of experiment design. And we developed the guideline through the result analyzing.

Discussion about the experiment performed in this study can be devided into two things. First, the vibrating motor used in Anti G-suit can differ in the intensity of stimulus according to the changes of air pressure. Therefore, it is thought that additional study about the difference of intensity of recognition is needed. Second, the shape of haptic can differ from the information that we provide. In other words, in the normal situations, it should be designed with highly satisfied factors. In case of urgent situations, it is thought that the intensity, location, rhythm should be designed to react promptly though the satisfaction is low. Therefore, additional process is needed to apply this study's result to the real aviation environment. For this, we designed five scenarios that would be applied to Haptic interface system and we will select an optimum scenario through expert analysis in further study. Nevertheless, the meaning of this study is that we presented the method as the application standard for the Haptic Interface design in aviation environment.

REFERENCES

Ercoline, W., & McKinley, A. (2008). Applied Research Review/Lessons Leatned. In J. B. F. v. Erp & B. P. Self (Eds.), *Tactile Displays for Orientation, Navigation and Communication in Air, Sea and Land Environments, TR-HFM-122* (pp. 91-106): NATO Research & Technology Organisation.

Iwata, H. (2008). History of haptic interface *Human haptic perception: Basics and applications* (pp. 355-361).

Koskinen, E. (2008). *Optimizing tactile feedback for virtual buttons in mobile devices.* Helsinki University of Technology.

Liu, X., Dodds, G., McCartney, J., & Hinds, B. K. (2004). Virtual DesignWorks - Designing 3D CAD models via haptic interaction. *CAD Computer Aided Design, 36*(12), 1129-1140.

Mayer, H., Nagy, I., Knoll, A., Braun, E. U., Bauernschmitt, R., & Lange, R. (2007). Haptic feedback in a telepresence system for endoscopic heart surgery. *Presence: Teleoperators and Virtual Environments, 16*(5), 459-470. doi: 10.1162/pres.16.5.459

Montfort, N. (2002). From PlayStation to PC, from http://www.technologyreview.com/computing/12770

Mouloua, M., Gilson, R. D., Deaton, J., & Brill, J. C. (2003). Pilot Interactions with Alarm Systems in the Cockpit. *Proceedings of the Human Factors and Ergonomics Society Annual Meeting, 47*(1), 199-201.

Robles-De-La-Torre, G. (2008). Principles of haptic perception in virtual environments *Human haptic perception: Basics and applications* (pp. 363-379).

Rupert, A. H. (2000). An instrumentation solution for reducing spatial disorientation mishaps. *Engineering in Medicine and Biology Magazine, IEEE, 19*(2), 71-80.

Sklar, A. E., & Sarter, N. B. (1999). Good vibrations: Tactile feedback in support of attention allocation and human-automation coordination in event-driven domains. *Human Factors, 41*(4), 543-552.

Van Erp, J., & Self, B. (2008). Tactile Displays for Orientation, Navigation and Communication in Air, Sea and Land Environments, TR-HFM-122: NATO Research & Technology Organisation.

Van Erp, J. B. F., & Van Veen, H. A. H. C. (2004). Vibrotactile in-vehicle navigation system. *Transportation Research Part F: Traffic Psychology and Behaviour, 7*(4-5), 247-256.

CHAPTER 77

Making Electronic Infographics Enjoyable: Design Guidelines Based on Eye Tracking

Leah Scolere, Brie Reid, Carolina Acevedo Pardo, Gilad Meron,

Jordan Licero and Alan Hedge

Department of Design and Environmental Analysis
Cornell University
Ithaca, NY 14853-4401, USA
Email address: lms43@cornell.edu

ABSTRACT

Infographics are designed graphic visual representations of information, data or knowledge that combine symbols, visual imagery, text, and data in innovative ways in order to quickly and clearly convey complex information to viewers. Increasingly infographics are being used on the web yet little systematic investigation has been undertaken on the most effective design guidelines. Based on an analysis of previous research 5 design categories were selected for study: Annotated Graphic, Poster Style, Flow Chart, Chosen Path, and Icon-grid. It was hypothesized that order will increase readability, guided sight-paths will aid narrative flow, and visual synthesis of text and graphic will improve user-perceived quality. The present study used a sophisticated eye movement tracking system to investigate how various graphic elements, layouts and organizational schemes influence a user's viewing times, their interest in and comprehension of the information, and their ratings of overall quality of these electronic infographics.

Keywords: eyetracking, electronic infographic design

1   INTRODUCTION

Infographics are designed graphic visual representations of information, data or knowledge that combine symbols, visual imagery, text, and data in innovative ways in order to quickly and clearly convey complex information to viewers. The increasing relevance of infographics in our post-industrial information-based economy has been brought to light in recent decades, and acted as an impetus for this study. Meta-data has shown a dramatic and significant increase in the prevalence of infographics from the late 1970's onward. A study of infographics in 114 newspapers showed that graphic devices were already covering significantly larger percentages of newspaper front pages in the late 1980s than they were 20 years earlier (Kenney & Lacy, 1987). Since 1984 there has been a steady and continual increase in the occurrence of the word infographic in books (Google Ngram Viewer, 2003), and since 2004 there has been a dramatic increase in Google searches that include the word infographic (Google Insights for Search, 2012). The past ten years have also seen significant growth in the number of websites and blogs devoted solely to infographics, coupled with the emergence of dozens of new graphic design and digital media firms that specialize in producing infographics. Large progressive firms have also begun investing more heavily in services and research related to infographics, notably Facebook and Google (Fast Company, 2012a). These examples are not meant to demonstrate an exhaustive survey of the increasing prevalence of infographics, but rather to highlight the increasing relevance of the infographic in our culture and the need for further critical study.

Further study is even more critical when the role of infographics is contextualized within the so called "information age" of a data-driven economy. Acclaimed author and historian George Dyson aptly states "information is cheap, meaning is expensive...access to raw information has grown exponentially, our time to process it has declined rapidly, placing an unprecedented premium on the act of meaning-making" (The European, 2012). The simple fact that the term terabyte (equal to one million million (10^{12}) bytes) is now commonplace underscores the necessity to condense, synthesize and communicate information in modes and mediums that allow for rapid internalization, comprehension and interpretation. This need is what prominent information designer Francesco Franchi calls "infographic thinking",- a narrative language of representation that invites reader interpretation (Fast Company, 2012b). Neuroscientist Douglas Nelson's work in the late 1970's first opened the door for the infographic paradigm by demonstrating that our brains are essentially hard-wired for visuals- "the very architecture of our visual cortex allows graphics a unique mainline into our consciousness" (Nelson, 1976). Paivio's dual coding theory further corroborated this by showing that imagery has the unique ability to stimulate both verbal and visual mental representations, and activate multiple neural pathways for memory recall (Paivio, 1969, 1986). The need for further study of infographics is highlighted by: a- the neurological framework that supports the unique cognitive potency of graphics, b- the surge in quantity, use and application of infographics, and c- the growing importance of meaning-making in an increasing data-rich world.

Segel and Heer (2010) analyzed the design space of infographics identifying visual narrative tactics or devices that assist in visual narrative. One of these tactics was defined as highlighting or the visual mechanisms that help direct a viewer's attention to particular elements in the display such as color, size, and boldness (ibid.). There have been extensive studies by psychologists that address the notion of visual salience, illustrating how outliers or rare elements in a certain context can attract attention (Itti and Koch, 2001). However, we have been unable to find any studies that use eyetracking with infographics to understand how we read infographics and the relationship between text and image in directing our attention.

More recently information designer Francesco Franchi asserts that there is a "non-linearity" of reading a strong infographic that invites the viewer to join the process of interpretation (Fast Company, 2012b). This background research made our team interested in the ability to use eyetracking methods in tracking and measuring the techniques that draw our attention to various parts of the infographic. One of the most significant previous studies on infographics using eyetracking examined the role of visual embellishment on comprehension and memorability of charts and found that recall after a two-to-three week gap was significantly better than charts without visual embellishment (Bateman et al., 2010). While this study examined recall and gaze related to embellishment, the study only looked 'plain' charts vs 'embellished' infographics and did not propose design guidelines for infographics. Our study seeks to fill this gap by being able to critically examine categories or typologies of infographics and propose design guidelines for designing infographics that viewers enjoy reading.

Our study builds on the identification of infographic categories from previous related research (Segel & Heer 2010), which concluded that different designs/layouts of infographics were influential in how the information contained was "digested" by the reader. After conducting an extensive review of current infographics on the web, our research team decided upon five major categories, which we believe encompass a large majority of infographics. Each of the five categories is defined by certain organizational structure. The categories included: Annotated Graphic, Poster, Flow Chart, Chosen Path, and Icon-grid. For our study, two examples of each category were selected, one with low text content, and one with high text content. The variability of text content was employed not as a direct offshoot of our hypothesis but as a control to further understand how various factors effected reader perception.

The present study will highlight specifically the need for building an understanding of how participants read infographics; a process which we suggest is largely affected by the relative quantity, structure, and graphic-interface of text. We use eyetracking as the method to analyze the locations, timing, and sequence of how viewers read infographics. This study suggests a series of guidelines related to typologies of infographics that help make the viewing of infographics more enjoyable and contributes to our knowledge of successful structures for the combination of information and graphics. This study also tests a "rare-elements" theory, which attempts to explain how individuals commonly read and digest infographics and why the relative amount of text is a key factor.

2   METHODS

2.1 Sample Profile

Six men and 6 women between the age of 19-27 volunteered to participate in the study. All participants (Ps) were non-designers, having no formal training in design or graphics. Ps were paid $15 for the study. The project was approved by Cornell University's Institutional Review Board for Human Participants.

2.2 Apparatus & Materials

The stimulus display was a series of high text content and a low text content infographics on a range of topics for each of five design categories. (Figure 1).

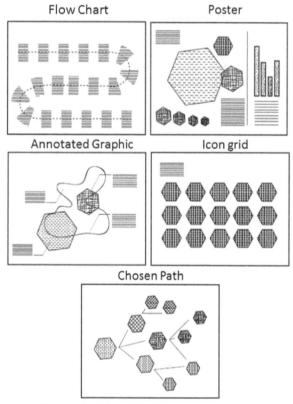

Figure 1 Five infographic design categories tested.

A remote infrared eyetracking system (Facelab 4.3) placed beneath the LCD screen was used to record gazetrails and times. The eye movement data were post-processed using software (Gazetracker v9.0).

2.2 Procedure

Ps were tested individually in a windowless room, the Cornell Human-Computer Interaction Usability Laboratory. Following gaze calibration, which involved viewing each of 9 target circles (3 rows of 3), each P began a 25 minute experimental session in which they viewed ten infographics presented on an LCD computer display (Dell 20"). Two images of landscapes were randomly interspersed to serve as "resting images". Ps were allowed to click through the infographics at their own pace and their eye movements were recorded. Upon completon of this test Ps were shown each infographic again on the computer screen and the experimenter asked them for verbal responses to the following questions: "How easy or difficult is it to know where to start reading this infographic?" where 1 = very difficult and 4 = very easy; "How easy is it to follow the sequence of where to look next in this infographic?" where 1 = very difficult and 4 = very easy; "How enjoyable is the content of this infographic?" where 1 = not at all enjoyable and 4 = very enjoyable; and "How would you rate this infographic in terms of the amount of text and images it contains?" where 1 = way too much text and 5 = way too many images. Ps were also asked what they liked most about an infographic and what they would change about the infographic design.

2.3 Data Analysis

All data were analyzed with a multivariate statistical package (SPSSv19). "Look zones" were defined for the text regions of each of the infographics and the percentage of total viewing time that participants viewed these look zones was calculated for each image. The number of words on each infographic was used to compute a percent viewing time per word. Repeated measures analysis of variance was used to test the effects of gender, infographic design and amount of text on total viewing time (seconds), percent text viewing time and percent viewing time per word.

3   RESULTS

3.1 Total Viewing Time

For the total viewing time there significant main effects of infographic ($F_{(4,40)} = 6.61$, $p = 0.001$) and the amount of text per image ($F_{(1,40)} = 26.67$, $p = 0.000$), and a significant interaction of infographic and text content ($F_{(4,40)} = 3.87$, $p = 0.009$) which is shown in Figure 2. No other effects were significant. The amount of text had no appreciable effect on total viewing time for the Flow Chart, Annotated Graphic or Poster designs, but a significant effect on the Icon-grid ($p=0.032$) and Chosen Path ($p=0.000$) designs where viewing times were longer with more text.

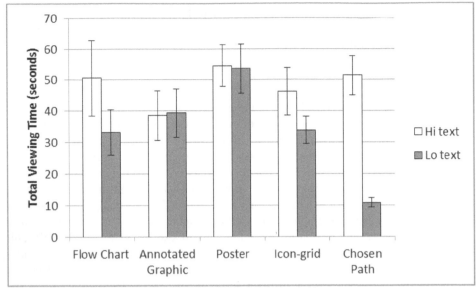

Figure 2 Percent text viewing time for the infographics (mean ± S.E.)

3.2 Percent Text Viewing Time

There was a significant main effect of infographic on the percentage of total viewing time spent viewing the look zones for text ($F_{(4,40)}$ = 12.32, p = 0.000) and Ps spent the least amount of time looking at text on the Icon-grid design followed by the Chosen Path design than the other designs (Figure 3).

There was a significant main effect of the amount of text on the infographic on the percentage of total viewing time spent viewing the look zones for text ($F_{(1,40)}$ = 17.64, p = 0.002) and Ps spent proportionally more time looking at text on the infographic designs with the most text (high text = 21.8 ± 0.6%; low text = 15.1± 1.3%). To further test this effect the percent viewing time per word was analyzed and this showed significant main effects of infographic ($F_{(4,40)}$ = 127.13, p = 0.000), and amount of text ($F_{(1,40)}$ = 127.27, p = 0.000), and an interaction between infographic and amount of text ($F_{(4,40)}$ = 121.11, p = 0.000). The difference between the percent viewing time per word for the high and low text versions of each of the 5 infographic designs is shown in Figure 4.

3.3 Survey

The mean ratings for answers to each of the survey questions are shown in Table 1. For Q1 there was a significant main effect of infographic ($F_{(4,40)}$ = 12.96, p = 0.000) and an interaction of infographic and the amount of text ($F_{(4,40)}$ = 3.65, p = 0.013).

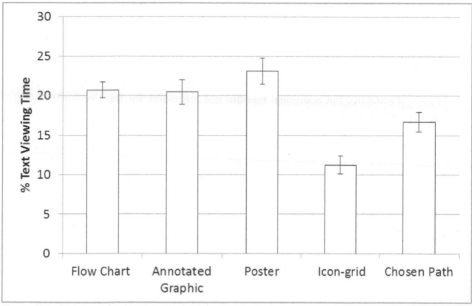

Figure 3 Percent text viewing time for the infographics (mean ± S.E.)

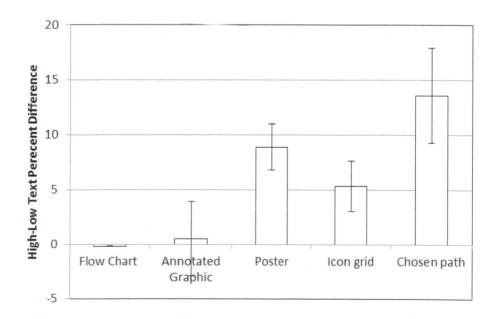

Figure 4 Differences in percent text viewing time per word for the infographics (mean ± S.E.)

The Annotated Graphic was rated as the hardest to know where to start reading but this was easier when more text was included, whereas the Chosen Path was the easiest to start reading and additional text did not substantially affect this.

For Q2 there was a significant main effect of infographic ($F_{(4,40)}$ = 16.19, p = 0.000) and the Chosen Path and Flow Chart designs were rated easiest and the Annotated Graphic and Poster were rated hardest to follow in a sequence. There was also an interaction of infographic and the amount of text ($F_{(4,40)}$ = 3.86, p = 0.01). Adding text to the Annotated Graphic, Poster and Flow Chart made it easier to follow the sequence but additional text did not affect this for the Chosen Path and actually made the sequence worse for the Icon-grid.

For Q3 there was a significant main effects of infographic ($F_{(4,40)}$ = 6.47, p = 0.000) and the Chosen Path was rated as the most enjoyable and the Flow Chart as the least enjoyable. There was an interaction of gender and amount of text ($F_{(1,40)}$ = 10.76, p = 0.008) and women rated less text as more enjoyable than more text (2.63 vs 3.00) whereas men preferred the opposite (3.2 vs 2.87). There was an interaction of infographic and the amount of text ($F_{(4,40)}$ = 2.74, p = 0.042) and adding text yielded higher enjoyment ratings for the Annotated Graphic, Poster and Flow Chart, but had no effect on Chosen Path and decreased enjoyment ratings for the Icon-grid.

For Q4 there were significant main effects of infographic ($F_{(4,40)}$ = 7.34, p = 0.000) with the Flow Chart (2.42) and Annotated Graphic (2.46), the Poster (3.13) and Icon-grid (2.96) being about balanced, and the Chosen Path (3.67) having somewhat too many images. There was a significant main effect of the amount of text ($F_{(1,40)}$ = 24.04, p = 0.001), and low text infographics were rated higher than high text infographics (3.27 vs 2.43) and there was an interaction of infographic type and the amount of text ($F_{(4,40)}$ = 3.57, p = 0.014).

Table 1 Opinions on the infographic images (mean ratings)

	Text	Q1. How easy or difficult is it to know where to start reading this infographic?	Q2. How easy is it to follow the sequence of where to look next in this infographic ?	Q3. How enjoyable is the content of this infographic?	Q4. How would you rate this infographic in terms of the amount of text and images it contains?
Flow Chart	Hi	3.75	3.67	2.58	2.08
	Lo	3.42	2.92	2.25	2.42
Annotated Graphic	Hi	2.50	2.50	2.67	2.58
	Lo	1.75	1.58	2.25	4.50
Poster	Hi	2.75	2.00	3.08	2.67
	Lo	2.33	1.75	3.17	3.17
Icon-grid	Hi	2.83	2.83	2.58	2.50
	Lo	3.50	3.17	3.33	3.33
Chosen Path	Hi	3.75	3.67	3.67	2.33
	Lo	3.92	3.67	3.67	2.92

4   DISCUSSION

Our results have led to multiple conclusions for understanding the role that text plays in the reading of infographics and developing guidelines for designing infographics. It appears as though the measures we have used in this study don't support our hypothesis of the 'rare elements' theory that essentially predicts that graphic or textural elements of an infographic will be viewed longer and more critically when they are "rare" relative to the rest of the infographic (i.e. a single small block of text amid a slew of graphics). However, when text is visually displayed in "digestible" chunks (i.e. condensed into small defined blocks rather then spread across the infographic) viewers are more likely to read the text and therefore more deeply engage with the infographic. Our second major hypothesis was related to synthesis of text and graphic; we predicted that when text was overlaid on graphics viewers would spend more time focusing on those areas. This concept was built off of the notion of "infographic thinking", a phrase that embodies the underlying message of numerous influential information designers such as Francesco Franchi, who proposes the need to move beyond simply pairing text with graphic, and instead work towards a narrative visual language which synthesizes text, graphic, icon and image to communicate something greater than the sum of their parts (Fast Company, 2012b). In the Poster style category, text was more frequently overlaid on the graphic and the participants rated this design as more enjoyable and balanced than Flow Chart, Annotated Graphic and Icon-grid. However, some found it more difficult to find the start point and follow a sequence for reading information in the Poster style.

Hierarchy of text plays an important role in readers' time spent as well. Our data shows that users are more likely to spend increased periods of time looking at text when a hierarchy of importance is conveyed visually. Generally this means smaller text is perceived as less important, and text highlighted, bolded, enlarged or called out in some other manner is often read first. In summary, the results seem to show that adding text to infographics generally increases viewing time, although the size of this effect varies with the design of the infographic. However, adding text doesn't improve the enjoyment; if anything it lowers this for some of the categories of infographics. The exception seems to be the Chosen Path design which was subjectively rated the most highly and the addition of text to this infographic substantially increases the viewing time although it doesn't change the ratings of enjoyability. The Chosen Path category allows for guided choices or exploration, achieving a balance between a singular path and the ambiguity of path. In addition, the Chosen Path designs were the only designs that were self-reflexive, asking the viewer questions about themselves, allowing the viewer to see themselves in the infographic. These two elements of the Chosen Path start to explain the rating of most enjoyable. One limitation of this study was that we used viewing time to test the 'rare elements' hypothesis. Perhaps a future study would examine the gazetrails as a better way to test this hypothesis.

Based on our findings and various responses from participants we formulated several preliminary guidelines for the design of infographics as a means to better convey information to readers. Overall, results confirmed that order does help increase readability, and visual synthesis of text and graphic will improve user-perceived quality, but visual sight-paths do not necessarily aid in narrative flow. A visually conveyed intended order, e.g. an Icon-grid, can significantly impact visual navigation. Several of the components of the Chosen Path typology may serve as guidelines for the development of future infographics; guided choice and self-reflexive content, which by engaging the viewer, contribute to the overall enjoyment of the infographic. Providing the viewer with guided choices to reading and experiencing the infographic achieves a balance between the organizational structure and the ability for the viewer to explore.

ACKNOWLEDGMENTS

This research was supported by funds from the College of Human Ecology and the Human Factors and Ergonomics Laboratory, Cornell University

REFERENCES

Bateman,S., R. L. Mandryk, C., Gutwin, A. Genest, D. McDine, C. Brooks. et al. 2010. Useful Junk? The Effects of Visual Embellishment on Comprehension and Memorability of Charts. CHI' 10.

Djamasbi, S., Siegel, M., Tullis, T.et al. 2010. Generation Y, web design, and eye tracking. Int. J. Human-Computer Studies 68: 307-323.

Fast Company 2012a. "Google Opens Up Infographic Tools for Everyone's Use," Accessed Febuary 16, 2012,http://www.fastcodesign.com/1663247/google-opens-up-infographic-tools-for-everyones-use.

Fast Company 2012b."Why Infographic Thinking is the Future, Not A Fad," Accessed Febuary 08, 2012, http://www.fastcodesign.com/1668987/why-infographic-thinking-is-the-future-not-a-fad.

"Google Ngram Viewer- search for 'infographic' 1980-2003," Accessed Feburary 27, 2012, http://books.google.com/ngrams/graph?content=infographic&year_start=1800&year_end=2000&corpus=0&smoothing=3.

"Google insights -search for 'infographic'," Accessed Feburary 27, 2012, http://www.google.com/insights/search/#q=infographic%2C&cmpt=q.

Itti, L. and C.Koch. 2001. Computational Modeling of visual attention. Nature Reviews Neuroscience 2 (3): 194-203.

Kenney, K. and S.Lacy. 1987. Economic Forces behind Newspapers' Increasing use of color and graphics. Newspaper Research Journal 8 (3): 33-41.

Nelson, D. 1987. Pictorial Superiority Effect. Journal of Experimental Psychology 2(5): 523-228.

Paivio, A.1969. Mental Imagery in Associative Learning and Memory. Psychological Review 76(3): 241-263.

Paivio, A. 1986. Mental Representations: A Dual Coding Approach, Oxford, England: Oxford University Press.

Segel, E. and J. Heer. 2010. Narrative Visualization: Telling Stories with Data. IEEE TVCG: vol.16.

The European 2011. "Information Is Cheap, Meaning is Expensive," Accessed October 10, 2011,http://theeuropean-magazine.com/352-dyson-george/353-evolution-and-innovation.

CHAPTER 78

Verbalization in Search: Implication for the Need of Adaptive Visualizations

Kawa Nazemi[1], Oliver Christ[2]

[1]Fraunhofer Institute for Computer Graphics Research
Darmstadt, Germany
kawa.nazemi@igd.fraunhofer.de

[2]Technische Universität Darmstadt
Darmstadt, Germany
christ@psychologie.tu-darmstadt.de

ABSTRACT

Interactive information visualization enables human to interact with huge and complex data and gather implicit information. Different visualization strategies allow solving visualization tasks e.g., exploring information, making decision or searching explicit information. The process of searching information premises the human verbalization ability. The solving probability of a search problem increases with the precise ability of formulating the query. The query formulation depends on pre-knowledge and verbalization ability of the subject. In this paper we show that, beside the interactive information visualization technique like bottom up or top down, the need of additional ideas e.g. adaptive visualization to increase the verbalization abilities of subjects should be implemented. This is endorsed by an evaluation study of users with significant differences in previous subjective ratings of high or low values of self-assurance in working with personal computers. The results of this evaluation let us assume that personalized or adaptive visualization will help to enhance the verbalization ability and therewith the search and exploration efficiency in subjects with low values of self-assurance. The paper concludes with a short description of an adaptive semantics visualization model.

Keywords: Visualization Evaluation, Adaptive Semantic Visualization, Information Exploration, Verbalization in Visualization

1   INTRODUCTION

Information visualization systems help users in heterogeneous tasks to interact in a comprehensible way with complex data. Especially the complexity of semantic data and the process of searching for implicit information require high acceptance of the user. The design of the visualization is essential for the acceptance and therewith for their usage and exploitation.

Most of today's visualizations investigate a top-down methodology to interact with data. Based on Shneiderman's *Visual Information Seeking Mantra* (Shneiderman 1996) users starts with interacting in an overview-level of the data. With zooming and filtering a special part of the underlying data is chosen, which can further be expanded by asking for more details. This process requires the recognition of certain motifs, structures and entities for gathering the needed information. On the other hand the process of search requires the formulation and verbalization ability of the users as an initial point. The search process begins with the verbalization of a term-of-interest, based on an intention. Different stages of the searching process involve the verbalization ability of the users in different manner. The process of information search is strongly related to the verbalization ability of users during the search process.

For visualizing information, especially semantic annotated visualization two inverse processes should be investigated, first the search process, which has more a bottom-up characteristic and second the *Visual Information Seeking Mantra* with a top-down characteristic.

In this paper we describe an evaluation study of users with significant differences in previous subjective ratings of high or low values of self-assurance. In this study subjects participated and worked with different visualization types as visualization cockpits (K. Nazemi et al. 2010) to fulfill a visualization task. The INTUI (Ullrich & Diffenbach 2010) questionnaire was used for measuring the intuitivism of the different visualization cockpits. The INTUI measures intuitive interaction containing 16 seven-point semantic differential items on the four subscales *Effortlessness*, *Gut Feeling*, *Verbalization Ability* and *Magical Experience*. Further the INCOBI (Richter et al. 2001) questionnaire was used to gather the self-assurance of the users.

The result of the study showed that regardless of the two visualization cockpits, different levels of self-assurance showed significant differences in the verbalization ability of the subjects. Because the process of search is coupled directly to the verbalization ability of the users and the verbalization ability is essential for the acceptance of visualizations, we assume that the acceptance of visualizations and the self assurance can be improved by adapting the visualizations.

This also includes the possibility that the visualization acts as either a top-down or bottom-up process for supporting the verbalization abilities.

2 SEARCH AND INFORMATION VISUALIZATION

Information visualization aims at visualizing data and information in a comprehensible way to understand the context and gather *implicit knowledge* from the underlying data. The implicit knowledge is both, the information that are not formal modeled by the data and the knowledge which may not be formulated by the user explicitly. Different disciplines have already investigated this aspect of search and the efficient representation of data and information respectively.

The existing approaches for gathering this implicit knowledge can be classified in bottom-up and top-down approaches. The standard search model (M. Hearst 2009), e.g. is simplification of a bottom-up approach. The approach attempts to formalize the iterative search process a three-stepped model of Query Formulation, Query refinement and Result Processing. This model assumes that the search begins with the formulation of query of known knowledge. During the search process the subject gets more knowledge about a certain topic to refine his query and gather more knowledge about the certain topic. The main aspect of this model is that the search process starts with ability to formulate a query and to reformulate the query during the search. During the search process new knowledge is adopted, which leads to a reformulation of the query. A more complex example for a bottom-up information gathering model is Marchionini's eight phases of information seeking (G. Marchionini 1995). This model encloses the internalized problem solving of subjects too and shows in a very comprehensible way the importance of the verbalization ability. Marchionini's model consists of eight phases in search: *Recognize and accept an information problem, Define and understand the problem, Choose a search system, Formulate a query, Execute search, Examine results, Extract information* and *Reflect/iterate/stop.* (Marchionini 1995)

The most famous example for a top-down information gathering model is Shneiderman's *Visual Information Seeking Mantra.* (B. Shneiderman 1996) This model proposes the opposite of the *bottom-up* approach and is designed for the visual information seeking. The three-stepped model propagates to *Overview* the data first, than *Zoom and Filter* the relevant parts and finally gather *Details on Demand.* Beginning with the overview of data, this model premises not the verbalization ability, here the focus is on the recognition ability. If a subject detects in the overview step an area-of-interest, he can zoom into the area or filter this information out. After he gets enough information to recognize a seeking problem, details about the information can be fetched. The described seeking models show that there are two different human abilities required for solving a seeking problem. In a bottom-up approach the verbalization and formulation of the searched topic is essential, whereas the recognition ability plays the key-role in top-down approaches. The mentioned top-down approaches are primary information visualization approaches, thus the overview of information and recognition of area-of-interest can be more supported with visualization systems. Figure 1 illustrates the two approaches based on the standard search model and the *Visual Information Seeking Mantra.*

726

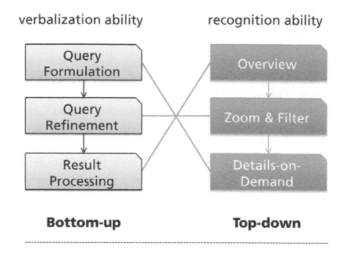

verbalization ability   recognition ability

Query Formulation

Query Refinement

Result Processing

Overview

Zoom & Filter

Details-on-Demand

Bottom-up   **Top-down**

Figure 1 Top-down versus bottom-up seeking approaches, based on the three-stepped Visual Information Seeking Mantra (B. Shneiderman 1996) and the standard search model (M. Hearst 2009).

3   SEMANTICS VISUALIZATION COCKPIT

Semantically annotated data provides complex structures for seeking information in different ways. Both methods (top-down and bottom-up) of information seeking are supported by the formal structure of semantic data. A specific query on semantic data would provide a very specific result from a domain of interest, whereas the schema of the semantic structure enables viewing an abstracted overview of the domain. The abstracted view is possible due the structure is often modeled with formal models by using concepts, instances, relations and a schema level of the underlying data. Based on the given search problem different visualization or representation methods would lead to successful solutions.

We proposed in previous works (Nazemi et al. 2009, Nazemi et al. 2010, Nazemi et al. 2011) different visualizations for semantic data and introduced the knowledge and visualization cockpit metaphor. The cockpit metaphor is rampant and indicates that different information systems are arranged as a visualization board. Our visualization cockpit separates information attributes from each other and visualizes this information in separate visualization units. The advantage of the separation of complex information units is obvious; the user of a cockpit is able to perceive the same information from several perspectives juxtaposing visualization techniques. With this approach both, bottom-up and top-down approaches are supported. A bottom approach starts with the query formulation. If the formulated query is precise enough to provide an information instance, this instance and the semantic neighborhood is presented. Otherwise, if the query is not specific or the user wants to have an overview, the abstracted schema of the semantics is presented with concepts as categories. This second approach follows the Visual Information Seeking Mantra and supports the three mentioned steps. Whereas the bottom-up approach follows the search model of Marchionini and provides visual query refinements for reformulating the query through the semantic relationships.

Bottom-up   Top-down

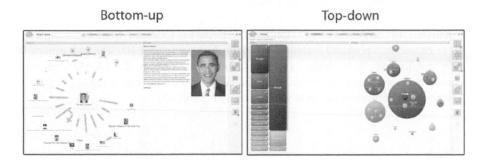

Figure 2 The semantics visualization cockpit with different types of visualizations. On the left the support of bottom-up approach, based on a precise query. On the right a top-down visualization based on un-precise query. The top-down visualization provides an abstracted schema visualization of the semantics, whereas the bottom-up provides an entity plus semantic relationship visualization.

4   USER STUDY

4.1   Method

18 subjects with a median age of 23 years participated in the experiment. The participants completed a demographic questionnaire and items related to PC- user behavior. The participants were divided into two groups of eight and ten persons. The groups were counterbalanced so that they did not differ concerning frequency of search engine usage. Both groups used SeMap (Nazemi et al. 2009), a facetted visualization for hierarchical data, and a representation of giving written information on the search criterion. One Group also used SemaGraph (Nazemi et al. 2009), a network visualization for exploring linked information, and a second group SemaSpace (Bhatti 2008), a set visualization showing connections between data sets, additionally. SeMap was placed at the left top of the page, The content representation at the right top of the page and SemaGraph respectively SemaSpace was placed at the bottom of the page. Icons, other visualization types and the adaptive version of SemaVis were not used in this experiment. 25 tasks concerning life data of different well known psychologists were generated. Answers could be found in either in the content representation or SemaGraph respectively SemaSpace, so both types of visualizations had to be used. The participants were instructed to answer as many questions as possible only by means of SemaVis engine within 25 minutes. Usage data such as timestamp, action, applied visualization, information

and data type were tracked during the period of search. Afterwards, they were asked to fill-in two questionnaires. The INTUI (Ullrich & Diefenbach, 2010) measures intuitive interaction containing 16 seven-point semantic differential items on the four subscales *Effortlessness*, *Gut Feeling*, *Verbalization Ability*, and *Magical Experience*. The COMA questionnaire is a subscale of the INCOBI questionnaire (Richter et al. 2001) concerning the self-assurance in using the computer containing eight items. The participants were tested simultaneously, each seated in front of a Windows 7 PC with a LG 22'' Monitor, the SemaVis (Nazemi, Stab and Kuijper 2011) engine was prepared including an individual participant ID. The two groups were subdivided in a high (group 1) and low (group 2) COMA group (counterbalanced regarding SemaGraph and SemaSpace). Inference statistics using t-tests with bonferroni adjustment were done with SPSS 17.

Figure 3 The evaluation scenarios: Left the SeMap visualization juxtaposed to a content visualizer and SemaGraph; the right scenario uses SemaSpace instead SemaGraph (low positioned visualizations.

4.2   Results

Subjects working with SemaGraph and SemaSpace visualizations showed no significant differences in the frequency of search engines use time. Also no significant differences between SemaGraph or SemaSpace Group regarding INTUI-scales were found. After subdividing, group 1 showed significant lower self assurance values than group 2. Subjects with lower self-assurance showed regardless of the use of different types of visualization no significant differences in solving the given tasks correctly (see fig 5) but significant lower skills in verbalization. One main factor for the ability of verbalization, as gathered with the INTUI questionnaire was the type of visualization in the two scenarios. The results showed no significant differences in the verbalization ability during search with different visualization techniques but significant differences in the verbalization ability of users with different levels of self-assurance (see fig 4).

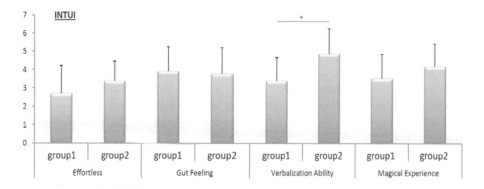

Figure 4 Significant differences (marked with a star) in the subscale "verbalization ability" between the groups with high and low self assurance.

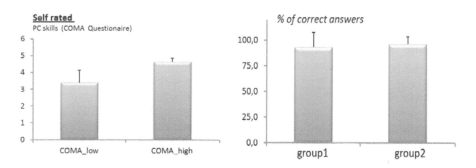

Figure 5 Left: Self-rated computer skills between subjects with high (COMA_high) and low self-assurance (COMA_low); right: correct given answers: no significant differences.

5   VISUALIZATION ADAPTATION

The evaluation evidences that one factor for enabling the verbalization ability in subjects is the higher self-assurance but not the visualization technique or visualization orchestration. The efficiency of a search task is strongly related to users' abilities in verbalization and recognition. The verbalization ability is in turn related to users' pre-knowledge. This covers the pre-knowledge in a specific domain, in a certain language or in usage with computer systems. As certain visualization techniques may help users in given seeking tasks for precise queries other visualization helps in discovering knowledge.

We assume that considering users' pre-knowledge and self assurance for the visualization selection and adaptation would provide a more efficient system for differentiating between a top-down and bottom-up search. The recognition of the given search is an initial step to provide visualization techniques that may help to formulate a query or to recognize visual patterns. Based on users' interaction history and incorporating machine learning algorithms the pre-knowledge in a specific domain can be gathered and the user can be supported during the task. If a user reaches a certain level of expertise, the formulation of his queries changes and the recommended visualization changes too.

Existing visualization adaptations consider either the visualization structure (Ahn and Brusilovky 2009) or the visualization type (Gotz and Wen 2009) to provide a personalized visualization for search tasks. The existing systems supports both, the visual pattern recognition in visualization tasks and the dedicated and recommendation based personalized search. But none of the existing systems supports the top-down and the bottom-up search approaches. For an enhanced adaptation it is more than necessary to classify the visual search tasks. Existing classifications (C. Kuhlthau 1991) or (Fluit et al. 2006) do not distinguish between a top-down and bottom-up search, although it may be obvious to take this important aspect into account.

Furthermore the stages of attention (Ware 2004) and their mapping to the visual variables (Bertin 1983) are not considered in the adaptation process of visualizations. As studies in visual attention showed (Treisman and Gelade 1980) and (Wolfe 2007), the visual features play an essential in the different stages of preattentive and attentive stages. Also the working memory seems to have an important impact here because some visual features may facilitate different ways of encoding information. There is evidence that in some context the working memory is able to be "trained" (Christ et al, 2011). In this case adaptation may give the user the change to be trained "online", which could enhance the verbalization process and acceptance in subjects with lower values in verbalization ability. We could find out that different visualization types support the search process, but which of the visual criteria enhances the lack of verbalization ability in some subjects could not be found out in our study. Future adaptive visualizations should consider the visual features and use them to support the users during the different attention stages.

5   CONCLUSIONS

In this paper we described a user study with significant differences in previous subjective ratings of high or low values of self-assurance. Subjects have participated and worked with different visualization cockpits (K. Nazemi et al. 2010) to fulfill a visualization task. To measure the intuitive behavior of the visualization the INTUI questionnaire was used, which also considers retrospectively the verbalization ability of the proceeded steps. Further the INCOBI test was used to measure how subjects are rating their own skills.

We assume that considering users' pre-knowledge and self assurance for the visualization selection and adaptation would provide a more efficient system for differentiating between a top-down and bottom-up search. The different searching methods were introduced in this paper too. The recognition of the given search is an initial step to provide visualization techniques that may help to formulate a query or to recognize visual patterns.

ACKNOWLEDGMENTS

This work has been carried out within the project FUPOL 287119: Future Policy Modeling, partially supported by the European Commission. The FUPOL project proposes a comprehensive new governance model to support the policy design and implementation lifecycle. The innovations are driven by the demand of citizens and political decision makers to support the policy domains in urban regions with appropriate ICT technologies. The results will be considered in the development of our visualization framework SemaVis.

REFERENCES

Ahn, J.-w., and Brusilovsky, P. 2009, 'Adaptive visualization of search results: Bringing user models to visual analytics', Information Visualization, 8/3: 167-179.

Bhatti, N. 2008:Web Based Semantic Visualization to Explore Knowledge Spaces - An Approach for Learning by Exploring. In: Luca, Joseph (Ed.) u.a.; Association for the Advancement of Computing in Education (AACE):Proceedings of ED-Media 2008: World Conference on Educational Multimedia, Hypermedia & Telecommunications. Chesapeake, 2008, pp. 312-317

Bertin, J. 1983. Semiology of graphics. University of Wisconsin Press, 1983.

Christ, O. Schwarz, K., Ohlmes, S. Berger, H.(2011). The effect of cognitive remediation with a biotic designed computer based training on depressive patients: A pilot study. European Psychiatry, 26 (S1), p.687

Fluit, C.; Sabou, M. & Harmelen, F. V. 2006. Ontology-Based Information Visualization: Towards Semantic Web Applications. In Geroimenko, V. and Chen, C. (eds.): Visualizing the Semantic Web. XML-Based Internet and Information Visualization. Springer-Verlag London Limited, 2006.

Gotz, D., and Wen, Z. 2009, 'Behavior-driven visualization recommendation', in IUI '09: Proceedings of the 13th international conference on Intelligent user interfaces (New York, NY, USA: ACM), 315-324.

Hearst, M. A. (2009): Search User Interfaces. Cambridge, UK: Cambridge University Press, 2009.

Kuhlthau, C. 1991. Inside the search process: Information seeking from the user's perspective. Journal of the American Society for Information Science 42, 5, 361-371

Maeno, Y. and Ohsawa, Y. 2008. Reflective visualization and verbalization of unconscious preference. CoRR, 2008, abs/0803.4074.

Marchionini, G. 1995. Information Seeking in Electronic Environments. Cambridge University Press, New York.

Nazemi, Kawa; Stab, Christian; Kuijper, Arjan 2011: A Reference Model for Adaptive Visualization Systems.In: Jacko, Julie A. (Ed.):Human-Computer Interaction: Part I : Design and Development Approaches.Berlin, Heidelberg, New York : Springer, 2011, pp. 480-489 DOI 10.1007/978-3-642-21602-2 52(Lecture Notes in Computer Science (LNCS) 6761).

Nazemi, K.; Breyer, M.; Hornung, C. 2009: SeMap: A Concept for the Visualization of Semantics as Maps. In Proceedings of the 5th International Conference on Universal Access in Human-Computer Interaction. Part III: Applications and Services,

Constantine Stephanidis (Ed.). Berlin, Heidelberg, New York : Springer, 2009, LNCS 5616, pp. 83-91.

Nazemi, K.; Breyer, M.; Burkhardt, D. and Fellner, D. W. 2010. Visualization Cockpit: Orchestration of Multiple Visualizations for Knowledge-Exploration. In: International Journal of Advanced Corporate Learning. 3.

Nazemi, K.; Stab, C.; Fellner, D. W. 2010. Interaction Analysis: An Algorithm for Interaction Prediction and Activity Recognition in Adaptive Systems.In: Chen, Wen (Ed.); Li, Shaozi (Ed.): IEEE International Conference on Intelligent Computing and Intelligent Systems. Proceedings : ICIS 2010. New York : IEEE Press.

Nazemi, K.; Breyer, M.; Forster, J.; Burkhardt, D.; Kuijper, A.: Interacting with Semantics: A User-Centered Visualization Adaptation based on Semantics Data. In: Jacko, Julie A. (Ed.): Human-Computer Interaction: Part I: Design and Development Approaches. Berlin, Heidelberg, New York: Springer, 2011, LNCS 6771, pp. 239-248.

Richter, T., Naumann, J., & Groeben, N. (2001). Das Inventar zur Computerbildung (INCOBI): Ein Instrument zur Erfassung von Computer Literacy und computerbezogenen Einstellungen bei Studierenden der Geistes- und Sozialwissenschaften [The Computer Literacy Inventory (INCOBI): An instrument for the assessment of computer literacy and attitudes toward the computer in students of the humanities and social sciences]. Psychologie in Erziehung und Unterricht, 48, 1-13.

Shneiderman, B. 1996. The Eyes Have It: A Task by Data Type Taxonomy for Information Visualization. In Proceedings of the 1996IEEE Symposium on Visual Languages, pp. 336-343. Washington, DC: IEEE Computer Society.

Treisman, A. M. and Gelade, G. A 1980. Feature-Integration Theory of AttentionCognitive Psychology, 12:1, 97-136

Ullrich D.; Diefenbach S. 2010: INTUI. Exploring the Factes of Intuitive Interaction. In Ziegler J., Schmidt A. (Eds): Mensch & Computer 2010. Oldenbourg Verlag, Duisburg, 251-260.

Ware, C. 2004. Information Visualization. Perception for Design Morgan. Kaufmann Publishers.

Wolfe, J. M. 2007. Guided Search 4.0: Current Progress with a model of visual search W. Gray (Ed.), Integrated Models of Cognitive Systems, 2007, 99-119.

CHAPTER 79

Learning to Use a New Product: Augmented Reality as a New Method

Deise Albertazzi, Maria Lucia Okimoto, Marcelo Gitirana Gomes Ferreira

Universidade Federal do Paraná
Curitiba, Brasil
dalbertazzig@gmail.com

ABSTRACT

When someone starts using a product for the first time, the user has many ways to learn how to use the product. It was developed a new method to help the user learn how to use a product. This method uses an augmented reality serious game to guide the user during the first interaction. Six metrics were evaluated, being compared the results from learning by using the augmented reality application and learning by using three other methods: direct manipulation, reading of the instructions manual and watching of the instructional video.

Keywords: augmented reality, serious games, instructional design, usability test

1   UNDERSTANDING THE BASIS: AUGMENTED REALITY AND SERIOUS GAMES

There are many methods people can choose to start learning how to use a new product. Each method has its own positive and negative aspects. There is nothing as a perfect method that suits any situation. Trying to offer a new and approachable method, it was developed an alternative that guides the user during an augmented reality serious game.

Augmented reality is a technology that mixes the virtual and real worlds by placing virtual elements in the real world (Milgram, Takemura, Utsumi, & Kishino, 1994). As in the AR the main scene comes from the real world - the one people already know, by using our own eyes - it is more natural to interact in an AR environment than in a virtual reality environment, that is entirely virtual and not so natural to the user of the technology (Kirner & Tori, 2006).

To create an augmented reality environment, it is needed the following items: a camera to capture the real world image; AR markers, that might be graphic patterns or a known 3D object; a software to identify the AR markers and to place the virtual elements in the right position; an AR visualization system (i.e. a head mounted display or a computer monitor) to visualize the AR environment.

A game is a play activity apart from the real world, where a player interacts by facing conflicts, attacking opponents, following rules and gaining experience to achieve goals, obtaining quantifiable results and satisfaction - the main objective of a game (Adams & Rollings 2007; Bobany 2008; Crawford 2003; Huizinga 2003; Salen & Zimmerman 2004).

There is a kind of game, however, that has a different main objective. The serious games are games focused on learning instead of having fun. The rules must simulate the reality and the game must be simple, avoiding what is not really needed to focus on learning (Michael & Chen, 2006).

2 TESTING THE AUGMENTED REALITY APPLICATION

The main goal of the test was to evaluate if the use of an augmented reality serious game helps the user learning how to use a product for the first time. To achieve the objective, the augmented reality application was compared to three other methods of learning how to use a product and six usability metrics were evaluated in a test that evaluated the first use of a portable vacuum cleaner.

2.1 Variables

The independent variable (IV) was the process used to learn how to use the product. There were four groups: direct manipulation, instructions manual, instructions video and augmented reality application, as seen on the Figure 1. The dependent variable (DV) was the usability, subdivided in six metrics: learnability, ease of use, satisfaction, task time, task success and errors. There were 40 participants (10 to each group) and it was required the previous use of a vacuum cleaner as the control variable (CV). To be sure that the age of the participant would not affect the results, it was settled two different age groups as the test factor (TF): below 30 and above 31 years old.

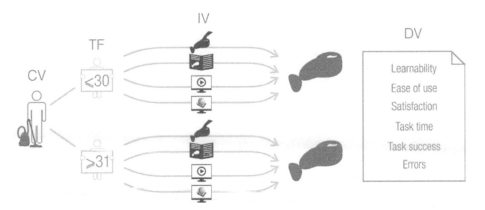

Figure 1: Variables

2.2. Tools

To learn how to use the product, the participants from the four groups used the following tools seen on the Figure 2: (A) a portable vacuum cleaner, that was the product used in the test; (B) a real use simulator, that simulates the small spaces and corners an user would have to clean in a real situation, and paper balls in two colors - the participants should aspire the blue ones and avoid aspiring the yellow ones. Also, all participants had to answer an after task questionnaire.

The group that learned by direct manipulation used only the tools described above.

The group that learned by the instructions manual also used the instructions manual provided by the manufacturer. It includes one page of general tips, two pages of precautions, one page of non recommended uses, one page of instructions for cleaning the filter, one page of instructions for cleaning the dust bag and one page of warnings.

Figure 2: Tools used by all the groups

The group that learned by the instructions video also used a display and the instructions video . The instructions video was made for this research, as seen on the Figure 3. It is a mute video that shows how to turn the vacuum cleaner on, how to use it and change its cleaning brushes, how to open it and remove the dust bag, how to clean the dust bag and place it again, how to close the vacuum cleaner.

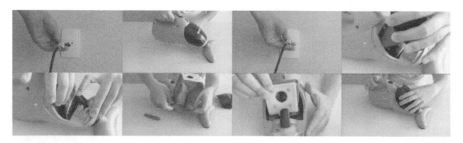

Figure 3: Scenes from the instructions video

The group that learned by the augmented reality application also used a display as the visualization system for the AR, a webcam to capture the image from the real world, the augmented reality serious game, AR fiducial markers as seen on the Figure 4 and an AR serious game evaluation questionnaire. It was required the presence of the researcher to advance the serious game through the stages, as it was an experimental study that requires further development.

It was used the Method for Developing an Augmented Reality Serious Game (MARSEGA) (Albertazzi, 2012) to develop the AR serious game. It resulted in an puzzle type serious game. In the beginning of the game, three situations are presented to the user, and he must choose among the cleaning brushes the right one to clean the object showed in a 2 dimensions photo. If he chooses a wrong cleaning brush, an audio instruction says it is not the right one, explaining where it is usually used and asks the user to try another cleaning brush. If he chooses the right one, an audio instruction congratulates him while explaining when to use that cleaning brush. When all the situations were presented the vacuum cleaner is "full" of dust, and the user has to clean it. Being a puzzle, the user must think about what he is doing and discover the right way to use the product by hearing the given tips. When he tries to open the lid, he must disconnect the electric plug. If he forgets to disconnect it, an audio warning asks if it is safe to open a powered electrical equipment. The user is given audio instructions to learn when to use each cleaning brush, how to safely open the lid, how to remove the dust bag and clean it, how to place the dust bag and how to close the lid.

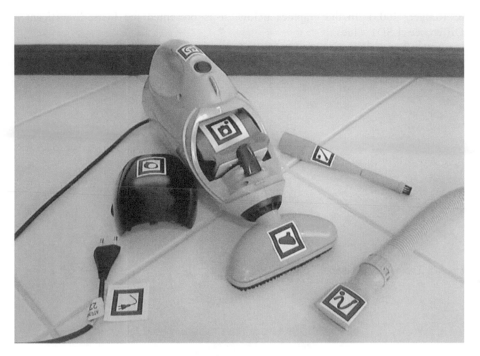

Figure 4: Vacuum cleaner with AR fiducial markers

2.3 Questionnaires

There were two developed questionnaires: the past task questionnaire, that evaluated the learnability, ease of use and satisfaction metrics, and the AR serious game evaluation questionnaire.

The past task questionnaire was developed based on the USE (Lund, 2001) and the SUS (Brooke, 1996) questionnaires. It consist of 12 Likert scales, with sentences from the USE questionnaire. Some of the sentences were converted to negative sentences. Based on the SUS questionnaire, it was given an usability score from 0 to 10 according to the answer in each 5-points scale.

The AR serious game evaluation questionnaire asked to mark the adjectives/ sentences to which the participant agreed. It was given 20 mixed positive and negative options. Also, the participant could write his positive and negative reviews about the process of learning using the method.

3   RESULTS

The data for the metrics learnability, ease of use and satisfaction came from the past task questionnaire. The data for the metrics task time, task success and errors came from the analysis of the video recordings of the test. It was used the analysis of variance in the metrics with a normal distribution of the residuals (Montgomery

& Runger, 2003). In the graphics below, it is presented the Least Significant Difference (LSD) values.

In the first set of metrics - learnability, ease of use and satisfaction - there was no significant difference between the AR serious game and the other methods, as seen on the figures below:

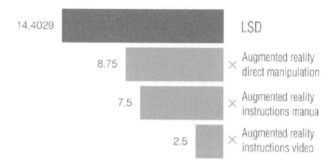

Figure 5: Learnability (Albertazzi, 2012)

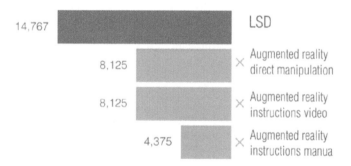

Figure 6: Ease of use (Albertazzi, 2012)

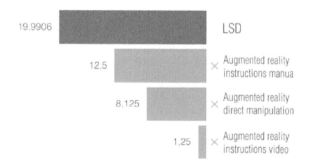

Figure 7: Satisfaction (Albertazzi, 2012)

In the second set of metrics - task time, task success and errors - there was significant differences between the AR serious game and the other methods, as seen on the figures below:

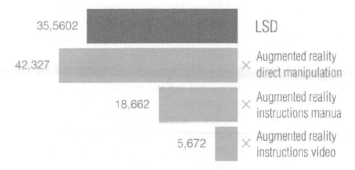

Figure 8: Task time (Albertazzi, 2012)

Figure 9: Task success (Albertazzi, 2012)

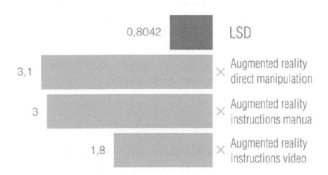

Figure 10: Errors (Albertazzi, 2012)

While in the first set of metrics there was no significant difference between the groups, the opposite happened in the second set. When the participants learned to use the vacuum cleaner by interacting with the serious game, they discovered their mistakes in the moment they were learning how to use the product. When they used other methods, many times they had not even noticed that an error was made.

A very common error was opening the vacuum cleaner without disconnecting it from the power outlet. The participants had not noticed this error when learning using one of the other three methods evaluated, but they did notice it while interacting with the AR serious game. Other common error was placing the dust bag in the wrong position after cleaning it. In a real use situation, the user would think he made everything right, but after using the vacuum cleaner for the second time and have the equipment full of dust - and not the wrong placed dust bag - he would discover his error and have a bad experience with the product.

What was answered in the questionnaire was the perceptions from the participants. As seen above, the errors were not clear in some situations, but they were there. Because of this difference between what was noticed by the participants and what they really did, the metrics that came from the questionnaire differ this much from the metrics that came from the video recordings.

Finally, the AR serious game evaluation questionnaire resulted in positive reviews from the participants: 70% marked it helped to learn and it was interesting, 60% marked it had clear instructions. The main concern was with the time spent to learn by the method: 20% though it was slow to use the method.

4   CONCLUSIONS

The developed augmented reality serious game was a better method to teach the user how to use the portable vacuum cleaner than the other methods evaluated. It has resulted not only in better results for the task time, task success and errors metrics, but also it showed that the users does not notice their errors while learning using the other methods. An error is an important tool for learning when it is noticed. If an error stays unknown, it will result in problems and bad experiences in the future.

Further works are required to evaluate the use of an AR serious game ready to be used by the general public, not requiring the researcher presence anymore. The augmented reality technology evolves quickly, being required testing other possibilities for the AR implementation: AR using mobile devices or head mounted displays, recognizing a 3D object as the AR marker and the use of the AR serious games in other areas, from other products to education.

ACKNOWLEDGMENTS

The authors would like to acknowledge Professor Stephania Padovani, Professor Romero Tori and Professor Susana Domenech for their assistance and also to acknowledge all the 40 participants that made the research possible.

REFERENCES

Adams, E., & Rollings, A. (2007). *Fundamentals of Game Design*. Upper Saddle River: Pearson.

Albertazzi, D. (2012). *Implantando a realidade aumentada em instruções de uso sobre um novo produto*. Universidade Federal do Paraná.

Bobany, A. (2008). *Videogame Arte*. Teresópolis: Novas Idéias.

Brooke, J. (1996). SUS: A quick and dirty usability scale. *Usability evaluation in industry* (pp. 189–194). London: Taylor & Francis. Retrieved from http://scholar.google.com/scholar?hl=en&btnG=Search&q=intitle:SUS+-+A+quick+and+dirty+usability+scale#0

Crawford, C. (2003). *Chris Crawford on Game Design*. Indianapolis: New Riders.

Huizinga, J. (2003). *Homo Ludens. A Study of the Play Element in Culture* (p. 220). Routledge. Retrieved from http://books.google.com/books?id=ALeXRMGU1CsC&pgis=1

Kirner, C., & Tori, R. (2006). Fundamentos de Realidade Aumentada. In R. Tori, C. Kirner, & R. Siscoutto (Eds.), *Fundamentos e Tecnologia de Realidade Virtual e Aumentada* (pp. 22-38). Porto Alegre: SBC.

Lund, A. M. (2001). Measuring Usability with the USE Questionnaire. *Usability & User Experience, 8*(2). Retrieved from http://www.stcsig.org/usability/newsletter/0110_measuring_with_use.html

Michael, D., & Chen, S. (2006). *Serious Games: Games that Educate, Train and Inform*. Boston: Thomson Course Technology.

Milgram, P., Takemura, H., Utsumi, A., & Kishino, F. (1994). Augmented reality: A class of displays on the reality-virtuality continuum. *SPIE 2351-34 (Proceedings of telemanipulator and telepresence technologies, 2351*(34), 282–292. Citeseer. Retrieved from http://citeseerx.ist.psu.edu/viewdoc/download?doi=10.1.1.83.6861&rep=rep1&type=pdf

Montgomery, D. C., & Runger, G. C. (2003). *Estatística Aplicada e Probabilidade para Engenheiros* (2nd ed., p. 478). Rio de Janeiro: LTC. Retrieved from http://books.google.com/books?id=hkt0AAAACAAJ&pgis=1

Salen, K., & Zimmerman, E. (2004). *Rules of Play: Game Design Fundamentals*. Cambridge: MIT.

CHAPTER 80

Visualizations Encourage Uncertain Users to High Effectiveness

[1]Matthias Breyer, [2]Jana Birkenbusch, [1]Dirk Burkhardt, [2]Christopher Schwarz, [1]Christian Stab, [1]Kawa Nazemi, [2]Oliver Christ

[1]Fraunhofer Institute for Computer Graphics Research
Darmstadt, Germany
matthias.breyer@igd.fraunhofer.de

[2]Technische Universität Darmstadt
Darmstadt, Germany
christ@psychologie.tu-darmstadt.de

ABSTRACT

Users have to handle a lot of information in order to fulfill their current task. For achieving an appropriate time and level of quality the users' motivation plays a key role. In this paper we present a user study which aimed to evaluate if the self-rated expertise of the subjects in their computer system skills has an impact on their task completion effectiveness using visualizations. The results reveal that regardless of the self-rated assurance of the users, no significant difference in the effectiveness of task completions using visualizations could be registered. Furthermore the participants indicate in the questionnaire that using visualization their individual satisfaction level had no significant differences when compared to the users' self-assurance levels. This indicates even users feeling not confident in interacting with computer systems they may feel confident in interacting with visualizations. Thus if visualizations are applied for tasks of information search and exploration, the user is encouraged to higher effectiveness.

Keywords: visualization, evaluation, effectiveness and satisfaction

1   INTRODUCTION

In the frequent interactions with computer systems users have to deal with a lot of information they have to interpret and apply to their current task. In order to fulfill these tasks in an appropriate time and level of quality the users' motivation plays an important role. There are several reasons why this motivation may decrease, one of which is the satisfaction and confidence of the user with the system.

User interface design guidelines propose principles which shall increase the comprehensibility and decrease potential frustration by unpredictable reactions of the system. But still unskilled users do not dare to try to click somewhere in order to retrieve the systems' reaction. They expect this could be a wrong action and in worst case harm the system. Or the user may not anticipate the systems' reaction correct and thus could not find the right action(s) to perform in order to fulfill his task. The overall result is the decreasing users' motivation.

In this paper we present a user study which aimed to evaluate whether the self-rated expertise of the subjects in their computer system skills has an impact on their task completion effectiveness using visualization technologies. The participants completed a questionnaire containing demographic aspects as well as items related to computer usage behaviors. The participants were divided into two counterbalanced groups so that they did not differ concerning frequency of search engine usage. Furthermore the participants were asked to complete another questionnaire to assess their self-assurance for their skills in working with computer systems. Therefore the COMA questionnaire, a subscale of the INCOBI questionnaire, had been used. Based on these outcomes the two groups were subdivided in significantly varying high and low COMA subgroups. In the main evaluation part the participants had to fulfill tasks in a group specific visualization cockpit. Thus in each of the demographically and usage-specifically counterbalanced group there were subjects feeling skilled (high self-assurance) and subjects feeling unskilled (low self-assurance). In the study the visualization cockpit, timestamps and actions were tracked. Afterwards we used the INTUI questionnaire to measure the intuitive effortlessness. Recent literature discusses the importance of the relationship between effortlessness and individual satisfaction (Orsinghe et al. 2011).

The results of the study reveal that independent of the self-rated self-assurance of the users there was no significant difference in the effectiveness of task completions using visualizations. Furthermore the users indicate in the questionnaire that using visualization their individual satisfaction level had no significant differences when compared to the users' self-assurance levels.

1.1   VISUALIZATION COCKPIT

The increasing amount of information is a well-known phenomenon in the current information age (Keim and Mansmann 2008). Information Visualization aims to provide visualization techniques to present data in an efficient and effective way (Keim 2002; Nazemi et al. 2011). But the visualization of complex structures with various details en masse tends to result in visualization with reams of graphs, lines and icons. For this reason the usage of a single visualization is not adequate for all tasks or all users. It is necessary to combine different visualization techniques and reduce the complexity of information by splitting in different separated areas of visualization (Nazemi et al. 2010). For the evaluation in this paper two visualization cockpits had been used, consisting of the visualization SeMap, SemaSpace, concentric radial graph visualization and a text-based detailed view.

SeMap is a combination of the Shneiderman's Treemap and Treeview. The Semantic Map (SeMap) uses the two graphical metaphors, Treemap and Treeview, to combine the surpluses for a special case: the usage of annotated data and the implicit impartation of knowledge. Graphical primitives like color, order and size are used to communicate relevant information in a way the user can fast and proper percept it. Color indicates user specific relevance whereas the order and size are determined by a combination of user- and data-based relevance. Order arranges the most relevant element next to the selected element of the last row. SeMap is a visualization that visualizes the concepts and their hierarchy. It is possible to navigate through the hierarchy, where different graphical primitives indicate the relevance of the concept (Nazemi et al. 2009; Nazemi et al. 2010).

SemaSpace is a visualization of knowledge spaces supporting different aspects, e.g. thematic, co-occurrences, spatial, clusters, or configurable domain-specific representations. It provides different knowledge domains (ontology concepts) visualized as circles containing the instantiation of the knowledge domain as smaller circle. SemaSpace offers a sophisticated way to explore knowledge spaces. It offers concepts and related knowledge items to them as factual knowledge and interrelation between knowledge spaces. Awareness knowledge is acquired, when the user explores knowledge spaces and makes decisions to follow different branches or chooses alternative branches in the visualization. Users can also reorganize the visualized knowledge spaces to put the important or relative knowledge spaces in the focus just like working on the desktop (Nazemi et al 2010; Bhatti 2008).

1.2 EVALUATION METHODS FOR VISUALIZATIONS

To design computer systems making people with high self-assurance and a high frequency of computer usage feel confident and comfortable while working with them seems not to be a challenging task. But if people with low self-assurance and low usage frequency are in the same target audience, the design process becomes more difficult. It is in the interest of users and designers to generate computer systems that encourage high and low self-confident users to high effectiveness and satisfaction. Therefore, it is important to evaluate new computer systems with regard to their fitness for both high and low self-confident users using appropriate evaluation methods.

The COMA questionnaire is a subscale of the INCOBI (Richter et al. 2009) that is an instrument for the assessment of attitude towards and competence with the computer. The COMA subscale is a very useful and accurate tool to measure the self-confidence of users working with computer systems. Therefore the COMA questionnaire is used to collect self-assurance data to assign the participants into high and low self-assurance groups.

Another influential factor in the interaction with computer systems is whether users perceive the system as being intuitive and satisfying. Even if a system facilitates to solve tasks effectively it might be designed not intuitively, thus the interaction with the product is perceived as complicated, unforeseeable or inefficient. A lack of intuitive interaction might lead to dissatisfaction. As a result users may avoid using the system or may not feel comfortable in working with it. Therefore it is very important to evaluate, how intuitive users perceive the interaction with a computer system. The INTUI questionnaire (Ullrich & Diefenbach, 2010a,b) is a measurement tool to collect data on how intuitive the interaction with computer systems and software is assessed by users. It contains the subscales Effortlessness, Gut Feeling, Magical Experience and Verbalizability.

A third dimension is the users' presence Immersion or immersive tendencies. This dimension describes the state of presence the user is in while interacting with the computer system. If a user easily shifts his or her attention from the actual physical to the computer system environment, he has a high degree of immersion and is able to focus and be aware of the entire task in contrast to a user with low presence (Fontaine 1992). Especially in the interaction with new software, higher attention to the task and a more detailed overview on the characteristics of the application is important. The attention facilitates more immersive users to solve the given tasks better than users with low presence. The ITQ (Immersive Tendencies Questionnaire, Witmer and Singer, 1998) measures immersive tendencies on the subscales tendency to become involved in activities, tendency to maintain focus on current activities and tendency to play video games.

To evaluate the computer system a typical scenario of usage is necessary. The evaluation setting should be as realistic as possible, in order to observe a test users experience very similar to a real users experience. Therefore specific tasks are generated that users typically solve using the system.

2   EVALUATION SETTING AND PROCEDURE

The participants were tested simultaneously, each seated in front of a Windows 7 PC with a LG 22'' Monitor, the visualization cockpit was prepared including an individual participant ID. 18 subjects (w=14), all students with major in psychology at the Technische Universität Darmstadt, with a median age of 23 years participated in the experiment and received course credits for their participation.

At first, all participants completed the COMA questionnaire of eight items concerning the self-confidence in using a computer system. Based on the results of this questionnaire two groups of each nine persons were generated by median split, see Figure 1.

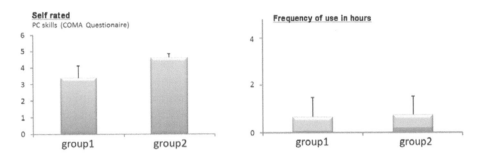

Figure 1 Left: participants self-confidence of computer system usage. Right: frequency of computer system usage in hours.

Afterwards, the participants completed the ITQ with 29 items on the subscales tendency to become involved in activities, tendency to maintain focus on current activities and tendency to play video games.

In the main part of evaluation the participants were instructed to solve typical search and exploratory tasks with the SemaVis system (Nazemi et al. 2011). Therefore, 25 questions concerning life data of different well known psychologists were generated. The participants were instructed to answer as many questions as possible only by means of SemaVis within 25 minutes. The questions were asked in English language to fit to the English data provided by the database. The short answers, e.g. date and place of birth, nationality or religion, should be written down on a sheet of paper, on which the questions were presented.

Both groups used SeMap and the text-based detail view giving information on the search results. Another type of visualization, either concentric radial graph visualization, a network visualization for exploring linked information (Figure 2), or SemaSpace (Figure 3), was randomly chosen and used by the participants. SeMap visualization was placed at the left top of the page, SemaContent at the right top of the page. The concentric radial graph visualization respectively SemaSpace was placed at the bottom of the page. Icons, other visualization types and adaptive functions of SemaVis were deactivated. It had been ensured all questions could be answered using the displayed information in either the text-based detail view or the concentric radial graph visualization respectively SemaSpace, so both types of visualizations had to be used.

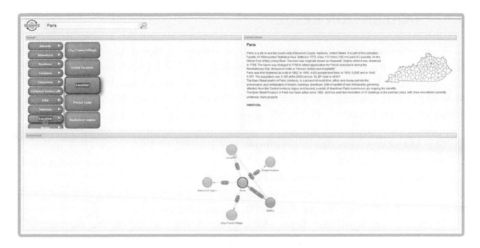

Figure 2 Evaluation scenario 1: SeMap visualization, text-based detail view and concentric radial graph visualization juxtaposed to a visualization cockpit.

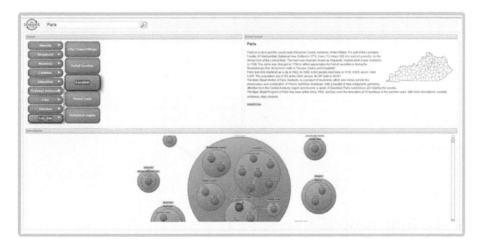

Figure 3 Evaluation scenario 2: SeMap visualization, text-based detail view and SemaSpace juxtaposed to a visualization cockpit.

Usage data such as timestamp, action (left-click, right-click, double-click), applied visualization, information and data type were tracked during the period of search.

At the end the participants were asked to complete the INTUI questionnaire. The INTUI measures intuitive interaction containing 16 seven-point semantic differential items on the four subscales *Effortlessness*, *Gut Feeling*, *Verbalizability*, and *Magical Experience*.

3   EVALUATION RESULT

The self-assessed self-assurance of both groups differed significantly. The overall mean of correct answered questions was 11.61 with 9.98 for the low self-assurance (LSA) group and 13,33 for the high self-assurance (HSA) group. A MANOVA was conducted to examine the effect of LSA resp. HSA (independent variable) on the number of correct answered questions and the INTUI scales (dependent variables).

The MANOVA revealed a significant main effect of LSA resp. HSA on the INTUI subscale verbalization, $F(1,16)=5.699$, $p=.03$, with a mean of 3.53 for the LSA group and a mean of 4.89 for the HSA group.

There was a significant correlation between the frequency of computer use and self-rated self-assurance of users (r=.58**), see also Figure 1.

There was no significant main effect of LSA resp. HSA on the number of answered questions. The results were tested on correlations. There was no significant correlation between the self-rated self-assurance of the users and the number of answered question (r=.43), see Figure 4.

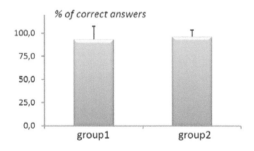

Figure 4 Percentage of correct answered questions in the groups.

5   CONCLUSIONS

In this paper a user study was presented which depicts that regardless of the self-rated self-assurance of the users no significant difference in the effectiveness of task completions in using visualization technologies can be registered. Furthermore the users indicate in the questionnaire that using visualization their individual satisfaction level had no significant differences compared to the users' self-assurance levels. This indicates even if users feeling not confident in interacting with computer systems, they may feel confident interacting with visualizations. Thus when applying visualizations for tasks of information search and exploration the user is encouraged to high effectiveness.

The satisfaction with the usability of the visualization measured on the subscales effortlessness, gut feeling, verbalizability, and magical experience is not related to the degree of self-assurance. This illustrates also users with low computer experience are encouraged to use visualization to fulfill tasks on information search.

These results indicate that even if users do not feel confident in interacting with computer applications, they may feel confident with visualizations. Visualizations seem to be an appropriate way to encourage high and low self-confident users to fulfill tasks of information search with the same high degree of effectiveness and satisfaction. The results of the evaluation indicate that even users feeling not confident in their interaction with computer systems are able to solve specific information search tasks as efficient and satisfied as their high-confident colleges.

ACKNOWLEDGMENTS

This work has been carried out within the project FUPOL 287119: Future Policy Modeling, partially supported by the European Commission. The FUPOL project proposes a comprehensive new governance model to support the policy design and implementation lifecycle. The innovations are driven by the demand of citizens and political decision makers to support the policy domains in urban regions with appropriate ICT technologies. The results will be considered in the development of our visualization framework SemaVis.

REFERENCES

Bhatti, N. 2008: Web Based Semantic Visualization to Explore Knowledge Spaces - An Approach for Learning by Exploring. In: Luca, Joseph (Ed.) u.a.; Association for the Advancement of Computing in Education (AACE):Proceedings of ED-Media 2008: World Conference on Educational Multimedia, Hypermedia & Telecommunications. Chesapeake, 2008, pp. 312-317

Fontaine, G. (1992). The experience of a sense of presence in intercultural and international encounters. *Presence: Teleoperators and Virtual Environments, 1(4)*, 482-490.

Richter, T., Naumann, J., & Groeben, N., (2001). Das Inventar zur Computerbildung (INCOBI): Ein Instrument zur Erfassung von Computer Literacy und computerbezogenen Einstellungen bei Studierenden der Geistes- und Sozialwissenschaften. Psychologie in Erziehung und Unterricht, 48, 1-13

Keim, D. A. 2002. Information Visualization and Visual Data Mining. IEEE Transactions on Visualization and Computer Graphics (TVCG) 7(1): 1-8.

Keim, D.A., Mansmann, F., Schneidewind, J., Thomas, J., Ziegler, H.: Visual Analytics: Scope and Challenges. In: Simoff, S.J., Böhlen, M.H., Mazeika, A. (eds.) Visual Data Mining. LNCS, vol. 4404, pp. 76–90. Springer, Heidelberg (2008)

Nazemi, K.; Breyer, M.; Hornung, C. 2009: SeMap: A Concept for the Visualization of Semantics as Maps. In Proceedings of the 5th International Conference on Universal Access in Human-Computer Interaction. Part III: Applications and Services, Constantine Stephanidis (Ed.). Berlin, Heidelberg, New York : Springer, 2009, LNCS 5616, pp. 83-91.

Nazemi, K.; Breyer, M.; Burkhardt, D. and Fellner, D. W. 2010. Visualization Cockpit: Orchestration of Multiple Visualizations for Knowledge-Exploration. In: International Journal of Advanced Corporate Learning. 3.

Nazemi, Kawa; Stab, Christian; Kuijper, Arjan 2011: A Reference Model for Adaptive Visualization Systems.In: Jacko, Julie A. (Ed.):Human-Computer Interaction: Part I : Design and Development Approaches. Berlin, Heidelberg, New York : Springer, 2011, pp. 480-489.

Orsingher, C., G. L. Marzocchi and S. Valentini 2011. Consumer (goal) satisfaction: A means-ends chain approach. Journal Psychology and Marketing 28(7): 730-748.

Ullrich, D., & Diefenbach, S. (2010a). From Magical Experience to Effortlessness: An Exploration of the Components of Intuitive Interaction. In Proceedings: NordiCHI, October 16-20.

Witmer, B.G., & Singer, M.J., (1998). Measuring Presence in Virtual Environments: A Presence Questionnaire. Presence, 7 (3): 255-240.

CHAPTER 81

Exploring Low-Glance Input Interfaces for Use with Augmented Reality Heads-up Display GPS

Edward Cupps, Patrick Finley, Britta Mennecke, Sunghyun Kang,

Iowa State University
Ames, IA, USA
ejc@iastate.edu, pmfinley@iastate.edu, ruth463@iastate.edu, shrkang@iastate.edu

ABSTRACT

An Augmented Reality Automotive GPS Navigation System (AR-HUD GPS) is designed to streamline the current state of automotive wayfinding. AR-HUD GPS can improve usability and reduce distractions by superimposing critical navigational information directly on the vehicle's windscreen. A necessary first step in developing this system is to design a low-glance input interface system. This project includes four primary activities – background research, development of a prototype, prototype testing and evaluation, and model refinement. These activities are a necessary precursor to the full development and testing of a complimentary windshield display. Preliminary research included reviewing existing GPS systems and contemporary papers on AR-GPS to establish baseline constraints. Two potential interface solutions were created from these findings and tested for the usability factors of icon/text relationship and navigation model. With an improved input interface that reduces glance time, future steps can be taken to further develop the windshield interface.

Keywords: GPS navigation system, interfaces, usability

1. INTRODUCTION

As with cell-phone use and SMS texting, automotive navigation systems such as GPS require that users divert their gaze from the road for significant periods of time. According to a report by the Highway Loss Data Institute, the distractions caused by interacting with electronic devices are a leading contributor to traffic accidents (Highway Loss Data Institute, 2009). For example, in a similar study it was found that truck drivers who texted while they drove had a 23 times greater risk of an accident, near accident, or unintentional lane drifting (McCartt, 2009). With this in mind, how can one minimize glance time while still delivering essential navigation information? Are there interface solutions that maximize understanding while minimizing distractions? One starting point is exploring the possibilities offered by the technology of augmented reality heads-up displays combined with GPS navigation or AR-HUD GPS.

An AR-HUD GPS works by projecting important navigation information in front of the driver on a glass medium or the windshield itself (i.e. augmented reality). By displaying current navigational data and visual markers such as an arrow for an upcoming turn, an AR-HUD GPS assists in keeping a driver's eyes on the road. Similar HUD devices in extant vehicles commonly deliver information on speed, rpm, or range (Chu, Brewer, & Joseph, 2008). An AR-HUD GPS, on the other hand, could be used for other functions such as delivering location details, presenting live traffic information, and displaying real-time road conditions. Yet, any device, especially a windshield-based system, needs a control interface. All control devices have the potential to be a distraction because they require that a driver takes his eyes off the road to inspect and interact with the device. The objective of this paper is to address this problem by examining how users react to various touch-screen interface options and offer suggestions for the design of these interfaces. To accomplish this, we evaluate a basic assumption; that is, driver willingness to use a system and enjoy its benefits will be positively related to the ease of use of the AR-HUD GPS interface. This assumption is predicated on designing a smoother input interface, discovering an effective icon/text ratio, and minimizing glance time from the road to the input device.

2. METHODOLOGY

In order to create a testable prototype interface solution, the team employed a design-thinking methodology. Step one was to research several GPS systems as well as review extant peer-reviewed research on AR-HUD GPS. Step two was to hold a design thinking ideation session based on constraints derived from this research. Step three was to develop and test a series of low-fidelity paper prototypes. Lastly, two of the protoypes were selected for usability testing.

2.1   Background Research

The proposed system should focus on simplicity to reduce glance time. With the navigation information of an AR-HUD GPS focused on the ego-centric display (i.e. the windshield), and only the process of changing that information remains for the exo-centric display, except for the occasional need of a bird's-eye-view (map) for orientation not turn-by-turn navigation (Medenica, Kun, Park, & Palinko, 2011).

Distractions can be minimized by restricting the input interface to essential information (Kruger, 2011). Specific distraction factors include a) information overload, b) cognitive capture due to details and c) avoiding perpetual tunneling – flashing or moving images/text that force attention (Toonis & Klinker, 2008). The typography and iconography should be understandable at a glance, whether icons, text or a combination in different ratios (Plavsic, Bubb, Duschul, Tonnis, & Klinker, 2009).

The main task of a driver is to drive. Only after safe-driving is considered can the three goals of any GPS system be applied: a) wayfinding – find a route, b) surveying – find places along that route, and c) stabilization – following the route (Toonis & Klinker, 2008).

Reviews of current systems such as Garmin Nuvi, Tom Tom, apps such as MotionX GPS Drive were conducted to discover current challenges, set design constraints and reveal insights usable in developing a flow chart of navigation tasks. Since simplicity is key, the user should have a method of "escaping" the system if necessary. Many of the current systems do not make this obvious. A quick search system should be included by default. All these factors set design and functionality constraints that were taken to ideation, flowcharting and rapid prototyping.

2.2   Ideation, Flowcharting, and Rapid Prototyping

The ideation sessions included the three principal investigators plus two lay persons outside the project. The participants considered different approaches to keeping clutter low, including exploring the balance between iconographic and typographic content and controls. They considered over a dozen variations on current solutions for input device navigation, developing seven distinct iterations of device task flowcharts. Lastly, the participants considered audio feedback and direction-complimenting visuals.

A testable flowcharting system was developed from the ideation session. Of the navigation concepts, two were selected for rapid prototyping. The *tabular* interface focused on dominant icons with subordinate reference text; whereas, the *linear* interface balanced these factors. The two interface concepts contained the same overarching information structure; however, the linear interface required a persistent back button that allowed the user to step-back on a task and add to it, change it, or start over. The tabular system, on the other hand, did not require a back feature; instead a user could flip between "pages" at any point.

2.4   Low Fidelity Testing

A series of three low-fidelity prototyping sessions for each interface further refined the two concepts. Each prototype interface was tested as paper models in turn by three volunteers using personas developed by the researchers. The participants were given a series of simple navigation input tasks to perform from the point of view of each persona. As the testing progressed, new paper interface options were presented to the volunteers to simulate a user interface experience.

The test results led to a final revised flowchart. They also revealed several additional factors to be considered. Volunteers noted that the ability to quickly determine alternate routes while driving, add via-points, and save favorite destinations was desirable. The alternate routes should be auto-generated by the AR-HUD GPS for the best speed and the shortest distance to minimize glance time. Universal cancel and home buttons should also added to both of the revised designs.

Based on the results, a prototype of the windshield display was created for the usability testing phase. Lastly, to remove color as a variable, it was additionally determined that the color schemes of the prototypes should be normalized.

2.5   Usability Testing

Throughout the prototyping, paper-testing, and revision process three primary questions arose concerning the usability testing AR-HUD GPS input interfaces:

1. Determine if the Tabular interface or the Linear interface is more effective in reducing glance time.
2. Determine if the Tabular interface or the Linear interface is easier to understand without training.
3. Determine if large icons with small text is more or less effective than a balanced approach.

Two semi-final prototypes, Tabular (Figure 1) and Linear (Figure 2) were tested using identical usability testing tasks with eight participants selected from a wide demographic (19 to 65) split 50/50 male and female. The participants were verbally given an introductory statement followed by a demographic pre-test.

The testing took place in the UX Lab in Howe Hall at Iowa State University. Two iMac computers using Morea software were employed: one as a recording and testing platform and another as an observation and tagging platform. Two experimenters accompanied each participant, one recording the session and the other offering instructions and prompting if necessary.

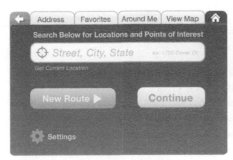

Figure 1 Tabular interface concept.

Figure 2 Linear interface concept.

In testing phase, the user was presented with a full-screen example of an AR-HUD GPS display with the prototype input interface placed on the lower right side of the screen to simulate the in-vehicle experience. Each was given a set of five tasks (or routes) to program into the interface. The first three tasks tested different initial input methods, while the final two tested the use of alternate route and via point functionality. The tasks are as follows:

Task 1: *"Set a route by entering the address of your preferred destination."* Enter goal address information and engage the route.

Task 2: *"Search for the closest Starbucks."* Use the search option (a field on the Tabular interface and a button on the Linear interface) to find the goal and then engage the route.

Task 3: *"Choose your favorite Starbucks that is also closest to you."* See the favorites button and use it to create a route to a saved location, and then engage this route.

Task 4: *"Find the closest Starbucks around you, then choose an alternate route that is the shortest distance."* Enter address information, and then select and engage an alternate route instead of the initial result.

Task 5: *"Choose your favorite Starbucks that is closest to you, then set a via point to find the closest gas station to you.* Enter address information, add a via point, and then engage the revised route.

The tasks were performed in an identical sequence for every user in order of assumed difficulty with the user employing a speak-aloud protocol. Each interface test was counterbalanced with the other to minimize experience-based preferences. The general observations, click paths, prompts, errors, mouse movements, time-on-task and usability challenges were recorded. The users were given an exit survey after testing each interface.

3.   FINDINGS

The Tabular interface had an error sum of 66, and the Linear interface had an error sum of 35. The Tabular interface had 79 total errors with the errors per task average being 13.167 at a standard deviation of 5.115. This result is nearly double the same figures for the Linear interface which stand at 45, 7.5 and 4.550 respectively. Each participant for the Tabular interface created an average of 8.778 errors with a standard deviation of 7.155, equating to an error rate of 2.19 errors per user per task. The Linear interface participants created 5.625 errors on average with a standard deviation of 2.72, equating to an error rate of 1.406 errors per Participant per task. The number of prompts for each interface was 13 for the Tabular interface and only 9 for the Linear interface. In addition, there were 25 discrete observations were made during the Tabular testing where only 15 were made during the Linear.

The confidence level on each interface varied significantly. The exit survey reveals that the Linear interface scored higher for participants on average (4.279 of 5 with a standard deviation of 0.422) than on the Tabular interface (3.952 of 5 with a standard deviation of 0.539). In fact, 75% of the participants explicitly favored the Linear interface approach.

The 25% of participants who favored the Tabular approach fell into the lower end of the User Mental Map Variable (UMMV) ratings. The UMMV was derived from the sum of the Participant's comfort level plus their computer use score. Those participants who scored 20 or more preferred the Linear interface. The two lowest UMMV scores preferred the Tabular model.

4.   DISCUSSION

In reviewing the results of the pre-tests, usability tests, and exit surveys, the majority of the metrics for the Linear interface scored higher than the Tabular interface. For example, in their exit surveys, 75% of the participants reported that they preferred the Linear interface. The UMMV scores of the participants who preferred the Tabular interface were the lowest. A reason for this variance is found by analyzing Task 2, which is the only task where the Tabular interface was scored consistently higher. This task included a request to "Search" for a Starbucks. The Linear interface had no clearly defined search bar, using a button instead. The Tabular interface, having this obvious search bar (which was primarily for address entry as designed), was found to make Task 2 easier despite any other challenges in the interface. Each of the two outlying participants first attempted Task 2 on the Linear interface with some difficulty; however, when they did the same on the Tabular interface, seeing the search bar, they completed the task quickly. A more obvious search bar is clearly an advantage and therefore a refined prototype should build upon the Linear interface with the aforementioned search preference of the Tabular interface.

The UMMV score revealed additional information about the interfaces in general. Participant H, having the lowest UMMV score, also had the largest click differential compared to the assumed path. Conversely, participants A and D, having among the highest UMMV scores of 25 and 22 respectively, completed all their tasks with the fewest number of clicks in comparison to the assumed number. Both of these participants selected the Linear interface.

Path analysis and click-through rates are biased toward the Linear interface in general. The number of required clicks clearly shows the graphical elements of the Linear interface are easier to use or easier to understand. Compared to the expected path scenarios, there were 8 additional total clicks for the Linear interface versus 17 additional total clicks for the Tabular interface. For the Linear interface, Tasks 1, 3, and 5 had a majority of participants taking the assumed path, while Task 2 and 4 were problematic. The data for the Tabular interface revealed near identical issues with Task 4; however, there was far more confusion on Tasks 1, 3, and 5, with less so for Task 2. Of note was the tendency for participants to go directly to search across many of the tasks if they became lost, or to use a "back" functionality to start over.

During the testing, it was observed that participants employing the Tabular interface were focused primarily on discerning system functionality and how to use the system effectively, and less so on their immediate task. Conversely, participants using the Linear interface focused on the basic characteristics of the interface and achieving their task, and less so on finding the correct controls. This concurs with the observed comfort level of the Linear interface over the Tabular interface (with the exception of Task 2 as noted above). This may be due in part to the balanced icon/text ratio of the Linear interface design. Several observations and the mouse tracking data showed participants hesitating on the icon-centric Tabular interface when searching for the correct icon versus readily identifying the larger text of the Linear interface.

It can be inferred that the Tabular interface's challenges lie primarily with the layout and style of the system itself in comparison to the Linear option. The tabs were not immediately understood, nor were the icons quickly acted upon. The Linear interface's issues were with interaction (e.g. one can find the object, but one then may not know what to do with it.) One additional observation of note was the consistency in which participants employed browser and smart-phone specific norms in their use of search fields, back and home buttons, virtual keyboards and scrolling interfaces.

4.1   Recommendations

With all the above in mind, it is believed that the input interface start page should feature a prominent combination search/address field. The data from Task 2 supports the inclusion of this feature from the Tabular interface. In addition, observations support that for the average user, an address entry field and a search field are effectively the same, and so should not be separated out by function (e.g. address, city, etc.). In addition, the landing page should also include functionality for "My Favorites" and "Around Me." In the testing data, search, favorites, and around me scored nearly even with each other in terms of initial clicks. While a search field by its nature will stand out in the hierarchy, useful optional functions should be considered nearly as important for the end user.

The balanced icon/typographic hierarchy of the Linear interface should be maintained. This recommendation is predicated by the observations of confusion when users were attempting to identify an icon for "Coffee" but took additional time in the Tabular interface, ostensibly because of the smaller text size and the focus on iconography. Further study of the size of the icon compared to consistent text may be warranted to refine the new prototype.

The revised design should employ standard UI norms such as a back button in the upper left, a home button, a virtual keyboard and scrolling results. This is supported by the numerous attempts by the users to employ these functions. In addition, textual or vocal feedback for the user should be offered after completing each critical part of the data entry. Observations showed that users on occasion noted that they were less confident that they had completed a particular part of a task without vocal feedback.

Several interaction changes are also recommended First, an "Along My Route" function should be added at the home screen to allow users to plan stops near their particular route or via points. "Around Me" can provide some of this functionality in the immediate vicinity of a driver, but for long-term planning it does not function in the same way. Next, the "Go" button should be more obvious. While not addressed conclusively by the testing, several outside commentators and colleagues noted that this button needed a stronger visual impact. Lastly, the interface should offer an option for vocal input. In order to maximize the "eyes-on-the-road" and minimize glance time a vocal input should be available as a value-added option.

Figure 3 Revised interface concept with additional features such as voice activation.

Figure 4 AR-HUD GPS windshield display.

CONCLUSION

Further testing should be conducted with the revised interface (Figure 3), based on the Linear model. These interface tests can further refine ideal icon size, ideal color scheme, vocal input options, and ideal interface positioning. In addition, these testing scenarios can explore age factors such as diminishing eyesight, speed/safety factors such as using the interface while driving versus at a stand still, and further explore knowledge factors such as computer interface familiarity.

AR-HUD GPS navigation systems may be commonplace in the near future if both the display and input parts of the system are designed and engineered in such a way as to be simple to understand with little experience working with this type of system. The next step after testing the revised design is to begin the testing and revision process of the windshield display itself (Figure 4). Taken to its logical conclusion, we believe the AR-HUD GPS research will succeed in revolutionizing the way we travel.

REFERENCES

Chu, K.-H., Brewer, R., & Joseph, S. (2008, 05 02). *Traffic and Navigation Support Through an Automobile Heads Up Display (A-Hud).* Retrieved 10 17, 2011, From University Of Hawaii At Manoa: Information & Computer Sciences:Http://Www.Ics.Hawaii.Edu /Research/Tech-Reports/Chu-Brewer-Joseph.Pdf/View

Huger, F. (2011). User interface transfer for driver information systems: a survey and an improved approach. *Proceedings of the AutomotiveUI 2011 Conference.* Salzburg.

Highway Loss Data Institute. (2009). Hand-Held Cellphone Laws and Collision Claim Frequencies. *Highway Loss Data Institute , 26* (17), 1.

McCartt, A. (2009, 11 04). *Driven to Distraction: Technological Devices and Vehicle Safety.* Retrieved 09 11, 2011, from Insurance Institute for Highway Safety: http://www.iihs.org/

Medenica, Z., Kun, A., Park, T., & Palinko, O. (2011, 08 30). *Augmented Reality vs. Street Views: A Driving Simulator Study Comparing Two Emerging Navigation Aids.*

760

    Retrieved 09 20, 2011, from Andrewkun.com: http://andrewkun.com/papers/ 2011/fp495-medenica.pdf

Plavsic, M., Bubb, H., Duschul, M., Tonnis, M., & Klinker, G. (2009). Ergonomic Design and Evaluation of Augmented Reality Based Cautionary Warnings for Driving Assistance in Urban Environments. *Contemporary Ergonomics: Proceedings of the International Ergonomics Association.* New York: Taylor & Francis.

Toonis, M., & Klinker, G. (2008). Augmented 3D Arrows Reach Their Limits in Automotive Environments. *Mixed Reality in Architecture, Design, and Constructions* , 185-202.

CHAPTER 82

An Augmented Interactive Table Supporting Preschool Children Development through Playing

Emmanouil Zidianakis[1], Margherita Antona[1], George Paparoulis[1] and Constantine Stephanidis[1,2]

[1]Foundation for Research and Technology – Hellas (FORTH), Institute of Computer Science, Heraklion, GR-70013, Greece
[2]University of Crete, Department of Computer Science
E-mail: cs@ics.forth.gr

ABSTRACT

This paper discusses the opportunities and challenges of Ambient Intelligence (AmI) technologies in the context of child development, and presents the methodology and preliminary results of the development of an augmented interactive table which offers to preschool children various AmI educative and entertaining applications. The overall objective of this work is to assess how AmI technologies can contribute to the enhancement of children's skills and abilities through common play activities during the various stages of their growth and development.

Keywords: Ambient Intelligence, human computer interaction for children, smart games, learning, skills monitoring, tangible interaction

1   INTRODUCTION

Ambient Intelligence (AmI) is an emerging field of research that is gaining the attention all over the world and especially in Europe (IST Advisory Group, 2003). Smart environments that combine a number of interoperating computing-embedded devices to facilitate everyday life in an unobtrusive and natural manner are no longer a futuristic concept. As a matter of fact, AmI is becoming a key dimension of the emerging Information Society, since many of the new generation industrial digital products and services are clearly already shifting towards an overall intelligent computing environment.

As a result, AmI is bringing major changes in the way people interact with interactive products and services, their content and functionality. People and their social situation, ranging from individuals to groups, and their corresponding environments (office buildings, homes, public spaces, etc.), are at the centre of the design considerations.

The work reported in this paper, which is conducted in the context of the Ambient Intelligence Programme of ICS-FORTH (Grammenos, Zabulis, et al, 2009), constitutes an initial step towards investigating the role of Ambient Intelligence technologies in the child's developmental stages.

According to Barnett (1990), play contributes to overall child's development. First, playing enhances the child's physical and motor development which involves height, weight, general appearance/tone of the body and coordination of large and finer muscles. Second, it contributes to the child's cognitive development which includes forming self-concept and forming concepts of size, shape and colors. Finally, playing contributes to the child's social and emotional development which includes establishing relationships, developing behavioral controls and social skills that make them acceptable in their family, school and community. It is common knowledge that child's play is very revealing. How children play, learn, speak, and act offers important clues about their development.

Children's physical and cognitive abilities increase over time as they go through the developmental milestones. The Swiss psychologist Jean Piaget (1970) showed that children not only lack knowledge and experience, but also perceive and understand the world differently than adults. Therefore, it is very challenging to design and develop Ambient Intelligence applications for young children that will cater for their interaction skills and capabilities.

This paper presents a preliminary attempt to develop an augmented interactive table called Beantable, for children in the age-range of 2 to 7, and the related AmI applications which aim at integrating AmI technologies during play time. The purpose of Beantable is to support children's development through the monitored use of appropriate smart games in an unobtrusive manner. Beantable aims to provide intuitive and seamless tools to monitor and enhance the child's playing experience, through appropriate smart games. Additionally, adaptation and personalization techniques in the domain of educational games (Vasilyeva, 2007) are revisited here in an AmI perspective, aiming at exploiting the interaction possibilities offered by AmI.

Section 2 of this paper discusses the background and the related work, followed by a detailed description of the Beantable design and characteristics, as well as the hardware and software set-up of the prototype in Section 3. Section 4 presents the usage scenario of Beantable and the developed games. Finally, Section 5 reports the results of an expert walkthrough and the preliminary outcomes of an informal testing with children. The paper ends with overall conclusions and future work.

2   BACKGROUND AND RELATED WORK

In the literature there are a few examples of work that focus on augmented interfaces geared towards young children. NIKVision (Marco, et al, 2010) is a tabletop prototype designed to be mainly used by kindergarten children. Interaction is carried out through the physical manipulation of conventional toys on the table surface suitable to be installed in nurseries, schools and public spaces such as museums. Mansor, et al. (2008) designed the DiamondTouch™ Tabletop game based on a traditional dolls house with a virtual reconstruction on the tabletop. Smart table, an interactive learning center, is a commercial interactive table designed for preschool and elementary age children (4 to 11 years old). Other applications which exploit some features of tangible interaction and augmented reality towards supporting children interaction, playing and learning, are based on tabletop interfaces. Examples include the Smart Jigsaw Puzzle Assistant (Bohn, 2004), which is a fully operational augmented jigsaw puzzle game using miniature RFID tags, as well as SIDES (Piper, et al, 2006), a cooperative tabletop computer game for social skills development.

However, none of the above applications have the capacity to be personalized to individual children's needs, skills, and abilities, and more importantly, none of them is capable of evolving in order to address different developmental stages of children as they grow up. The main difference of Beantable with respect to the mentioned applications is that it monitors the interaction level of the child with any given game application and adapts its functionality accordingly to different forms of play. For example, in the case of a game that does not capture the attention of the child, the system is able to efficiently change the mode of interaction or offer an alternative game that is more engaging to the child while it satisfies the same developmental goals. The main characteristics of Beantable can be summarized as follows:

- **Interaction monitoring**: Taking into account the way a child plays, the selection of materials and game themes, and the way the child takes part into an activity, the system can extract indications of the achieved maturity and skills.

- **Game adaptation**: Through the use of smart games and appropriate guidance based on scaffolding (Berk and Winsler, 1995), Beantable provides the child with many opportunities for games, in order to support overall development, empowerment of imagination and creativity, as well as strengthening of initiative (and not just knowledge).

Beantable aims at meeting not only the needs of the children, but also the needs of their parents and educators. For children, Beantable acts as an object that carries various toys. For parents, the Beantable system acts as a tool that provides them with general information on their child's physical and mental development progress. At the same time, for educators and development experts the system can actually act as a diagnostic agent, as it provides them with analytical data that can be extracted from the interaction history and can be used to examine whether the child is meeting all the necessary developmental milestones.

The actual interaction of the child with Beantable covers part of the child's needs for activity, exercise and pleasure, and enhances social interaction and communication skills. At the same time, this interaction supports the overall development and enhancement of the child's sensory, motor, cognitive, and social abilities based on the child's individual biological and cognitive maturation pace (rather than based on social and cultural trends).

3   DESIGNING AN AUGMENTED INTERACTIVE TABLE

3.1   Physical artifact design

In the context of AmI, interactive devices must be unobtrusive, hidden or embedded in traditional everyday objects and furniture augmented with ICT technology without compromising general health and safety requirements. It is also very important for the equipment to be easy to install, easy to move around in the room and it shouldn't take too much space. To this end, artifact-oriented approach has been adopted in the context of this work which introduces independent AmI augmented artifacts in the environment.

The first developed artifact is an augmented interactive table, Beantable, appropriate for use by children in age-range of 2 to 7 (Figure 1). The table, which has been custom built, is a wooden prototype with dimensions of 116cm (L) X 105cm (W) X 46cm (H), and has been designed to be robust yet transferrable. The height of the table can be adjusted to fit children's needs as they grow. The hardware infrastructure (i.e., desktop computer, projectors, cameras, infrared beacons, etc.), that is required for the operation of the AmI applications is embedded in a way that is invisible to the eye. A main display device with a standard resolution of 1024 X 786 pixels is located on the top side of the table surface and has dimensions of 56cm (L) X 40cm (W).

The main display device is enabled with multi-touch and force-pressure sensitive capabilities, and is able to recognize the location and the rotation of physical objects that are placed on it provided that each physical object carries at least one fiducial marker (visual pattern printed on paper which has topological characteristics that make the fiducial easy to detect and track by visual recognition algorithms) at its bottom. The minimum size of each fiducial marker is 3X3 cm. Beantable comes with a custom made chair, with dimensions of 68cm (L) X 51cm (W) X 78cm (H), which is able to detect when the child is sitting on it and captures body posture data. Both the table and the chair have a soft yellow color which according to pedagogical experts is neutral and easily recognizable to children.

In addition to the above, the Beantable setup will also include the following components:
- A wall-mounted sensor module that captures data used for:
  ○ body posture recognition
  ○ head position estimation
  ○ speech recognition

      ○ gaze tracking
      ○ gesture recognition

- A secondary device displaying a 3D animated model acting as a virtual child's partner during playing. The virtual character implements body animation, lip-synching, and shows emotions through facial expressions and voice.
- A smart pen that uses various sensors measuring the applied pressure weight, the position and orientation on the screen, and movement acceleration.

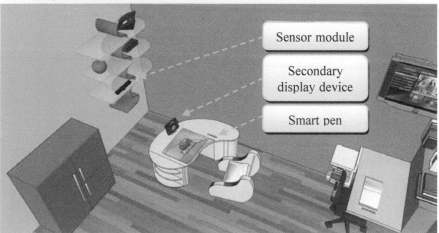

Figure 1 A room equipped with Beantable

3.2   Hardware and Software set up

An important feature of the interactive table is the use of vision-based back projection which facilitates the support of multi-touch interaction and object recognition, while ensuring gesture interaction quality under variable lighting conditions. Even though back project introduces space-related requirements, it eliminates the need for the installation of ceiling mounted cameras or hanging projectors, which would require for the fiducial markers to be placed on the top surface of each physical object and wouldn't allow a natural multi-touch interaction experience. The software used for the multi-touch and physical object recognition is an open-source software called reacTIVision (Kaltenbrunner and Bencina, 2007). The hardware set up includes (see Figure 2):

- Intel Core i7 PC
- A mini portable led projector located inside the artifact
- A mirror for reducing the projection distance
- 2 cameras located behind the screen with wide lens to maximize quality of the captured image
- 4 infrared beacons located behind the screen covered with thin fabric to diffuse the light smoothly

- 1 custom-designed high robust triplex glass covered by a window frosted glass film
- 2 cooling fans located at specific points that facilitate natural air flow and which are controlled by software running on a microcontroller
- 2 temperature sensors
- 1 stereo speaker set
- 4 pressure sensors located just behind the screen
- 4 pressure sensors located under the chair

Figure 2 Beantable hardware set up

4   USAGE SCENARIO AND GAMES

An important objective of Beantable is to ensure that the various technological artifacts that are developed have the playing process as their focal point. To this purpose, games involving physical objects, such as puzzles, were selected as a testing domain and common game practices were considered while building the actual usage test scenarios. In these scenarios, Beantable is the child's playmate during her / his growth. The interaction of the child with the Beantable begins the moment the child sits on the custom made chair. The sitting action initializes the starting screen that shows a menu (see Figure 3) of smart games that are appropriate for the child's age and skills based on previous interaction history.

Each smart game is composed of various micro games and it is built progressively by the child. As the child completes various levels in a micro game, more parts and features are added to the smart game. Consider, as an example, a variation of the classic game "Snake and Ladders" in which each building block (ladder, snake, etc.) is dynamically added as a reward from the successful completion of various levels in one of the relevant micro games.

Each micro game targets specific child skills and abilities according to the ICF classification (CAD, 2001), and supports their enhancement and development. It

Figure 3: Beantable menu

Figure 4: Winnie the Pooh

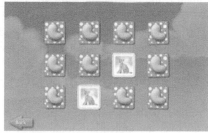

also acts as a monitoring and evaluation tool based on information given from pedagogical and child development experts and data captured during interaction. Eventually, combining appropriately designed micro games will result into the creation of motivating smart games that will attract the attention of the child throughout the age range targeted by the system.

As examples of the aforementioned micro games, two jigsaw puzzles ("Winnie the Pooh" and "The Three Little Pigs"), as well as a classic memory game (Pick & Match), were developed and tested with young children (see Figure 4, Figure 5, Figure 6). Jigsaw puzzles composed of physical parts were selected as a testing game because most children are tactile learners, i.e., they learn best by touching and manipulating physical objects with their hands (see Figure 8). Puzzle games are not only entertaining, but can also provide a variety of learning opportunities. Michalewicz and Michalewicz (2007) categorize this type of learning as the "Eureka factor".

During playing, Beantable is able to identify the location and rotation of any piece of the puzzle on the surface and provide scalable and personalized guidance. For the moment, auditory feedback is provided on each successful match between two pieces and the game ends with applauses when the child fully completes the puzzle. Furthermore, there is visual information about the completion progress of the puzzle game.

Figure 7 Interaction with custom made stamp

Figure 8 Playing with jigsaw puzzle

Memory games are an optimal way to stimulate a child's brain and help enhance mental power and strength. Pick & Match allows the child to interact either by using fingers in the context of natural user experience or by using a custom made stamp (see Figure 7), which is appropriate for very young children with difficulties using multi-touch devices. Auditory feedback is used in this game as well.

5 EVALUATION

Two levels of evaluation have been conducted so far on the preliminary Beantable prototype. The first evaluation level involved expert walkthroughs which were conducted by three accessibility and usability experts from the FORTH Human-Computer Interaction Lab. The second evaluation level involved prototype testing with three children of ages 3, 3.5, and 6. The main objective was to assess the overall system usability and provide recommendations on how to improve design. The findings of both evaluations are reported below.

5.1 Expert walkthroughs evaluation

The experts were asked to play the preliminary developed micro games in order to uncover any potential violations of usability standards in the design, as well as identify any areas of the design that could potentially cause problems specifically to children.

Overall, the experts found the design of the game applications intuitive and engaging and pointed out only minor problems with the presentation of the menu and the game information. Regarding the physical design of the table, the experts found it ergonomic and commented on how its circular design would facilitate cooperative gaming. Their only concern was that the chair was too heavy for a younger child to drag or lift to get closer to the table. The main suggestions that the experts made on the design and logic of the games were:

- Increasing the difficulty level of the memory card game by adding more cards every time the child solved the puzzle.
- Projecting a border on the screen the size of the actual puzzle. The border image will help the child understand the physical dimensions of the puzzle and where to place the actual puzzle pieces.
- Projecting the image of the solved puzzle upon which the child could build the actual puzzle. This feature could be used for younger children that need help with solving the puzzle.

Changes based on the above observations and suggestions were implemented before moving to the second level of evaluation, the informal user-based evaluation.

5.2 Informal user-based evaluation

Three children were invited to participate in the preliminary user-based evaluation of the Beantable prototype with the consent of their parents, who were present but did not play any specific role in the experiment. Two of the children were boys with ages 3.5 and 6, and one was a girl 3.5 year old. They were each asked by the evaluator to sit on the table and select a game from the menu to play. All three of them were able to open, close and play the games with very little given instructions by the evaluator. They all reached the highest level of difficulty of the memory card game (24 cards) and completed the two puzzles with ease. What was really impressive was the fact that even though the button labels that appeared in the dialog boxes were written in English, they knew which button to select based on its color, green for yes, red for no. They all expressed that they liked the games and that was also evident by the fact that they remained engaged throughout the evaluation and selected to play all three games.

Figure 9 From left to right: a girl 3.5 old and two boys 3.5 and 6 years old respectively

6 CONCLUSIONS

AmI technologies have the potential to enhance the child's playing experience and to be accepted by young children, provided that they are carefully designed and tested. The main characteristic of the proposed system is the capability to monitor the child's interaction during playtime and to adapt the game according to the child's performance in an unobtrusive and personalized way. This adaption and experience enhancement is based on the recorded interaction history data regarding the child's skills and abilities.

Currently, the prototype table has been assembled and all the necessary software building blocks have been installed. In cooperation with child development experts, fully functional smart games are being developed, along with the software required for supporting the monitoring functionalities using the wall mounted sensor. Upon full completion of the system, a large-scale evaluation experiment is being planned, involving children, pedagogical experts and parents.

Based on bibliography, a knowledge base (ontology) of children's skills, abilities and developmental milestones in the targeted age range is currently under development in order to support the whole system's functionality and share with other AmI applications in the context of AmI home environment, nursery or kindergarten. Moreover, further types of smart games are being designed in order to fully exploit the potential for cooperative interaction offered by Beantable.

Beantable will be installed and tested under almost real-life conditions in a complete AmI home environment located in the AmI Research Facility of ICS-FORTH.

ACKNOWLEDGMENTS

This work is supported by the FORTH-ICS internal RTD Programme 'Ambient Intelligence and Smart Environments'. The authors would like to thank Mrs. Ilia Adami for her contribution to the usability evaluation of Beantable.

REFERENCES

Barnett, L.A., 1990, Developmental benefits of play for children. Journal of Leisure Research, National Recreation & Park Assn

Berk, L.E., Winsler, A. 1995. Scaffolding children's learning: Vygotsky and early childhood education. National Association for the Education of Young Children Washington, DC

Bohn, J. 2004. The Smart Jigsaw Puzzle Assistant: Using RFID Technology for Building Augmented Real-World Games. Workshop on Gaming Applications in Pervasive Computing Environments at Pervasive 2004

CAD, P. 2001. International classification of functioning, disability and health (ICF), World Health Organization

Grammenos, D., Zabulis, X., Argyros, A.A., Stephanidis, C. 2009. FORTH-ICS internal RTD Programme 'Ambient Intelligence and Smart Environments, In the Proceedings of the 3rd European Conference on Ambient Intelligence, Salzburg, Austria, November 18-21.

IST Advisory Group, 2003. Ambient Intelligence: from vision to reality. Electronically available at: tp://ftp.cordis.lu/pub/ist/docs/istag-ist2003_consolidated_report.pdf

Javier Marco, Eva Cerezo, and Sandra Baldassarri. 2010. Playing with toys on a tabletop active surface. In Proceedings of the 9th International Conference on Interaction Design and Children (IDC '10). ACM, New York, NY, USA, 296-299. DOI=10.1145/1810543.1810596 http://doi.acm.org/10.1145/1810543.1810596

Kaltenbrunner, M. and Bencina, R. 2007. reacTIVision: a computer-vision framework for table-based tangible interaction. Proceedings of the 1st international conference on Tangible and embedded interaction. ACM

Mansor, E.I.; De Angeli, A.; De Bruijn, O. 2008. Little fingers on the tabletop: A usability evaluation in the kindergarten. Horizontal Interactive Human Computer Systems, 2008. TABLETOP 2008.

Michalewicz, Z. and Michalewicz, M., 2007, Puzzle-based learning. Proceedings of the 18th Conference of the Australasian Association for Engineering Education, Melbourne, Australia

Piaget, J. 1970. Science of Education and the Psychology of the Child. New York: Orion Press.

Piper, A. M., O'Brien, E., Morris, M. R., and Winograd, T. 2006. SIDES: a cooperative tabletop computer game for social skills development. In Proceedings of the 2006 20th Anniversary Conference on Computer Supported Cooperative Work.

Smart Table, http://smarttech.com/table

Vasilyeva, E., 2007, Towards personalized feedback in educational computer games for children, Proceedings of the sixth conference on IASTED International Conference Web-Based Education-Volume 2, ACTA Press

Author Index